Remodelers Handbook

A manual of professional practice
for home improvement contractors

Edited by Benjamin Williams

Craftsman
Book Company
542 Stevens Avenue
Solana Beach
California 92075

Library of Congress Cataloging in Publication Data

Williams, Benjamin.
   Remodelers handbook.

   Includes index.
   1. Dwellings--Remodeling. I. Title.
TH4816.W49   690'.8   76-53565
ISBN 0-910460-21-3
©1976 Craftsman Book Company

Contents

Chapter 1
Why Home Improvement?

About two-thirds of the 70 million homes in the U.S. are over 13 years old. Most of these older homes lack modern conveniences, do not have desirable design features or are seriously in need of repair. Even if we had the nearly unlimited resources necessary to demolish and rebuild the over 45 million older homes, it would be a mistake to do so. Many older homes in America have architectural features, spacious floor plans and a character which would be difficult or prohibitively expensive to reproduce using modern building techniques and materials. Yet most people want to live in a well lighted, energy efficient, well designed home with adequate bedrooms, bathrooms, and modern conveniences.

Fortunately, most of the homes we have been building in America for the last hundred years have foundations and frames substantial enough to last for several hundred years. In spite of this structural soundness, many older houses are being razed, abandoned, or left to a slower destruction by decay, insects, rodents, and the elements. Some of these houses have deteriorated to a point where rehabilitation would be impractical; but many could be restored to a sound condition and updated in convenience and comfort at a lower cost than that required to build a new house. In addition to monetary savings, rehabilitation has other advantages: The owner can stay in familiar surroundings; some older houses provide more space than can be achieved in a new house at reasonable cost; the work can usually be done as finances become available; and the character of the older house is often preferred to that of a new one.

The first chapters of this handbook are a guide to appraising the suitability of a house for remodeling. This relates only to the condition of the house itself, not to an area. Once a house is deemed suitable for rehabilitation, suggestions on how to proceed are covered in the later chapters. The last part of this handbook is a guide to operating a home improvement business.

Your Future In Home Improvement

The remodeling industry is sure to grow and change in the 1980's. Every year the average age of homes in the U.S. increases. Homes are growing old faster than new homes are being built. Yet improvements in materials and design, and changes in the economics of home maintenance make having a modern home much more attractive. Over 80% of the homes in America have little or no insulation. Insulating these homes is a national goal because each dollar spent on insulation will save more than a dollar in fuel cost over a relatively short period. Installing new aluminum or vinyl siding that will last 20 years or more is much less expensive than repainting wood siding several times over a 20 year period. Modernizing kitchens and bathrooms, adding floor space, or making better use of existing space makes older homes more convenient, often lowers maintenance and probably will add more than the construction cost to the value of a home. New asphalt and fiberglass shingles are available that have the look and texture of expensive cedar shakes yet are guaranteed to give good service up to 25 years. All this means that the home improvement industry will continue to grow at a rate in excess of 10% a year, faster than the economy in general and faster than the construction industry in all but a few boom years.

In the 1980's the home improvement industry faces the challenge of changing to meet the growing and changing need for home improvement. Many remodelers agree that the industry must adapt itself in three key areas.

1. More professionalism is needed. Contractors cannot be content to do a good job while others about them in the industry continue to abuse the public trust. The ethical contractor must joint forces in his community with others like him. Local organizations of these professionals must be formed to develop industry standards and to work to eliminate misrepresentation, shoddy workmanship and inferior materials being applied to the homes of their communities.

2. The professional remodeler must advertise and promote his business more than ever before. Thousands of home owners have the need to remodel. They need to be told where in their community they can find quality workmanship at fair prices. Survey after survey has shown that home owners who know where to get good home improvement services at a reasonable price will choose to remodel their homes rather than sell

their problems to someone else.

3. The professional remodeler must make his voice heard in local, state and national government where decisions affecting his business are made every day. The voice of the home improvement contractor in the land has been weak indeed. As a result he is fingerprinted in at least one area of the country and over-regulated in many more. His image is besmirched on national television down to his local newspaper. This happens less where there exists an organized remodeling group to protect such mistreatment and misrepresentation. Where a local group of professionals has the ear of community leaders there is a noticeable improvement in publicity about the industry, and consumer confidence goes up.

It is inconceivable that today's home improvement businessman wants to return to the days and the reputation of the "blue-suede shoe" operator. Yet, many are moving too slowly to grasp the opportunity that presently exists for remodeling. Those who will miss out through lethargy or lack of interest should be prepared to regard with some envy the competitor who chose to move into the mainstream of the future remodeling industry.

For a lot of obvious reasons the home improvement industry is going to keep on growing. And because home improvement is just starting to be defined as an industry, it presents a unique opportunity for any individual to go into the business and succeed. This will be one of the relatively few areas where young people can start a business for themselves with limited capital and develop strong, profitable organizations in their own lifetimes. And that's good. Though there will be larger and larger remodelers, some regional and even national, there will be room for young people to come in because the market is growing broad and is taking many forms. No matter how the industry changes, there will always be an independent remodeling contractor. His business in the 1980's will be better organized, more professional, better represented in government and promoted more consistently than it is today. The independent remodeling contractor is the only man who can pull it all together and make the concept of home improvement a reality across the country.

Chapter 2

Evaluating a Deteriorated Dwelling

How a house should be remodeled can be determined only by systematically inspecting the house and evaluating any necessary repairs in comparison with the value of the finished product. If the foundation is good and the floor, wall, and roof framing are structurally sound, the house probably is worth rehabilitating; but a thorough inspection is still in order. Even if only one part or a small portion of the house is to be remodeled, you should be aware of other potential problems that may become evident as work progresses. Use the checklist at the end of this chapter for evaluating the job before work begins.

Examination of the house is treated here by parts to provide a systematic approach to evaluation. Although no publication could possibly cover every condition that might exist in houses of varying age, style, and geographic location, several general considerations should cover most situations. Generally, suggestions are also given for examining heating, plumbing, and electrical systems; but in these areas subcontractor assistance may be desirable. Also, code requirements vary and what seemed to be a minor addition or repair may become a major revamping of the entire structure for code compliance. In most communities a building official will inspect the house and indicate code violations.

Foundations

The most important part of a house from a rehabilitation standpoint is the foundation. It supports the entire house, and failure can have far reaching effects. Check the foundation for general deterioration that may allow moisture or water to enter the basement and may require expensive repairs. More importantly, check for uneven settlement. Uneven settlement will distort the house frame or even pull it apart. This distortion may rack window and door frames out of square, loosen interior finish and siding, and create cracks that permit infiltration of cold air. A single localized failure or minor settling can be corrected by releveling of beams or floor joists and is not a sufficient reason to reject the house. Numerous failures and general uneven settlement, however, would indicate a new foundation is required or, more critically, that the house is probably unsuitable for rehabilitation.

Many old houses have stone or brick foundations, and some may be supported on masonry piers. Check the masonry foundation for cracks with crumbling mortar, a common defect that can usually be repaired, depending on its extent. More extensive deterioration may indicate the need for major repair or replacement.

Crawl space houses usually have a foundation wall or piers supporting the floor joists. These supports must be checked for cracks and settlement the same as the perimeter foundation.

Occasionally houses will be on pier-type wood post foundations. These are more common in certain areas of the country. Such foundations give good service if the wood is properly treated with preservatives. However, in the inspection they must be checked for decay and insect damage.

Most foundation walls of poured concrete have minor hairline cracks that have little effect on the structure; however, open cracks indicate a failure that may get progressively worse. Whether a crack is active or dormant can be determined only by observation over several months.

Basements

Damp or leaky basement walls may require major repair, especially if the basement space is to be used. Possible causes of the dampness are clogged drain tile, clogged or broken downspouts, cracks in walls, lack of slope of the finished grade away from the house foundation, or a high water table. Check for dampness by examining the basement a few hours after a heavy rain.

The most common source of dampness is surface water, such as from downspouts discharging directly at the foundation wall or from surface drainage flowing directly against the foundation wall. Therefore the cardinal rule is to keep water away from the foundation, and this is best accomplished by proper grading.

A high water table is a more serious problem. There is little possibility of achieving a dry basement if the water table is high or periodically high. Heavy foundation waterproofing or footing drains may help but, since the source cannot be controlled, it is unlikely they will do more than minimize the problem.

Masonry

Uneven settlement of the foundation will cause cracks in brick or stone veneer. Cracks can be grouted and joints repointed, but large or numerous cracks will be unsightly even after they are patched. The mortar also may be weak and crumbling, and joints may be incompletely filled or poorly finished. If these faults are limited to a small area, regrouting or repointing is feasible. For improved appearance, the veneer can be sandblasted.

It is important to prevent water from entering the masonry wall or flowing over the face of the wall in any quantity. Examine flashing or caulking at all projecting trim, copings, sills, and intersections of roof and walls. Plan to repair any of these places where flashing or caulking is not provided or where need of repair is apparent.

Porous or soft brick or stone should be coated with a clear water repellent after care has been taken to see that no water can get behind the veneer.

The most obvious defects to look for in a chimney are cracks in the masonry or loose mortar. Such cracks are usually the result of foundation settlement or the attachment of television antennas or other items that put undue stress on the chimney. These cracks are a particular hazard if the flue does not have a fireproof lining.

The chimney should be supported on its own footing. It should not be supported by the framework of the house. Look in the attic to see that ceiling and roof framing are no closer than 2 inches to the chimney. Either of these defects are fire hazards and should be corrected.

Check to see if the fireplace has an operating damper. Where no damper exists, one should be added to prevent heat loss up the flue when the fireplace is not in use. A fireplace that looks like it has been used a lot probably draws well; however, you can check this by lighting a few sheets of newspaper on the hearth. A good fireplace will draw immediately; a usable one will draw after about a minute.

Floor Supports

In a basement house, interior support is usually provided by wood or steel girders supported on wood or steel posts. Wood posts should be supported on pedestals and not be embedded in the concrete floor, where they may take on moisture and decay. Examine the base of the wood posts for decay even if they are set above the floor slab. Steel posts are normally supported on metal plates. Check the wood girders for sag and also for decay at the exterior wall bearings. Sag is permanent deflection that can be noted especially near the middle of a structural member. See Figure 2-1. Some sag is common in permanently loaded wood beams and is not a problem unless parts of the house have obviously distorted. Sag is usually an appearance problem rather than a structural problem. Some deflection is normal

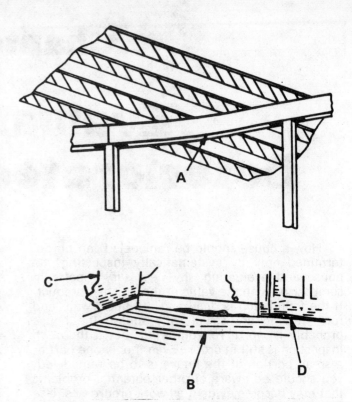

Sagging horizontal member, A, has resulted in: B, uneven floor; C, cracked plaster; and D, poorly fitting door
Figure 2-1

and about 3/8 inch deflection in a 10 foot span girder is acceptable in design.

Floor Framing

The sill plates, or joists and headers where sill plates are not used, rest on top of the foundation. Thus they are exposed to moisture and are vulnerable to decay or insect attack. Examine these members specifically, as well as the entire floor framing system for decay and insect damage, particularly if the basement or crawl space is very damp.

Joists, like girders, should be examined for sag. Here too some sag can be expected and is not a sign of structural damage. It is usually not a serious problem in floor joists unless the foundation system has settled unevenly, causing excessive deflection in parts of the floor system. Look for local deflection due to inadequate support of a heavy partition load that runs parallel to the joists. Sag might be considered excessive if it is readily apparent from a visual appraisal of the levelness of a floor.

A floor may be noted to be excessively springy when walking across it. This may be remedied by adding extra joists or girders to increase stiffness.

Another point of particular concern is the framing of the floor joists around stair openings. Some builders estimate that 50 percent of the houses have inadequate framing around stairs.

8

Check floors around the opening for levelness. Where floors are sagging, the framing will have to be carefully leveled and reinforced.

Wall Framing

The usual stud wall normally has much more than adequate strength. It may be distorted, however, for reasons covered in preceding sections. Check openings for squareness by operating doors and windows and observing fit. Some adjustments are possible but large distortions will require new framing. Also check for sag in headers over wide window openings or wide openings between rooms. Where the sag is visually noticeable, new headers will be required.

Roof Framing

Examine the roof for sagging of the ridge, the rafters, and the sheathing. This is easily done by visual observation. If the ridge line is not straight or the roof does not appear to be in a uniform plane, some repair may be necessary. The ridge will sag due to improper support, inadequate ties at the plate level, or even from sagging of the rafters. Rafters will sag due to inadequate stiffness or because they were not well seasoned. Sheathing sag may indicate too wide a spacing between rafters or strip sheathing, or plywood that is too thin or has delaminated.

Siding and Trim

The main problems with siding and trim stem from excessive moisture, which can enter from either inside or outside. One of the main contributors to the problem is the lack of roof overhang, allowing rain to run down the face of the wall. Moisture may also enter from the inside, because of the lack of a vapor barrier, and subsequently condense within the wall.

Look for space between horizontal siding boards by standing very close and sighting along the wall. Some cracks can be caulked, but a general gapping or looseness may indicate new siding is required. If the boards are not badly warped, renailing may solve the problem. Check siding for decay where two boards are butted end to end, at corners, and around window and door frames.

Decorative trim is sometimes excessive and presents unusual decay and maintenance problems, particularly where water may be trapped.

Good shingle siding appears as a perfect mosaic, whereas worn shingles have an all over ragged appearance and close examination will show individual shingles to be broken, warped, and upturned. New sidings will be required if shingles are badly weathered or worn.

Windows

Windows usually present one of the more difficult problems of old wood-frame houses. Note Figure 2-2. If they are loose fitting and not weatherstripped, they will be a major source of uncomfortable drafts and cause high heat loss. Check the tightness of fit and examine the sash and the sill for decay. Also check the operation of the window. Casement windows should be checked for warp at top and bottom.

Signs of excessive water damage are evident in the paint peeling off this window sill and sash and broken caulking around the window
Figure 2-2

When replacement of windows is planned, check the window dimensions. If the window is not a standard size or if a different size is desired, the opening will have to be reframed or new sash must be made, both of which are expensive.

In cold climates windows should be double glazed or have storm windows, both to reduce heat loss and avoid condensation. Again, if the windows are not a standard size, storm windows may be expensive.

Doors

Exterior doors should fit well without sticking. They should be weatherstripped to avoid air infiltration, but this is a very simple item to add. Difficulties in latching a door can usually be attributed to warping. A simple adjustment of the latch keeper will solve the problem in some instances, but badly warped doors should be replaced.

Storm doors are necessary in cold climates, not only for heat saving and comfort, but also to avoid moisture condensation on or in the door and to protect the door from severe weather.

If the door frame is out of square due to foundation settlement or other racking of the house frame, the opening will probably have to be reframed.

The lower parts of exterior doors and storm doors are particularly susceptible to decay and should be carefully checked. Also observe the condition of the threshold, which may be worn, weathered, or decayed, and require replacement.

Exessive paint peeling on siding
Figure 2-3

Porches

One of the parts of a house most vulnerable to decay and insect attack is the porch. Since it is open to the weather, windblown rain or snow can easily raise the moisture content of wood members to conditions for promoting growth of wood-destroying organisms. Steps are often placed in contact with soil, always a poor practice with untreated wood.

Check all wood members for decay and insect damage. Give particular attention to the base of posts or any place where two members join and water might get into the joint. Decay often occurs where posts are not raised above the porch floor to allow air to dry out the base of the post. It may be worthwhile to replace a few members, but the porch that is in a generally deteriorated condition should be completely rebuilt or removed.

Finishes

Failure of exterior finishes on siding or trim results most commonly from excessive moisture in the wood. This may result either from direct rain or from moisture vapor condensing in the walls. Finish failures may also be caused by poor paints, improper application of good paints, poor surface preparation, or incompatible successive coatings.

Excessive peeling (Figure 2-3) may require complete removal of the paint. Since this can be very expensive, re-siding may be considered.

Roofing

If the roof is actually leaking, it should be obvious from damage inside the house. A look in the attic may also reveal water stains on the rafters, indicating small leaks that will eventually cause damage. Damage inside the house is not always attibutable to roofing, but could be caused by faulty flashing or result from condensation.

Asphalt shingles are the most common roof covering and are made in a wide range of weights and thicknesses. The most obvious deterioration of asphalt shingles is loss of the surface granules.

The shingles may also become quite brittle. More important, however, is the wear that occurs in the narrow grooves between the tabs or sections of the shingle or between two consecutive shingles in a row. This wear may extend completely through to the roof boards without being apparent from a casual visual inspection. A good asphalt shingle should last 18 to 20 years.

Wood shingles also find considerable use for covering of pitched roofs and are most commonly of durable woods such as cedar in No. 1 or No. 2 grades. A good wood shingle roof should appear as a perfect mosaic, whereas a roof with worn shingles has an all over ragged appearance. Individual shingles on the worn roof are broken, warped, and upturned. The roof with this worn appearance should be completely replaced even though there is no evidence of leaking. Excessive shade may cause fungus growth and early shingle deterioration. A good wood shingle roof will last up to 30 years under favorable conditions.

Built-up roofing on flat or low sloped roofs should be examined by going onto the roof and looking for bare spots in the surfacing and for separations and breaks in the felt. Bubbles, blisters, or soft spots also indicate that the roof needs major repairs. Alligatoring patterns on smooth surface built-up roofs may not be a failure of the roof. The life of a built-up roof varies from 15 to 30 years, depending on the number of layers of felt and the quality of application.

Flashing should be evident where the roof intersects walls, chimneys, or vents, and where two roofs intersect to form a valley. Check for corroded flashing that should be replaced to prevent future problems. Likewise, check for corroded gutters and downspouts, which can be restored by repainting unless severely corroded.

If the house was built with no roof overhang, the addition of an overhang should be considered in the remodeling plan. It will greatly reduce maintenance on siding and window trim, and prolong the life of both.

Flooring

In checking wood floors look for buckling or cupping of boards that can result from high moisture content of the boards or wetting of the floor. Also notice if the boards are separated due to shrinkage. This shrinkage is more probable if the flooring boards are wide. If the floor is generally smooth and without excessive separation between boards, refinishing may put it in good condition; however, be sure there is enough thickness left in the flooring to permit sanding. Most flooring cannot be sanded more than two or three times; if it is softwood flooring without a subfloor, even one sanding might weaken the floor too much. Sanding of plywood block floors should also be quite limited. If floors have wide cracks or are too thin to sand, some type of new flooring will have to be added.

Floors with resilient tile should be examined

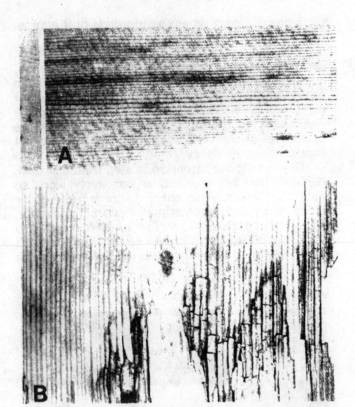

A, Discolored wood showing advanced stages of
decay; B, cubical checking and collapse
Figure 2-4

for loose tile, cracks between tile, broken corners, and chipped edges. Look to see if any ridges or unevenness in the underlayment are showing through. Replacement of any tile in a room may necessitate replacing the flooring in the whole room because tile change color with age and new tile will not match the old.

Walls and Ceilings

The interior wall covering in old houses is usually plaster, but may be gypsum board in more recently built homes. Wood paneling may also be found but is usually limited to one room or to a single wall or accent area.

Plaster almost always has some hairline cracks, even when it is in good condition. Minor cracks and holes can be patched, but a new wall covering should be applied if large cracks and holes are numerous, if the surface is generally uneven and bulging, or if the plaster is loose in spots. The same general rule applies to ceilings.

If walls have been papered, check the thickness of the paper. If more than two or three layers of paper are present, they should be removed before applying new paper, and all wallpaper should be removed before painting.

The paint on painted surfaces may have been built up to excessive thickness. It may be chipped due to mechanical damage to incompatibility between successive layers, or to improper surface preparation prior to repainting. Old kalsomine surfaces may require considerable labor to

recondition, so a new wall covering should be considered. Paint failures may be due to application of paint over kalsomine.

Trim, Cabinets, and Doors

Trim should have tight joints and fit closely to walls. If the finish is worn but the surface is smooth, refinishing may be feasible. If the finish is badly chipped or checked, removing it will be laborious regardless of whether the new finish is to be a clear sealer or paint. Trim or cabinetry of plain design will be less difficult to prepare for refinishing than that having ornately carved designs.

If any trim is damaged or if it is necessary to move doors or windows, all trim in the room may have to be replaced as it may be difficult to match the existing trim. Small sections of special trim might be custom made but the cost should be compared with complete replacement. Check with the building supply dealer to see if the particular trim is still being made. Also check some of the older cabinet shops to see if they have shaper knives of this trim design.

The problems with interior doors are much the same as those for exterior doors except there are no decay or threshold problems.

Lightening of wood with fine black
lines showing decay
Figure 2-5

Recognizing Damage by Decay

Look for decay in any part of the house that is subject to prolonged wetting. Decay thrives in a mild temperature and in wood with a high moisture content.

One indication of decay in wood is abnormal color and loss of sheen. The brown color may be deeper than normal (Figure 2-4A), and in

Fungal surface growth in crawl space area under joint. [Pen shows comparative size.]
Figure 2-6

advanced stages cubical checking and collapse occur (Figure 2-4B). The abnormal color may also be a lightening which eventually progresses to a bleached appearance (Figure 2-5). Fine black lines may be present with the bleached appearance.

Fungal growths appearing as strandlike or cottony masses on the surface of wood indicate excessive water and consequently the presence of decay (Figure 2-6).

The visual methods of detecting decay do not show the extent of damage. The two strength properties severely reduced by decay are hardness and toughness. Prod the wood with a sharp tool and observe resistance to marring. To determine loss of hardness compare this resistance with that of sound wood. Sound wood tends to lift out as one or two relatively long slivers, and breaks are splintery. To determine loss of toughness use a pointed tool to jab the wood and pry out a sliver. If toughness has been greatly reduced by decay, the wood breaks squarely across the grain with little splintering and lifts out with little resistance.

Decay may exist in any part of the house, but some areas are particularly vulnerable. Special attention should be given to the areas described in the following paragraphs.

Decay often starts in framing members near the foundation. It may be detected by papery, fanlike growths that are initially white with a yellow tinge, and turn brown or black with age. Look for these growths between subfloor and finish floor and between joists and subfloor. They may become exposed by shrinkage of flooring during dry weather. These growths may also exist under carpets, in cupboards, or in other protected areas that tend to stay damp.

Where siding is close to the ground, look for discoloration, checking, or softening. Also check

for signs of decay where siding ends butt against each other or against trim.

Observe wood shingles for cubical checking, softening, and breakage of the exposed ends. Asphalt shingles have deteriorated if they can be easily pulled apart between the fingers. Edges of roofs are particularly vulnerable if not properly flashed. If the roofing is deteriorating, check the underside of the roof sheathing for evidence of condensation or decay.

Give particular attention to step treads or deck surfaces that are checked or concavely worn so they trap water. Also check joints in railings or posts. Enclosed porches may have condensation occuring on the underside of the deck and framing. Check the crawl space for signs of dampness and examine areas where these signs occur.

Stain and decay may occur at such joints in a window frame
Figure 2-7

Look for brown or black discoloration near joints or failure of nearby paint (Figure 2-7). Both are signs of possible decay. Also check the inside for water stains on the sash and sill resulting from condensation running down the glass. Where these stains exist, check for softening and molding.

Insect Problems
The three major kinds of wood-attacking insects that cause problems in wood-frame houses are termites, powderpost beetles, and carpenter ants. Methods of recognizing each of these are discussed under separate headings. Where there is any indication of one of these insects, probe the wood with a sharp tool to determine the extent of damage.

There are two main classifications of termites: (1) subterranean termites, which have access to the ground or other water source, and (2) nonsubterranean termites, which do not require direct access to water.

Termite shelter tubes on foundation
Figure 2-8

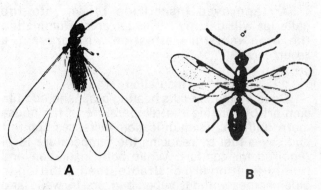

Differences between winged termite, A;
and winged ant, B.
Figure 2-9

Termite damage to interior of a pine 2 by 4
Figure 2-10

Examine all areas close to the ground for subterranean termites. One of the most obvious signs is earthen tubes (Figure 2-8) built over the surface of foundation walls to provide runways from the soil to the wood above. Termites may also enter through cracks or voids in the foundations or concrete floors. They do not require runways to the soil where there is a source of water such as a plumbing leak.

Another sign of the presence of termites is the swarming of winged adults early in the spring or fall. Termites resemble ants, but the termites have much longer wings and do not have the thin waist of an ant (Figure 2-9). Where there is an indication of termites, look for galleries that follow the grain of the wood, usually leaving a shell of sound wood (Figure 2-10).

Nonsubterranean termites live in damp or dry wood without outside moisture or contact with the ground. Look for these only in Hawaii and in a narrow coastal strip extending from central California to Virginia. One of the early signs of these termites is sandlike excretory pellets that are discarded outside the wood. Nonsubterranean termites can also be identified by the presence of swarming winged forms. They cut freely across the grain of the wood rather than following the grain as the subterranean termites do.

To combat termites, soil poisoning is often recommended, but this is generally done by a professional exterminator.

Powder-post beetles are most easily recognized by their borings, which are about the consistency of flour. Many borings remain inside the wood. The adults leave the wood through a hole about the diameter of a pencil lead, giving the wood the appearance of having been hit by birdshot. Such holes may be just the result of a previous infestation, so check for fresh, clean sawdust as a sign of current activity. Activity may also be recognized by the rasping sound the beetles make while tunneling.

Look for powder-post beetles in humid locations such as near the ground. Sometimes the homeowner may destroy them with an approved insecticide, but in severe cases, fumigation by a professional exterminator is required.

The presence of carpenter ants is often discovered by their chewed wood, which resembles coarse sawdust and is placed in piles outside the wood. They do not eat the wood, but only nest in it. Working ants may be as much as half an inch long. They make a rustling noise in walls, floors, or woodwork. Look for signs of carpenter ants in softwood in high humidity locations.

An approved insecticide blown into the galleries will destroy carpenter ants. Eliminating the high moisture situation will prevent a recurrence.

Insulation

Good insulation cuts heating costs and adds to comfort by making the temperature in the house more uniform. Humidification increases comfort and saves fuel by reducing the temperature level required for comfort. While both insulation and humidification are desirable, their addition to older homes without vapor barriers in walls and ceilings may create moisture condensation problems. Where large differences exist between indoor and outdoor temperatures, pressure forces water vapor out through the walls.

In the uninsulated house this vapor usually moves on to the outside without any problem. Where insulation is added, the dewpoint often occurs within the insulation, so water vapor condenses into free water with consequent wet insulation and siding. In some instances where indoor relative humidities are low and the outside covering material allows moisture in the walls to escape readily, no moisture problems may result. However, mechanical humidification, in addition to normal moisture from cooking, bathing and respiration, amplifies moisture problems; the water vapor pressure drive is increased and consequently the rate of moisture movement into the walls. Vapor barriers in walls and ceilings reduce the rate of moisture movement into these areas and thus help to control the moisture problems otherwise created by adding insulation or humidification.

Look in the attic to determine the amount of ceiling insulation present. The ceiling represents the greatest source of heat loss on cool days as well as the greatest source of heat gain on warm days. At least 6 inches of insulation should be provided for homes in mild climates and 9 to 12 inches for those in cold climates. To find out if the walls are insulated, some siding and sheathing or interior covering must be removed. Insulation in walls should be included in any house rehabilitation in cold climates and in warm climates where summer cooling is essential. Insulation is also needed under floors of crawl space houses in cold climates.

Vapor Barriers

Vapor barriers should be provided on the warm side of all insulation. Most houses built before the mid 1930's do not have vapor barriers. If the ceiling insulation is in blanket form with a covering around it, the covering material may resist the passage of moisture. However, if the ceiling is loose fill, look under it for a separate vapor barrier of coated or laminated paper, aluminum foil, or plastic film. The same thing is true of insulated walls, where the vapor barrier should be on the inside of the walls.

Check in crawl spaces for a vapor barrier laid on top of the soil. If there is none and the crawl space seems quite damp, a vapor barrier could be added.

There is no convenient way to determine if there are vapor barriers under floor slabs. If the floor seems damp most of the time, there probably is no vapor barrier. A barrier would then have to be added on top of the slab, with a new finish floor applied over it, to have a dry finish floor.

Ventilation

The two major areas where good ventilation is required are the attic, or roof joist spaces in the cases of a cathedral ceiling or flatroofed house, and the crawl space. The general adequacy of existing ventilation can be observed just from the degree of dampness.

Moisture passes into the attic from the house and condenses as the air cools down or where the moist air contacts the cold roof members. Both inlet and outlet vents must be located properly for good circulation of air through all the attic area. These vents not only help keep the attic dry in winter, but keep hot air moving from the attic during summer and help to cool the house.

Observe the size and location of crawl space vents. There should be at least four vents located near building corners for optimum cross ventilation and minimum dead air space.

Mechanical

Because many of the plumbing, heating, and wiring systems in a house are concealed, it may be difficult to determine their adequacy. For the same reason it is difficult to make major changes without considerable cutting of wall surfaces and, in some situations, even structural members.

In a very old house the mechanical systems may have to be replaced and this will become a major cost item. At the same time, however, properly installed new systems will be responsible to a large extent for comfort and convenience in the house. One bonus can be the dramatic recovery of space and improvement in the appearance of the basement when an old "octopus" gravity warm air heating system is replaced by a modern forced-air system.

Plumbing

Water pressure is important. Check several faucets to see if the flow is adequate. Low pressure can result from various causes. The service may be too small or it may be reduced in diameter due to lime, particularly with very old lead service pipes. The main distribution pipes should be ¾ inch inside diameter but branch lines may be ½ inch inside diameter. Sizes can be checked easily. Copper pipes ½ inch inside diameter are 5/8 inch outside diameter, and ¾ inch inside diameter pipes are 7/8 inch outside diameter. Galvanized pipes ½ inch inside diameter are 7/8 inch outside diameter and pipes ¾ inch inside diameter are

1/18 inch outside diameter.

The supply pressure may be inadequate. If the house has its own water system, check the gauge on the pressure tank, which should read a minimum of 20 and preferably 40 to 50 pounds. Anything less will indicate the pump is not operating properly, or the pressure setting is too low. If the supply is from a municipal system, the pressure in the mains may be too low, though this is unlikely.

Check shutoff valves at the service entrance and a various points in the system to determine if they have become frozen with age or little use.

Check for leaks in the water supply system. Rust or white or greenish crusting of pipe or joints may indicate leaks.

Water hammer may be a problem. This results from stopping the water flowing in the pipe by abruptly closing a faucet. Air chambers placed on the supply lines at the fixtures usually absorb the shock and prevent water hammer. If there is water hammer, air chambers may be waterlogged. If there are no air chambers, they may be added.

The water from any private well should be tested even though the well has been in continuous use.

The drainage system consists of the sewer lateral, the underfloor drains, the drainage pipes above the floor, and the vents. Pipes may have become clogged or broken or they may be of inadequate size. Venting in particular may be inadequate and far below code requirements.

Flush fixtures to see if drains are sluggish. If so, check the following:

Old laterals are commonly of vitreous bell tile. These may have been poorly installed or have become broken, allowing tree roots to enter at the breaks or through the joints. Roots can be removed mechanically but this operation may have to be repeated every few years.

The underfloor drains may be of tile or even of steel and could be broken or rusted out. They may have become clogged and only need cleaning.

The drainage system above the basement floor or within the house should be checked for adequacy and leaks.

Vents may be inadequate or may have become clogged; in extreme cases they may cause the water in the traps to be siphoned out, allowing sewer gas to enter the house. Note any excessive suction when a toilet is flushed.

Additional supply and drain lines may be desirable in modernizing a house. New lines may be required for automatic washers, added baths, adequate sill cocks, or in reorganizing the layout.

With a hot water heating system, water may also be heated satisfactorily for cooking, bathing, and other personal needs. However, in a hot air furnace, the water heating coil seldom provides enough hot water. Furthermore, during summer months when the hot air heating is not needed, a separate system is required to provide hot water. A gas water heater should have at least a 30 gallon capacity and preferably more. An electric water heater should have a capacity of 50 gallons or more.

Plumbing fixtures that are quite old may be rust stained and require replacement, or it may be desirable to replace them just for the improved appearance.

Drainage may not be adequate. Run water for a few minutes to check for clogged drain lines between the house and the sewer main. If a private sewage system exists, consider the adequacy of the drain field. If a new drainage field is needed, some codes require percolation tests of soil.

Heating

Heating system advances and concepts of comfort outdate the heating systems in most old houses. Central heating with heat piped to all rooms is considered a necessity in all but very small houses.

The only way to satisfactorily check the adequacy of the heating system is through use. If the system is adequate for the desired degree of comfort, check the furnace or boiler for overall general condition.

Gravity warm air systems are common in older homes. Some gravity warm air furnaces may heat a house relatively well, but it is doubtful that the temperature control and heat distribution will be as good as with a forced circulation system.

If a warm air system is exceptionally dirty, there may be bad smudges above the registers. This will indicate some repair work is required; if the furnace is old, it may need to be replaced. Rusty ducts may need replacement.

One-piece gravity steam heating systems are common in older homes. The system is similar in appearance to hot water. This is an extremely simple system and, if properly installed, it will provide adequate heat but with no great speed or control. It can be modernized without basic changes merely by replacing standing radiators with baseboard heaters.

A one-piece gravity steam system can be made more positive in action by converting to a two-pipe system. This requires adding traps and return lines.

A two-pipe steam system can be modified to a circulating hot water system. Circulating pumps must be added but this results in greater speed of heat distribution and excellent control.

Radiant heat from hot water flowing through coils embedded in concrete floors or plastered ceilings is less common but may provide excellent heating. Such systems may become airlocked and require an expert to restore proper operation. Breaks in ceiling coils can be repaired fairly easily but repairing breaks in floors is extremely difficult. If floor breaks are extensive, the system will probably need to be replaced. The electric heating panels have no moving parts to wear out and should be in good condition unless a heating element has burned out.

Electrical

The electrical service is the first thing to check. So many new electrical appliances have come into common use in recent years that old houses may not have adequate wiring to accommodate them, particularly if air conditioning is installed. The service should be at least 100 amperes for the average three bedroom house. If the house is large or if air conditioning is added, the service should be 200 amps. If the main distribution panel has room for circuits, additional circuits can be added to supply power where there is a shortage. Otherwise another distribution panel may be added.

Examine electrical wiring wherever possible. Some wiring is usually exposed in the attic or basement. Wiring should also be checked at several wall receptacles or fixtures. If any armored cable or conduit is badly rusted, or if wiring or cable insulation is deteriorated, damaged, brittle or crumbly, the house wiring has probably deteriorated from age or overloading and should be replaced.

At least one electrical outlet on each wall of a room and two or more on long walls is desirable but may not always be necessary. Ceiling lights should have a wall switch, and rooms without a ceiling light should have a wall switch for at least one outlet.

Other Considerations

Some very subtle and not so obvious conditions should be evaluated before completing the house inspection. Attention in this handbook is focused on the building itself, with no mention of problems that relate to a particular site. It is beyond the scope of this handbook to get into such problems—important as they might be—as the condition and future of the neighborhood, availability of transportation, distance to shopping and schools, or the vulnerability to such natural disasters as floods. Some homes, though structurally sound and not badly deteriorated, do not lend themselves to remodeling. You are not doing anyone a service to undertake an expensive project which will not materially increase the value of the property. Some points to consider are relationship and convenience of areas to each other, traffic circulation, privacy, and adequacy of room size.

Circulations

Observe circulation of traffic patterns. For good circulation the general rule is to keep through traffic from all rooms or at least keep traffic at one side of a room rather than through its center. If circulation is a problem, it can sometimes be improved simply by moving doors to the corners of rooms or by placing furniture in a manner to direct traffic where it will be least objectionable. Traffic considerations will be discussed further as they relate to specific areas.

Layout

Ideally, houses should have rooms arranged in three areas—the private or bedroom area, the work area consisting of kitchen and utility rooms, and the relaxation area consisting of dining and living rooms. A family room, a den, or a recreation room may exist in or between these general areas. The den should be out of the general circulation areas and, if it is part of the bedroom area, may double as a guest room. A recreation room in the basement can serve some of the same functions as a family room. The relationship of areas will be treated further as each area is discussed.

The location of the kitchen in relation to other areas of the house is one of the most critical. It should have direct access to the dining area and be accessible to the garage or driveway for ease in unloading groceries. Being near the utility room is also convenient as the housewife often has work in progress in the kitchen and utility room at the same time. Traffic should not pass through the kitchen work area, i.e., the range-refrigerator-sink triangle.

The size of the kitchen is important. There was a period when kitchens were made very small with the idea that this was convenient, but the many modern appliances that now commonly go into a kitchen, as well as the inclusion of a breakfast area, require much more space. If the kitchen is too small, a major addition or alteration will be necessary.

A coat closet near the kitchen entrance and some facility for washing up near the work area are desirable. However, in the small house neither may be feasible.

The bedroom and bathroom area should be separated as much as possible, both visually and acoustically, from the living and work areas.

Every bedroom should be accessible to a bathroom without going through another room, and at least one bathroom should be accessible to work and relaxation areas without going through a bedroom. One of the basic rules of privacy is to avoid traffic through one bedroom to another. If this privacy is not presently provided, some changes in layout may be desirable.

Check the size of bedrooms. It is desirable to have a floor area of at least 125 square feet for a double bed and 150 square feet for twin beds. Smaller bedrooms can be very usable, but furnishings will be limited.

The relaxation area is usually at the front of an older house, but rooms at the side or rear may be used, particularly if this provides a view into a landscaped yard. If the house has a small parlor or a living room and dining room separated by an arch, consider removing partitions or arches to give a more spacious feeling. The main entrance is usually at or near the living area. Check for a coat closet near this entrance, and a passage into the work area without passing through the living room or at least not more than a corner or end of the room.

Home Evaluation Checklist

Address_____ **Date**_____

Phone_____ **Evaluated by**_____ **Owner**_____

Year built_____ **Year last remodeled**_____

1. Foundation. Type _____
☐ Evidence of deterioration _____
☐ Water in basement _____
☐ Uneven settlement _____
  ☐ Windows or doors cracked _____
  ☐ Cracks evident _____
  ☐ Frame distorted _____
☐ Mortar or grout loose _____
☐ Support or pier settlement _____
☐ Decay in wood supports _____
☐ _____

2. Basement. Average depth below floor joists _____
☐ Water present _____
☐ Walls cracked _____
☐ Walls damp _____
☐ Surface water entering basement _____
☐ Waterproofing needed _____
☐ _____

3. Masonry veneer. Where located _____
☐ Cracked _____
☐ Repointing needed _____
☐ Mortar joints deteriorated _____
☐ Sandblasting needed _____
☐ Flashing rusted or missing _____
☐ Caulking needed _____
☐ Separating from wall _____
☐ _____

4. Masonry fireplace and chimney. Type _____
☐ Cracks present _____
☐ Repointing needed _____
☐ Mortar joints deteriorated _____
☐ Sandblasting needed _____
☐ Flashing rusted or missing _____
☐ Caulking needed _____
☐ Cap deteriorated _____
☐ Foundation missing or cracked _____
☐ Separating from wall _____
☐ Evidence of rain entry or smoke exist _____
☐ Fireproof lining needed _____
☐ Wood frame closer than 2'' to chimney _____
☐ Damper needed or needs repair _____
☐ Draws poorly _____
☐ _____

5. Floor supports
☐ Girders deteriorated at bearings _____
☐ Posts not on pedestals _____
☐ Posts decayed _____
☐ Girders deflected over ½'' _____
☐ _____

6. Floor framing
☐ Sill plates deteriorated _____
☐ Decay or insect damage _____
☐ Joists deflected excessively _____
☐ Floor ''springy'' under load _____
☐ Floor deflection around stair openings _____
☐ _____

7. Wall framing
☐ Distorted _____
☐ Headers over windows or doors distorted _____
☐ _____

8. Roof framing
☐ Ridge deflected _____
☐ Rafters deflected _____
☐ Sheathing deteriorated _____
☐ Surface ''springy'' under load _____
☐ Inadequate internal support _____
☐ _____

9. Siding and trim. Type of siding _____
☐ Inadequate roof overhang _____
☐ Decayed or deteriorated _____
☐ Vapor barrier needed _____
☐ Cracked _____
☐ Loose or warped _____
☐ Needs caulking _____
☐ Needs renailing _____
☐ Needs painting or resurfacing _____
☐ _____

10. Windows. Type _____
☐ Loose fitting _____
☐ Need weatherstripping _____
☐ Inoperative _____
☐ Decayed sash or sill _____
☐ Warped _____
☐ Hardware missing or broken _____
☐ Windows not standard sizes _____
☐ Need storm windows _____
☐ Need double glazing _____
☐ Inadequate window area _____
☐ Inadequate ventilation area _____
☐ Need resurfacing _____
☐ _____

11. Doors
☐ Fit opening poorly _____
☐ Warped or damaged _____
☐ Hardware missing or broken _____
☐ Not weatherstripped _____
☐ Obsolete design _____
☐ Need resurfacing _____
☐ Fail to latch _____
☐ Need storm doors _____
☐ Frame out of square _____
☐ Threshold decayed or worn _____
☐ Trim missing or damaged _____
☐ _____

12. Porch
☐ Flooring decayed _____
☐ Supports decayed _____
☐ Steps in contact with soil _____
☐ Insect damage _____
☐ Trim damaged or missing _____
☐ Structural members damaged or missing _____
☐ Needs refinishing _____

13. Roofing. Type _____
☐ Evidence of leaks present _____
☐ Asphalt shingles. Year of application _____
  ☐ Loosing surface granules _____
  ☐ Shingles missing or broken _____
  ☐ Worn spots _____
  ☐ Nails loose or missing _____
  ☐ Valley worn or inadequate _____
  ☐ Ridge missing or deteriorated _____
  ☐ Poorly applied _____
☐ Wood shingles. Year of application _____
  ☐ Worn or deteriorated _____
  ☐ Shingles missing or broken _____
  ☐ Shingles loose _____
  ☐ Nails loose or missing _____
  ☐ Poorly applied _____
  ☐ Valley deteriorated _____
  ☐ Ridge missing or deteriorated _____
  ☐ Warped or upturned _____
  ☐ Fungus growth _____
☐ Built up roofing. Year of application _____
  ☐ Blistered _____
  ☐ Cracked _____
  ☐ Water accumulation _____
  ☐ Gravel surface deteriorated _____
☐ Flashing deteriorated _____
☐ Caulking missing or deteriorated _____
☐ Gutters and downspouts deteriorated _____
☐ Attachments or decorations on roof _____
☐ _____

14. Flooring
☐ Wood flooring. Type _____
  ☐ Buckling or cupping _____
  ☐ Separated or loose _____
  ☐ Rough surface _____
☐ Resilient. Type _____
  ☐ Loose _____
  ☐ Cracked _____
  ☐ Broken _____
  ☐ Missing _____
  ☐ Uneven surface _____
  ☐ Discolored _____
☐ Carpeting. Type _____
  ☐ Worn _____
  ☐ Spotted _____
  ☐ Needs cleaning _____
☐ _____

15. Walls and ceilings
☐ Plaster or wallboard. Type _____
  ☐ Cracks _____
  ☐ Holes _____
  ☐ Uneven surface _____
  ☐ Needs new finish _____
☐ Paneling. Type _____
  ☐ Chipped or cracked _____
  ☐ Discolored _____
  ☐ Needs new finish _____
☐ Wallpaper. Type _____
  ☐ Peeling or worn _____
  ☐ Discolored _____
  ☐ Needs to be removed _____
☐ _____

16. Cabinets and trim
☐ Needs refinishing _____
☐ Obsolete style or design _____
☐ Inadequate space _____
☐ Hardware needed _____
☐ Trim damaged or missing _____
☐ Racked or distorted _____
☐ _____

17. Insulation. Type _____
☐ Needed in walls _____
☐ Needed in crawl space _____
☐ Needed in attic _____
☐ _____

18. Vapor Barrier. Type _____
☐ Needed on inside of walls _____
☐ Needed above ceiling _____
☐ Needed in crawl spaces _____
☐ _____

19. Ventilation
☐ Attic needs vents _____
☐ Crawl space needs vents _____
☐ _____

20. Plumbing
☐ Faucet flow inadequate _____
☐ Supply inadequate _____
☐ Valves frozen _____
☐ Leaks apparent _____
☐ Water hammers _____
☐ Waste drains poorly _____
☐ Vent stacks inadequate _____
☐ Inadequate traps on fixtures _____
☐ Inadequate water heater _____
☐ Fixtures need replacement _____
☐ _____

21. Heating. Type _____
☐ Capacity inadequate _____
☐ Poor heat distribution _____
☐ System repairs needed _____
☐ Controls obsolete _____
☐ System obsolete _____
☐ _____

22. Electrical. ☐ 100 amps, ☐ 200 amps, ☐ 110 volts, ☐ 220 volts
☐ Wiring insulation deteriorated _____
☐ Inadequate electrical outlets _____
☐ Distribution equipment inadequate _____
☐ Fixtures need replacing _____
☐ Switches need replacing _____
☐ _____

23. Community considerations
☐ Adjoining properties deteriorated _____
☐ Future neighborhood improvement unlikely _____
☐ Local transportation system inadequate _____
☐ Remote from schools, shopping, etc. _____
☐ Flood or storm damage likely _____
☐ Inadequate privacy _____
☐ Offensive activities in community _____
☐ _____

24. Design considerations
Number of bedrooms _____
Number of bathrooms _____
Approximate square footage _____
Size of garage _____
Unusual features _____

☐ Room size inadequate _____
☐ Poor traffic patterns _____
☐ Bedrooms not remote from kitchen _____
☐ Kitchen poorly located or designed _____
☐ Kitchen too small or too large _____
☐ Bathrooms poorly located _____
☐ Bathrooms too small _____
☐ Inadequate closet space _____
☐ Inadequate ceiling height _____
☐ Excessive ceiling height _____
☐ Architectural style unattractive _____
☐ Cluttered, unattractive exterior _____
☐ Landscaping inadequate or unattractive _____
☐ _____

Appearance

Taste is largely a personal matter so only basic guidelines will be given. Simplicity and unity are the major considerations. A good "period" style may be worth preserving and a house possessing the quality commonly referred to as "charm" may be appraised $2,000 to $3,000 higher than a plainer house of the same size and general description. Style or charm assume less importance, however, for anyone who has to live with inconvenience, discomforts, and constant repairs.

Simplicity is one of the first principles. Observe the main lines of the house. Some variety adds interest, but numerous roof lines at a variety of slopes present a busy, confused appearance. Strong horizontal lines are usually desirable in a conventional residence to give the appearance of being "tied to the ground." Strong vertical lines tend to make a house look tall and unstable.

List the number of materials used as siding. There should never be more than three, and not more than two would be preferable. Look at the trim and see if it seems to belong with the house or is just struck on as ornamentation.

Unity is as important as simplicity. The house should appear as a unit, not as a cluster of unrelated components. Windows and trim should be in keeping with the style of the house. Windows should be of the same type and of a very limited number of sizes. Shutters should be one-half the width of the window so that, if closed, they would cover the window. Porches and garages should blend with the house rather than appear as attachments.

If the dwelling appears unattractive, consider how paint and landscaping may affect it. Even an attractive house will not look good without being properly painted or landscaped.

Final Evaluation

After the house has been completely examined, the information should be carefully evaluated. Some general guides for evaluating the information are presented here. Judgment will be required to draw conclusions from these guides.

Major remodeling and repair work is not advisable if your evaluation discloses any of the following:

1. The foundation is completely unrepairable. Houses are occasionally moved onto a new foundation, but this is generally not economical unless the house is otherwise in extremely good condition.

2. The entire frame of the house is badly out of square, the framing is generally decayed or badly termite damaged.

3. Numerous costly replacements (or major repairs and replacements combined) are required. In general, the rehabilitation cost should not exceed two-thirds of the cost of a comparable new house.

A particularly good location would be justification for spending more; a generally undesirable or deteriorating location would indicate much less than two-thirds the new house cost should be spent. If there are sentimental attachments, the value of the rehabilitated house must be decided by the individual concerned. However, neither the finance company nor a prospective buyer will allow anything for sentimental value.

Chapter 3
Planning the Job

After the house has been examined, plans should be made to improve the layout, provide more space, add modern conveniences, and improve the appearance. Here is where you can use your imagination to adapt the existing structure to modern living patterns, conveniences, and concepts of comfort. This increases the value of the house and gives a continuous benefit in convenience and liveability.

General Layout

In the layout of a house, it is desirable to provide separate zones for various family functions and to provide good traffic circulation through and between areas. Where cost is the most important consideration, this may not be practical. Even where cost is not a major limitation, there will have to be compromises in the layout. The considerations presented here are goals to aim for where practical; inability to fully satisfy them should not prevent remodeling the home to create a sound, comfortable dwelling.

The layout of the house should be zoned to provide three major family functional areas for relaxation, working, and privacy. The relaxation zone will include recreation, entertaining, and dining. In the small house these may all be performed in one room, but the larger house may have living room, dining room, family room or den, study, and recreation room. The latter is frequently in the basement. The working zone includes the kitchen, laundry or utility room, and may include an office or shop. The privacy zone consists of bedrooms and baths; it may include the den which can double as a guest bedroom. In some layouts a master bedroom and bath may be located away from the rest of the bedrooms so the privacy zone is actually in two parts.

Zones within the house should be located for good relationship to outdoor areas. If outdoor living in the backyard is desirable, perhaps the living room should be at the back of the house. The working zone should have good access to the garage, the dining room, and outdoor work areas. The main entrance to the house should have good access to the driveway, or usual guest parking area, which may be on either the front or side of the house. In considering where to locate rooms and entrance, conventional arrangement of the past should not be binding, but convenience in the particular situation should govern.

Traffic Circulation

One of the most important items in layout is traffic circulation. Ideally there should be no traffic through any room. This is usually difficult to accomplish in the living and work areas. A more feasible plan is to keep traffic from cutting through the middle of the room. Many older houses have doors centered in the wall of a room; this not only directs traffic through the middle of the room, but also cuts the wall space in half, making furniture arrangement difficult. Study the plan and observe where a door might be moved from the middle of a wall space to the corner of the room; however, movement of doors is costly and should be limited in the moderate remodeling job. Also consider where doors might be eliminated to prevent traffic through a room. Figure 3-1 shows examples of improved layout through relocation of 4 doors.

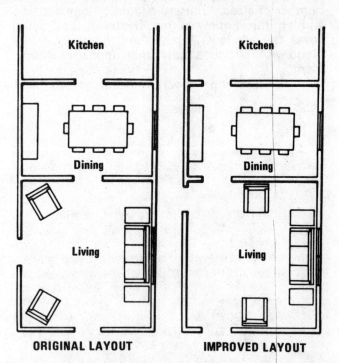

ORIGINAL LAYOUT   IMPROVED LAYOUT

**Relocation of doors to direct traffic
to one side of rooms
Figure 3-1**

Changing Partitions

Often rooms are not the desired sizes, and it may be necessary to move some partitions. This is not difficult if the partition is nonload-bearing and plumbing, electrical, or heating services are not concealed in the partition. It is possible to move even load-bearing partitions by adding a beam to support the ceiling where the partition is removed.

To determine whether a partition is load-bearing or not, check the span direction of ceiling joists. If joists are parallel to the partition, the partition is usually nonload-bearing (Figure 3-2); however, it may be supporting a second floor load so this should be checked. In most constructions where the second floor joists are perpendicular to the partition, they do require support so the partition is load-bearing (Figure 3-2). An exception would occur when trusses span the width of the building, making all partitions nonload-bearing.

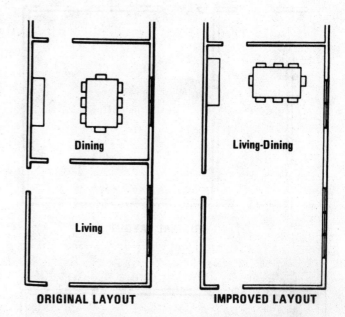

ORIGINAL LAYOUT   IMPROVED LAYOUT

**Removal of partition for better space utilization
Figure 3-3**

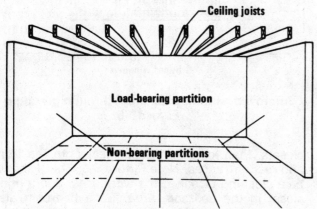

**Load-bearing and nonload-bearing partitions.
[A second-floor load may place a load
on any partition]
Figure 3-2**

Although removal of a non-bearing partition will not require a structural modification, the wall, ceiling, and floor will require repairs where the partition intersected them.

If rooms are small, it may be desirable to remove partitions to give a more open spacious feeling. A partition between living and dining rooms can be removed to make both seem larger and perhaps make part of the space serve a dual purpose (Figure 3-3). In some instances unneeded bedrooms adjacent to the living area can be used to increase living room or other living space by removing a partition. In some rehabilitation, removing a partition between a hallway and a room in the living area gives a more spacious feeling even though traffic continues through the hallway (Figure 3-4).

Window Placement
Windows influence general arrangement and so should also be considered in the layout. Moving windows is costly, involving changes in studs, headers, interior and exterior finish, and trim. The number of windows relocated should be very

limited in the moderate-cost remodeling job but, where changes are practical, properly placed windows can enhance the livability of a house. Where possible, avoid small windows scattered over a wall, as they cut up the wall space and make it unusable. Attempt to group windows into one or two large areas and thus leave more wall space undisturbed (Figure 3-5). Where there is a choice of outside walls for window placement, south walls rank first in cold climates. Winter sun shines into the room through a south window and heats the house, but in the summer the sun is at a higher angle so that even a small roof overhang shades the window. In extremely warm climates, north windows may be preferable to south windows to avoid heat gain even in the winter. West windows should be avoided as much as possible because the late afternoon sun is so low that there is no way of shading the window.

Windows provide three major functions: They admit daylight and sunlight, allow ventilation of the house, and provide a view. Points to consider in planning for each of these three functions are discussed below.

Some general practices to insure adequate light are:

1. Provide glass area equal to 10 percent of the floor area of each room.

2. Place principal window areas toward the south except in warm climates.

3. Group window openings in the wall to eliminate undesirable contrasts in brightness.

4. Screen only those parts of the window that open for ventilation.

5. Mount draperies, curtains, shades, and other window hangings above the head of the window and to the side of the window frame to free the entire glass area.

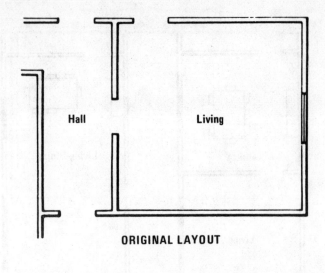

Hall   Living

ORIGINAL LAYOUT

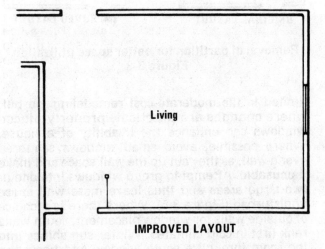

Living

IMPROVED LAYOUT

Removal of partition for more spacious feeling
Figure 3-4

To insure good ventilation, some practices to follows are:

Provide ventilation in excess of 5 percent of the floor area of a room.

2. Locate the ventilation openings to take full advantage of prevailing breezes.

3. Locate windows to effect the best movement of air across the room and within the level where occupants sit or stand. (Ventilation openings should be in the lower part of the wall unless the window swings inward in a manner to direct air downward.)

To provide a good view:

1. Minimize obstructions in the line of sight from sitting or standing, depending on the use of the room.

2. Determine sill heights on the basis of room use and furniture arrangement.

Closets

An item sometimes overlooked in planning the layout is closets. Most older houses have too few

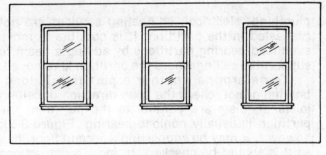

Spaced windows

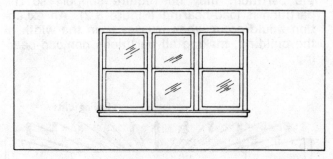

Grouped windows

Improved window placement through grouping
Figure 3-5

closets. Plan for a coat closet near both the front and rear entrances. Note Figure 3-6. There should be a cleaning closet in the work area, and a linen closet in the bedroom area. Each bedroom also requires a closet. If bedrooms are large, it will be a simple matter to build a closet across one end of the room. The small house will present more difficulty. Look for wasted space, such as at the end of a hallway or at a wall offset. If the front door opens directly into the living room, a coat closet can sometimes be built beside or in front of the door to form an entry (Figure 3-6). In the story-and-a-half house, closets can often be built into the attic space where the headroom is too limited for occupancy.

Closets used for hanging clothes should ideally be at least 24 inches deep, but shallower closets are also practical. Other closets can vary in depth depending on their use, but a depth greater than 24 inches is usually impractical. The exception is the walk-in storage closet, which is very useful and should be considered if space is available. Where existing closets are narrow and deep, rollout hanging rods can make them very usable. To make the best use of closet space, plan for a full-front opening.

In many remodeling situations, plywood wardrobes may be more practical than conventional closets that require studs, drywall, casing, and doors. Very simple plywood wardrobes are illustrated in a later chapter. More elaborate closets can be built by dividing the wardrobe into a variety of spaces for various types of storage and installing appropriate doors or drawers.

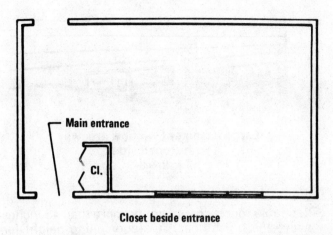

Closet beside entrance

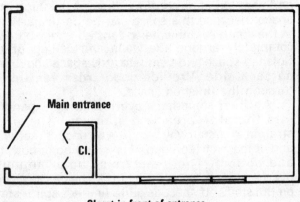

Closet in front of entrance

Entry formed by coat closet
Figure 3-6

Porches

Porches on older houses are often very narrow, have sloping floors, and cannot easily be enlarged. They do not lend themselves well to the type of outdoor living, such as dining and entertaining, usually desired today. Consider this in determining whether or not to retain an existing porch and in planning a new porch that will be used.

Expansion Within The House

Regardless of the size of a house, there always seems to be a need for more space. Seldom is there enough storage space. Work space for a shop or other hobby area is nearly always in demand. More recreational and other informal living space are usually desired. Family rooms are seldom found in older homes. Furthermore, in some houses rooms may be small or additional bedrooms may be needed. Many older houses have only one bathroom, so bathroom additions are often required. Often there are possibilities for expansion by using existing unfinished space such as the attic, basement, or garage. If expansion into these areas is not practical, an addition can be built.

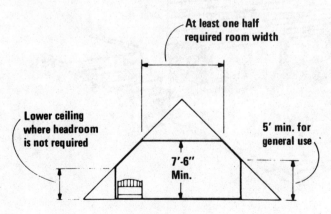

Headroom requirements for attic rooms
Figure 3-7

Attic

The house with a relatively steep roof slope may have some very usable attic space going to waste. It can be made accessible for storage space by the addition of fold-down stairs. If this space is usable for living area and has an available stairway, it may be finished to form additional bedrooms, a den, a study, a hobby room, or an apartment for a relative who lives in the house but is not a member of the immediate family.

Where the attic is at the third floor level, local codes should be checked. Some codes do not permit use of the third floor for living areas; others requires a fire escape.

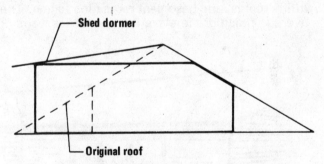

Shed dormer for additional attic space
Figure 3-8

The head room requirements for attic rooms are a minimum ceiling height of 7 feet 6 inches over at least one-half the room, and a minimum ceiling height of 5 feet at the outer edges of the room (Figure 3-7). The space with a lower ceiling height could be used for bunks or other built-in furniture or as storage space.

If there is sufficient head room only in a narrow strip at the center of the attic, consider building a large shed dormer to increase the usable space (Figure 3-8). Dormers may be required for windows even though they are not needed for increasing space.

It is important that insulation, vapor barriers,

and ventilation be considered when finishing attic space, because this space can be particularly hot in the summer. Insulate and install vapor barriers completely around the walls and ceiling of the finished space and ventilate attic space above and on each side. Provide good cross ventilation through the finished space.

An item sometimes overlooked in expanding into the attic area is a stairway. The usual straight-run stairway requires a space 3 feet wide and at least 11 feet long plus a landing at both top and bottom. There must also be a minimum overhead clearance of 6 feet 8 inches at any point on the stairs. If space is quite limited, spiral stairs may be the solution. Some spiral stairs can be installed in a space as small as 4 feet in diameter; however, code limitations should be checked. While these are quite serviceable, they are a little more difficult to ascend or descend, and so may not be suitable for the elderly. Spiral stairs will not accommodate furniture, so there must be some other access for moving furniture into the upstairs space.

Basement

An unfinished basement may be one of the easiest places for expansion, although certain conditions must be met if the basement is to be used for habitable rooms. A habitable room is defined as a space used for living, sleeping, eating, or cooking. Rooms not included and therefore not bound by the requirements of habitable rooms include bathrooms, toilet compartments, closets, halls, storage rooms, laundry and utility rooms, and basement recreation rooms. The average finish grade elevation at exterior walls of

**Large basement window areaway
with sloped sides
Figure 3-9**

habitable rooms should not be more than 48 inches above the finish floor. Average ceiling height for habitable rooms should be not less than 7 feet 6 inches. Local codes should be checked for exact limitations. Other basement rooms should have a minimum ceiling height of 6 feet 9 inches.

Dampness in the basement can be partially overcome by installing vapor barriers on the floor and walls if they were not installed at the time of construction. In an extremely damp basement, plan to use a dehumidifier for summer comfort.

One of the main disadvantages of basement rooms is the lack of natural light and view. If the house is on a sloping lot or graded in a manner to permit large basement windows above grade, the basement is much more usable than in the house where the basement has only a few inches of the top of the wall above grade. However, even the completely sunken basement can have natural light if large areaways are built for windows, and the walls of the areaways are sloped so that sunlight can easily reach the windows (Figure 3-9). At least one window large enough to serve as

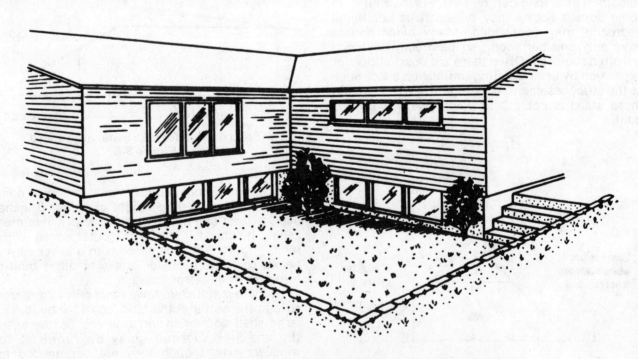

**Sunken garden forming large basement window areaway
Figure 3-10**

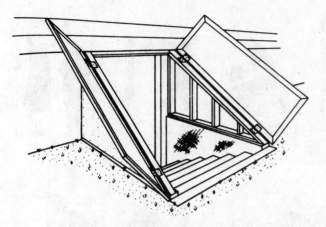

Areaway type of basement entrance
Figure 3-11

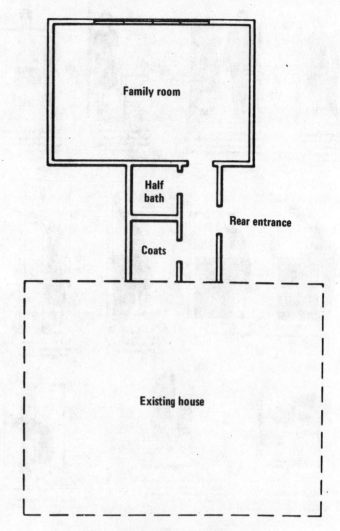

Addition using the satellite concept
Figure 3-12

a fire exit is recommended and often required by code. It may be convenient to extend the areaway to form a small sunken garden (Figure 3-10), but adequate drainage must be included.

The usefulness of the basement may also be increased by adding a direct outside entrance. This adds to fire safety by giving an alternate exit and is particularly desirable if the basement is to be used as a shop or for storing lawn and garden equipment. If an outside stairway is provided, try to place it under cover of a garage, breezeway, or porch to protect it from ice, snow, and rain. Otherwise, use an areaway-type entrance with a door over it (Figure 3-11).

Garage

Another place for expansion is the garage built as an integral part of the house. If the garage is well built, the only work is in finishing, which is much less costly than adding onto the house. The main consideration is whether the additional finished space is needed more than a garage. The garage is often adjacent to the kitchen, so that it is an ideal location for a large family room. It could also be used for additional bedrooms and possibly another bathroom.

Walls and ceiling of the garage can be finished in any conventional material. The floor will probably require a vapor barrier, insulation, and a new subfloor. It may be convenient to use the existing garage door opening to install a large window or a series of windows; otherwise the door opening can be completely closed and windows added at other points.

Additions

After considering all the possibilities of expanding into the attic, basement, or garage, if space requirements are not met, the only alternative left is to construct an addition. The house on a small lot may have minimum setback limitations that will present problems. Local zoning and lot restrictions should be checked. The distance from the front of the house to the street is

usually kept the same for all houses on a particular street, so expanding to the front may not be permitted. Often the house also has the minimum setback on the sides, preventing expansion on either side. Thus, the only alternative may be to add on behind the house. A house in the country usually can be expanded in any direction without restriction.

Use the addition for most critically needed space. If the need is for more bedrooms, but a much larger living room is also desirable, maybe the present living room can be used as a bedroom and a large living room can be added. If the main requirement is a large modern kitchen, add on a new kitchen and use the old kitchen as a utility room, bathroom, or some other type of living or work space. The important thing for good appearance is that the addition be in keeping with the style of the house. Rooflines, siding, and windows should all match the original structure as closely as possible to give the house continuity, rather than giving the appearance that something has just been stuck on.

25

Walking between two high walls (space adequate for both men and women)

Two people passing (figure derived; twice the space for one person to walk between two high walls)

Walking between high wall and 30" high table (space adequate for both men and women)

Walking with elbows extended (space adequate for both men and women)

Kneeling on one knee (woman only)

Man bending at a right angle

Reaching, maximum height

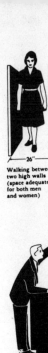

Reaching over obstruction, 24" deep and 36" high

Reaching over obstruction, 12" deep and 36" high (women only)

Maximum reach to back of shelf 12" deep (women only)

Using a conventional range

Using a wall oven

Using a refrigerator

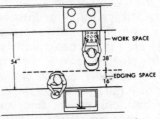
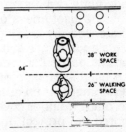

Using a base cabinet

Using a front-opening dishwasher requires 4 inches more space than using other appliances in a kitchen

Using a cleaning closet

Minimum space (allowing for edging) for two people working at cabinets and appliances opposite each other (except a front-opening dishwasher)

Liberal space (allowing for walking) for two people working at cabinets and appliances opposite each other (except a front-opening dishwasher)

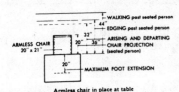

WALKING past seated person 44"
EDGING past seated person 32"
ARMLESS CHAIR 20" x 21" — ARISING AND DEPARTING CHAIR PROJECTION 20" 36" (seated person)
20" MAXIMUM FOOT EXTENSION

Armless chair in place at table

Rising from table, armless chair (armchair 2" more)

Walking past seated person

Dining areas for eight persons with free-standing table 72" x 40", one armchair, and seven armless chairs (calculated on basis of edging space on sides where there is not serving space, so that everyone can leave his place without disturbing others)

Serving space on one side and one end

Foot extension, knees crossed, not at table

WALKING past seated person 38"
EDGING past seated person 34"
ARMCHAIR 22" x 23" — ARISING AND DEPARTING CHAIR PROJECTION 22" (seated person)
20" MAXIMUM FOOT EXTENSION

Armchair in place at table

Edging past seated person

Serving space on two sides and one end

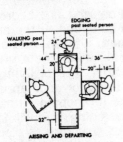

EDGING past seated person 24"
WALKING past seated person 44" 20" 36" 20" 16"

ARISING AND DEPARTING

Using tables and chairs in free area

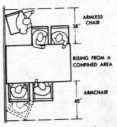

ARMLESS CHAIR 38"
RISING FROM A CONFINED AREA
ARMCHAIR 40"

Using tables and chairs in confined area

Arising from a card table

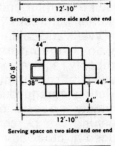

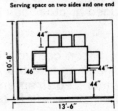

Serving space all around table

Space standards
Figure 3-13

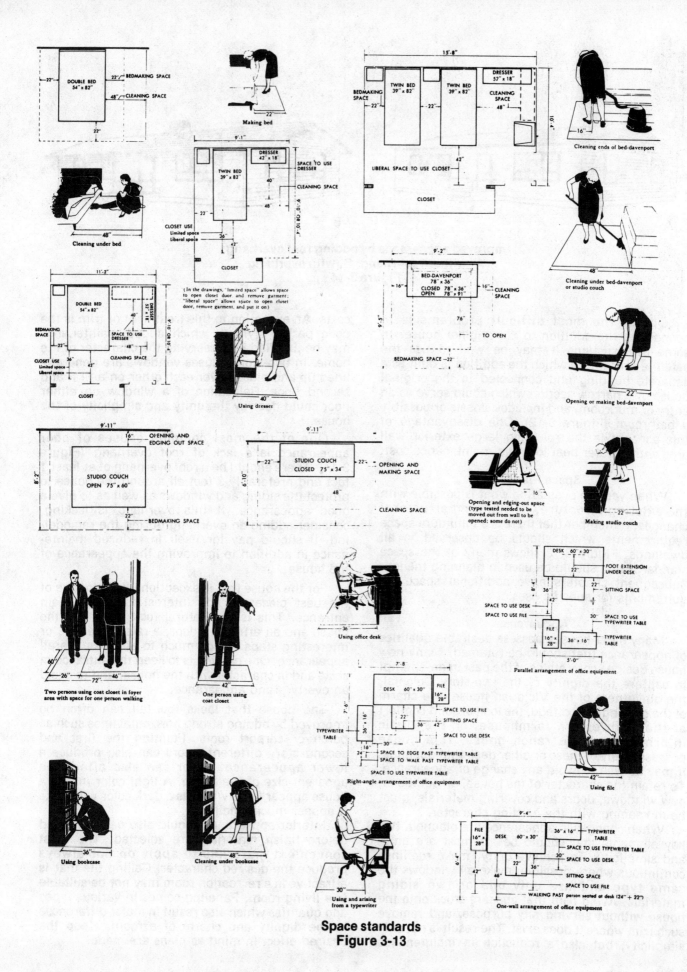

Space standards
Figure 3-13

27

A   B

**Improved appearance by adding roof overhang;
A, without overhang; B, with overhang
Figure 3-14**

One of the most difficult problems is in connecting the addition to the original house. In some constructions it may be well to use the satellite concept, in which the addition is built as a separate building and connected to the original house by a narrow section which could serve as an entry or mudroom, and includes closets or possibly a bathroom (Figure 3-12). One disadvantage of this concept is the resulting large exterior wall area with greater heat loss and maintenance cost.

Space Planning

When you are evaluating what is possible with the existing structure, or planning structural changes, remember that there are minimum space requirements which should be observed in all dwellings. Figure 3-13 shows many of the space standards that should be used in planning the job. Subsequent chapters have additional space requirements for specific uses.

Appearance

Many older houses possess desirable qualities of appearance that should be retained. Many new house designs copy styles of the past in an attempt to capture the dignity of the two-story Colonial, the quaintness of the Victorian house, the charm of the old English cottage, the look of solid comfort of the Midwestern farmhouse, or the rustic informality of the ranch house. If a house possesses any of these or other desirable qualities, it may be well to avoid any change of appearance. To retain the character of the house, all additions, new windows, doors and covering materials, must be in keeping with the existing character.

When changes in appearance are planned, two key elements that should be foremost are unity and simplicity. To achieve unity, make rooflines continuous where possible, make all windows the same type and use only one or two siding materials. Avoid trim that appears stuck onto the house without serving any purpose, and remove such trim where it does exist. The result is not only simplicity, but also a reduction in maintenance

costs. An exception to this treatment of trim is the good period style in which added maintenance may be justified in preserving the character of the home. In two-story houses windows are generally lined up and placed over each other on a first and second floor. Relocation of a window on either floor could destroy the unity and simplicity of the house.

One of the most common causes of poor appearance is a lack of roof overhang (Figure 3-14). There should be a roof overhang of at least 1 foot and preferably 2 feet all around the house to protect the siding and windows as well as to give a good appearance. If this overhang is lacking, consider adding an overhang during the remodeling. It should pay for itself in reduced maintenance in addition to improving the appearance of the house.

For the house that is exceptionally plain, one of the best places to add interest is at the main entrance. This is the natural focal point for the house and an attractive door, a raised planter, or interesting steps can do much to enhance overall appearance. One caution is to keep the entrance in scale and in character with the house, and to avoid an overly grand appearance.

The house that looks too tall can often be improved by adding strong horizontal lines such as porch or carport roofs. Painting the first and second story different colors can also produce a lower appearance. Color can also affect the apparent size of the house. A light color makes a house appear large whereas a dark color will make it appear much smaller.

Interior appearance should also be considered before finish materials are selected. The most convenient materials to apply do not always produce the desired character. Ceiling tile that is attractive in a recreation room may not be suitable in a living room. Paneling comes in various types and qualities which also result in major differences in the dignity and charm of a room. Keep the desired effect in mind as plans are made.

Chapter 4

Making Structural Repairs

With the information from a thorough examination of the house and a complete plan for changes to be made, the actual work can begin. The order in which components of the house are repaired or remodeled will vary with each situation; however, the first step with a house that is not level should be to level the house and floor system. A level base to work from is essential. Roof repairs and changes in windows and exterior doors should be made before interior work begins. All changes in plumbing, electrical wiring, and heating should also be completed before interior finish work is started.

Certain general information will be useful in selecting materials for repairing or remodeling jobs:

(1) Moisture content of framing material should not exceed 19 percent, and a maximum of 15 percent would be better in most areas of the United States.

(2) Recommended moisture content for interior finish woodwork varies from 6 to 11 percent depending on the area of the United States. A map showing recommended average moisture contents is shown in Figure 4-1.

(3) Plywood should be exterior type anywhere

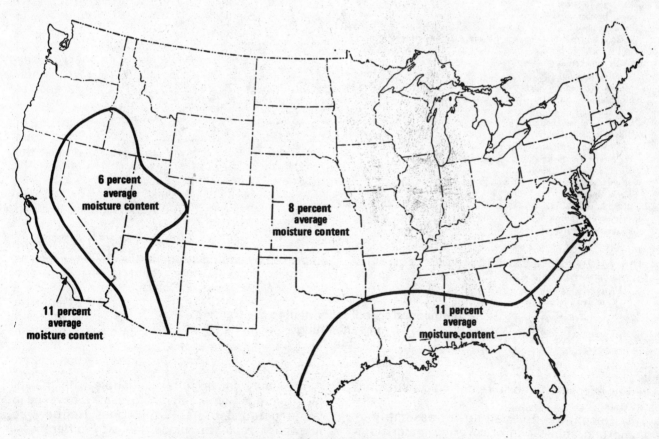

Recommended average moisture content for interior finish woodwork in different parts of the United States
Figure 4-1

Joining	Nailing method	Number	Size	Placement
Header to joist	End nail	3	16d	—
Joist to sill or girder	Toenail	2	10d	—
			or 8d	
Header and stringer joist to sill	Toenail	—	10d	16 in. on center
Bridging to joist	Toenail each end	2	8d	—
Ledger strip to beam, 2 in. thick	—	3	16d	At each joist
Subfloor, boards:				
1 by 6 in. and smaller	—	2	8d	To each joist
1 by 8 in.	—	3	8d	To each joist
Subfloor, plywood:				
At edges	—	—	8d	6 in. on center
At intermediate joists	—	—	8d	8 in. on center
Subfloor (2 by 6 in., T&G) to joist or girder	Blind nail (casing) and face nail	2	16d	
Soleplate to stud, horizontal assembly	End nail	2	16d	At each stud
Top plate to stud	End nail	2	16d	—
Stud to soleplate	Toenail	4	8d	—
Soleplate to joist or blocking	Face nail	—	16d	16 in. on center
Doubled studs	Face nail, stagger	—	10d	16 in. on center
End stud of intersecting wall to exterior wall stud	Face nail	—	16d	16 in. on center
Upper top plate to lower top plate	Face nail	—	16d	16 in. on center
Upper top plate, laps and intersections	Face nail	2	16d	—
Continuous header, two pieces, each edge	—	—	12d	12 in. on center
Ceiling joist to top wall plates	Toenail	3	8d	—
Ceiling joist laps at partition	Face nail	4	16d	—
Rafter to top plate	Toenail	2	8d	—
Rafter to ceiling joist	Face nail	5	10d	—
Rafter to valley or hip rafter	Toenail	3	10d	—
Ridge board to rafter	End nail	3	10d	—
Rafter to rafter through ridge board	Toenail	4	8d	—
	Edge nail	1	10d	—
Collar beam to rafter:				
2-in. member	Face nail	2	12d	—
1-in. member	Face nail	3	8d	—
1-in. diagonal let-in brace to each stud and plate (four nails at top)	—	2	8d	—
Built-up corner studs:				
Studs to blocking	Face nail	2	10d	Each side
Intersecting stud to corner studs	Face nail	—	16d	12 in. on center
Built-up girders and beams, three or more members	Face nail	—	20d	32 in. on center, each side
Wall sheathing:				
1 by 8 in. or less, horizontal	Face nail	2	8d	At each stud
1 by 6 in. or greater, diagonal	Face nail	3	8d	At each stud
Wall sheathing, vertically applied plywood:				
3/8 in. and less thick	Face nail	—	6d }	6-in. edge and 12-in. intermediate
1/2 in. and over thick	Face nail	—	8d }	
Wall sheathing, vertically applied fiberboard:				
1/2 in. thick	Face nail	—	(1) }	3-in. edge and 6-in. intermediate
25/32 in. thick	Face nail	—	(2) }	
Roof sheathing, board, 4-, 6-, 8-in. width	Face nail	2	8d	At each rafter
Roof sheathing, plywood:				
3/8 in. and less thick	Face nail	—	6d }	6-in. edge and 12-in. intermediate
1/2 in. and over thick	Face nail	—	8d }	

1 1 1/2-in. roofing nail.
2 1 3/4-in. roofing nail.

**Recommended schedule for nailing the framing
and sheathing
Table 4-2**

it will be exposed to moisture during construction or in use.

(4) Recommended nailing for assembly of framing and application of covering materials is listed in Table 4-2. Sizes of common nails are shown in Figure 4-3.

(5) Actual lumber sizes have changed and new lumber may not be fully compatible with that of old framing. Some older homes may have been constructed using full thickness lumber, so allowance must be made for size differences.

Halting Insect Damage
Subterranean termites are the only wood-

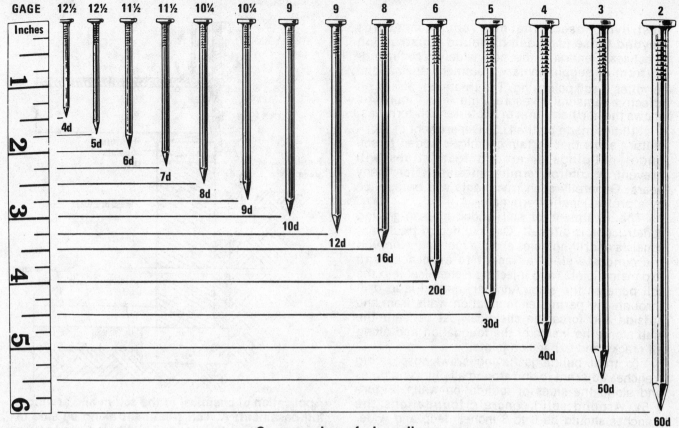

Common sizes of wire nails
Figure 4-3

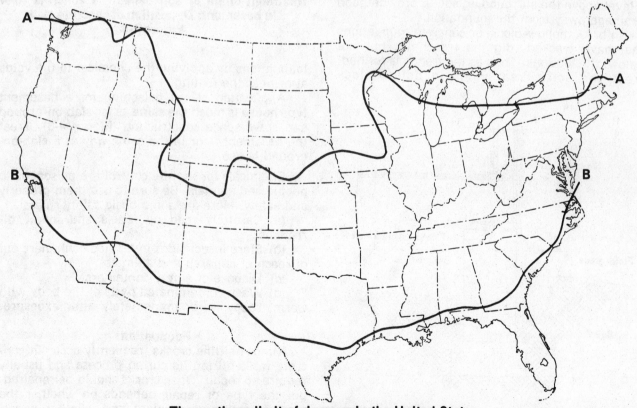

**The northern limit of damage in the United States
by subterranean termites, line A; by
dry-wood or nonsubterranean termites, line B**
Figure 4-4

31

destroying insects that may require measures beyond those provided by sound construction practices. Generally, the most widely recommended form of supplementary treatment against the termites is soil poisoning. The treatment also is an effective remedial measure. Line A on Figure 4-4 shows the northern limit of subterranean termites.

Studies made by the U.S. Department of Agriculture show that certain chemicals added to soil under buildings or around foundations will prevent or control termite infestation for many years. Generally such treatments will be applied by a professional exterminator.

The treatment of soil under slab-on-ground construction is difficult. One method of treatment consists of drilling holes about a foot apart through the concrete slab, adjacent to all cracks and expansion joints, and injecting a chemical into the soil beneath the slab. Another method is to drill through the perimeter foundation walls from the outside and force the chemical just beneath the slab along the inside of the foundation and along all cracks and expansion joints.

To treat buildings having crawl spaces, dig trenches adjacent to and around all piers and pipes and along the sides of foundation walls (Figure 4-5). Around solid concrete foundations the trenches should be 6 to 8 inches deep and wide. A chemical is poured into the trench and, as the excavating soil is put back into the trench, it also is treated. The soil is tamped, and the trench filled to a level above the surrounding soil to provide good drainage away from the foundation.

In brick, hollow-block, or concrete foundations that have cracked, dig the trench to, but not below, the footings. Then as the trench is refilled, treat the soil. Treat voids in hollow-block

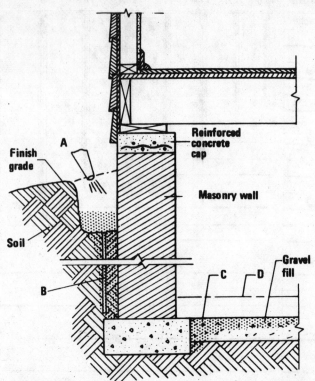

Application of chemical to the soil in and around a full basement: A, soil treatment along outside of the foundation; B, pipe and rod hole from bottom of trench to the top of the footing to aid distribution of the chemical; C, drill holes for the treatment of fill or soil beneath a concrete floor in basement; D, position of concrete slab
Figure 4-6

foundations by applying the chemical to the voids at or near the footing.

Application of soil poisoning for a basement type house is much the same as for slab-on-ground and crawl space construction (Figure 4-6). Treat the basement floor in the same way as a slab-on-ground house.

Chemicals for termite control are poisonous to people and animals. Be sure to use them properly and safely. Here are some basic safety rules:

(a) Carefully read all labels and follow directions.

(b) Store insecticides in labeled containers out of reach of children and animals.

(c) Dispose of empty containers.

(d) Wash contaminated parts of the body with warm, soapy water immediately after exposure.

Foundations

Minor hairline cracks frequently occur in concrete walls during its curing process and usually require no repair. Open cracks should be repaired, but the type of repair depends on whether the crack is active or dormant and whether waterproofing is necessary. One of the simplest methods of determining if the crack is active is to place a mark at each end of the crack and observe

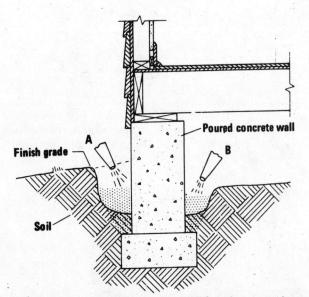

Application of chemical to crawl-space construction soil treatment: A, along outside wall; and B, inside foundation wall
Figure 4-5

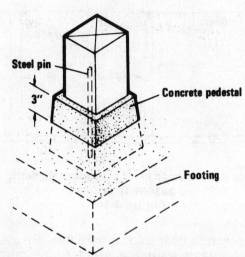

Basement post on pedestal above the floor
Figure 4-7

at future dates whether the crack extends beyond the marks.

If the crack is dormant, it can be repaired by routing and sealing. Routing is accomplished by following along the crack with a concrete saw or chipping with hand tools to enlarge the crack near the concrete surface. The crack is first routed ¼ inch or more in width and about the same depth; then the routed joint is rinsed clean and allowed to dry. A joint sealer such as an epoxy-cement compound should then be applied in accordance with manufacturer's instructions.

Active cracks require an elastic sealant. These should also be applied in accordance with manufacturer's instructions. Sealants vary greatly in elasticity, so a good quality sealant that will remain pliable should be used. The minimum depth of routing for these sealants is ¾ to 1 inch, and the width is about the same. The elastic material can then deform with movement of the crack. Strip sealants which can be applied to the surface are also available, but these protrude from the surface and may be objectionable.

Where masonry foundations or piers have crumbling mortar joints, these should be repaired. First chip out all loose mortar and brush thoroughly to remove all dust and loose particles. Before applying new mortar, dampen the clean surface so that it will not absorb water from the mixture. Mortar can be purchased premixed. It should have about the consistency of putty and should be applied like a caulking material. For a good bond, force the mortar into the crack to contact all depressions. Then smooth the surface with a trowel. Provide some protection from sun and wind for a few days to keep the mortar from drying out too fast.

Uneven settlement in a concrete foundation due to poor footings or no footings at all usually damages the foundation to the point that precludes repair. In a pier foundation the

individual pier or piers could be replaced or, if the pier has stopped settling, blocking could be added on top of the pier to level the house. In either situation, the girder or joists being supported must be jacked and held in a level position while the repairs are being made.

Basement Posts

Any type of basement post may have settled due to inadequate footings, or wood posts may have deteriorated due to decay or insect damage. To correct either problem, a well supported jack must be used to raise the floor girder off the post in question. This releveling must be done slowly and carefully to avoid cracking the plaster in the house walls. Steel jack posts are convenient replacements for the post removed. If a wood post is used, a pedestal should be built to raise the base of the post slightly above the floor surface (Figure 4-7). This allows the end of the post to dry out if it becomes wet.

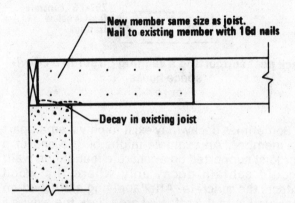

Repair of joint with decay in end contacting the foundation
Figure 4-8

Floor System

If examination of the floor framing revealed decay or insect damage in a limited number of framing members, the members affected will have to be replaced or the affected sections repaired. Large scale damage would probably have resulted in classifying the house as not worth rehabilitation. Damaged members should be replaced with preservative treated wood if exposure conditions are severe. To accomplish this, the framing supported by the damaged member must be temporarily supported by jacks in a crawl space house or jacks with blocking in a basement house. A heavy crossarm on top of the jack will support a width of house of 4 to 6 feet. Where additional support is necessary, more jacks are required. Raise the jacks carefully and slowly, and only enough to take the weight off the member to be removed. Excessive jacking will pull the building frame out of square. After the new or repaired member is in place, gradually take the weight off the jack and remove it.

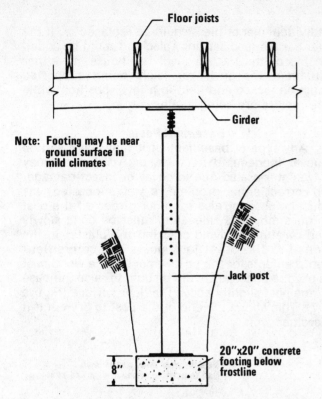

Note: Footing may be near ground surface in mild climates

Jack post

8"

20"x20" concrete footing below frostline

Jack post supporting a sagging girder in a crawl-space house
Figure 4-9

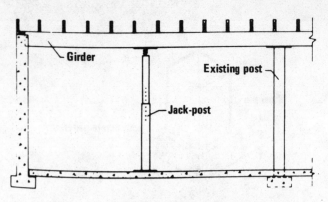

Jack post used to level a sagging girder in a basement house
Figure 4-10

Sometimes decay may exist in only a small part of a member. An example might be the end of a floor joist supported on a concrete foundation wall; it could contain decay only where the wood contacts the concrete. After applying a brushed-on preservative to the affected area, jack the existing joist into place and nail a short length of new material to the side of the joist (Figure 4-8).

After the foundation repairs have been properly made, the support points for the floor should be level; however, the floor may still sag. Where the floor joists have sagged excessively, permanent set may have occurred and little can be done except by replacement of the floor joists. A slight sag can be overcome by nailing a new joist alongside alternate joists in the affected area. If the new joists are slightly bowed, place them crown up. Each joist must be jacked at both ends to force the ends to the same elevation as the existing joist. The same treatment can be used to stiffen springy floors.

Girders that sag excessively should be replaced. Excessive set cannot be removed. Jack posts can be used to level slightly sagged girders or to install intermediate girders, but unless the space is little used or jack posts can be incorporated into a wall, they are generally in the way Methods for installing jack posts are shown in Figures 4-9 and 4-10.

When jack posts are used to stiffen a springy floor or to carry light loads, they can be set directly on the concrete floor slab. Where they are used to support heavy loads, a steel plate may be necessary to distribute the load over a larger area of the floor slab. The jack post should not be used for heavy jacking. Where heavy jacking is involved, use a regular jack to carefully lift the load and then put the jack post in place.

Squeaks in flooring frequently are caused by relative movement of the tongue of one flooring strip in the groove of the adjacent strip. One of the simplest remedies is to apply a limited amount of mineral oil to the joints.

Sagging floor joists often pull away from the subfloor and result in excessive deflection of the floor. If this is the cause of squeaks, squeeze construction mastic into the open joints. An alternate remedy is to drive small wedges into the spaces between joints and subfloor (Figure 4-11). Drive them only far enough for a snug fit. This method of repair should be limited to a small area.

Undersized floor joists that deflect excessively are also a major cause of squeaks. The addition of girders (described previously) to shorten the joist span is the best solution to that problem.

Strip flooring installed parallel to the joists may also deflect excessively. Solid blocking nailed between joists and fitted snugly against the

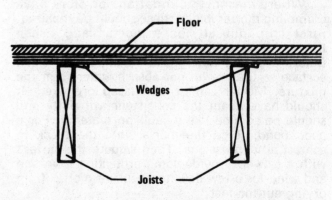

Wedges driven between joists and subfloor to stop squeaks
Figure 4-11

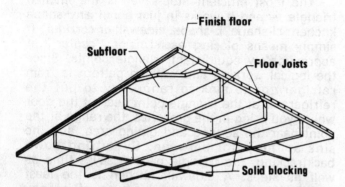

Solid blocking between floor joists where finish floor is laid parallel to joists
Figure 4-12

subfloor (Figure 4-12) will prevent this deflection if installed at relatively close spacing.

One of the most common causes of squeaking is inadequate nailing. To correct this, drive a nail through the face of the flooring board near the tongue edge into the subfloor—preferably also into a joist. Set the nail and fill the hole. A less objectionable method from the standpoint of appearance is to work from under the floor using screws driven through the subfloor into the finish floor. This method will also bring warped flooring into a flat position.

Roof System

The first steps in repairing a roof system are to level sagging ridge poles and straighten sagging rafters. Then the new roof covering can be applied to a smooth, flat surface.

The sagging ridgepole can sometimes be leveled by jacking it at points between supports and installing props to hold it in a level position. When this is done, the jack must be located where the load can be traced down through the structure so the ultimate bearing is directly on the foundation. Where there is no conveniently located bearing partition, install a beam under the ridge and transfer the load to bearing points. After the ridgepole is jacked to a level position, cut a 2 by 4 just long enough to fit between the ceiling joist and ridgepole or beam and nail it at both ends (Figure 4-13). For a short ridgepole, one prop may be sufficient. Additional props should be added as needed. In some repairs the addition of collar beams may be sufficient without requiring props.

Where rafters are sagging, nail a new rafter to the side of the old one after forcing the new rafter ends into their proper position. Permanent set in the old rafters cannot be removed.

Sheathing may have sagged between rafters, resulting in a wavy roof surface. Where this condition exists, new sheathing is required. Often the sheathing can be nailed right over the old roofing. This saves the labor of removing old roofing and a big cleanup job. Wood shingles which show any indication of decay should be completely removed before new sheathing is applied. Where wood shingles are excessively cupped or otherwise warped, they should also be removed. Wood shingle and slate roofs of older houses were often installed on furring strips rather than on solid sheathing.

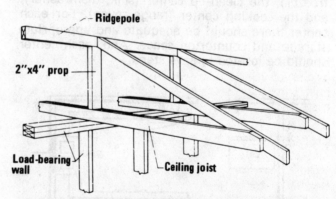

Prop to hold sagging ridgepole in level position
Figure 4-13

Sheathing nailed over existing sheathing or over sheathing and roofing must be secured with longer nails than would normally be used. Nails should penetrate the framing 1¼ to 1½ inches. Nail edges of plywood sheathing at 6 inch spacing and to intermediate framing members at 12 inch spacing. Apply the plywood with the length perpendicular to the rafters. For built-up roofs, if the plywood does not have tongued-and-grooved edges, use clips at unsupported edges. Clips are commercially available and should be installed in accordance with the fabricator's instructions. For 16 inch rafter spacing, 3/8 inch plywood is the minimum thickness to be used and ½ inch thick plywood is preferable.

Chapter 5

Remodeling Kitchens

New appliances and present concepts of convenience have revolutionized the kitchen in recent years. More space is required for the numerous appliances now considered necessary.

In spite of changing requirements, certain principles of kitchen planning will always apply. The four generally recognized arrangements for kitchens are: the ''U'' and ''L'' types, the corridor, and the sidewall (Figure 5-1). The arrangement selected depends on the amount of space, the shape of the space, and the location of doors. If the kitchen is going to be a new addition, select the layout preferred, and plan the addition in accordance with the layout.

No matter how you dress it up, a kitchen is essentially a workroom where efficiency is never out of place. The kitchen is a series of work ''centers,'' each with a major appliance as its hub: the fresh and frozen food center (refrigerator-freezer), the clean-up center (sink, dishwasher), and the cooking center (range, oven). For each center there should be adequate and appropriate storage and countertop space. And, each center should be located to save steps.

The most efficient step-saver is the kitchen triangle — and it works in just about any shape kitchen: U-shape, L-shape, sidewall or corridor. It simply means placing each major ''center'' at approximately equidistant triangle points. Since the logical working and walking pattern is from refrigerator to sink to range, try to put the refrigerator at the triangle point nearest the door where you bring in the groceries, the range at the point near the serving and dining area, and the sink at the point between. By cutting down backtracking, the triangle pattern saves time as well as steps. A sidewall kitchen is the least satisfactory arrangement from an efficiency standpoint but is preferred where space is quite limited. The work triangle is smallest in the ''U'' and corridor layouts. The ''L'' arrangement is used in a relatively square kitchen that must have a dining table in it.

Storage And Counters

Logical storage patterns also save steps. Keep cooking and cleaning items in the work center where they're most often used. Lots of storage

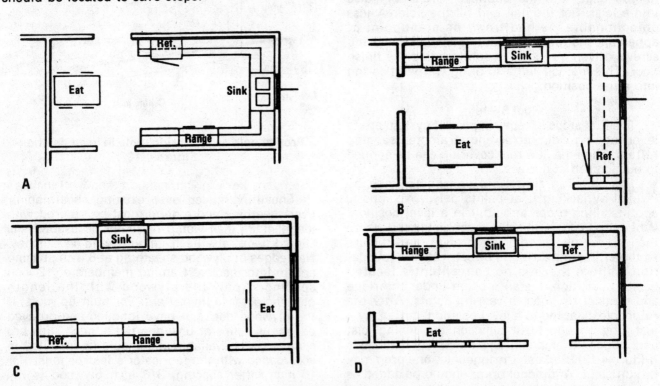

Kitchen arrangements: A, U-type; B, L-type; C, corridor; D, sidewall
Figure 5-1

space for canned foods is a virtual mileage-saver — it cuts trips to the supermarket. In addition to storage, each kitchen center needs its own countertop space. Provide generous amounts on the latch side of the refrigerator-freezer to set down food, on both sides of the sink for food preparation and dish washing, and alongside the range for pots and pans.

The basis for determining the correct amount of storage space necessary in a kitchen is the number of bedrooms in the house. One person for each bedroom, plus an additional person for the master bedroom is considered the number of people occupying the house. Studies have proved that 6 square feet of storage space in wall cabinets is required for each permanent resident of a house. To this must be added 12 square feet for entertaining. Base cabinet storage is measured on a lineal foot basis. As a general rule, base cabinets should occupy all space beneath the wall cabinets not already used by the range, sink, refrigerator or dishwasher. If there is sufficient wall storage space, there will usually be enough base storage. Accordingly, in a two-bedroom house there would be 6 square feet of wall cabinet space for each person or 18 square feet. This, plus the 12 square feet for entertaining and accumulation, total 30 square feet of wall cabinet storage space. This much is needed for such a home.

In some cases window locations limit the amount of wall space available although there is plenty of base cabinet space. In such cases the difference should be made up by installing another wall cabinet, a tall shelf cabinet or a storage closet elsewhere in the room without a base cabinet below. The correct balance of wall storage space above the counter is important because substandard space reduces efficiency, makes for poor housekeeping and creates awkward storage requirements.

It has been found that there are definite lengths advisable for all counter areas. There should be not less than 4½ feet nor more than 5½ feet of counter surface between the refrigerator and sink. There should be no less than 3 feet nor more than 4 feet between the sink and range. There are, of course, variations and additions to these basic principles. Snack bars are often extended out from the wall into the kitchen or at the end of an installation. Here rounded corners are necessary to provide a radius counter top and base end shelves. In large kitchens, islands are sometimes used to save steps. These are assembled by backing base cabinets against each other and covering with a one-piece top.

A clearance of about 18 inches is considered desirable between the counter surface and bottom of wall cabinets. Base cabinets are 36 inches high to correspond with the established height of ranges and sinks. Wall cabinets located over the range should have a clearance of 22 inches to 30 inches above the unit for best results.

Doors and windows should always be located in

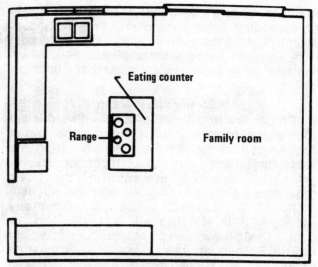

Island counter dividing kitchen and family room
Figure 5-2

relation to corners of rooms and equipment to be installed. Most appliances and cabinets are from 24 inches to 28 inches in depth. To clear such equipment, a door opening should be not less than 30 inches from the corner. In cases where a range is next to a door, a clearance of 48 inches is desirable. Generally doorways in corners should be avoided, and door swings should avoid conflict with the use of appliances, cabinets, or other doors. If swinging the door out would put it in the path of travel in a hall or other activity area, consider using a sliding or folding door. Sliding doors and installation are expensive, but may be worth the expense in certain situations.

Windows should be adequate to make the kitchen a light cheerful place, because the homemaker spends much of her time there. The current trend toward indoor-outdoor living has fostered the "patio kitchen," with large windows over a counter which extends to the outside to provide an outdoor eating counter. This is particularly useful in the warmer climates, but is also convenient for summer use in any climate. The distance between a window opening and a corner should not be less than 15 inches because standard flat wall cabinets are about 13 inches in depth. If this clearance is allowed, wall cabinets can be installed all the way to the corner. Because backsplash at the back of counter tops vary from about 3 inches to 8 inches in height, the underside of window sills should be not less than 44 inches from the floor.

The placement of the sink in relation to windows is a matter of personal preference. Many women like to look out the window while they are at the sink, but installing the sink along an interior plumbing partition is usually less costly than on an outside wall.

If the kitchen is quite large, it may be convenient to use part as a family room. The combined kitchen-family room concept can also be met by removing a partition to expand the kitchen or by adding on a large room. One method of arranging workspace conveniently in such a room is by using an island counter (Figure 5-2), which can also serve as an eating counter for informal dining.

Planning The Kitchen

Proper layout of work centers with sufficient counter space and specialized storage, the kind of appliances that can do the best job within the space limitations, and enough light for ease of seeing and working are some of the basic requirements for planning an efficient kitchen. Although a kitchen may be used for many activities, it is still basically a work room for certain food preparation processes. But the type of use the kitchen will receive should determine its design. Will the kitchen be used as a living area - a place where the family gathers? Should the kitchen be isolated behind closed doors to separate it from children's areas or should it be open where all can participate in activities? Figure 5-3 is a questionnaire which will help planning of equipment and space utilization.

The primary concern is convenience and ease in performing the basic processes of storage, food processing, and clean-up. Planning a kitchen to provide this convenience involves placement of equipment, cabinets, and counter spacing so that there will be a smooth sequence of work. Follow the paragraphs below to design a kitchen that can be used efficiently.

Heading the list is the proper location of major appliances to form a good kitchen work triangle. The total distance should not exceed 22 feet, nor be less than 13 feet. Most kitchen work is performed at the sink; therefore, the sink should be placed between the refrigerator and the range. If a separate cooktop and oven are used, the oven is not included within the work triangle. The flow of work through the kitchen proceeds from the refrigerator, to the sink, to the range, then to the dishwasher for clean-up. A minimum distance of four feet should be maintained between work areas. The flow of household traffic through the kitchen should be kept out of the kitchen work triangle.

There are three primary work centers in a well planned kitchen, each of which includes a major appliance with its related storage cabinets and countertop. All equipment, supplies and utensils should be stored in the area where they are first and most often used. The cooking center should have adequate counter space on both sides of the range or cooktop. A chopping block insert can be added, or ceramic tile installed for landing hot cooking utensils without fear of burns. Avoid installing the range next to a vertical wall return, as this is not recommended by Underwriters Laboratories and major manufacturers. A range should be installed on an exterior wall, if possible, to facilitate venting of the exhaust hood directly to the outside. If wall ovens are used, locate them at the end of a counter run with ample space adjacent to them for placement of hot foods from the ovens. This planning also applies to microwave ovens that can be built-in.

The refrigerator center should be near an exterior doorway. Since this center will be used for storage of staple foods, it should be conveniently located for unloading groceries. Locating the refrigerator at the end of a counter improves the kitchen aesthetically by having this large appliance at the "end" of the kitchen. Be sure, however, the refrigerator door handle is next to the counter, allowing it to be opened away from the counter so items can be easily transferred. Install side-by-side models with counter space on both sides. Avoid placing the refrigerator next to a range or cooktop.

The preparation/clean-up center includes the sink, dishwasher, food waste disposer and trash compactor. The dishwasher should be installed next to the sink. A 24 x 21 inch single bowl sink is usually adequate with a dishwasher. If a double bowl sink is specified, there are many styles available that are adequately sized for installation with a dishwasher. The bowl next to the dishwasher should contain the disposer. The trash compactor is best located near the sink. Plan for one base cabinet 12 or 15 inches wide and a drawer for cutlery and flatware storage beside the dishwasher.

Adequate electrical circuits and outlets must be provided in a functional kitchen design. The kitchen should be provided with heavy circuits for large appliances, such as the range or refrigerator, and with at least three dual convenience outlets on separate circuits. If a "cooking island" is planned, install convenience outlets in its ends below the countertop.

A good kitchen design includes proper lighting. An adequate ceiling fixture is needed to create a soft and pleasing atmosphere with general illumination. Luminous ceilings are desirable for kitchens with little or no window space. False beams with strip light inserts can create an interesting appearance. Special lighting to "spotlight" work areas is required over the sink, and the preparation center.

Proper storage is a must. Base and wall cabinets should provide ample storage for items where they are first used, including accessories such as small appliances, pots, pans or trays. Storage drawers are more functional if they are not over 18 inches wide; larger drawers tend to become too heavy when full. To make use of "dead" corner spaces, corner cabinets can be incorporated. Tall storage should be provided in the kitchen area for brooms and mops along with specialized pantry storage if space is available. Staple food storage needs to be easily accessible and close to the refrigerator-freezer location. It is important that pantry storage and cabinet storage of dry food stuffs be located in a cool dry area of

Kitchen Planning Questionnaire

Street address _____ Owners _____

Number of bedrooms _____ Phone _____ Date _____

1. How many in your family? Adults:_____ , Children:_____ , Ages of children:_____
2. How tall is the person who will use the kitchen most?_____. Right or left handed?_____
3. How much cooking is to be done in the kitchen? ☐One meal a day or less, ☐About two meals a day, ☐Three meals a day.
4. How often do you entertain adult guests? ☐More than once a month, ☐Occasionally, ☐ Seldom.
5. What activities would you like to have in the kitchen besides cooking? ☐Eating, ☐Desk and phone, ☐Bar, ☐Children's play area, ☐Laundry.
6. Do you plan to do most of your eating in the kitchen?
7. Which features in the kitchen you now have do you want to retain? List the brand name.

   ☐Oven _____ ☐Cabinets_____
   ☐Range_____ ☐Lighting _____
   ☐Dishwasher _____ ☐Floor covering _____
   ☐Food waste disposer_____ ☐Eating area _____
   ☐Range hood _____ ☐ _____
   ☐Refrigerator _____ ☐ _____
   ☐Sink _____ ☐ _____

8. What new features do you want planned into the kitchen? ☐ New ceiling finish, type_____, ☐New wall finish, type_____, ☐New floor finish, type_____, ☐Improved lighting, type_____, ☐New cabinets, type_____, ☐Range exhaust hood, type_____, ☐Food waste disposer, brand_____, ☐Dishwasher, brand_____, ☐Trash compactor, brand_____, ☐Microwave oven, brand_____, ☐Oven and range, brand_____, ☐Refrigerator, brand_____, ☐Sink, type_____, ☐Breakfast nook, ☐Breakfast bar, ☐Built-in mixer or powered food preparation center, ☐Butcher block sandwich area, ☐Lazy susan shelves, ☐Pantry, ☐Additional space, ☐Greenhouse window, ☐Barbecue area, ☐Garbage compactor, brand_____, ☐Counter top, type_____, ☐Air conditioner, brand_____, ☐Additional heating, type_____, ☐Open shelving, ☐Small appliance area, ☐Other_____
9. Which cabinet treatment do you prefer? ☐Cabinets extend to ceiling, ☐Cabinets installed on a lowered ceiling (soffit), ☐Open space above cabinets.
10. How much of the remodeling work do you want to do yourself (painting, wallpaper, etc)?_____
11. What is wrong with the kitchen as it is now? _____
12. Would you consider changing the location of windows? _____
13. Would you consider changing the location of doors? _____
14. Would you consider opening a new doorway or closing an existing one? _____
15. Would you consider opening a new window or closing an existing one? _____
16. Would you consider removing a partition wall? _____
17. Would you consider extending the kitchen beyond the existing house perimeter? _____
18. Would you consider changing the location of the range _____, the refrigerator_____, the sink _____
19. What major changes would you like to see made in your kitchen? _____
20. What color appliances would you prefer? _____
21. What color and style of cabinets would you prefer? _____
22. What style or theme would you prefer for the kitchen?
    ☐Contemporary, clean lines, openness, dramatic colors, bold natural woods and chrome.
    ☐Early American, traditional woods, copper utensils, pewter, handmade patterns.
    ☐Spanish/mediterranean, massive wood, heavily carved, rich fabrics in reds, wrought iron.
    ☐Traditional, flexible style with no single motif, simple and informal lines.
23. What wall and floor colors should predominate? _____
24. Do you have a picture of the type of kitchen you would like to have?
25. Which of the following would you like to exclude from the remodeling plan at this time?
    ☐Cabinets, ☐Counter, ☐Flooring, ☐Wall and ceiling cover, ☐New appliances
    ☐Structural changes, ☐Heating, ☐Plumbing, ☐Electrical work
26. Is there a dollar limit which is the prime consideration in developing the new plan?What is that limit?

Figure 5-3

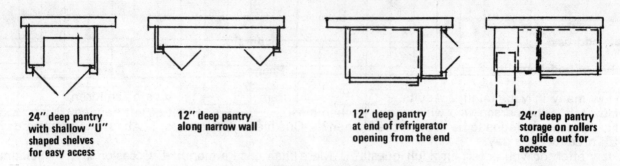

24" deep pantry
with shallow "U"
shaped shelves
for easy access

12" deep pantry
along narrow wall

12" deep pantry
at end of refrigerator
opening from the end

24" deep pantry
storage on rollers
to glide out for
access

Wall cabinet design
Figure 5-4

the kitchen. This helps in the preservation of the foods. Pantries should be accessible and have shallow shelved storage so items at the back of shelves are visible (Figure 5-4) without climbing on a chair.

Secondary kitchen centers should be considered as the remodeling budget allows. A trend that is becoming a "must" is the incorporation of an "eating center" in the kitchen area — a 30 inch high table is preferred. In a well planned layout, the eating center is near as possible to the kitchen, but separate from it. Other secondary centers are mixing, planning, bar and glass storage, barbecue facility, flower arrangement space (for the greenhouse look), and even space for pets and their beds, dishes or toys. If ample space is available, a freezer center also can be located in the kitchen.

A key point to remember is that basic kitchen planning concepts — the work triangle, work centers, proper storage and the rest — apply regardless of the type of kitchen a homemaker wants, or the way it is to be decorated. You approach the problem of designing a Colonial kitchen or a contemporary kitchen in the same way you approach any other kitchen.

Solving Specific Problems

The first step in evaluating the job is identifying the problems in the existing kitchen. Make a rough appraisal of what will have to be repaired or replaced because it is deteriorated or obsolete. Check the condition of the floor, wall and ceiling surfaces. If cabinets are to be moved, a new floor and wall surface are almost certainly needed. Check the condition of the appliances and decide what is to be replaced. You should plan on installing an exhaust hood, food waste disposer, dishwasher and fluorescent lighting if these are not in the present kitchen. Consider a trash compactor, barbecue, microwave oven, built-in mixer or power operated food preparation equipment, built-in butcher block, sandwich area, lazy susan shelves, pantry cabinet, laundry center, breakfast bar, bar sink, kitchen desk, and additional wall and base cabinets.

Now find the total distance of the three legs of the kitchen triangle: 1) from the center of the refrigerator to the center of the sink, 2) from the center of the sink to the center of the range, 3) from the center of the range back to the center of the refrigerator. If the total is less than 13 feet, the kitchen may be too small for efficient use by more than one person. If the three triangle sides are over 22 feet, time and effort is being wasted in food preparation. Many older homes may have a work triangle of 30 feet or more because they are designed for a refrigerator in an adjoining room or an entryway. The refrigerator should be near the entry where groceries arrive but should also be closer to the food preparation area.

If the work triangle falls between 13 and 22 feet, a major replanning of equipment locations may not be needed. But watch out for other problems. Is the kitchen eating area convenient to the range and dishwasher? Are the 3 legs of the kitchen triangle located across an area of heavy traffic? Is the sink located in front of a window so the homemaker has a pleasant view while working at the sink?

If the major work centers are going to be relocated, it is often better to start with the basic room dimensions and fill the space in the most convenient manner regardless of the existing equipment locations. Use Figure 5-5 to sketch the room size, appliances, fixtures, and opening locations while you are at the job site.

Let each large square represent one square foot. When you begin your planning, make another floor plan of just the existing walls and openings. On this floor plan place appliances and fixtures in several locations until an ideal arrangement is found. Now consider what compromises can be made in this plan to avoid relocating major appliances or utility connections while still preserving the best features of the plan. Be sure to consider opening or closing window and door spaces if you have trouble arriving at a satisfactory plan. Generally the window area should be 10 percent to 15 percent of the floor area. More than two entries leading to the food preparation area are usually unnecessary. Finally, review your plan from the standpoint of the space requirements and "do's and dont's" given earlier in this chapter.

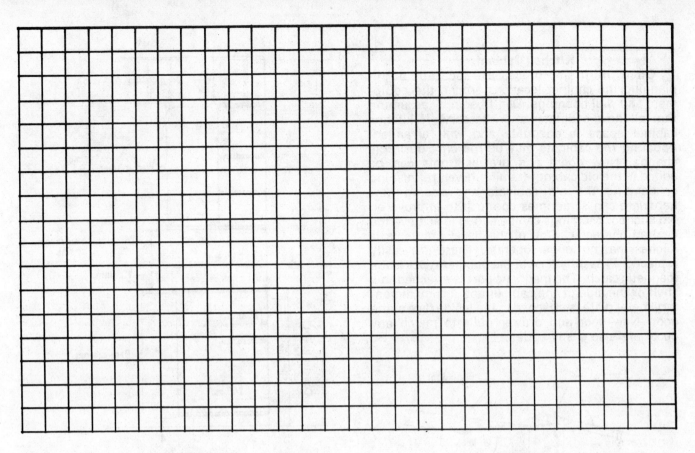

Dimensions

1. Indicate the length and width of kitchen from wall to wall.
   Length: _____ ft. ___ in. _____ Width ___ ft. ___ in.
2. Height of room from floor to ceiling:___ft.___in.
3. Indicate the location of all windows and doors in kitchen:
   (a) show distance from corners of room
   (b) indicate direction in which doors open
   (c) indicate if location of any can be changed
4. Indicate all offsets in the walls such as chimneys, exposed pipes and other obstructions to installation.
5. Show the location of present sink. Indicate the distance from center line of sink to nearest corner of the room._____ ft._____in.
6. Give location and height from floor of all electrical outlets and switches; also for ceiling and wall fixtures.
   Item____Ht._____ in._____ Item_____ Ht._____ in.
   Item____Ht._____ in._____ Item_____ Ht._____ in.
7. Show location, height and width of all registers and heaters.
   Item_____ Height_____in._____ Width_____ in.
   Item_____ Height_____in._____ Width_____ in.
8. If there is a ventilating fan, show the location; give the dimensions of the fan and the distance from the floor. Size_____ in. Distance_____
9. Include the width and height of all present appliances and indicate the distance of each from nearest corner of the room.

Openings

Item _____Width_____Height_____Distance_____
Item _____Width_____Height_____Distance_____

NUMBER each window using the same number for windows of the same size. Fill in dimensions in charts below. Be sure numbers on the sketch correspond to sizes in chart below.

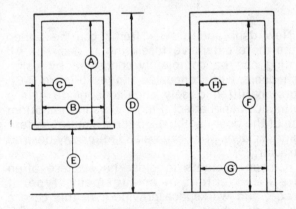

Indicate dimensions in inches

WINDOW	A	B	C	D	E	DOOR	F	G	H
No. 1						No. 1			
No. 2						No. 2			
No. 3						No. 3			
No. 4						No. 4			

Figure 5-5

Kitchen Cabinets

Once the major pieces are located, begin planning the cabinet locations around the equipment and wall openings. Don't begin by assuming that all new cabinets are necessary. If kitchen cabinet space is adequate and well arranged, updating the cabinets may be the only desirable change. New doors and drawer fronts can be added to the old cabinet framing. Even refinishing or painting the old cabinets and adding new hardware can sometimes do much to improve an old kitchen. Moldings can be added to achieve the desired character. One of the usual problems is latches that no longer operate. These are easily replaced. Presently, one of the popular types is the magnetic catch. The magnetic part is attached to a shelf or the side of the cabinet and a complementary metal plate is attached to the inside face of the door. New door and drawer pulls to match any decor are also easily added.

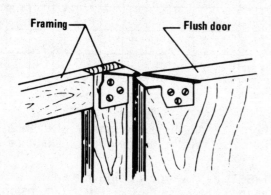

Concealed hinge used with flush cabinet door
Figure 5-6

New door and drawer fronts can be added where more extensive facelifting is desired. All framing can be completely concealed by using flush doors with concealed hinges (Figure 5-6). Doors are fitted closely edge to edge to give a continuous panel effect. Finger slots in the bottom edge of the door can be used for a simple modern design, or door pulls can be added for any desired character.

Kitchen cabinets in older houses are often quite inconvenient in arrangement, type of storage, and workspace provided. In this case a new set of cabinets will be in order. Figure 5-7 shows the normal range of kitchen cabinet dimensions. Note that cabinets over a refrigerator can be a full 21 inches deep.

A wide variety of stock kitchen cabinets are available in widths increasing in three inch increments from 12 inches to 48 inches. Some units have filler strips which can be cut on the site to fit any desired width. Occasionally custom made cabinets may be required to carry out a particular design element. Factory made cabinets may arrive at the site as assembled units or knocked down.

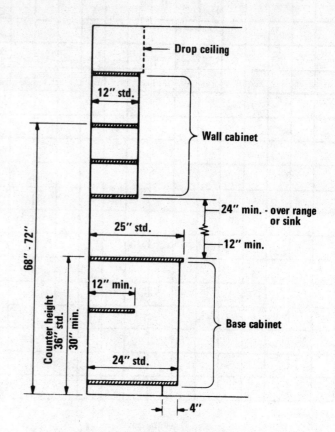

Kitchen cabinet proportions
Figure 5-7

Generally the manufacturer supplies (or will supply on request) installation instructions. Regardless of the type of cabinet, certain principles should be followed.

1. Many kitchen floors in older homes are not level. It is essential that the cabinets be installed level even if the floor is not level. If a new floor surface is being installed, level the floor with shims or new underlayment before installing the cabinets. If a new surface is not part of the job, use scrap lumber to provide a level surface on which the cabinets can rest.

2. Mark the location of the studs in the walls where both wall and base cabinets will be installed. Be sure that you will be able to hit a stud with each screw that you use to attach the cabinet to the wall.

3. Place the base cabinets in position on the leveled floor and check how the cabinets fit against the wall. If the wall is not straight or has concave or convex areas, install furring to bring the cabinet backs into line. Use a straight edge 2 x 4 to check the wall where the wall cabinets will be installed. Place furring on the wall where necessary.

4. Mark a level line on the wall 84 to 89 inches above the floor wherever wall cabinets will be installed. You will use this line to locate the position for the top of each wall cabinet. The top of wall cabinets is usually 84 inches above the floor and a soffit is built to enclose the area between the cabinet top and the ceiling. Rather than frame a

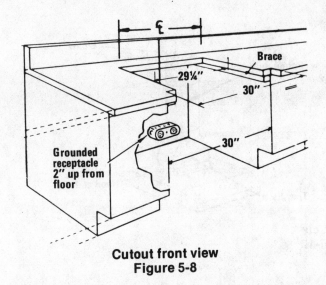

Cutout front view
Figure 5-8

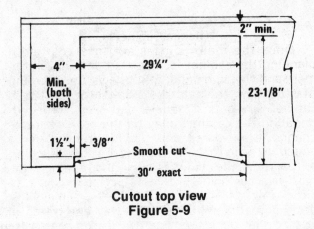

Cutout top view
Figure 5-9

soffit, plywood held in place by strips of molding can close off the opening above the cabinets. In many jobs the entire kitchen ceiling is lowered to 89 inches (with recessed fluorescent fixtures) and the cabinets extended to the ceiling.

5. The base cabinets are usually installed first. Begin with a corner unit and then attach the adjoining units to both the wall and the other cabinets. Use a C-clamp to hold adjoining cabinets together while you are working on them. Check carefully the openings for range, refrigerator and dishwasher as you work.

6. Next install the wall cabinets. Usually hanging cleats or strips are used to attach the cabinets to the wall. Don't drive the screws down tight until all the cabinets are in place, level and plumbed. A racked, out of plumb cabinet installed with one end higher than the other will be a source of annoyance as long as it remains in use.

Counter Tops

Where cabinets are adequate, the one best way to update a kitchen is to apply new counter tops. The most common type is plastic laminate.

Several sheet and roll materials that can be glued to clean, smooth backing are available. These include laminated melamine, laminated polyester, vinyl, and linoleum. They are normally applied to ¾ inch exterior-type plywood. Linoleum and vinyl are flexible enough to be shaped to a covered backsplash on the job. Other materials can only be applied flat, so the backsplash is covered separately and a metal strip is used to cover the joint between the backsplash and the countertop. The cabinets usually are predrilled to receive the counter top. Use screws, not nails, to attach the top. If the wall is uneven at the back of the backsplash, a metal cove molding will cover any opening.

Marble is shop-fabricated and requires no backing material. It must be precut to size in the shop because special tools are required. It is self-edged. There are some objections to marble

because of its hardness, and it is expensive, so it is usually limited to bath counters.

Ceramic tile can be set in a mortar bed or applied with adhesive. It is the only material that must be applied at the building site rather than in a shop or factory. It is available in a variety of sizes. The smaller 1 inch square tile are often preassembled on a mesh backing in 1 foot square units. It was once very popular and is quite attractive, but is quite hard for kitchen counters and the joints become a maintenance problem.

Installing Drop-In Ranges

Drop-in ovens and cook tops should be installed after the counter top is in place. Range dimensions vary somewhat but the dimensions listed in Figures 5-8 and 5-9 will be about right for any drop-in unit 30 inches wide overall. The

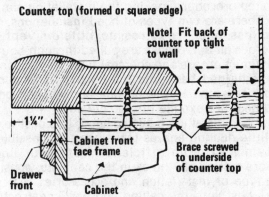

Counter and cabinet detail
Figure 5-10

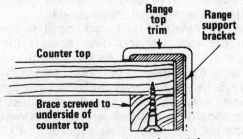

Cross section view of top trim
Figure 5-11

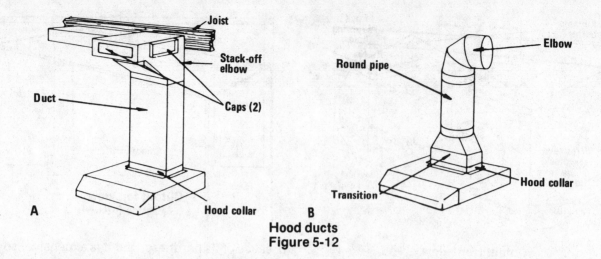

Hood ducts
Figure 5-12

cabinets on each side of the range must be level and plumb because the range will be supported by the counter top and the cabinets. The counter top should lap over the face of the cabinets 1¼ inches as in Figure 5-10. Make the gas or electrical connection before installing the unit. Slide the range into place with the range top trim resting on the counter top. Note Figure 5-11. Complete the installation by raising the cooktop and attaching the range to the adjoining cabinets with screws.

Installing Range Hoods

These instructions will give you an idea of the various requirements for installing a hood. Complete instructions will be included with the hood you order. They will explain in detail the step by step procedure making the installation simple.

There are two types of hood installations. The ductless hood requires no outside venting. Odorous air and grease are pulled through both an aluminum grease and charcoal filter. The clean air is discharged back into the kitchen. The ducted hood is vented to the outside and kitchen odors and heat are expelled to the outside.

When installing a ducted hood, it is best to use as little ducting and as few elbows as possible to obtain maximum efficiency from your hood. Before cutting any holes in the cabinets, consider the type of installation and accessories that will work best with the ceiling and wall construction. Note that the vent in Figure 5-12A allows ducting in four directions. The circular elbow in Figure 5-12B can be used only if the ceiling joists are parallel to the sides of the cabinets.

Before installing duct work, position the hood under the cabinets and flush against the wall. Mark the holes (four) in each corner that will be used later for fastening the hood in position. (See Figure 5-13.) If the wall cabinets have recessed bottoms, strips of wood the same thickness as the recess should be used to form a flush mounting surface for the hood. (See Figure 5-14.) These should be firmly fastened to the cabinet bottom with screws.

Refer to the installation instructions for the dimensions that will determine where the hole for the hood collar and other ducts should be cut in the cabinets. This will also give the location of the electrical knockout holes. Select the location most convenient for your installation.

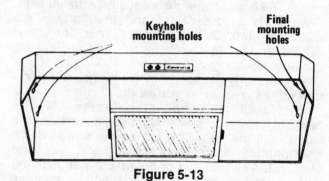

Figure 5-13

Cabinet with recessed bottom
Figure 5-14

The hood is usually assembled at the factory to vent the vapors through the top. If you wish to vent out the rear, the blower may have to be removed, reversed and reinstalled to close the top opening and vent out the rear opening.

To install the hood collar to the hood, insert the

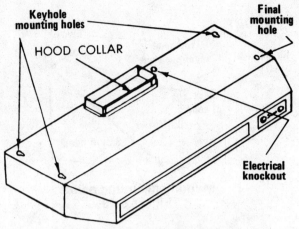

Hood collar for vertical venting
Figure 5-15

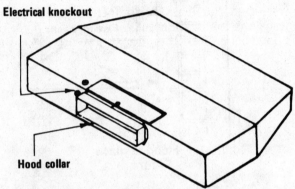

Hood collar for horizontal venting
Figure 5-16

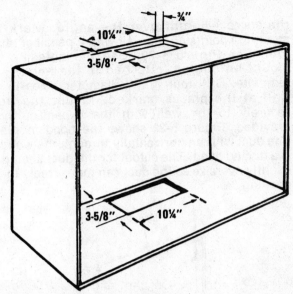

Cabinet cutouts necessary for vertical venting
Figure 5-17A

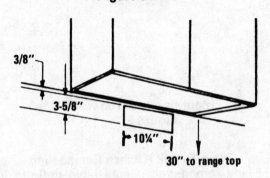

Cabinet cutouts necessary for horizontal venting
Figure 5-17B

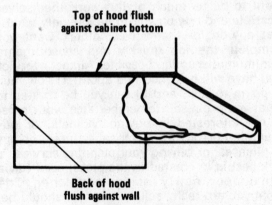

Ductless hood installation
Figure 5-18

flange (without holes) in the opening in the top (or rear) and fasten the opposite flange with the screws furnished. See Figures 5-15 and 5-16.

Now make the necessary cutouts in the cabinets or wall. See Figures 5-17A and 5-17B. Be sure these are positioned so that the ductwork will line up with the hood collar (now fastened to the hood) when the hood is installed. The hood collar will slide inside the duct.

After the cutouts are complete, proceed with the installation of the duct work. When installing the outside cap, be sure it is properly sealed against the outside of the home. To be sure of an air tight connection when installing two pieces of pipe, use 2 inch wide duct tape to seal the joint.

A ductless hood can be fastened in the same way as a ducted hood. With this type of installation, the cooking center and hood can be installed anywhere in the kitchen with no planning necessary for duct work. Figure 5-18 shows how a ductless hood can be installed. Because of the absence of duct work, practically any type of installation is possible.

If there are no cabinets above the hood, the hood can be hung from the wall with a wall mounting kit available from most hood manufacturers. Expansion bolts are included. If it should ever become necessary to remove the hood, the bolts can be unscrewed and the sleeves remain in the wall.

Attach the mounting angles from the wall mount kit to the inside of both ends of the hood with the nuts and bolts furnished, as illustrated in Figure 5-19. Drill holes in the back of the hood using the mounting angles for templates. Position the hood on the wall, making sure it is level and at the correct height over the angle. Mark the mounting points on the wall with a pencil or sharp tool. See Figure 5-20. Use the holes in the mounting angles (attached to the hood) as a template. Drill four 7/16'' diameter holes in the wall at the places marked. Mount the hood securely to the wall with the expansion bolts provided. Figure 5-21 shows the hood in place. The duct will run horizontally through the wall if it is a ducted hood. The cutout for the duct should be slightly oversize so the duct can move freely in the opening.

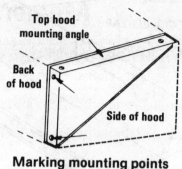

Marking mounting points
Figure 5-20

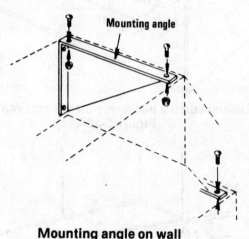

Mounting angle on wall
Figure 5-19

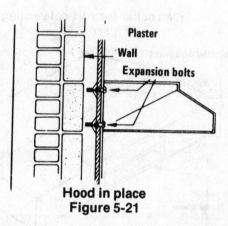

Hood in place
Figure 5-21

How To Sell Kitchen Remodeling

Many remodeling clients have definite ideas about what they want and how much they want to spend. Other clients may come to you hoping that you will supply the ideas. Some prospects may want to do as much of the work themselves as possible and rely on you to handle only what they can not do for themselves; others want you to complete the job quickly and cause them the minimum disruption possible. Almost every client you have will be concerned about: 1) how much it is going to cost, and 2) can you be trusted to do what you promise. Remember also, your prospect is not interested in buying cabinets, or buying plumbing or buying electrical fixtures, or buying appliances, or buying your planning service. Your prospect is interested in the pleasure of having a convenient, newly remodeled kitchen. That is what you are really selling and that should be the focus of your sales pitch.

When you meet your prospects, convince them that you are a ''pro.'' Offer references if that will help sales. Many remodelers join the Better Business Bureau to distinguish themselves from less reliable operators. Develop and use a form such as Figure 5-3 to get your client's ideas about what they need. If you meet them in their home, listen carefully to their explanations of what is needed and then use your knowledge of kitchen planning to confirm that they have a problem that can be remedied. If you meet them in your office,

use a display kitchen and perhaps several small vignette displays to illustrate efficient kitchen planning. If you don't have a display, several sample floor plans and a few color pictures will work almost as well. After you get a feel for what they want (but only if the subject comes up), answer the question about cost. Explain that the typical kitchen remodeling job costs between $6,000 and $8,000 and that their job would be more, or less, or about that much. Then explain that you will give them a bid on the job before you begin work and that the bid will be based on the final plan which you will develop to suit their needs. If low cost seems to be the most important consideration, offer to develop a plan based on a certain maximum cost. Be sure to add that your bank can arrange financing and that the cost of the job will increase the value of the house significantly.

By this time you have qualified your client as being willing and able to pay for your services, established yourself as a ''pro,'' gained some information about their likes and dislikes that will help you close the sale, and eased some of their fears about the cost and your reliability. Now try out a few ideas on your client. From the moment you saw the kitchen (or had it described to you) you probably had several suggestions for improvement. After learning about your client's concept of what is needed, you should offer a few ''How would you feel about . . .'' or ''Have you considered . . .'' suggestions. Take notes on their

Memorandum Of Agreement

The undersigned agree to have _____ prepare a floor

plan, elevation view, construction cost estimate and outline specifications for a proposed remodeling

of the _____ in the residence located at _____ at a cost of

_____ . The sum of _____

is hereby acknowledged as received and the balance is due upon completion of the plans, estimate

and specifications. If _____ is authorized to complete all or a

substantial portion of the work proposed, the fee for preparing the plan, estimate and specifications

will be credited in full against the cost of construction.

Owner_____ Builder _____

Owner_____ Date _____

Figure 5-22

answers and make a preliminary sketch of a proposed layout. Don't give out all your good ideas. Save a few for the plan you prepare. Don't make any concrete "take it or leave it" proposals. Just open up a few possibilities for discussion. Your prospects will eliminate some suggestions and become enthusiastic about others. Focus on what they like and build your plan around that. But make the plan practical and efficient. You want satisfied customers, not ex-clients who have to live with a plan you knew was impossible. Explain the cost implication of their ideas when appropriate. Be sure to ask about colors and a decorating theme. This may be important in concluding the sale.

By now you have learned almost enough to start planning. If you are meeting in your prospect's home, take down the kitchen dimensions on a form such as Figure 5-5 so you can develop your plan. If you have been meeting at your office, schedule a meeting at your prospect's home. Now is the time to make your offer. Propose that you develop a plan for their kitchen and an estimate for the job for a certain fee. If your plan is accepted and the agreement signed, the fee is credited against the job. If your prospects are

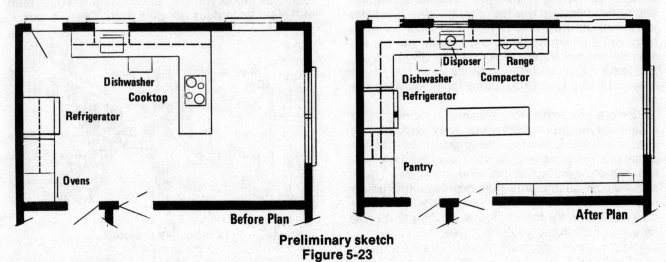

Preliminary sketch
Figure 5-23

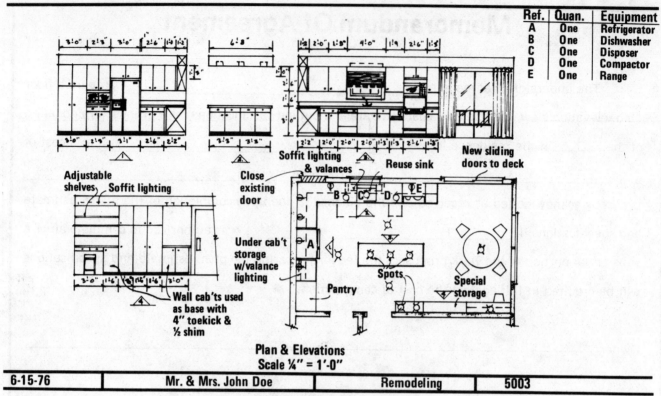

Ref.	Quan.	Equipment
A	One	Refrigerator
B	One	Dishwasher
C	One	Disposer
D	One	Compactor
E	One	Range

Adjustable shelves
Soffit lighting
Close existing door
Soffit lighting & valances
Reuse sink
New sliding doors to deck
Under cab't storage w/valance lighting
Wall cab'ts used as base with 4" toekick & ½ shim
Pantry
Spots
Special storage

Plan & Elevations
Scale ¼" = 1'-0"

| 6-15-76 | Mr. & Mrs. John Doe | Remodeling | 5003 |

Floor plan and elevations
Figure 5-24

serious about the job and have confidence in you and your ideas, they will probably accept your proposal. Set the fee high enough so that it pays you for the work you plan to do. Fifty dollars is probably a minimum for a simple floor plan estimate and a recommendation on colors and cabinets. A floor plan, elevation and perspective view together with outline specifications and a detailed estimate should cost between $150 and $250. You should charge at least $25 per hour for your time in planning. If the proposal takes 4 hours a reasonable fee would be $100. The higher the fee, the more likely you will eliminate prospects who are "just shopping" and the more likely you will get the job once the plan is submitted because the fee is credited against the job if you do the work. If your prospects agree to have you develop a plan and estimate, have them sign the memorandum of agreement (Figure 5-22) and leave a copy with them. Make an appointment a week to 10 days in the future for presenting the plan.

Before you leave your prospect's home, try to anticipate all the problems that may arise during construction. Examine the location points for proposed windows and doors. Consider how you can match existing colors and textures. Think about electrical and plumbing problems (including extra electrical runs, vent stack relocation) you may have. The estimating forms in Chapter 19 will help you find items you might overlook.

Next, develop your plan. The planning information in this chapter should help you evaluate the plan as it is completed. Turn your rough sketch (Figure 5-23) into a more detailed floor plan. This will be essential to your crew as work is done but an elevation view will be more useful in concluding the sale. The floor plan and the elevation view can be combined on a single sheet as in Figure 5-24. Then have a blueprinter make two prints of the plan. The prints go to your customer. You retain the original drawings in your file.

You can visualize the kitchen from the plans, but your customer may not be familiar with working drawings and will not be as excited by your drawing as you are. Even a detailed floor plan with elevations does not give that "lived-in look."

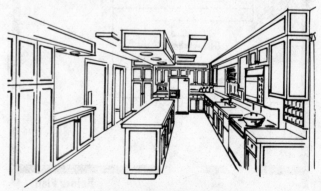

Perspective drawing
Figure 5-25

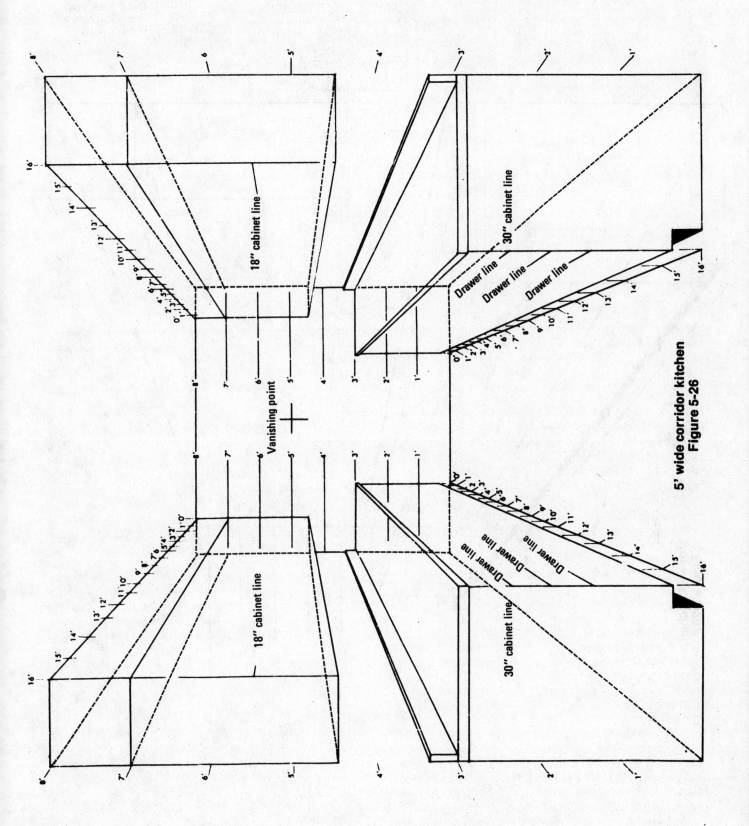

5' wide corridor kitchen
Figure 5-26

49

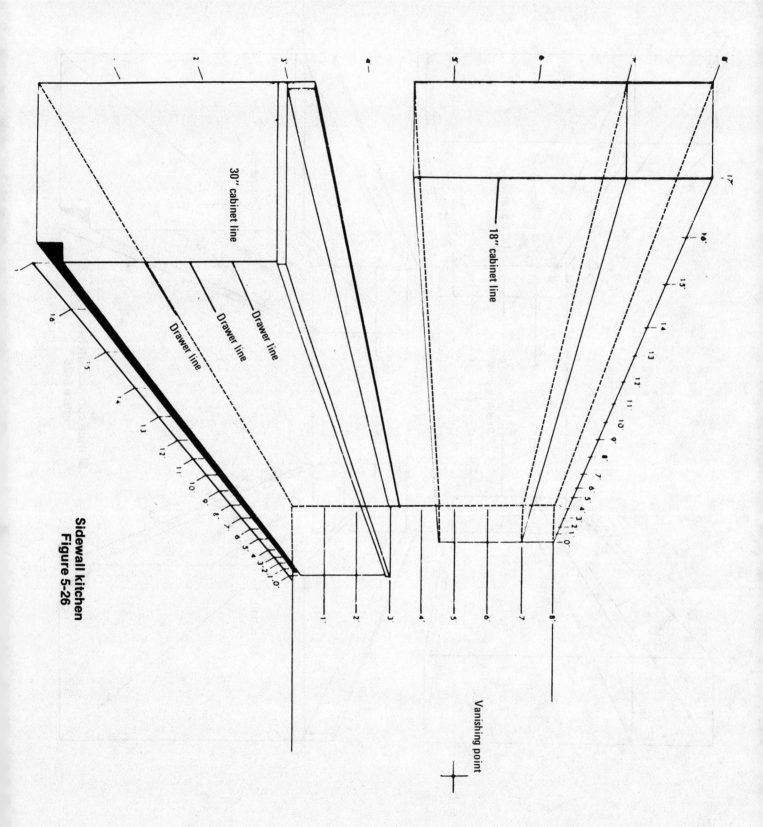

30" cabinet line

18" cabinet line

Drawer line

Drawer line

Drawer line

Vanishing point

**Sidewall Kitchen
Figure 5-26**

50

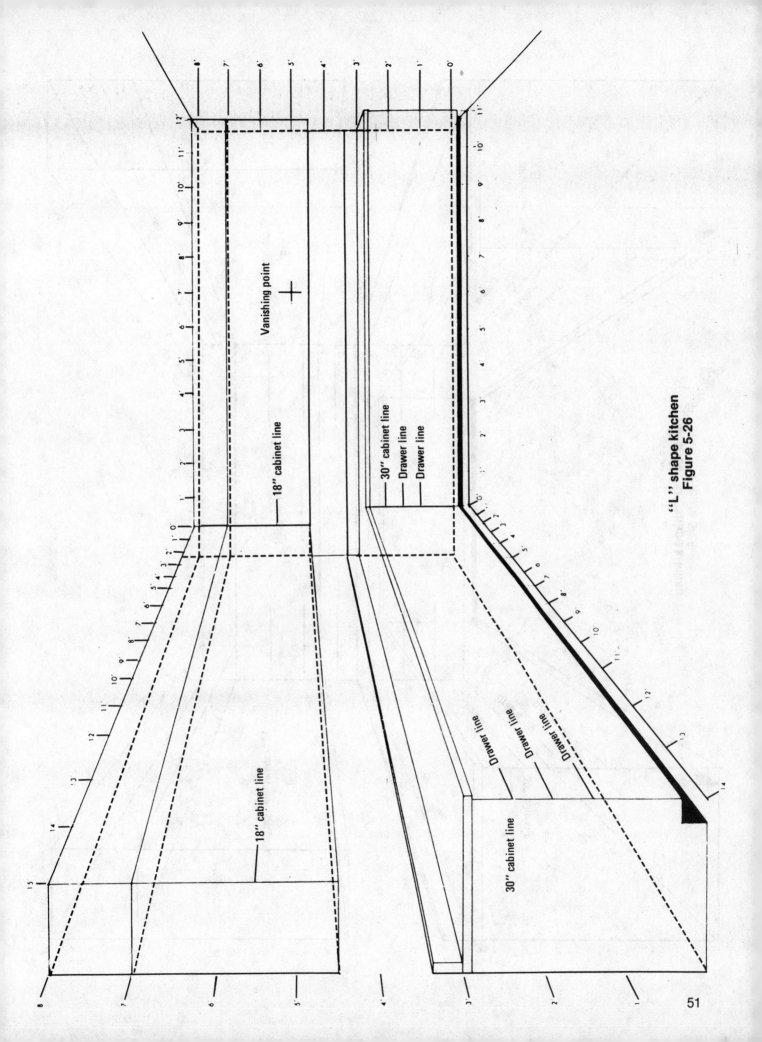

Vanishing point

18" cabinet line

30" cabinet line
Drawer line
Drawer line

18" cabinet line

30" cabinet line

Drawer line
Drawer line
Drawer line

"L" shape kitchen
Figure 5-26

51

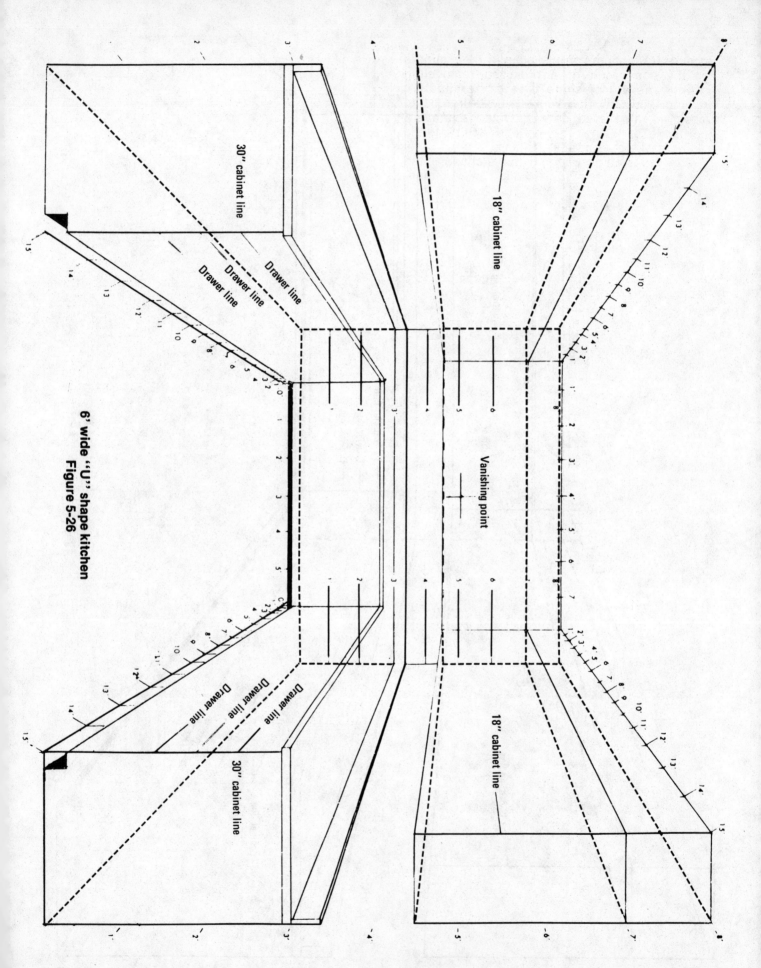

6' wide "U" shape kitchen
Figure 5-26

30" cabinet line

Drawer line

Drawer line

Drawer line

18" cabinet line

Vanishing point

18" cabinet line

Drawer line

Drawer line

Drawer line

30" cabinet line

52

Because few people can visualize in their mind's eye the end result of a proposed remodeling without the aid of some type of perspective drawing, you may find yourself hard-pressed to close the sale based only on your reputation, knowledge of kitchen design and ability to do free-hand sketches and floor plans.

The visual aid needed to help merchandise the sale is the rendering, Figure 5-25, a three-dimensional perspective of the floor plan and elevation that can create a "visual handshake" with the potential customer.

How involved you should get in producing the rendering depends on how important it is to helping close the sale. If deemed critical, the services of a professional designer should be employed to produce the rendering, showing the location of all kitchen features and major appliances.

To produce a rendering that helps sell a remodeling job may appear difficult, but don't let it scare you away from trying to develop one yourself. If you can draw a floor plan and decent elevations, then you're a good candidate to learn the art. Practice makes perfect, so don't give up if the first attempt is not impressive.

The tools of the trade are basically the same ones used when doing floor plans and elevations. You'll need a drawing board, T-square, architectural scale, at least one triangle, ellipse template, French curve, push pins, paper, vellum tracing paper, 2H pencils, and don't forget the eraser. These items can be obtained from any art or architectural supply house.

There are many techniques that can be pursued in accomplishing the finished rendering,

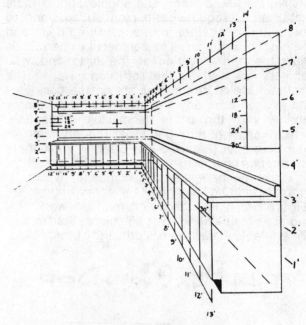

**Using perspective chart
Figure 5-27**

but the least complicated and quickest involves use of perspective charts (Figure 5-26). These charts are available in a variety of shapes and sizes, including such standard kitchen plans as sidewall, L-shaped with right and left walls, corridor and the U-shaped. These charts contain vertical dimensions, cabinet dimensions and the vanishing point, the most important influence in achieving a successful rendering. The vanishing point represents the eye-level and location the

Figure 5-28

kitchen is viewed from, and should approximate that of an average height person. All side-angled lines will run together at this vanishing point and seemingly disappear. The perspective charts also show the end elevation of the base and wall cabinets, as well as the soffit and walls. The "actual" scaled dimensions are on the back wall and you can draw any size after you have become familiar with the art of perspective dilineation. Select a chart (Figure 5-27) that roughly matches your floor plan (Figure 5-23). Chapter 6 has more information on how to make perspective drawings.

Once the perspective is complete, it can be set-off by adding shading or the appropriate color, (Figure 5-28). Felt pens and markers work best and are obtainable in over 50 colors. Starter sets are available in basic pen widths in 12 basic colors.

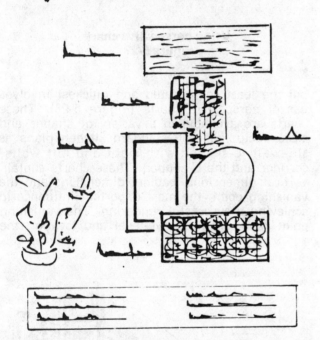

Swatch board
Figure 5-29

If samples of the actual materials to be used in the new kitchen are available, a swatch board also can be made (Figure 5-29). The samples contained will enable the customer to "feel" the kitchen in addition to seeing it.

You have now created that "visual handshake" with the prospective customer. The rendering represents the finished product as you and your customer agree it should be. Aside from eliminating doubts about any details, and being used to "sell" the job to your client, the rendering also enables subcontractors to bid more accurately since all details are clear. It also can serve as a tangible object for the client to show when applying for a loan. The rendering is an invaluable merchandising tool. It is the ultimate answer to your customer's question: "What will my kitchen look like?" It establishes you as a "pro" as well as a careful craftsman and will discourage delays, indecision and suggestions for changes when you ask for the job.

Now complete your estimate and outline specifications. Finally prepare a contract for your customers' signature. Use your own documents or the sample in Figure 6-44. The sample is a binding document prepared especially for this book by an attorney. It is fair and complete and protects you against most contingencies that may arise. Read over the fine print on the back to see the ways you are protected.

When you meet your prospects again to close the deal, their biggest fear will be that they are about to make a mistake, a $6,000 mistake! Have several options in cabinet and equipment grade written into your plan so you can show some flexibility as to price. Convince your prospects that you are going to give them good value for their money and that you have created a plan which minimizes the cost without sacrificing on quality. Give your prospects the option of supplying the stove and refrigerator themselves. Have some samples of hardware or pictures of range hoods along so you can focus attention on making small decisions that assume the big decision — that they will give you the go-ahead to begin work. Usually you will want to let your clients select the wallpaper and paint themselves. They may want to do the painting and paperhanging themselves also. Finally, present the contract and make it clear that you are asking for the job now so that buying materials and scheduling crews can begin right away. Explain the contract to them as may be required and leave one copy with them when it has been signed.

Finishing The Job

Kitchen remodeling is about the largest inconvenience any family can endure in their home. Ten days is too long for your clients to do without full use of their kitchen if you can get the job done in five days. Many successful kitchen remodelers try to finish each job in the same week it is begun. Where several trades are involved, scheduling can be a major problem. However, most experienced remodelers would agree that the operation that finishes jobs most promptly usually operates most profitably.

Chapter 6
Bathrooms

Every year over 2½ million American bathrooms are remodeled or added. More bathroom remodeling jobs are sold each year than any other single category of remodeling.

Over 45 million housing units in this nation are beyond 13 years of age. Most of these homes need remodeling or repair to a bathroom or an additional bathroom. Most important from a remodeler's standpoint, people seldom complain about the cost of a bathroom job if it is done right. The most frequent criticism is of the contractor who calls himself a bathroom remodeler but who, in truth, can complete only part of the job. This leaves the homeowner in the unfamiliar role of designer and general contractor. People want to buy a complete "Bathroom Package," and they want to buy it from one source. Their feelings toward a skilled remodeling contractor somewhat resemble a patient/doctor relationship. They know something is wrong, but they don't know how to correct it. When a sink, tub or toilet needs replacement, it logically follows that everything else in the bathroom needs an overhaul because none of the bathroom is more durable than the plumbing fixtures.

Design

Old bathrooms, like old kitchens, were poorly planned. People with old bathrooms have become accustomed to living with poor design (like having towels stored in a hall closet rather than in the bathroom). As a result, they respond positively to the creative salesman who suggests additional vanities and storage cabinets. The ability to sell and design what people really want is what can make you a leading bathroom remodeler in your community. As a skilled custom bathroom specialist, you should have a good understanding of harmonious wall, ceiling and floor coverings; plumbing fixtures and fittings; cabinets and vanities of all types; medicine cabinets; boutique items; lighting and ventilation. Offer creative design with an innovative approach, use sales aids offered by most plumbing fixture and vanity manufacturers to inspire and excite your prospects, offer to take on the whole job at a fair price, and then do exactly what you propose in good time and in a workmanlike manner. It's a success formula that can be repeated over and over again.

Planning Considerations

Today, a bathroom can be anything your clients want it to be. With a little imagination and an understanding of the possibilities that exist, you can make decisions that will make their lives more enjoyable. Whether a bathroom is being planned as an addition or your clients are just remodeling or improving an existing bathroom, it is important to examine their personal needs, desires, and finances before making a proposal. Extra energy spent in thinking before doing will be well worthwhile.

The first consideration to be faced concerns functions of the bathroom. A bathroom need not be just a small, white-tiled room with three fixtures. It might also include a dressing and grooming area, a space for sunbathing and exercising, a sauna, or even an area to relax, read, listen to the radio, or watch television. Free your clients' minds from conventional ideas, and they will probably be happier with the results.

The bathroom, even in its conventional uses, is a very important place. No homeowner can do without a bathroom. The bathroom is the first place we go in the morning, and the last place at night. We probably use the bathroom more often than any other room in the house. These are good reasons for making it a pleasant and comfortable place to be and use.

In planning for a remodeled bath or new bath in an existing house, the size of the family will largely determine the features needed. Particular individual needs and desires must also be taken into account. And of course the financial situation of the family will have an effect on final decisions.

If your clients are adding or remodeling a second or third bathroom, each bathroom can be more specialized. The standard two-bathroom solution is to have a family bathroom and a guest bath or powder room, perhaps with only a lavatory and toilet.

However, if there are older children in the family, a better solution might be a separate bath for them. It could be compartmented with a personal lavatory for each, connecting to a shared shower and water closet area. If there are small children in the family, a "mud room" by the back door rather than a powder room by the front door might be a good solution.

If your clients are considering the addition of a master bath, it should be thought of as part of an entire suite, related to the bedroom and to dressing and clothes storage areas. You should not have to go through the bedroom itself, however, to get to this bathroom.

A master bath should be a large and comfortable room. Designing it can be an excellent opportunity to make it a very individual place, reflecting your clients' tastes and ideas. For example, many plants grow well in the moist, warm bathroom climate and could give the room a distinctive, unusual appeal. The master bath might also be split into various compartments, or have his and her tubs. On the other end of the scale, you might suggest having one of the new large tubs for two.

If there is a family bathroom upstairs, it might be sensible to locate the laundering facilities next to it. Then soiled laundry would not have to be taken downstairs for washing, and back upstairs again for storage.

Equipment and Materials for the Bath

An enormous variety of bathroom equipment is available, in a wide range of colors, styles, and prices. Careful planning is essential to be sure your clients get the best value for their investment, and to give the bathroom the specific character and style they want. In general, a higher price will not only mean better quality and better design, but most importantly, a wider selection.

Showers and Tubs

The safest rule-of-thumb in deciding on the shower or a tub is: if in doubt, choose a combination tub-shower. Although most tubs are used almost exclusively as shower receptors, a tub is adaptable. If the house has more than one bathroom, a shower might serve well in the second bathroom.

One-piece fiberglass shower modules answer several needs. They are easy to install, clean, and maintain. Because they are integral units, there are no seams to leak. They also solve the question of wall-coverings around the inside of the shower. Note, however, that most one-piece fiberglass shower-tub units won't fit through most doorways. Keep this in mind when considering one of these modern modules.

The smallest comfortable shower is 42 inches by 30 inches, although models which include a built-in seat are available up to five feet long. A shower stool is a good selling point in any case. A built-in ledge or shelf for soap and shampoo is another essential feature.

For safety, a shower with a non-skid floor should be chosen. There should also be a grab bar located at the entry to the shower. The shower controls should not be located under the shower head, nor in such a place where they are difficult to reach from outside the shower. If doors are used for closure, by law they must be either plastic or safety glass.

If a shower is installed, it is important to provide an exhaust fan. This will protect the walls and ceiling from excess moisture, and provide necessary ventilation.

Bathtubs come in a variety of sizes and shapes. The length of a standard rectangular tub ranges from 4 feet to 6 feet and its height ranges from 12 inches to 16 inches. The height of the tub is normally computed from the floor to the top of the rim. The actual depth of water the tub will provide, however, is governed by the distance from the bottom of the tub to the overflow outlet, a dimension not always printed in literature and sometimes obtainable only by measuring a sample tub.

Bathtubs are usually made of either cast iron or formed steel with a porcelain enamel finish, or fiberglass made of reinforced polyester resins with a gel coat surface. Cast iron tubs are the most rugged and durable and the least susceptible to damage. They are the heaviest, weighing from 250 pounds to as much as 500 pounds. A thick glossy coat of porcelain (approximately 1/16 inch thick) provides a beautiful, durable and sanitary surface.

Enameled formed steel tubs are much lighter, 100 pounds or so, and consequently may be easier to install, especially in upper story bathrooms. Formed steel is also very durable and somewhat less expensive than cast iron.

Fiberglass units are a relatively recent development. The appearance is sleek and attractive. The walls and tub are one-piece, eliminating seams and grout lines. Fiberglass, however, is not as durable as cast iron or steel. Greater care must be exercised during cleaning because the gel coat surface is more susceptible to damage than porcelain enamel.

The most frequently installed tub is rectangular and is recessed into a niche, which is then tiled for showering. Corner tubs are for installations where the back and only one end of the tub will be against walls. This model is less popular and more expensive. If you want to create a unique looking bathroom, consider a large square tub or one of the other unique shapes on the market, all of which are premium priced. Small square tubs work well in areas where space is a problem, although they are not large enough for adult tub bathing. They are ideal for a children's bath or as a luxury shower base.

Because tub baths are generally enjoyed for relaxation and soaking, the tub chosen should be comfortable to lie back in. The slope and height of the back of the tub is therefore an important feature. Advise your clients to consider a five and one-half or six-foot long tub, rather than the standard five-foot length.

Many convenience features for tubs are available. Hand showers which attach to the tub-filler spout or the shower head can be purchased. These are useful for shampooing, child-bathing, and cleaning the tub after use. Consider also dual shower heads, one at 60 inches height for women and another at 66 inches for men.

For safety, tub controls should be higher than is customary, and near the entry side of the tub. Forty inches is a good height. The shower head should be at the opposite end of the tub from the controls.

A non-skid surface is necessary on the tub bottom. Grab bars and hand grips should be securely and strategically located for help in stepping into and out of the tub.

A sunken tub may be elegant, but it is also a great safety hazard and exceedingly awkward to clean. If your clients are interested in a sunken tub effect, install the tub in a 10 to 12 inch raised platform built into the end of the room where the tub is located.

Lavatories [Sinks]

There are five basic kinds of lavatories. The "flush-mount" requires a metal ring or frame to hold it in place. It can be expensive. The metal frame is objectionable to some who maintain it is difficult to keep clean where it rests on the countertop.

The "self-rimming" lavatory requires no metal ring or frame. It holds itself in place with an integral rim which rests on the countertop. Normally this rim overlaps sufficiently to permit the mounting hole in the countertop to be "rough cut." Properly sealed, the seam between lavatory rim and counter is easy to keep clean.

The "under the counter" lavatory mounts beneath an opening in the marble of a simulated marble countertop. The fittings are mounted through the countertop, creating a high style decorative effect. The seam where the lavatory meets the underside of the countertop may be difficult to keep clean.

The one-piece "integral lavatory and countertop" is seamless, attractive, easy to clean and rests on a vanity or cabinet. It can be made of plastic or some type of synthetic marble material.

The "wall-hung" lavatory is attached to and juts out from the wall. It works well where space is at a premium, where storage is not needed, or where there is no concern for exposure of the plumbing.

Lavatories can be made from cast iron, formed steel, vitreous china or plastic. The use of plastic for this purpose is fairly recent. The casting and molding of cast iron and china permits wide variations in shape and design including ovals, shell shapes, angles and swirls. Punching or stamping of steel does not permit those design variations.

For maximum convenience and adaptability, choose the largest basin built into the largest countertop (set at a height of 34 to 36 inches) that the space can accommodate. Space beneath the sink should be planned as a storage area. Wall hung sinks on legs or wall brackets provide no counter space and waste the space underneath—a luxury very few homes can afford today.

Faucet controls should be easily reached and easily cleaned and maintained. Cartridge-type controls are especially long-lived. The faucet itself should be as long as possible, and should be the mixing type.

The lavatory can be mounted off center in its vanity, and in fact this arrangement actually is preferable in most situations. An off-center lavatory leaves room for larger bathroom appliances or plants to one side. However, the mirror should be centered above the lavatory if only a small wall mirror is planned.

Water Closets [Toilets]

The toilet is a major part of the cost of any bathroom remodeling job. Your client should understand why toilets are expensive, especially good quality toilets, and why quality is an important consideration when selecting this fixture. The toilet is one fixture in the bathroom where design and quality control in manufacture are of critical importance. Unlike the tub and lavatory which function primarily as basins to hold water, the toilet is a much more sophisticated mechanism. It is semi-automatic in its function. Such things as proper relation between water volume and the interior design of the bowl must combine to create an efficient waste disposal system with automatic protection against sewer gases and unsanitary conditions.

Most residential toilets consist of a bowl and a tank. The function of the tank is to provide storage for sufficient water to create a proper flushing action. In commercial buildings, where large pipes are used for water lines, an adequate amount of water is provided directly from these supply lines, and all that's needed is a flush valve which shuts off automatically after the proper amount of water has been provided. In residences, however, the water lines are too small to provide the amount of water needed, as fast as it is needed. Therefore, a toilet tank (or flush tank) is required.

There are three different flushing actions commonly used for residential construction. These different actions result from bowl design and, although they have been available for years, few homeowners are familiar with their differences or the advantages of one over the other. Consequently, price has frequently been the determining factor in bowl selection, resulting later in dissatisfaction in the function of the toilet. These three flushing actions are illustrated in Figure 6-1.

Washdown. The washdown toilet bowl discharges into a trapway at the front and it is most easily recognized by a characteristic bulge on the front exterior. It has a much smaller exposed water surface inside the bowl, with a large flat exposed china surface at the front of the bowl interior. Since this area is not protected by water, it is subject to fouling, contamination and staining. The trapway in the washdown bowl is not round and its interior is frequently irregular in shape due to the exterior design and method of manufacturer. Characteristically, the washdown bowl does not flush as well or as quietly as other bowls.

The washdown bowl is no longer accepted by many municipal code authorities and several manufacturers have deleted it from their manufacturing schedule.

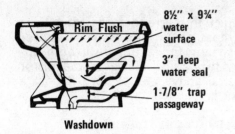
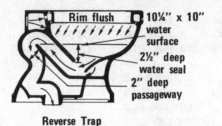
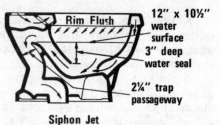

Washdown   Reverse Trap   Siphon Jet

Toilet types
Figure 6-1

Reverse Trap. The reverse trap bowl discharges into a trapway at the rear of the bowl. Most manufacturers' models have a larger exposed water surface, thereby reducing fouling and staining of the bowl interior. The trapway is generally round, providing a more efficient flushing action.

Siphon Jet. The siphon jet is similar to the reverse trap in that the trapway also discharges to the rear of the bowl. All models must have a larger exposed water surface, leaving less interior china surface exposed to fouling or contamination. The trapway must be larger and it is engineered to be as round as possible for the most efficient flushing action. Low profile toilets are usually siphon action one piece units. They are very quiet in operation, have almost no dry areas in the bowl interior to collect contaminants and are the most expensive to buy.

Most toilets are available with either round or elongated bowl rims. The elongated bowl (sometimes referred to as "extended rim") is 2 inches longer from the front edge of the rim to the back of the toilet. Most remodelers recommend the more modern elongated bowl. It is more comfortable, more attractive, and, since it provides a larger interior water surface, more sanitary and easier to keep clean. It costs very little more than a round bowl. Wall-mounted toilets are efficient and easy to clean under, but they are more expensive and must be specially mounted. This is not always possible when remodeling.

Toilet seats should be easy to clean. The hinges should be sturdy. Plastic seats are the longest-lasting. Painted wooden seats chip, and become unsightly sooner than the all-plastic type.

Toilets are also available with self-venting exhaust systems which carry off odors.

For safety and convenience, especially for older people, grab bars may be installed on either side of the water closet.

Bidets

Bidets provide the ultimate in personal hygiene. Widely used for decades in Europe and South America, they are now gaining popularity in the United States. A companion to the toilet, a bidet (bee-day) is a fixture for use in cleaning the perineal area of the body after using the toilet.

Many doctors believe that this washing practice prevents skin infections and irritations in the genito-urinary area. The thermal effect and soothing action created by water under pressure striking the body is also advantageous in the care of post-operative patients or elderly people.

The user sits astride the bowl of the bidet facing the controls which regulate water volume and temperature. Water enters the bidet either via the rinse-spray or through the flushing rim which helps maintain bowl cleanliness. A mechanical stopper permits water to accumulate in the bowl when desired.

A bidet seat which fits on the water closet and includes a water jet and a drying jet might also be considered, most especially by those who suffer from hemorrhoids.

Storage

The homeowner will know that you know what you're talking about when you suggest adequate storage facilities. Any homemaker will respect the contractor who recognizes how important it is to have space to store towels, tissues, cleansing materials, cosmetics and shaving equipment. Storage space is equally as important in small bathrooms as in a large one. Lack of adequate storage space is the most common characteristic of obsolite bathrooms. An abundance of storage space is the biggest advantage to modern, well-designed bathrooms.

The space above the toilet can be utilized wisely by building-in a cabinet and giving it character by using shutters as louvered doors. Consider storage cabinets on either side of the mirror above the lavatory. Shutters or louvered doors add simple finishing touches. Large vanity cabinets provide storage shelves and space for bottles and a waste basket. Some include extra storage in door shelves, similar to a refrigerator. Bathrooms in older homes are usually much larger than bathrooms built in modern dwellings. However older bathrooms typically have very little built in storage.

Make a list of all the items that logically could or should be kept in the bathroom, considering your client's needs, and plan accordingly.

This list might include the following: Electrical grooming aids, such as toothbrushes, water-pics, shaving equipment, hot combs, hairdryers, etc.; Beauty and grooming materials, such as lotions, powders, perfume, deodorant, brushes and combs; Medicines, drugs, sick-room and first aid supplies, such as heating pads, hot water bottles, and bandages; Cleaning supplies and spare paper goods; Soiled laundry, and clean underwear; Linens, including towels, face and hand cloths.

To this list might be added such amenities as a telephone, radio, or television, a clock, sunlamp, or exercise equipment, for example. Older people and young children will have added storage needs, perhaps for such things as water toys, diapers, scales, lotions and creams.

In planning storage equipment, consider whether it should be open or closed storage. A medicine cabinet should be placed out of reach of small children for safety.

When planning for storage, try to have the storage space located near the place where the stored items will be used. For example, provide storage for towels near the lavatory and bath.

A minimum of two lineal feet per person should ordinarily be planned for towel and washcloth racks. Hooks and rings result in bunched up towels, and inadequate drying. Hooks should be provided for hanging clothes and robes. The bathroom scale might be wall-mounted, but in all cases it should not be left out where it can present a hazard.

Lighting

Most bathrooms do not have enough light. Those without windows need special consideration. Bathroom lighting falls into two basic categories: general illumination of the room as a whole, and lighting for specific areas.

General lighting can range from a ceiling fixture to a luminous ceiling. Different fixtures have been developed for the various areas. Waterproof lights are available for use in the tub or shower area, but do not place light switches where they can be reached by someone in the tub or shower. If the toilet is in a separate compartment, it should have its own light fixture. The vanity area is the section of the bathroom that needs the most light. A single fluorescent fixture over a mirrored medicine cabinet is the budget solution, but has definite disadvantages. It creates badly placed shadows on the face of a man who is trying to shave and is not flattering to the woman who is applying makeup. A better choice would be a pair of lights on each side of the medicine cabinet, either wall bracket or hanging lights. Each fixture should be capable of utilizing a 100-watt bulb. Theatrical lighting (exposed bulbs mounted around the cabinet) or soffit lighting (lights mounted in the ceiling or bulkhead concealed by a translucent panel) are also possibilities for lighting the vanity area.

Solutions include obscure windows or plastic skylights to improve lighting during daylight hours. Insulated plastic skylights can be added at reasonable cost to give a dramatic, open, well lighted feeling. If you are adding or changing windows, strip windows placed high up on the wall give both good air circulation and good lighting.

When you plan a mirror running the entire length of a vanity or long counter, use a double row of fluorescent tubes and be sure to include an equal amount of warm and cool light tubes to attain proper color balance.

For safety, a night light or illuminated switch plate should always be installed at the entrance to the bathroom.

When looking for bathroom lighting, don't confine yourself to the bathroom section of lighting catalogs. Select any fixtures that will accommodate a 100 watt bulb and use warm incandescent lights where possible. The idea is style, not sterility.

Colors

Many bathrooms have no windows. Since these inside bathrooms are lighted completely by artificial light, it is always wise to select the bathroom colors under the same type of light that you will install in the bathroom. Some bathrooms have both fluorescent and incandescent lights and also natural light from a window. Three fixtures whose colors are perfectly matched will then appear to be different in color if the toilet, for instance, is under the daylight coming in the window, the lavatory basin is in a vanity with fluorescent light and the tub is lit by an incandescent ceiling light.

Color is an area where a creative remodeling salesman can make points with the homeowner. All of those beautiful bathrooms you see in consumer magazines have been created by the greatest decorating talent around. Take advantage of them. Create your own file of bathroom photographs and ideas. Out of it will come color schemes—plans, ideas and solutions to any decorating problems you have.

You can also impress the homemaker by buying a color wheel which is available at most art supply stores. The wheel shows which colors harmonize with others and which ones contrast pleasingly. Little points like this make you a true professional in the eyes of your customer.

Electrical Outlets

While you are planning the lighting, be sure to consider all the electrical appliances your clients might want to use in their bathroom. There are electric shavers, electric toothbrushes, hair dryers, curlers and setters, facial saunas, patters, electric manicure implements, and exercisers. Think about storing these appliances at the same time you are planning the electric outlets. Will these appliances be stored in the vanity, the medicine cabinet or in a linen closet? The planning stage is also the time to think about electric ventilating or heating units. Some of these units combine several functions. For instance, a single unit is available that will heat, ventilate and light the bathroom. Also consider the installation of sunlamps.

Vanities and Cabinets

Replacing the typical old wall hung lavatory with a modern vanity cabinet is about the easiest and least expensive way to modernize any bathroom. Women appreciate the convenience of additional concealed storage space and like the thought of establishing a theme or style to the bathroom with an attractive piece of built-in furniture. From a decorating standpoint, the vanity is probably the key element around which the remainder of the bathroom should be built. Most vanities are white because most bathroom fixtures are white, but wood tone and brightly colored vanities are also available. For a powder room, suggest a brightly colored vanity and color scheme to give a feeling of openness and space. Even if your prospect is more comfortable with a white or off white vanity, suggest bright accent tones in the top, doors, trim, walls, hardware or towels.

Generally you use space better if the vanity is placed adjacent to at least one side wall. Many vanities in larger bathrooms completely fill the space between two opposite walls so that the top can end against these side walls. A filler strip may be necessary to make a standard size vanity fit into the desired area. (Some vanities can be cut down 8 or 10 inches in length if necessary.)

Consider also a set of narrow drawers, perhaps only 6 inches wide, to help a vanity fill the space between two walls. Even very narrow drawers are useful in a bathroom. But keep the vanity in proportion to the rest of the bathroom. If a nice size vanity is a big improvement, a really big vanity isn't necessarily better yet. Show your prospect how big the unit will be in the bathroom. Make a paper cut-out the size of the vanity to avoid deciding on a unit that dwarfs the rest of the room. If space is at a premium and nearly any vanity is too large or would have to be placed in an awkward position, consider putting a vanity outside the bathroom in an adjacent master bedroom, separated from the sleeping area by a partition or filigree wood sight screen. This then becomes a makeup area separated from the moisture of the bathing area. Any woman would appreciate this convenience. The additional run of drain and supply piping may be only a foot or two.

The vanity base may be of wood or plastic laminate. Tops can be marble, plastic laminate, cultured marble, slate, pre-treated wood or ceramic tile. Styles range from French Provincial to Early American, Mediterranean or contemporary. If there is space for a long vanity, or for "his and her" vanities, consider making the man's vanity higher than the woman's. You could also include knee space (omitting cabinets underneath the vanity top) to create a makeup or cosmetic vanity.

If you are planning to recommend medicine cabinets, they should be considered at the time your clients choose their vanity. From the functional cabinet that is merely a box with a mirrored door, you can go to large mirrors that slide to expose storage behind, or decorative mirrors in rectangular or oval shapes that are actually the doors of a medicine cabinet. You might recommend a unit that has a mirror in the center mounted on the wall, with mirrored medicine cabinets at each side. When the mirrored doors of the cabinets are open, they form a three-sided mirror.

Most medicine cabinets are mounted over the lavatory, but some are designed to be mounted in the wall beside the vanity. These are usually used when the wall behind the vanity is completely mirrored. They range from an invisible door—one covered with the same wallcovering as the walls—to louvered or paneled doors.

Some medicine cabinets are equipped with lights, but lighting should be considered separately. Medicine cabinets are available in a choice of recessed or surface-mounted styles. Many remodelers recommend surface-mounted units to minimize wall piercing. This retains the natural sound barrier created by the wall surfaces, a factor frequently overlooked in bathroom design.

Other Equipment

Shower heads should be adjustable. The more expensive types, which have self-cleaning plastic parts inside, are preferable because they are more durable and do not become clogged, a common problem in hard water areas.

If grab bars are to be installed, they should be mounted to the wall studs or some similarly secure foundation. This requires forethought. Grab bars should be good quality, approximately one inch in diameter, and sturdy. Don't substitute towel rods for these bars. A filmsy bar could do more damage by giving way than having no bar at all.

Nothing makes a small bathroom look bigger and brighter than mirrors. A full length mirror is nearly essential if the bathroom is going to serve as a dressing area. In a small bathroom, consider covering the better part of two walls with mirrors rather than wall cover or tile. Many shower stalls and tub enclosures are available with optional mirror glass.

A built in scale, dressing table, clothes hamper or fold out stool for a small child are nice touches that you can add to your plan at small cost. If you notice the need for equipment such as this when you first call on your prospects, include it in your plan.

Bathroom fittings are probably the most constantly used items in the bathroom. They are also the plumbing products which most visitors observe closely. They help keynote the style of the bathroom. Because they are in constant use, they are subject to the most wear and resulting service problems. Consequently, quality of materials, workmanship, function, design, and ease of maintenance are also important. Advise your clients that the lower-priced fittings can frequently be the most expensive in the long term.

There are four types of valve controls. The two valves system uses two handles and two valves: One each for hot and cold water. The "single valve" system may be operated by either a lever or a knob. Movement right or left controls the temperature. Movement in or out, backwards or forwards, controls the water volume. A "pressure balancing" valve is usually preset as to volume. The user sets only the temperature. The valve maintains it automatically. A pressure-sensing device decreases the flow of hot or cold water when the pressure in the opposite line drops. Normally pressure balancing valves are used for controlling a shower head. In a "thermostatic control" valve a heat-sensing device automatically adjusts hot and cold water flow to conform with a preselected temperature. This permits the user to control volume as well as temperature.

Wall Covers

Ceramic tile has traditionally been the first choice in wall covers for the high-moisture area around the tub or shower. Tile can be anything from a 1 inch by 1 inch square to a huge Spanish-shaped tile, 9½ inch tip to tip. The most frequently used wall tile is 4½ inches by 4½ inches, usually in a bright shiny glaze. Colors cover the spectrum, but the prevailing trend is to light, pastel tones, and a simple treatment, rather than contrasting borders and trims. Decorative tile is available. If it is used, it becomes the most important design element in the room. Everything else must be subordinated to the design of the tile. Discretion should then be used in selection of accessories to avoid a clash of patterns or designs.

Marble, either man-made or natural, is a beautiful material to use in a tub recess, creating a luxurious effect. When marble is used, it is usually repeated on the vanity top.

A wide range of possibilities is offered by plastic laminates or plastic-finished hardboards, especially designed for bath and shower areas. These come in a broad spectrum of colors and patterns. Recent product improvements have made these materials quite satisfactory for shower walls and in tub areas.

Whatever material you recommend, think carefully about the color. If you cannot match the fixture color exactly (most difficult, since you are dealing with products from two different manufacturers) it is better to use contrast. Let the wall color be lighter or darker than the color of the bathtub, or use a completely different color that harmonizes with the tub color.

The material you use on the remaining walls doesn't have to be as durable and waterproof. These walls can be finished with any of the materials already mentioned, or they can be of wood with a special finish, painted plaster or a vinyl-on-fabric wallcovering.

Flooring

Floors can be ceramic tile, marble, sheet vinyl, vinyl tile, poured floors or carpet. Highly glazed wall tile is not usually used on the floor because it is less durable than the matte or crystalline glazed tile.

Unglazed mosaic tile makes a most durable floor. Another choice might be quarry tile. Floor tile comes in various shapes—hexagons, octagons, rectangles, small squares, larger squares and Spanish and Moorish shapes. Most tile companies manufacture floor tile which harmonizes with the colors of their wall tile.

Poured floors are also good-looking. These floors, actually formed on the job, are built up in layers, with polyurethane chips embedded in a polyurethane glaze. They are sometimes referred to as seamless floors and are easy to maintain because they require no waxing. Sheet vinyl is another choice for flooring. Most of these sheets are 6 feet or more wide, so that the average bathroom floor can be laid without a seam. Vinyl tiles are also a possibility for the bathroom, and they range in style from the very conservative marbleized patterns to large 12 inches or 18 inches tiles that resemble bamboo, marble or parquet flooring. Carpet can also be installed in the bathroom. Use either room-size rugs that can be easily removed to be cleaned or laundered, or the new permanent type that is cemented to the underfloor. (The former might be more easily maintained in a bathroom used by young children.)

Decorating the Bathroom

After your clients have chosen all the permanent and built-in furnishings for their bathroom, you may want to recommend things that may change from time to time: wall covering, for instance, for the walls other than those in the bath or shower recess. Vinyl-on-fabric is an excellent choice. It is washable and scrubbable and will stand up well under humidity. It is available in many different patterns and colors, from a delicate spray of flowers to a bold zigzag. Duck, seashell and fish patterns have been used for years in bathrooms and are somewhat trite. You may wish to recommend a big paisley, a large plaid or a vivid stripe instead. The wall covering is not as permanent as the other materials, so you can be bolder in color and design. Flocks and foils are more suitable for the powder room used by guests than they are for the main bathroom.

The powder room is a place to try bold decorating ideas. Since no one stays in the room very long, finishing touches can be much more dashing than in any other room in the house. This room is usually very small, so you can follow either of two courses in selecting the wall covering: use a see-through type pattern, such as a caning or

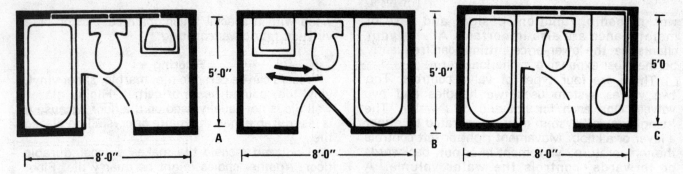

**A. Usual toilet location; B. Codes usually permit
reversed location; C. Run pipes under vanity
Fixture rearranging with one wet wall
Figure 6-2**

trellis design, or use a bold, zippy pattern that takes over the whole room. Mirrored walls are a great visual spacemaker, and overcome the claustrophobic effect of a small room.

Towels and small accessories are a very colorful part of bathroom decorating. Since they can be changed more easily and less expensively than any other item in the bathroom, here is the place to use brilliant colors, accent colors, the latest thing, a color that might be fashionable this year and out of date three or four years from now. Wastepaper baskets and tissue holders, drinking glasses and soap dishes, all are available in exotic colors. With beautiful fixtures as a pastel background, use brilliant dashes of color in these accessories, or match one of the dominant colors in the wallcovering with plain towels in a brilliant hue. When recommending accessories or selecting accessories for your own showroom bath, don't limit yourself to traditional bath accesory shops. Pottery canisters from a kitchen shop can be used to store bath salts, a dish from an antique shop can hold soaps, or an old hatstand rescued from the attic and spray-painted a brilliant color can be used to hold towels.

Don't forget pictures as bathroom accessories—flower prints in gold frames for a powder room, military prints for a boy's bathroom. Sculpture, a collection of seashells, framed butterflies, posters, a collection of small mirrors, unusual containers for soap or guest towels, all can add to the charm of a bathroom.

The last time you or your salesman calls on your client after finishing the job, make it a point to bring a gift of a set of towels or an attractive accessory specially selected to go well in their bath. It's a thoughtful gesture bound to create good feeling toward your company.

The treatment of any windows is also a definite part of bathroom decorating. Try to avoid having the tub beneath a window. It makes it difficult to open and close the window, and if the window is subject to water from the shower, there is danger that water will penetrate from the window into the wall below and destroy the plaster or other

construction materials. Normally, beaded or frosted glass is used for bathroom windows to preserve privacy. Louvered shutters, Roman shades, woven blinds, decorative window shades, or shutters with inserts of filigree, caning or fabric can be used, adding to the decorating scheme while reinforcing a feeling of privacy. If your clients insist on draperies or curtains, advise them that ruffly, delicate materials suffer from the humidity. Duck, sailcloth, denim and other tightly woven materials are better choices.

At the other end of the design spectrum is the bathroom visually open to the outdoors. The garden bath or patio bath has one wall that is all window. It looks out into a garden, patio or sundeck that is completely screened by a high wall or fence. Skylights, now available in the form of plastic bubbles, are another exciting way to expand the bathroom visually. What could be more glamorous than a beautiful bathroom with tub centered under a skylight?

One thing that is sometimes overlooked is the relation of the bathroom color scheme to other rooms nearby. Decorate the powder room in the same colors used in the adjacent foyer or entrance hall. Tie the master bathroom scheme in with that of the master bedroom.

Developing a Floor Plan

Those sunken Roman tubs equipped with slinking tigers and stunning models are nice to look at, but 95 percent of your customers aren't in that market. In fact the great majority of bathroom jobs you quote will be for areas measuring around 5 feet by 8 feet. As a remodeling professional, you have to show the homeowner how to convert this small area into a bathroom that has some of the appeal, some of the high style, and some of the feeling of luxury found in the larger more elaborate designs.

The basic plan of the bathroom must be a good one, one that will work even in the most restricted space. In many cases you will be working with one wet wall where construction economics place the plumbing pipes in one wall with the lavatory, toilet

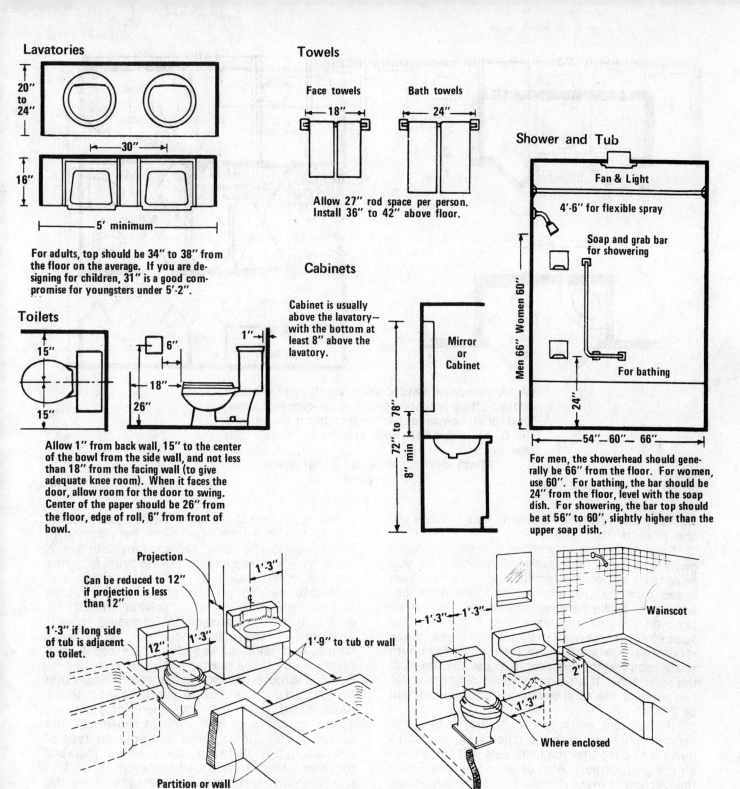

Lavatories

20" to 24"

30"

16"

5' minimum

For adults, top should be 34" to 38" from the floor on the average. If you are designing for children, 31" is a good compromise for youngsters under 5'-2".

Toilets

15"

15"

6"

1"

18"

26"

Allow 1" from back wall, 15" to the center of the bowl from the side wall, and not less than 18" from the facing wall (to give adequate knee room). When it faces the door, allow room for the door to swing. Center of the paper should be 26" from the floor, edge of roll, 6" from front of bowl.

Towels

Face towels

18"

Bath towels

24"

Allow 27" rod space per person. Install 36" to 42" above floor.

Cabinets

Cabinet is usually above the lavatory—with the bottom at least 8" above the lavatory.

Mirror or Cabinet

72" to 78"

8" min

Shower and Tub

Fan & Light

4'-6" for flexible spray

Soap and grab bar for showering

Men 66" Women 60"

For bathing

24"

54"— 60"— 66"

For men, the showerhead should generally be 66" from the floor. For women, use 60". For bathing, the bar should be 24" from the floor, level with the soap dish. For showering, the bar top should be at 56" to 60", slightly higher than the upper soap dish.

Projection

Can be reduced to 12" if projection is less than 12"

1'-3"

1'-3" if long side of tub is adjacent to toilet.

12"

1'-3"

1'-9" to tub or wall

Partition or wall

1'-3"—1'-3"

Wainscot

2"

1'-3"

Where enclosed

**Minimum space requirements
Figure 6-3**

and supply/drain end of the bathtub all lined up against that wall. Your job is to take this somewhat static arrangement and develop variety and design without a major change in plumbing. Sometimes variety in design can be achieved, without a major change in plumbing, by constructing a vanity countertop or cabinet along the end wall opposite the bathtub. Supply pipes and drain can be run under the countertop to the wet wall at the rear. (Note Figure 6-2). Normally the toilet is installed in the center of the wall between the tub and the lavatory, but if local codes permit back-venting the toilet to the stack, you might try reversing the position of the lavatory and toilet. Where ceiling height permits, an 8 inch raised platform floor can be built in a portion of the room to allow running

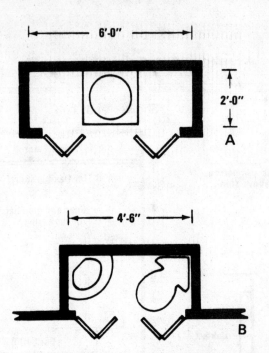

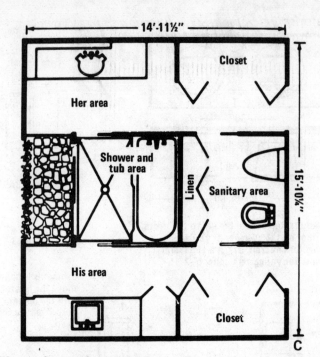

A. A grooming center with vanity and storage, closed off by louvered doors. B. A corner lavatory and toilet, convenient arrangement for the aged or ill. C. An island bath with "his" and "hers" compartments
Bath ideas for little and lots of space
Figure 6-4

waste to the new toilet location. This reduces the cost in most situations.

Each 12 inches of additional bathroom space gives you room for creative variations. All you have to do is become familiar with the minimum space each fixture requires and the minimum amount of space permissible between fixtures or between fixtures and walls. See Figure 6-3 for space requirements. If two lavatory bowls are requested in a vanity, five feet is the minimum space required for two people to use the bowls at the same time. If a sit-down vanity is planned for two persons, the total counter length required will be 7 feet.

If you want extra counter space, extend the vanity countertop over the toilet tank, but don't make it so deep that the toilet seat will not stay in an upright position. Also be sure to either hinge that section or make it removable to permit access to the interior of the toilet tank. Where space is at a premium, and if codes allow, consider a corner lavatory and triangle toilet. An island bathroom or compartmentalized "his" & "hers" dressing areas are some of the many innovations you can introduce when plenty of room is available. See Figure 6-4.

Proper planning and utilization of a bathroom's floor area can do a lot to improve its traffic handling ability. A properly designed small bath can be more serviceable than a poorly designed large bath. The conventional bathroom, consisting

of tub, toilet and lavatory, can only accommodate one person at a time. The same bathroom with twin lavatories, however, can accommodate two. A partition which separates the lavatories from the tub and toilet area adds additional privacy and convenience. To go one step further, some bathrooms have separate and private "zones" for each fixture, with a common "dressing room." This design works extremely well for a large family, and adds an especially nice touch to a luxury bath off the master bedroom.

Look through the sample bathroom layouts that follow to find the design features and space requirements that meet the needs of your job. These designs work well (both as shown and the mirror image) and can help you avoid the type of problem that is difficult to see in a floor plan but becomes obvious once the bathroom goes into service. Whenever possible, resist relocating the toilet to avoid moving the plumbing stack. The lavatory, tub and shower can be relocated relatively easily in most bathrooms.

The figures on page 72 show three "before" and "after" floor plans for remodeled bathrooms. The existing bathroom in Figure 6-5 measured 7 feet by 5 feet, about as small as a three fixture bathroom can be. All three fixtures are along one wall, leaving little room for convenient storage space and making use by more than one person at a time fairly difficult. If the door swings in against the right wall, less than 2 feet of the wall backing

64

Small Bathrooms For Family Use

When space is fixed or limited, you want to get the most out of every square foot. Set your priorities and plan around them—whether it's storage space, twin lavatories, countertop area, or whatever.

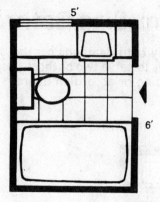

Inexpensive two-wall plumbing with new countertops and cabinet storage.

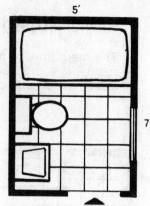

Neat, compact arrangement, one-wall plumbing, efficient use of space.

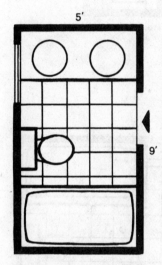

Inexpensive two-wall plumbing, luxurious twin lavatory countertop.

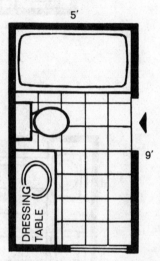

Efficient, inexpensive one-wall plumbing with dressing table convenience.

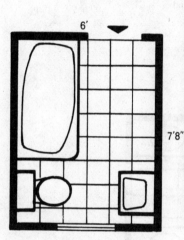

A corner tub can be enclosed in mirrored glass to make the room seem larger.

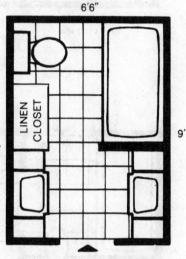

Twin lavatory convenience, double storage convenience below, plus a closet.

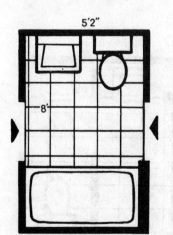

Providing access from two bedrooms with plenty of room for walk-through.

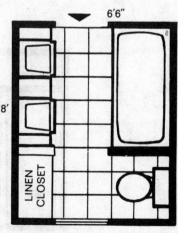

Privacy of a compartmented toilet and lots of storage space.

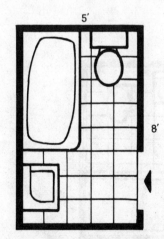

A handsome contour corner tub lets you modernize without adding a wall.

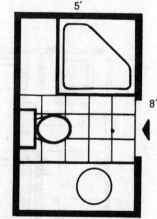

A square receptor-bath offers showering and bathing in minimum space, opens room.

Medium-Size Bathrooms For Family Use

The medium-size family bath demands careful space planning. Increase efficiency by dividing the room into compartments to serve two or three persons at the same time. Consider a bathtub in its own closed-off area and sliding or accordion doors.

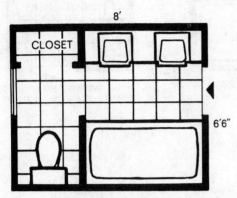

Compartmenting the toilet and twin lavatories.

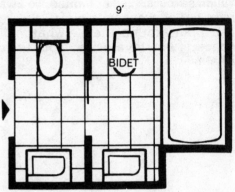

Twin lavatories and the luxury of a bidet for all.

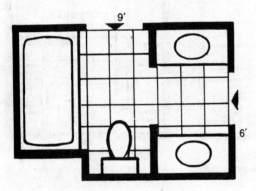

Separating the lavatories for semi-private use.

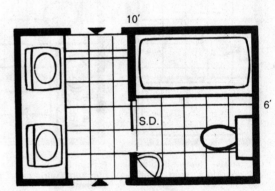

Accessibility from both sides; compartmented bathroom.

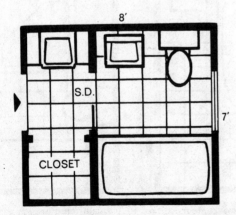

Complete bathroom privacy for one, shaving or makeup area for the other.

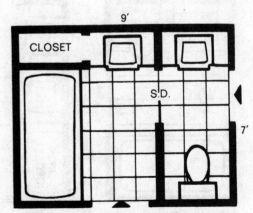

For a large family, by closing off the bottom door and adding another toilet, an extra powder room in a bath.

Powder Rooms

You can install a powder room just about anywhere there are 16 square feet of unused space...a closet, under stairs, etc. Use mirrors, a window, a skylight or bright colors to open up the room and make it look larger than it really is.

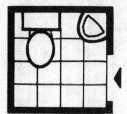

4' X 4'
It's tight but tidy so why not consider under a stairway, a large closet or just make space?

4' X 4'6''
How about the convenience of a countertop? For needed storage, for guests.

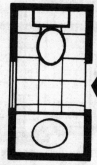

3' X 6'
Ideal for under a stairway. The decor can accent the hall. Consider flocked paper or fabric.

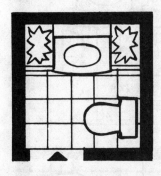

5' x 5'-3''
Guests will be impressed with built-in planters flanking the lavatory.

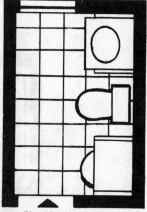

5' x 7'-6''
Plenty of room for busy families, primping guests. A separate vanity.

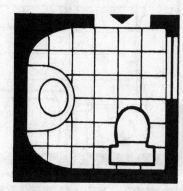

6' x 6'
Build in the dramatic, with curved walls from flexible plywood cornered with a dramatic fabric or foil paper.

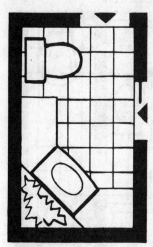

5'-8'-3''
Use a corner for the lavatory-dressing table . . . gain storage space.

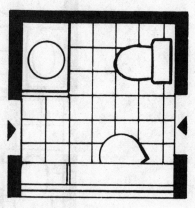

6'-6'' x 6'-6''
Square room with one whole wall for storage and make up.

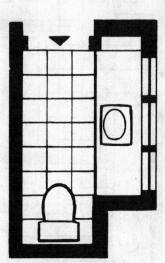

5' x 8'
Lots of room and lots of convenience in storage and counter top space.

"Zoned" Bathrooms

These plans can make use of the larger bathroom areas that are found in older homes. Partitions separate the privacy areas so the room can be used by two or more at the same time.

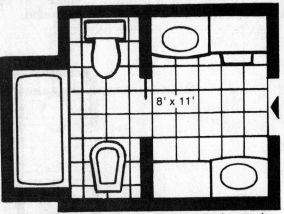

8' x 11'

The bathing area here may open into a small private garden.

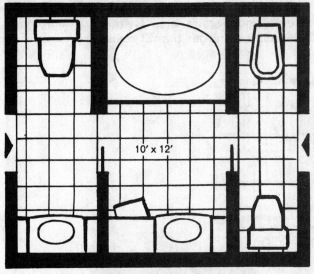

Double entry bath built around an oval tub.

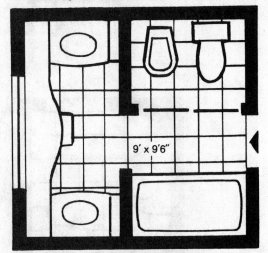

9' x 9'6"

Partitioned toilet-bidet privacy compartment.

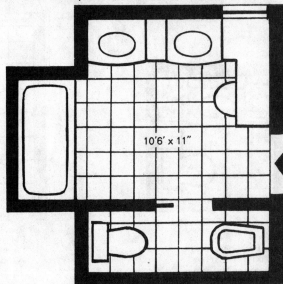

10'6' x 11"

Separate, personal toilet-bidet area for multiple use of a spacious room.

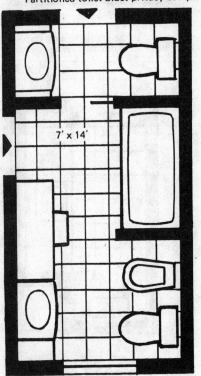

7' x 14'

Two separate entrances in a long floor plan.

10' x 11'6"

Two rooms in one, with a powder room off a hallway or accessible from bedroom.

Larger Bathrooms For Family Use

With more space, you have more flexibility. You can use larger, more luxurious fixtures and still have plenty of room. However, it's not necessarily the amount of space you have, but the way you use it that makes an ordinary bath into a real client pleasing job.

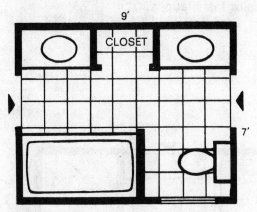

Deluxe recess tub offers wall for private toilet area.

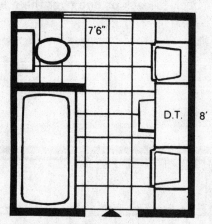

Twin lavatories frame a dressing table with semiprivate toilet.

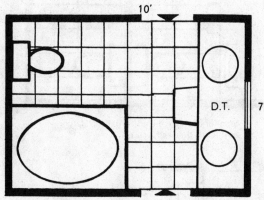

Oval corner bath with twin lavatories.

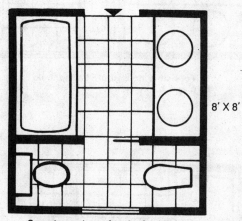

Complete privacy for the fastidious with toilet and bidet.

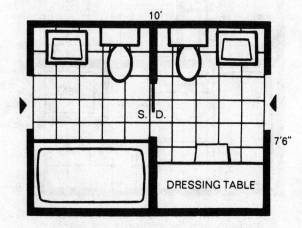

A double-door bath with double duty, double privacy.

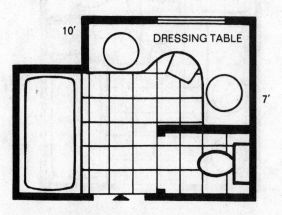

Unusual twin lavatory installation for the ideal grooming area.

Adjoining Or Auxiliary Baths

Nearly all home-owners today require a minimum two bathroom house if at all possible. These layouts illustrate an excellent way of providing two bathrooms in a compact area. Bathrooms built back-to-back or one over the other are the most economical kinds of installations.

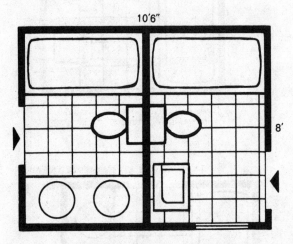

In a one-story house, both baths should be placed back-to-back.

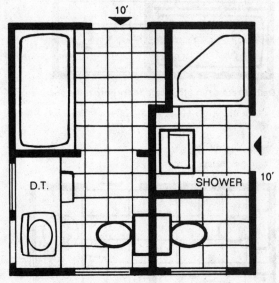

Smaller bath with traditional tub; larger bath with deluxe fixtures.

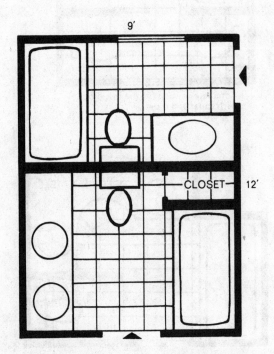

His and Her twin lavatories in the master bath; smaller countertop in second bath.

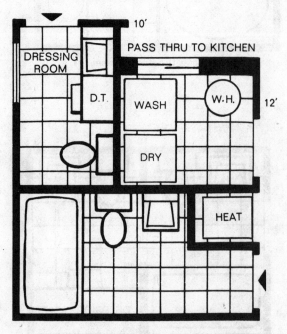

This all-encompassing 10' x 12' area includes the family bath, heating, laundry and water heater facilities, dressing room.

Dressing Room Bathrooms

The dressing room can be moderate or expansive. Borrow space from the master bedroom, adjacent room, hallway or large closets.

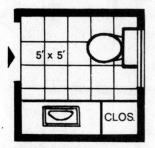

5' x 5'
CLOS.

Everything for privacy. Substitute a closet for the toilet if more space is needed.

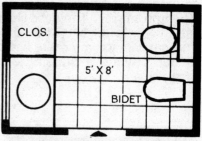

CLOS.
5' X 8'
BIDET

The finest continental dressing room convenience includes the luxury of a bidet.

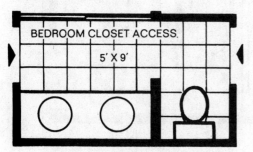

BEDROOM CLOSET ACCESS.
5' X 9'

Mother and daughter might share this convenient dressing room that joins their bedrooms.

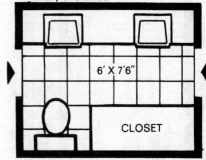

6' X 7'6"
CLOSET

Double lavatories for two people. Plenty of storage space in the cabinet and in the closet.

Convenience Bathrooms

Adding a bath can spare wear and tear on the main bath...and ease traffic on busy mornings and evenings.

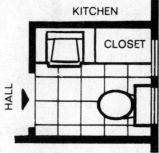

KITCHEN
CLOSET
HALL

4'-6'' X 5'

A sink and toilet next to the kitchen. A great convenience for the busy homemaker.

URINAL MAY BE USED
BACK DOOR
HALL
SHOWER STALL
LIVING AREA

4'-6'' X 6'

A wash-up mud room area for the whole family . . . especially children. Confines tracks to outside and here.

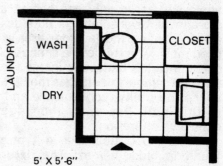

WASH
DRY
LAUNDRY
CLOSET

5' X 5'-6''

Next to the laundry in the basement, tap into the main plumbing lines for another convenience facility.

up to the closet can be used for towel racks or storage. The suggested change borrows space from an adjacent closet (or perhaps from below a stairway) for a vanity alcove and linen storage cabinet. The space between the tub and toilet will now accept a full height louvered door cabinet or full length mirror. The door should be swung in against the toilet or outward. The remodeled bathroom can accommodate two comfortably now and has become a dressing area as well as a bathroom.

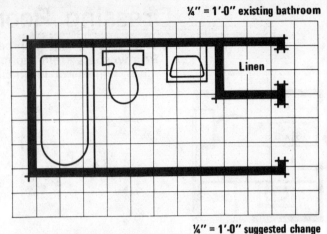

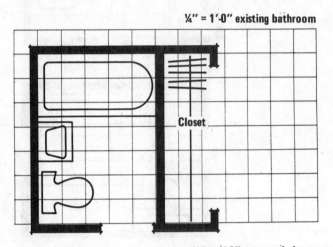

¼" = 1'-0" suggested change

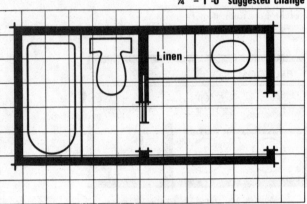

Figure 6-6

¼" = 1'-0" suggested change

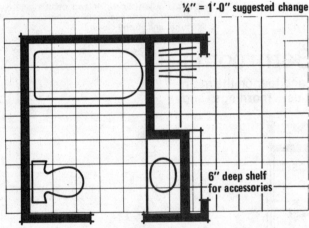

Figure 6-5

Figure 6-6 shows a long, narrow bathroom that again will serve only one person at a time. The storage of towels, etc. is outside the bathroom rather than in the lavatory area where they are used most. The suggested change brings the linen closet into the bath, eliminates the useless entry hall and compartmentalizes the bath with a sliding door. The utility, convenience and attractiveness of this bath has more than doubled without extensive alteration.

Figure 6-7 illustrates a typical larger bath found in older homes. The fixtures are widely spaced but little though went into making this bathroom serve as the primary bathroom for a larger family. There probably is little built-in storage space and only the most basic lighting.

¼" = 1'-0" existing bathroom

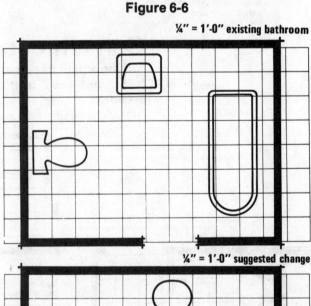

¼" = 1'-0" suggested change

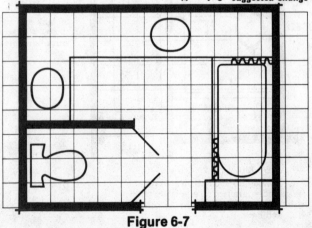

Figure 6-7

72

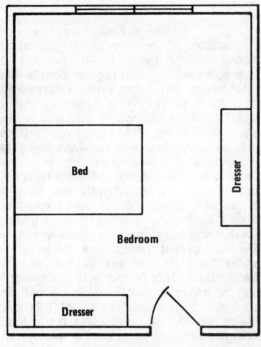

Original layout

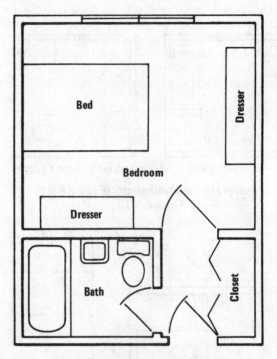

Bath and closet added

Portion of a large bedroom used to add a bath
Figure 6-8

The advantage you have here is plenty of space. An 8 feet by 10 feet bath such as this can serve three people at once with a little planning. The toilet in the remodeled bathroom was moved two feet and built into a compartment with swinging doors. Both walls above the lavatories were covered with a combination of mirrors and cabinets. Naturally, the cabinets below the lavatories became linen and appliance storage area, over 12 linear feet in this case. A family with 3 or 4 children would really appreciate the improvements in the changed bathroom.

Adding a Bath

Finding a convenient location to add a bath in an existing house is often more difficult. Adding a room to the house is seldom a good solution because it is usually best to have access to the bath from a bedroom hallway. A half bath near the main entrance or in the work area is usually a good solution. One consideration in locating a bath economically is to keep all piping runs as short as possible. Also, all fixtures on one plumbing wall can use a common vent.

One common mistake in adding a bath in an older house has been to place it in any unused space without regard to convenience of location. Consequently, many bathrooms have been placed in what was formerly a pantry, a large closet, or under a stairway. This usually means the only access to the bathroom is through the kitchen or a bedroom, or that the bathroom is totally removed from the bedroom area. If this mistake has been made in a house, it is important to add another bath in a good location.

In a house with large bedrooms, a portion of one bedroom can be taken for a bath (Figure 6-8). Such a bedroom should be at least 16 feet in one dimension so that it will still be no less than 10 feet in least dimension after the bath is built. If the bedrooms are small and all are needed, there may be no choice except to build an addition. It may be wise to make a small bedroom into a bath (or two baths) and add on another bedroom. A possibility in the one-and-one-half-story house is to add a bath in the area under the shed dormer. When this is done, remember that the wall containing the plumbing must have a wall below it on the first floor through which piping can run. This applies to two-story houses also.

The minimum size for a bathroom is 5 feet by 7 feet (Figure 6-9), and larger sizes are certainly

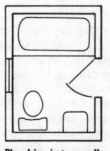

Plumbing in two walls

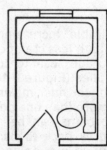

Common plumbing wall

Minimum size bathroom [5 ft by 7 ft]
Figure 6-9

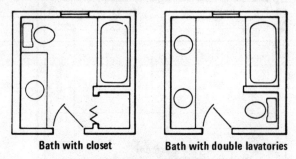

Bath with closet   Bath with double lavatories

Moderate size bathroom [8 ft by 8 ft]
Figure 6-10

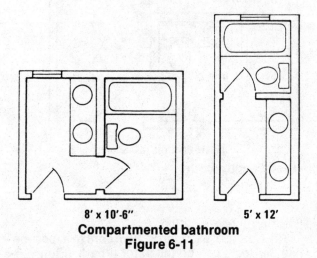

8′ x 10′-6″   5′ x 12′

Compartmented bathroom
Figure 6-11

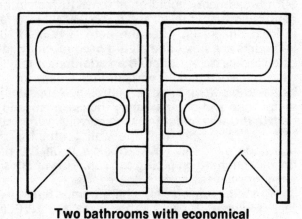

Two bathrooms with economical
back-to-back arrangement
Figure 6-12

desirable. Increasing the size slightly would make the bath less cramped and could provide space for a storage closet for towels, cleaning equipment, or supplies (Figure 6-10). If the plan is for only one bath, consider making it in compartments for use by more than one person at a time (Figure 16-11). Two baths can be most economically built with fixtures back to back (Figure 6-12), but do not sacrifice convenience of location to accomplish this. Bathrooms built on both floors of a two-story house are most economically built with second floor bath directly over the first floor bath.

Selling Bathroom Jobs

Chapter 20 outlines what you should know about selling remodeling services. In many respects, selling bathrooms jobs is like selling kitchens or any other home improvement work. You have to find and identify your prospect, qualify him as financially able and willing to buy your service, determine what your prospect needs and will buy, and then convince him that you are the bathroom specialist who can do a professional job at a reasonable price. Naturally, it helps if you are knowledgeable, friendly and have the references and testimonials to prove you are reliable. Like in kitchen remodeling, you are going to follow certain steps and use several sales aids in making the sale. Most remodelers follow a procedure something like is set out on the following paragraphs. This certainly is not the only way to sell bathroom remodeling, but most remodeling contractors use some or all of this system and many remodeling contractors could benefit from using or adapting some part of this system in their sales work.

Examine the Problem

When you arrive at your prospects' home, the focus of your attention and the attention of your potential clients will be on the existing inadequate, deteriorated or unpleasant bathroom. In the first few minutes you are in their home, your prospects are forming an opinion of your character and professionalism. These may be awkward minutes for both you and your prospects. The best way to establish a good working relation with your prospects and impress them favorably is to go right to work, listening to what they don't like about their bathroom and probing for what they want in the new bathroom while taking down the measurements and fixture locations. You should have a tape measure, pencil and pad of survey forms something like Figure 6-14. Fill in the fixture dimensions and make a floor plan of the existing bathroom. While you are taking down these dimensions, make notes on what your clients want in their new bathroom. Use a form such as Figure 6-13 to check off what your prospects want or don't want. An "instant printer" can print several hundred of these forms with your company name at modest cost and make them into a pad for easy use. Have Figure 6-13 printed on the front and Figure 6-14 printed on the back of each sheet.

While you are taking down dimensions, answer your clients questions in a general way and avoid making any specific recommendation or quoting anything more than broad cost estimates (such as "a completely remodeled bathroom generally runs between $3,000 and $6,000"). Once you have made the measurements you need, question your client about what the new bathroom should have. Tell them you will get their ideas of what they want and make up a proposal and cost estimate.

Bath Survey

Contractor's Name _____

Address _____

City _____ State _____ Zip _____

Phone _____ Job Number _____

Survey By _____

Owner's Name _____

Address _____

City _____ State _____ Zip _____

Job Address _____ Phone _____

Date _____ Job Phone _____

Vanity ☐New ☐Use existing ☐None
☐Remove existing _____
☐Wood finish ☐Plastic finish
  Style _____ Color _____
☐Hamper door ☐Knee space
  Length _____
  Drawers _____ Doors _____
  Style of pulls _____
_____

Tops ☐New ☐Use existing ☐None
  Material _____
  Style _____ Color _____
☐Hinge over toilet ☐Post formed
  Edge treatment _____
  Splash height _____
☐Intergal splash ☐Separate splash
  Length _____ Width _____
  Number of cut-outs _____

Lavatories ☐New ☐Existing
☐Wall hung ☐One piece ☐Self-rim
☐Flush mount ☐Undermount
☐Round ☐Oval ☐Other _____
  Size _____ Color _____
_____

Medicine cabinet ☐New ☐Existing
☐Recessed ☐Surface ☐Decorative
  Type _____ Shape _____
  Size _____
  Mirrors _____

Toilet ☐New ☐Reset ☐No change
  Color _____ Style _____
☐1 piece ☐2 piece

Tub ☐New ☐Use existing ☐None
☐Whirlpool
  Color _____ Style _____
  Size _____
☐"Sunken" tub in platform
_____

Bidet ☐New ☐None
  Color _____ Style _____

Lavatory Faucets ☐New ☐No change
☐Chrome ☐Brass ☐Crystal
☐China ☐Other _____
☐Two valve ☐Single valve
_____

Bath Fittings ☐New ☐No change
☐Single valve ☐Thermostatic
☐Shower head ☐Pressure balancing
☐"His and hers" heads ☐Personal
☐Diverter type _____

Ventilating ☐New ☐No change
☐Wall fan ☐Ceiling fan
☐Vent to _____ Distance _____
☐Heat lamp - light
☐Fan, lamp, heater ☐Timer
☐Humidstat

Enclosures ☐New ☐No change
☐Remove _____
☐Shower stall and receptor
☐Shower-tub module
  Size _____
☐Stall door, size _____
☐Tub door, size _____
  Door style _____
☐Glass ☐Plastic ☐Mirror
☐Shower rod

Storage ☐New ☐No change
☐Remove _____
☐Closets _____ Style _____
  Height _____ Door _____
☐Linen ☐Laundry ☐Scale
☐Clothes drying ☐Cleaning supplies
☐Electric appliances ☐Open shelving
_____

Floor ☐New ☐No change
☐Remove existing ☐Repair
☐Resilient ☐Underlayment
☐Carpet ☐Tile
☐Other _____
  Color _____ Style _____
☐New sill, type _____
☐New cove, type _____

Walls ☐New ☐No change
☐Remove existing _____
☐Repair existing _____

Tub area: ☐Plastic ☐Tile
  Style _____ Color _____
  Tub area size _____

Other areas:
☐Repaint ☐Paper
_____

Ceiling ☐New ☐No change
☐Repair _____
☐Remove _____
☐Drop luminous, size _____
☐Skylight, size _____

Accessories
☐Remove _____
☐Match tile, type _____
☐Paper holder
☐_____
☐Paper holder ☐Tissue dispenser
☐Hoods ☐Grab bar ☐Soap dish
☐Grab bar - soap dish ☐Towel bars
☐Tumbler - toothbrush holder
☐Fold out stool ☐Other _____
  Type of finish _____
☐Mirrors, size _____
☐Decorative items _____

General Construction
  Bulkheads at: ☐End of tub
☐Over vanity ☐Other _____
☐Rehang door _____
☐New entry door _____
☐Reset window _____
☐New window _____
☐Wood shutters _____
☐Remove _____

☐Partitions _____

☐Sliding door ☐Folding door
☐Raised floor level
☐Close off _____
☐Open up _____
☐Expand to _____
☐Wallboard ☐Panel ☐Plaster

Plumbing
☐Remove _____ fixtures
☐Rough-in for _____
  Feet of waste _____
  Feet of vent _____
  Feet of supply _____
  Pipe access from _____
☐Remove radiator ☐Relocate radiator
☐Shower piping _____
☐Steam piping _____
☐Install _____ fixtures
☐Other _____

Heating ☐New ☐No change
☐Relocate registers _____
☐Extend duct from _____
☐New ducting from _____
☐Close off _____
☐Baseboard ☐Radiant
☐Other _____

☐**Electric** ☐New _____ amp service
  Feet of wiring _____
  Type of wiring _____
☐Fan switch ☐Heater switch
☐Additional outlets _____

Lighting
☐Fluorescent ceiling _____
☐Wall _____
☐Medicine cabinet _____
☐Over bulkhead _____
☐Theatrical ☐Drop ☐Swag
☐Waterproof ☐Other _____

Owner to provide the following:
☐Materials _____
☐Labor _____

Figure 6-13

Who is going to use this bathroom: _____

What are the problems with the existing bathroom: _____

What colors and theme or style do you want in this bathroom: _____

How do you want to improve this bathroom: _____

How many bathrooms in this residence excluding the bath now being considered: _____

How many tubs in other bathrooms: _____

How many showers in other bathrooms: _____

How many toilets in other bathrooms: _____

Access is available below the floor by a ☐Crawl space ☐Basement ☐Lower room ☐Other

Access is available above the ceiling by an ☐Attic ☐Upper room ☐Other

Lines from the basement, attic, garage or crawl space can be concealed in _____

Tub access door location _____

Condition of walls   ☐Good ☐Fair ☐Poor

Condition of floor   ☐Good ☐Fair ☐Poor

Condition of ceiling   ☐Good ☐Fair ☐Poor

Condition of door and trim   ☐Good ☐Fair ☐Poor

Signs of structural repair needed _____

Existing fixture dimensions

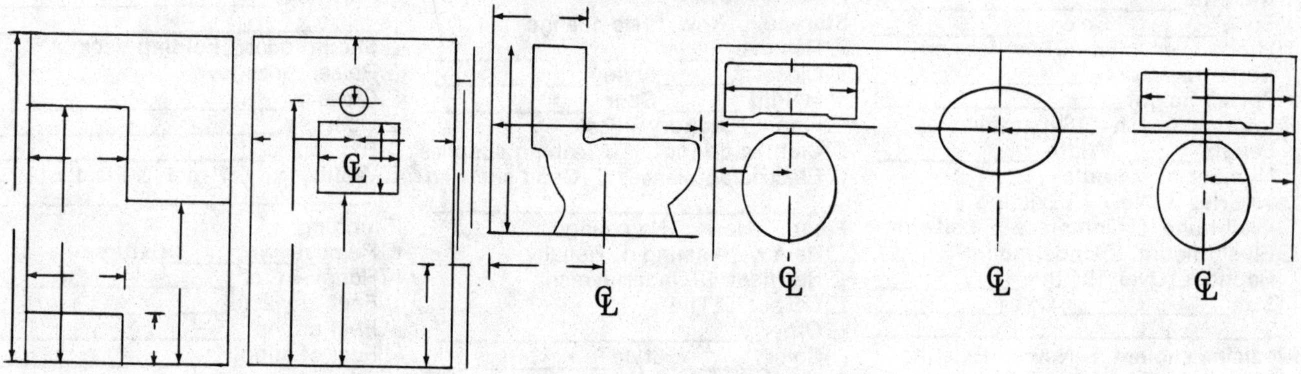

Existing bathroom layout

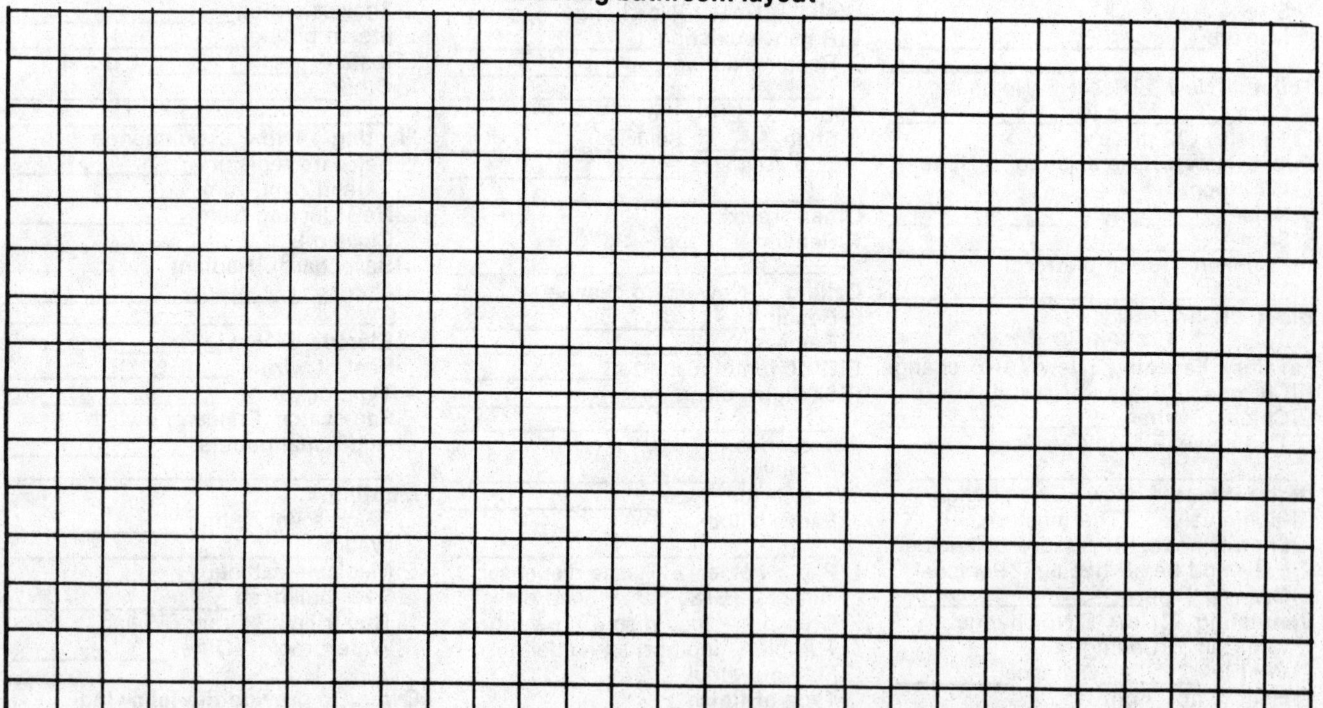

Figure 6-14

Figure 6-13 includes many improvements that might not have occurred to your clients. Get their reaction to a new tub, toilet, vanity, shower, floor, wall or ceiling cover, etc. Explain the advantages of each improvement and give them the benefit of your experience with some of the modern materials that are available. You should have some samples or pictures available to convey the beauty and attractiveness that you are going to build into their new bathroom. Get their reaction to an improved floor plan, expanding the bath into adjacent little used space, a particular decorating theme or adding a partition to compartmentalize the bath. Suggest a few luxury features such as dual lavatories, a bidet, pastel colored fixtures, a whirlpool bath or genuine marble vanity top if they seem appropriate. Mention that they probably want to improve the lighting, add storage space and an electrical outlet or two, and install a modern vanity. Make notes, either in your mind or on paper, about what your clients seem to approve and reject. Later you will try to incorporate into your proposal the ideas they seem to like.

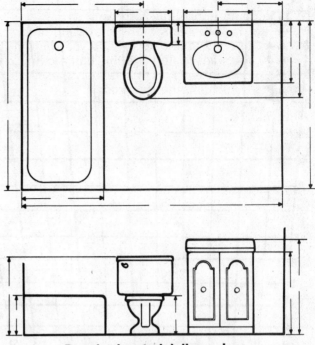

**Required material dimensions
Figure 6-16**

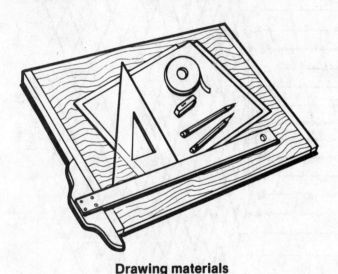

**Drawing materials
Figure 6-15**

**Typical perspective drawing
Figure 6-17**

When you have completed your survey you know pretty well what your clients want and will buy. If you are confident that they are ready to buy your services, and the job doesn't require major changes or a complete proposal, offer then and there to let them select the actual fixtures and materials they want. You will need pictures and samples they can select from and should be able to suggest approximate costs (but not give a firm estimate). After they have made their selections, make another appointment for the purpose of giving them your estimate and closing the sale.

In most cases, your prospects won't be ready to buy your services after the first meeting and the job will require more planning than you can do in the few minutes you are in their home the first time. You are going to have to make up a proposal and estimate. Advise your clients that you can make up a plan and estimate in a few days and set a time and date for presenting the proposal to them. If you sense that your clients are "just shopping" or if the job involves major changes, set a reasonable fee for preparing the plan and estimate. Collect the fee from them on the spot or at least get their signature on an agreement to pay the planning fee (such as in Figure 5-22). If they won't agree to pay $10 or $20 for professional planning, you probably are wasting your time to draw up a plan and estimate. Naturally, the fee is credited against the cost of the job if your proposal is accepted.

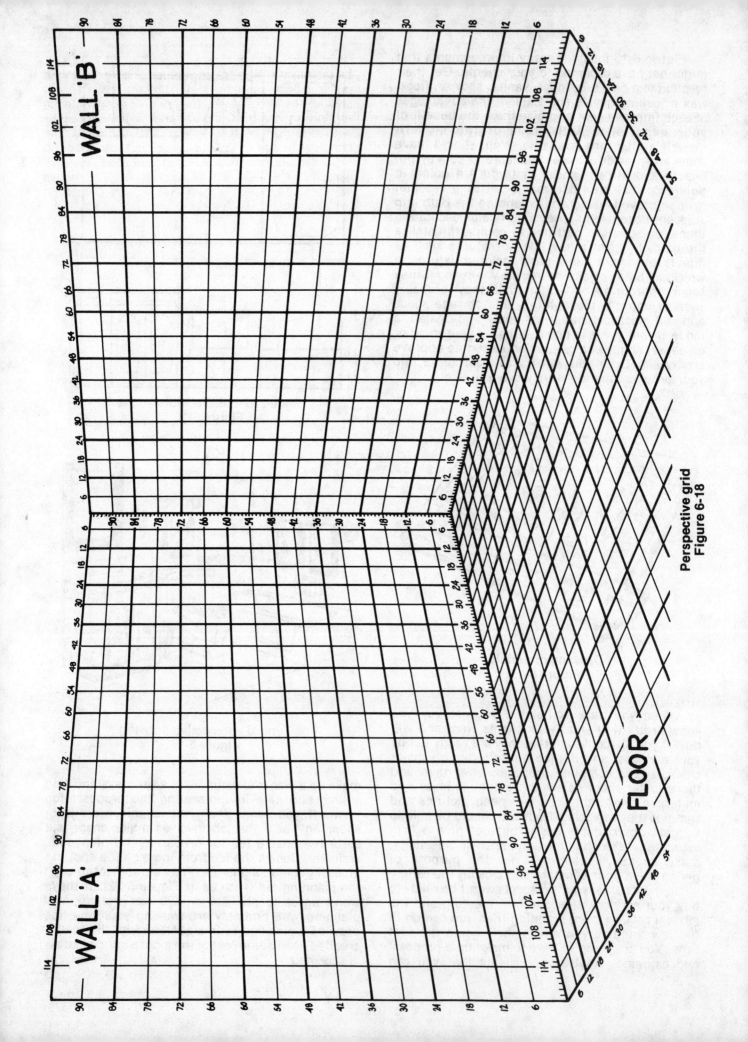

Perspective grid
Figure 6-18

Developing Your Proposal

Now draw on your experience and knowledge as a professional home improvement contractor. Sketch out several floor plans on a grid such as Figure 6-14 until you have found the plan that makes the most sense for your clients. Remember that you are going to have to execute the plan after you sell it to your clients. Avoid anything that is going to be difficult, overly expensive, wasteful, or problem prone from a construction standpoint. Be thinking of reasons why your plan is the best one for this bathroom and why all others rate as second best.

Once you have the floor plan, prepare a perspective drawing of the remodeled bathroom. Even if you don't have any previous drawing experience, you should have no trouble making a pleasing, attractive drawing in less than an hour if you use the method recommended on the following pages. The completed drawing will command the attention of your clients and provoke a "yes" answer to your question "Is this what you want?" The drawing is a physical object, a real thing that serves far better to convey the attractiveness of the new bathroom than you standing in the bathroom motioning with your hands and trying to describe materials and treatments that may be unfamiliar to your clients. Most important, the perspective drawing is a representation of "their bathroom." No picture cut from a magazine can be as impressive or as close to the completed job as your perspective drawing.

With a little practice you or someone in your office can prepare one of these drawings in less than an hour. You need some inexpensive drafting materials to do the job right: a drawing board, T square, triangle, tracing paper, tape, pencils, and gum eraser. See Figure 6-15.

The first step in preparing a perspective rendering is collecting the necessary dimensions. These can be found on your plan view of the bath layout and in the manufacturers' catalogs. You will need the actual fixture dimensions and the dimensions which locate the fixtures in the room as well as dimensions of the vanity, medicine cabinets, etc. Figure 6-16 shows some of the dimensions you will need.

The next step is deciding on the best view to draw for your customer. You will want to illustrate the most pleasing angle of the bathroom. In most cases this will be a view which focuses on the lavatory and toilet area, unless you are planning something special like a sunken tub. One of the more common layouts is illustrated in Figure 6-17.

Prepare for drawing by taping down the perspective grid (Figure 6-18) to the drawing board. Using your T square and triangle be sure that the vertical lines are lined up with the triangle as illustrated in Figure 6-19. Then place a sheet of tracing paper over the grid, taping it to the drawing board as well. The perspective grid will be visible thru the tracing paper. The grid is laid out in 6 inch squares. Note that the squares get smaller the farther away they are. For convenience in plotting, the grid is numbered every six inches with one inch marks within each square. The simplest fixture to lay out is the bathtub. The tub in its simplest form is a rectangular box. For our sample, we will use a tub 60 inches long, 30 inches wide and 14 inches high and it will be placed on the end wall of a 5 feet by 8 feet bathroom.

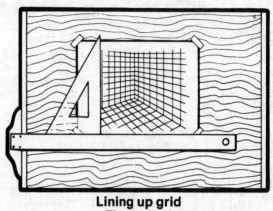
Lining up grid
Figure 6-19

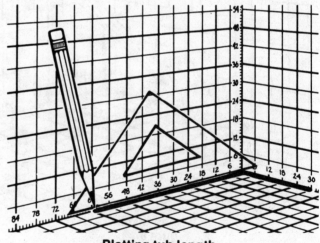
Plotting tub length
Figure 6-20

On the floor grid, plot the floor plan outline of the tub in its proper location. Since the tub will be against the end wall, measure 30 inches from the corner for the width of the tub and 60 inches from the corner for the length of the tub. Note: All dimensions will originate from this corner. Draw the lines, using a straight edge as a guide. See Figure 6-20.

Complete the outline of the tub by drawing a line which originates at the 60 inches length of the tub. Draw another line originating at the 30 inches width of the tub. These lines should follow the corresponding grid line at this measurement. See Figure 6-21. Note that these two lines will intersect to form the front corner of the tub

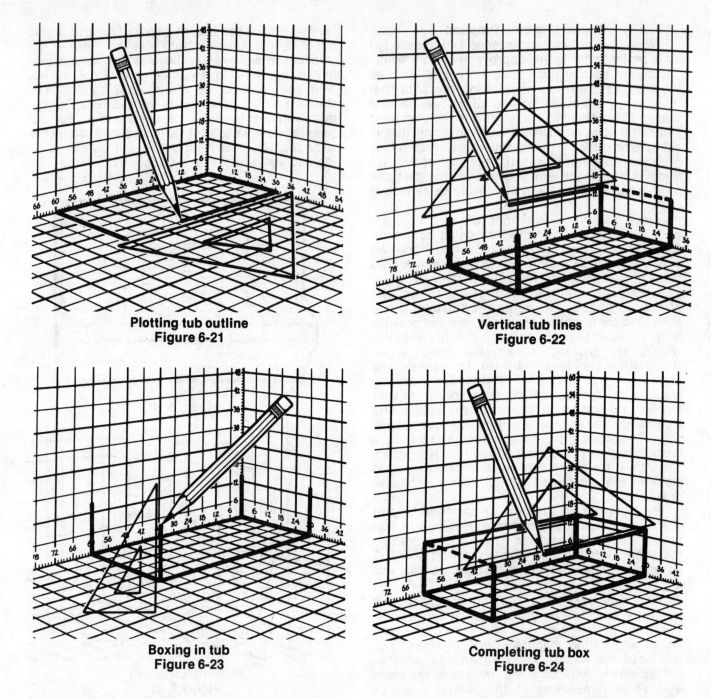

Plotting tub outline
Figure 6-21

Vertical tub lines
Figure 6-22

Boxing in tub
Figure 6-23

Completing tub box
Figure 6-24

outline. This completes the perspective floor plan outline of the tub.

Next, draw 4 vertical lines approximately 14 inches high at each corner of the tub using your T square and triangle, as illustrated in Figure 6-22. The top surface of the tub is drawn by first plotting the 14 inches height at each corner on the walls and drawing a line between each. The front edges of the tub are located by placing your straight edge thru the corner at this 14 inches height, against the grid wall and aligning the straight edge with the nearest grid line. When both front edges have been drawn, the box shape of the tub is complete. Note Figures 6-23 and 6-24.

All other bathroom fixtures will be basically the same boxlike structures. Before detailing the

tub, finish the rough shape of the other fixtures to be located in the bathroom.

Draw the lavatory or vanity cabinet next. This, too, is a simple rectangular box in its rough form. As we did with the tub, plot the outline of the cabinet on the floor grid. In determining the outline and location of the cabinet, refer to your plan layout, roughings and cabinet manufacturer's specifications. The vanity to be drawn here is 25 inches wide and 22 inches deep. Since these dimensions do not fall exactly on a grid line, remember to align your straight edge with the closest grid line when drawing the outline. See Figure 6-25.

Now plot the height of the vanity against the back wall. The height is 31 inches. Draw the two

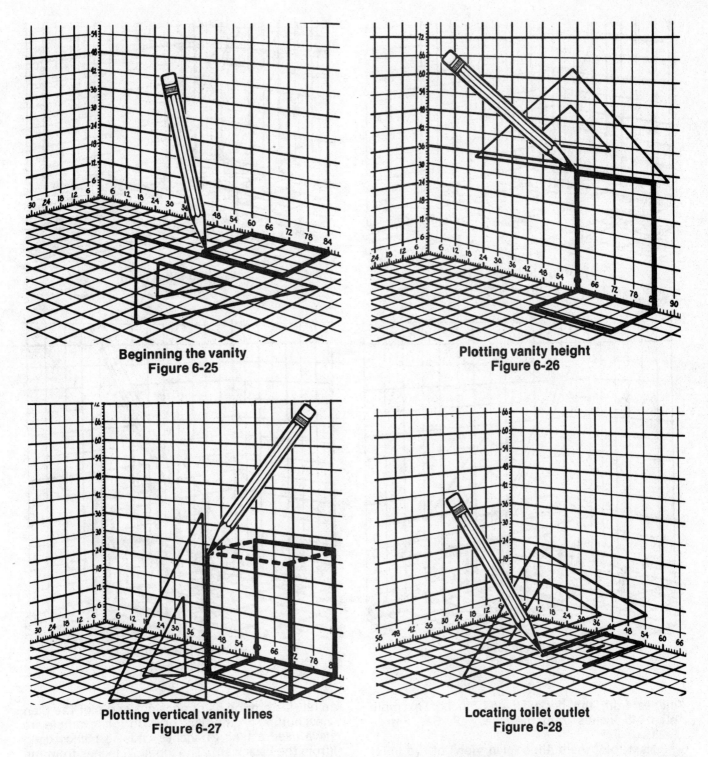

Beginning the vanity
Figure 6-25

Plotting vanity height
Figure 6-26

Plotting vertical vanity lines
Figure 6-27

Locating toilet outlet
Figure 6-28

vertical lines and the top edge. Since the 31 inches does not fall exactly on a grid line, remember to align your straight edge with the closest grid line. Figure 6-26. As with the tub, draw vertical lines at each corner to scale, approximately 31 inches high.

To complete the vanity box, draw the top surface of the box as was done with the tub, place your straight edge thru each of the top back corners, keeping it aligned with the closest grid line. Draw a line long enough to intersect the front vertical corner lines. Complete the vanity box by

drawing the front edge of the vanity. This line should intersect the two front corners of the box.

The final fixture is the toilet. Picture it, too, in terms of a box. First, refer to the rough sketch you made in your client's home to determine its location. You should locate the toilet by its rough-in dimension of the outlet. Let's say our toilet bowl is a 12 inches rough, 14 inches wide, 14 inches high and its front edge is 27 inches from the wall. The next step is locating the outlet on the grid floor. This is determined from your plan view and roughing drawing. The outlet should be located 12

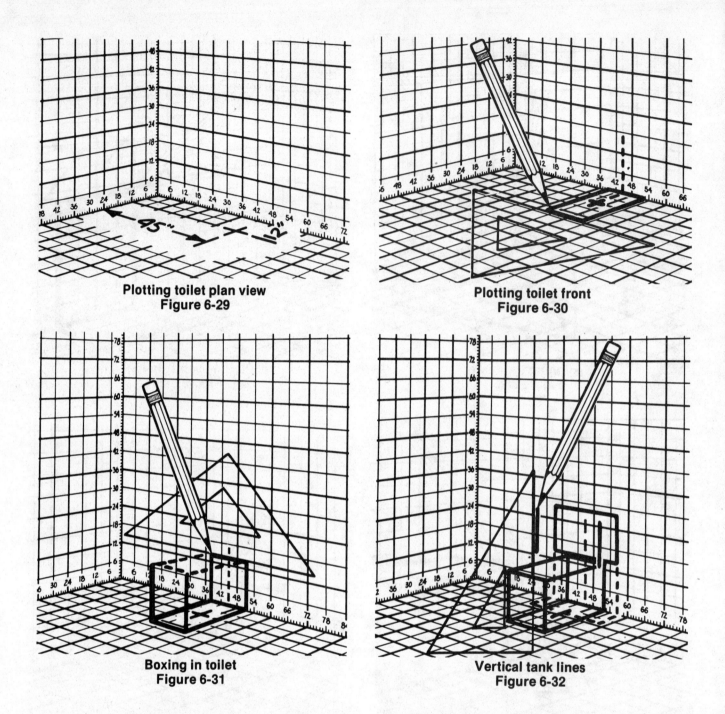

Plotting toilet plan view
Figure 6-29

Plotting toilet front
Figure 6-30

Boxing in toilet
Figure 6-31

Vertical tank lines
Figure 6-32

inches from the backwall and, in our example plan, 45 inches from the side wall. See Figure 6-28.

Next, plot the outline (plan view) of the toilet box on the grid floor. Since the toilet is 14 inches wide, plot 7 inches on either side of the outlet centerline to locate the "sides" of this box. See Figure 6-29. Complete the plan view outline by drawing the front edge 27 inches from the back wall. Remember to keep your straight edge aligned with the closest grid line. Figure 6-30.

Next determine the 14 inches height on the back grid, Wall B; draw the two vertical lines at each corner. Complete this box as you did the other fixtures. Figure 6-31.

The toilet tank sits to the rear on top of the

toilet. Determine its dimensions and plot the plan view outline on the floor grid. In this example we have used a tank 21 inches wide, 8 inches deep (from the back wall). The top is 28 inches from the floor and the base rests atop the bowl at the 14 inches height.

Using your T square and triangle, project 4 vertical lines, one from each corner of your floor plan view. These lines will form the corners of the tank. See Figure 6-32.

In the same manner as previous fixtures, complete the tank box; the top at 28 inches height, the bottom at 14 inches height. (Remember to keep your straightedge parallel to closet grid line.) Note Figure 6-33.

At this stage of the rendering, your perspective

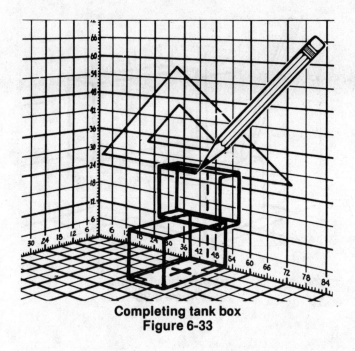

Completing tank box
Figure 6-33

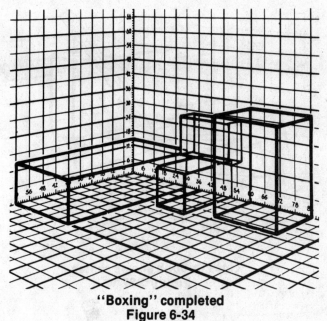

"Boxing" completed
Figure 6-34

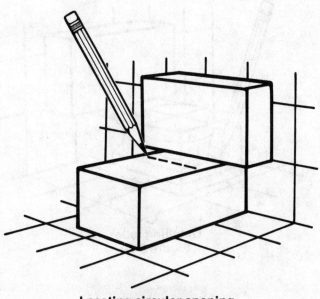

Locating circular opening
Figure 6-35

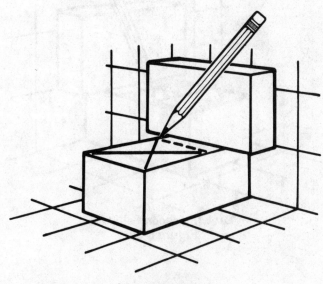

Locating center
Figure 6-36

drawing should look like Figure 6-34. All that's left is to add the details and give shape to the fixtures. The bathtub and lavatory are quite simple, while the toilet bowl is a bit difficult.

To turn the box shape into a toilet bowl, you must add some curves. Keep in mind that a circle in perspective appears as an ellipse. Begin by dividing the top surface of the toilet, as shown in Figure 6-35, to locate the circular opening of the bowl. Then draw lines diagonally from corner to corner to locate the center. Figure 6-36.

With the center located, draw two lines through it as shown in Figure 6-37. These lines should be parallel to the grid lines.

Now, if you have any ellipse template, place the appropriate size ellipse within the square and trace the ellipse. If you do not have an ellipse template, draw an ellipse freehand within the square, touching the 4 sides of the square as shown in Figure 6-38. This ellipse establishes the predominant shape of the toilet. Draw another ellipse just below this one to form the bottom edge of the rim. Blend the ellipse back into the tank as in Figure 6-39.

Remember that all toilets have different styling and you should attempt to represent the toilet you are installing when you detail it. This should be done with all fixtures you recommend. Figures 6-40 and 6-41 show you some of the details you can add to give each fixture some identify.

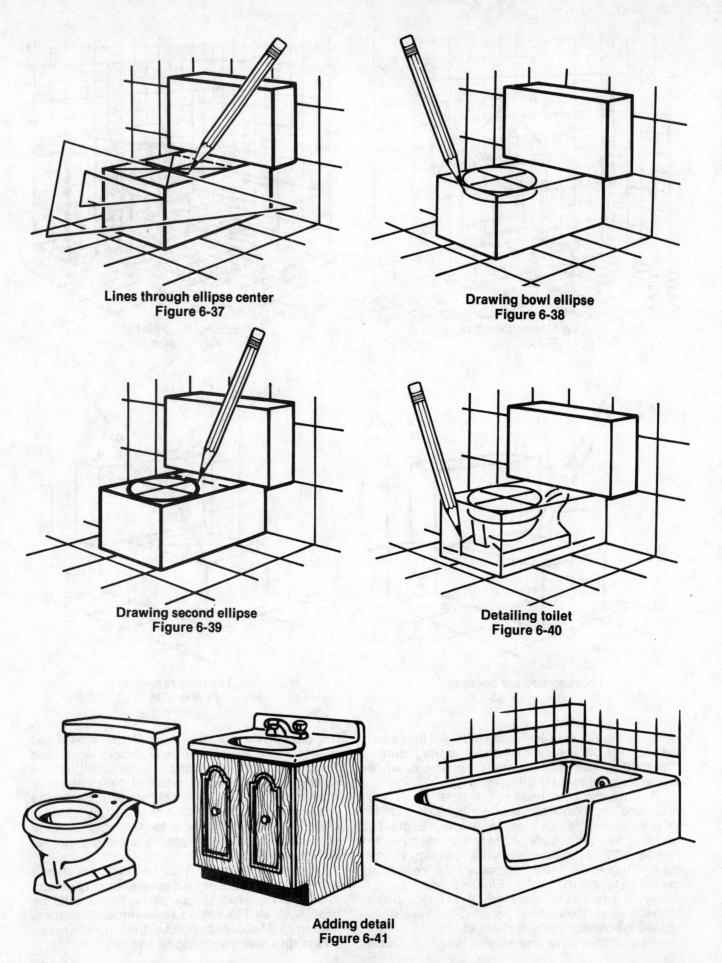

Lines through ellipse center
Figure 6-37

Drawing bowl ellipse
Figure 6-38

Drawing second ellipse
Figure 6-39

Detailing toilet
Figure 6-40

Adding detail
Figure 6-41

**Completed drawing
Figure 6-42**

For the final rendering you may wish to place a clean sheet of tracing paper over your drawing and trace only the finished lines. That is, the detail lines that describe each fixture, deleting all the guide lines you've drawn. Remember that in a perspective view the objects closest to you may conceal areas of other fixtures, so draw the objects nearest the front of the room first.

You will also add accessories to the bathroom, such as a medicine cabinet or mirrors, lights, towel racks, etc. This can be done simply; just remember that all lines should be parallel to the closest grid line.

Your final drawing may look like Figure 6-42. If your client has suggested a color combination, ink in the appropriate colors with a broad tip felt pen. The result will be a clear visual image which substantially increases your probability of making a sale without adding greatly to the cost.

Preparing The Plans And Specs

Now you should make up whatever floor plan and elevation plan you and your subs will need and prepare the specification sheet that outlines what you are including and omitting from your proposal. The specifications become part of your contract and must be clear enough to settle any dispute that may come up about what you are getting paid to do. The specifications sheet is also the basic planning document for preparing your

cost estimate. Make out the specification sheet carefully; it can guarantee a smooth, trouble-free, profitable job for you. Figure 6-43 is a sample specification sheet which follows the order of the bathroom survey sheet, Figure 6-13. You may want to use a material and labor take-off sheet such as Table 19-1 as a worksheet to list material quantities and costs, labor hours and costs and your mark-up. Your take-off sheet is your private record of the job. The job specifications is shown to your client and lists only the item by "sell price," which includes all your costs and markup. When you make your proposal, your clients will probably want to delete some items or cut costs in some areas. Giving your client an item by item cost summary makes it easier for you to figure changes on the spot so the job can be sold as revised without delay.

To make a profit on the job you have to estimate costs accurately and include all costs in your estimate. Chapter 19 of this book lays down good guidelines for preparing an estimate. As a bathroom remodeling specialists you should be able to make a fairly precise estimate of the labor cost on most jobs. Keep good records on the labor cost of completed jobs. Compare the actual costs with your estimates and make notes on the estimate about why the job was more or less difficult than expected. Review these estimates and job records when you estimate the cost of a

Job Specifications

Contractor's Name _____   Owner's Name _____

Address _____   Address _____

City _____ State _____ Zip _____   City _____ State _____ Zip _____

Phone _____   Job Address _____ Phone _____

Prepared By _____   Date _____ Job No. _____

Contractor proposes to provide the building permit, labor, materials and equipment necessary to complete installation of the following:

Vanity No. 1
Cabinet color _____
Material _____
Knob or pull No. _____
Back plate No. _____
Hinge No. _____

Vanity No. 2
Cabinet color _____
Material _____
Knob or pull No. _____
Back Plate No. _____
Hinge No. _____

Item No.	Quantity	Description	Shelves	Finish sides	Hinge	Cost

Total $ _____

Tops ( ) As per drawing attached
Material _____   Edge treatment _____
Style _____   Splash height _____
Color _____   Splash type _____
Size at back edge _____ X _____   Number of cut-outs _____
Total cost $ _____

Lavatories Number _____
Mount _____   Manufacturer & No. _____
Style _____   Color _____
Size _____   Total Cost $ _____

Medicine Cabinets Number _____
Mounted _____   Manufacturer & No. _____
Type _____   Mirror size _____ X _____
( ) Custom made per drawing attached.   Total Cost $ _____

Fixtures and Fittings	Color	Description	Cost
Toilet and seat			
Toilet, reset existing			
Tub			
Whirlpool			
Bidet			
Lavatory faucets			
Lavatory valves			
Lavatory legs			
Bath valve			
Shower head			
Personal shower			
Diverter			
Tub fittings and overflow			
Lavatory fittings			$ _____

Figure 6-43

Ventilating	Description	Cost
Heater		
Vent fan		
Timer		
Switch		
Humidstat		
	Total	$

Enclosures

Description_____   Manufacturer_____
Color_____   Size_____
Door size _____   Glass type_____
Rod length_____   Total Cost $_____

Storage

Type_____   Size_____
Finish_____   Hardware_____
Doors_____   Drawers_____
 Total Cost $_____

Floor

Underlay area_____   Total area_____
Description_____   Color_____
Manufacturer_____   Sill_____
Cove_____   Total Cost $_____

Walls

Tub area_____   Size_____
Description_____   Color_____
Manufacturer_____   Cost $_____
Other areas_____   Size_____
Description_____   Color_____
Manufacturer_____   Cost $_____
Finish_____   Total Cost $_____

Ceiling

Description_____   Size_____
Manufacturer_____   Cost $_____
Description_____   Size_____
Manufacturer_____   Cost $_____
Finish_____   Total Cost $_____

Accessories	Finish	Number	Description	Cost
Matched tile				
Tub trim				
Paper holders				
Tissue dispensers				
Hooks				
Grab bars				
Soap dishes				
Bar-soap dish				
Towel bars				
Tumbler holders				
Folding stools				
Mirrors				
Decorative items				

Figure 6-43 [continued]   Total $_____

General Construction

	Description	Cost
Bulkheads		
Platforms		
Doors		
Soffit		
Windows		
Tub access		
Partitions		
	Total	$

Plumbing & Heating

	Description	Cost
Waste		
Vent		
Supply		
Heating		
	Total	$

Electric & Lighting

	Description	Cost
Service section		
New 3 wire service		
Wire heater		
Wire lighting		
Wire fan		
Wire outlets		
Light switches		
Outlets		
Lighting		
Lighting		
	Total	$

Contractor proposes to do the following demolition and dispose of items removed:

( ) Vanity   ( ) Flooring _____ Square Feet
( ) Top   ( ) Wall cover _____ Square Feet
( ) Lavatory   ( ) Ceiling cover _____ Square Feet
( ) Medicine cabinet   ( ) Partitions _____
( ) Toilet   ( ) Doors _____
( ) Tub   ( ) Windows _____
( ) Bath fittings   ( ) Piping _____ Feet
( ) Shower enclosure   ( ) Register and ducting _____
( ) Radiator   ( ) Electric _____

Total Cost $ _____

Contractor proposes to undertake the following repairs:

Item	Description	Cost
	Total	$

Total of costs above $_____   Tax $_____   **Cost Complete** $_____

Owner will furnish labor and materials as follows:

Item	Description

These are the total and complete specifications for this job. Only the items checked or for which a cost is indicated are inlcuded in this job:

Approved: Owner _____

                       Owner _____

Date: _____   Designer _____

Additions or changes have been agreed to in writing as follows:

Item	Description	Cost of change

Figure 6-43 [continued]

Proposal and Contract

Date _____ 19 _____

To _____.

Dear Sir:

We propose to furnish all materials and perform all labor necessary to complete the following:

_____

_____

_____

_____

_____

Job Location:

All of the above work to be completed in a substantial and workmanlike manner according to the floor plan, job specifications, and terms and conditions on the back of this form for the sum of

Dollars ($_____)

Payments to be made as the work progresses as follows: _____

_____

_____

the entire amount of the contract to be paid within _____ days after substantial completion and acceptance by owner. The price quoted is for immediate acceptance only. Delay in acceptance will require a verification of prevailing labor and material costs. This offer becomes a contract upon acceptance by contractor but shall be null and void if not executed within 5 days from the date above.

By _____

" YOU, THE BUYER, MAY CANCEL THIS TRANSACTION AT ANY TIME PRIOR TO MIDNIGHT OF THE THIRD BUSINESS DAY AFTER THE DATE OF THIS TRANSACTION. SEE THE ATTACHED NOTICE OF CANCELLATION FORM FOR AN EXPLANATION OF THIS RIGHT."

You are hereby authorized to furnish all materials and labor required to complete the work according to the plans, job specifications, and terms and conditions on the back of this proposal, for which we agree to pay the amounts itemized above

Owner _____

Owner _____ Date _____

Accepted by Contractor _____ Date _____

Figure 6-44

1. The Contractor agrees to commence work within ten (10) days after the last to occur of the following, (1) receipt of written notice from the Lien Holder, if any, to the effect that all documents required to be recorded prior to the commencement of construction have been properly recorded; (2) the materials required are available and on hand, and (3) a building permit has been issued. Contractor agrees to prosecute work thereafter to completion, and to complete the work within a reasonable time, subject to such delays as are permissible under this contract. If no first Lien Holder exists, all references to Lien Holder are to be disregarded.

2. Contractor shall pay all valid bills and charge for material and labor arising out of the construction of the structure and will hold Owner of the property free and harmless against all liens and claims of lien for labor and material filed against the property.

3. No payment under this contract shall be construed as an acceptance of any work done up to the time of such payment, except as to such items as are plainly evident to anyone not experienced in construction work, but the entire work is to be subject to the inspection and approval of the inspector for the Public Authority at the time when it shall be claimed by the Contractor that the work has been completed. At the completion of the work, acceptance by the Public Authority shall entitle Contractor to receive all progress payments according to the schedule set forth.

4. The plan and job specification are intended to supplement each other, so that any works exhibited in either and not mentioned in the other are to be executed the same as if they were mentioned and set forth in both. In the event that any conflict exists between any estimate of costs of construction and the terms of this Contract, this Contract shall be controlling. The Contractor may substitute materials that are equal in quality to those specified if the Contractor deems it advisable to do so. All dimensions and designations on the plan or job specification are subject to adjustment as required by job conditions.

5. Owner agrees to pay Contractor its normal selling price for all additions, alterations or deviations. No additional work shall be done without the prior written authorization of Owner. Any such authorization shall be on a change-order form, approved by both parties, which shall become a part of this Contract. Where such additional work is added to this Contract, it is agreed that all terms and conditions of this Contract shall apply equally to such additional work. Any change in specifications or construction necessary to conform to existing or future building codes, zoning laws, or regulations of inspecting Public Authorities shall be considered additional work to be paid for by Owner as additional work.

6. The Contractor shall not be responsible for any damage occasioned by the Owner or Owner's agent, Acts of God, earthquake, or other causes beyond the control of Contractor, unless otherwise provided or unless he is obligated to provide insurance against such hazards. Contractor shall not be liable for damages or defects resulting from work done by subcontractors. In the event Owner authorizes access through adjacent properties for Contractor's use during construction, Owner is required to obtain permission from the owner(s) of the adjacent properties for such. Owner agrees to be responsible and to hold Contractor harmless and accept any risks resulting from access through adjacent properties.

7. The time during which the Contractor is delayed in his work by (a) the acts of Owner or his agents or employees or those claiming under agreement with or grant from Owner, including any notice to the Lien Holder to withhold progress payments, or by (b) any acts or delays occasioned by the Lien Holder, or by (c) the Acts of God which Contractor could not have reasonably foreseen and provided against, or by (d) stormy or inclement weather which necessarily delays the work, or by (e) any strikes, boycotts or like obstructive actions by employees or labor organizations and which are beyond the control of Contractor and which he cannot reasonably overcome, or by (f) extra work requested by the Owner, or by (g) failure of Owner to promptly pay for any extra work as authorized, shall be added to the time for completion by a fair and reasonable allowance. Should work be stopped for more than 30 days by any or all of (a) through (g) above, the Contractor may terminate this Contract and collect for all work completed plus a reasonable profit.

8. Contractor shall at his own expense carry all workers' compensation insurance and public liability insurance necessary for the full protection of Contractor and Owner during the progress of the work. Certificates of insurance shall be filed with Owner and Lien Holder if Owner and Lien Holder require. Owner agrees to procure at his own expense, prior to the commencement of any work, fire insurance with Course of Construction, All Physical Loss and Vandalism and Malicious Mischief clauses attached in a sum equal to the total cost of the improvements. Such insurance shall be written to protect the Owner and Contractor, and Lien Holder, as their interests may appear. Should Owner fail so to do, Contractor may procure such insurance, as agent for Owner, but is not required to do so, and Owner agrees on demand to reimburse Contractor in cash for the cost thereof.

9. Where materials are to be matched, Contractor shall make every reasonable effort to do so using standard materials, but does not guarantee a perfect match.

10. Owner agrees to sign and file for record within five days after substantial completion and acceptance of work a notice of completion. Contractor agrees upon receipt of final payment to release the property from any and all claims that may have accrued by reason of the construction.

11. Any controversy or claim arising out of or relating to this contract shall be settled by arbitration in accordance with the Rules of the American Arbitration Association, and judgment upon the award rendered by the Arbitrator(s) may be entered in any Court having jurisdiction.

12. Should either party bring suit in court to enforce the terms of this agreement, any judgment awarded shall include court costs and reasonable attorney's fees to the successful party plus interest at the legal rate.

13. Unless otherwise specified, the contract price is based upon Owner's representation that there are no conditions preventing Contractor from proceeding with usual construction procedures and that all existing electrical and plumbing facilities are capable of carrying the extra load caused by the work to be performed by Contractor. Any electrical meter charges required by Public Authorities or utility companies are not included in the price of this Contract, unless included in the job specifications. If existing conditions are not as represented, thereby necessitating additional plumbing, electrical, or other work, these shall be paid for by Owner as additional work.

14. The Owner is solely responsible for providing Contractor prior to the commencing of construction with any water, electricity and refuse removal service at the job site as may be required by Contractor to effect the improvement covered by this contract. Owner shall provide a toilet during the course of construction when required by law.

15. The Contractor shall not be responsible for damage to existing walks, curbs, driveways, cesspools, septic tanks, sewer lines, water or gas lines, arches, shrubs, lawn, trees, clotheslines, telephone and electric lines, etc., by the Contractor, subcontractor, or supplier incurred in the performance of work or in the delivery of materials for the job. Owner hereby warrants and represents that he shall be solely responsible for the condition of the building with respect to moisture, drainage, slippage and sinking or any other condition that may exist over which the Contractor has no control and subsequently results in damage to the building.

16. The Owner is solely responsible for the location of all lot lines and shall if requested, identify all corner posts of his lot for the Contractor. If any doubt exists as to the location of lot lines, the Owner shall at his own cost, order and pay for a survey. If the Owner wrongly identifies the location of the lot lines of the property, any changes required by the Contractor shall be at Owner's expense. This cost shall be paid by Owner to Contractor in cash prior to continuation of work.

17. Contractor has the right to subcontract any part, or all, of the work agreed to be performed.

18. Owner agrees to install and connect at Owner's expense, such utilities and make such improvements in addition to work covered by this Contract as may be required by Lien Holder or Public Authority prior to completion of work of Contractor. Correction of existing building code violations, damaged pipes, inadequate wiring, deteriorated structural parts, and the relocation or alteration of concealed obstructions will be an addition to this agreement and will be billed to Owner at Contractor's usual selling price.

19. Contractor shall not be responsible for any damages occasioned by plumbing leaks unless water service is connected to the plumbing facilities prior to the time of rough inspection.

20. Title to equipment and materials purchased shall pass to the Owner upon delivery to the job. The risk of loss of the said materials and equipment shall be borne by the Owner.

21. Owner hereby grants to Contractor the right to display signs and advertise at the job site.

22. Contractor shall have the right to stop work and keep the job idle if payments are not made to him when due. If any payments are not made to Contractor when due, Owner shall pay to Contractor an additional charge of 10% of the amount of such payment. If the work shall be stopped by the Owner for a period of sixty days, then the Contractor may, at Contractor's option, upon five days written notice, demand and receive payment for all work executed and materials ordered or supplied and any other loss sustained, including a profit of 10% of the contract price. In the event of work stoppage for any reason, Owner shall provide for protection of, and be responsible for any damage, warpage, racking, or loss of material on the premises.

23. Within ten days after execution of this Contract, Contractor shall have the right to cancel this Contract should it be determined that there is any uncertainty that all payments due under this Contract will be made when due or that an error has been made in computing the cost of completing the work.

24. This agreement constitutes the entire Contract and the parties are not bound by oral expression or representation by any party or agent of either party.

25. The price quoted for completion of the structure is subject to change to the extent of any difference in the cost of labor and materials as of this date and the actual cost to Contractor at the time materials are purchased and work is done.

26. The Contractor is not responsible for labor or materials furnished by Owner or anyone working under the direction of the Owner and any loss or additional work that results therefrom shall be the responsibility of the Owner. Removal or use of equipment or materials not furnished by Contractor is at Owner's risk, and Contractor will not be responsible for the condition and operation of these items or service for them.

27. No action arising from or related to the contract, or the performance thereof, shall be commenced by either party against the other more than two years after the completion or cessation of work under this contract. This limitation applies to all actions of any character, whether at law or in equity, and whether sounding in contract, tort, or otherwise. This limitation shall not be extended by any negligent misrepresentation or unintentional concealment, but shall be extended as provided by law for willful fraud, concealment, or misrepresentation.

28. All taxes and special assessments levied against the property shall be paid by the Owner.

29. Contractor agrees to complete the work in a substantial and workmanlike manner but is not responsible for failures or defects that result from work done by others prior, at the time of or subsequent to work done under this agreement.

30. Contractor makes no warranty, express or implied (including warranty of fitness for purpose and merchantability). Any warranty or limited warranty shall be as provided by the manufacturer of the products and materials used in construction.

31. Contractor agrees to perform this Contract in conformity with accepted industry practices and commercially accepted tolerances. Any cliam for adjustment shall not be construed as reason to delay payment of the purchase price as shown on the payment schedule. The manufacturers' specifications are the final authority on questions about any factory produced item. Exposed interior surfaces, except factory finished items, will not be covered or finished unless otherwise specified herein. Any specially designed, custom built or special ordered item may not be changed or cancelled after five days from the acceptance of this Contract by Contractor.

Figure 6-44 [continued]

Notice To Customer Required By Federal Law

You have entered into a transaction on _____ which may result in a lien, mortgage, or other security interest on your home. You have a legal right under federal law to cancel this transaction, if you desire to do so, without any penalty or obligation within three business days from the above date or any later date on which all material disclosures required under the Truth in Lending Act have been given to you. If you so cancel the transaction, any lien, mortgage, or other security interest on your home arising from this transaction is automatically void. You are also entitled to receive a refund of any down payment or other consideration if you cancel. If you decide to cancel this transaction, you may do so by notifying.

_____
(Name of Creditor)

at _____
(Address of Creditor's Place of Business)

by mail or telegram sent not later than midnight of _____. You may also use any other form of

written notice identifying the transaction if it is delivered to the above address not later than that time.

This notice may be used for the purpose by dating and signing below.

I hereby cancel this transaction.

_____   _____
(Date)   (Customer's Signature)

Effect of rescission. When a customer exercises his right to rescind under paragraph (a) of this section, he is not liable for any finance or other charge, and any security interest becomes void upon such a rescission. Within 10 days after receipt of a notice of rescission, the creditor shall return to the customer any money or property given as earnest money, downpayment, or otherwise, and shall take any action necessary or appropriate to reflect the termination of any security interest created under the transaction. If the creditor has delivered any property to the customer, the customer may retain possession of it. Upon the performance of the creditor's obligations under this section, the customer shall tender the property to the creditor, except that if return of the property in kind would be impracticable or inequitable, the customer shall tender its reasonable value. Tender shall be made at the location of the property or at the residence of the customer, at the option of the customer. If the creditor does not take possession of the property within 10 days after tender by the customer, ownership of the property vests in the customer without obligation on his part to pay for it.

Figure 6-44 [continued]

proposed job. Your experience and cost records should make estimating fairly easy for most types of work. You probably are going to buy most of your fixtures and materials from three or four suppliers and subcontract most work to three or four subs. They will help you stay posted on price changes for the trades, fixtures and materials you use most often. Keeping a current price book of key items by trade, product description or manufacturer will save estimating time and improve accuracy. Most important, remember that overhead, fees, commission, profit, supervision, taxes and insurance are nearly half the cost of any bathroom job. The cost of labor and materials will be only about 50 to 60 percent of your selling price. See the discussion of pricing in Chapter 19. If you cut costs to get the job you may be working for nothing or even loosing money for all your effort. You do far better to maintain high standards, work for fair prices and let others do the work for owners who insist on cut rates and minimum quality.

Figure 6-44 is a sample bathroom remodeling contract. Have the names of the parties and job information typed in before you make the return call on your prospects. They may make revisions in the specifications that change the cost, so leave the cost open until the job is sold. The contract refers to the specifications for a complete description of the job. These two documents, together with the working drawings, if any, are your agreement. Nothing else is included in the job no matter what your client assumed while you were making your sales presentation. Make the contract, job specifications and working drawings clear on what is included and excluded. Make sure your client understands that these three documents are the **only** agreement. Finally, be sure you live up to the agreement precisely. If you do this, and if you price your work correctly, you should make money on every job you have.

Making The Sale

Now you are ready to meet your prospects and close the deal. Chapter 20 spells out in detail a proven sales approach. It is enough here to emphasize that you are going to need samples or pictures of every major item in the job specification sheet. You want to get an affirmative reaction to each of your suggestions, if possible. If your first choice of tub, vanity, tile, lavatory, etc. is not accepted, offer several more possibilities at a range in price if possible. Know what the cost of each of these alternatives is so you can make the substitution on the job specification. You want a final decision, item by item on the specification sheet, not indecision or "I'll have to think about that." Your perspective drawing will act as a magnet drawing your prospects toward the choices you made when drawing up the plan. But have pictures and samples of second and third choices available.

Even before you begin to explain your proposal and uncover your perspective drawing, sell your company as a highly professional specialist in the field. Letters from satisfied customers, a bank reference, current balance sheet, and certificates of membership in the Better Business Bureau, a national remodeling association and the local business association should be displayed for your prospect's review even before you begin to talk about their bathroom. Chapter 20 has more on how to sell your company before you begin to sell the job. One good way to sell your company is to make the sales presentation in your showroom. It isn't essential, but a showroom can be used to advantage.

ABC BUILDERS
Bathroom Specialists
Address   City   State   Zip
Phone

Dear Mr. and Mrs._____:

Many thanks for the privilege of "doing" your new bathroom.

In developing your bath many ideas and possibilities have been considered. The enclosed specifications, drawing and contract cover your final selections of all materials, colors, hardware labor and services. These are our total firm agreement to be completed for a total firm price of $_____.

Payable as follows: _____
                        _____
                        _____

Please, let us avoid any misunderstanding. Will you thoroughly review our contract one more time? If we do not have a full and complete agreement please contact us by_____.

This will be the last day for changes without additional charge to you. After this date additional work must be a new contract - "changes" will be charged as follows:

    To change a catalog part - $25.00 plus 10% handling
    To change a custom part - $50.00 plus up to 100% handling
    To change installed part - $100.00 plus all labor, material and 20% overhead.

We have set a tentative date of_____to start your job. This is subject to the delivery of all materials. You will be advised of the starting day approximately one week in advance.

Please, one more "Look". Only with full understanding and agreement can we complete your job promptly and to your full satisfaction.

Thank you for your consideration.

                            Very truly yours,

                            ABC Builders

Figure 6-45

After the job is sold and countersigned by whoever has authority to obligate your company to do the job at the price quoted, a copy of the completed contract, job specifications and working drawings should be mailed to your clients with a letter something like Figure 6-45. This confirms your client's understanding of the job and gives

ABC BUILDERS
Bathroom Specialists
Address City State Zip
Phone

Thank you for your order. Our suppliers advise us that all materials will be ready to start your work approximately_____ _________. We will verify the exact starting day several days in advance.

Unless unforseen delays develop, we expect your work to take _________ working days (Saturday and Sunday not included) to have the contract substantially complete.

Will you please:

Assist us to proceed with your job installation without delay?

Unload all wall and base cabinets, clear all counter tops and walls and remove all of these items from the area before the day we start?

Help us protect your children by keeping them away from dangerous power tools?

Secure any pets away from areas needed for access such as halls, basement, utility room, etc?

Review the plans and specifications? The work will be done exactly as outlined in the plans and specifications. Let us avoid any misunderstanding.

Authorize some adult to be on the premises while the job is in progress?

If there are any questions at any time, for any reason, please call for assistance.

Very truly yours,

ABC Builders

Figure 6-46

them a final opportunity to make changes before you order your materials. About one week before work begins and about the time you have assembled all the materials necessary, your clients should receive a letter like in Figure 6-46. This will help prepare the bathroom for your work and eliminate potential problems.

Doing The Work

The more trades and craftsman involved in a bathroom job, the more supervision and control is necessary. There are a lot of things that can go wrong on a "simple" bathroom job, and many, if not most, will go wrong. Your client has to live with the bathroom and it will be a constant reminder of either your competence and prefessionalism or your neglect and carelessness. Expect that individual craftsman will occassionally lack the proper tools or materials, misread or misunderstand their instructions, and confuse your clients. Discover problems before your clients do whenever possible. If your client is the first to bring a problem to your attention, take his side immediately and assure him that he is right and there is something wrong and it will be fixed shortly. (Of course, avoid accepting criticism of your craftsmen in the presence of those men and don't do more than the job specifications call for without a written change order at a higher price.) Your objective is a satisfied client. You satisfy your client by producing what you promise and correcting anything that does not come up to standard. That sounds easier than it is, but a good set of job specifications, a good contract, and selling the job at a fair price are an excellent beginning. It is far easier to meet a standard if the standard is clear and obvious to everyone. Completing the job requires supervision and quality control. Someone has to check everything that is done and take steps to correct what was not done right. The point is that even the most professional bathroom specialists run into problems, but the successful, efficient operators correct the deficiences and produce uniformly satisfied clients.

Don't let your jobs become a good example of what can go wrong. One homeowner wrote the following about her experience with bathroom remodeling:

> Our bathroom was renovated by a contractor masquerading as a remodeler. The room had to be gutted and all of the existing fixtures replaced. A simple project as far as renovations go, but almost everything went wrong. In fact, everything did go wrong except the installation of the tile around the tub, which was done to perfection by a subcontractor.
>
> A partial listing of the mistakes would include:
>
> 1. A countertop was fitted over the toilet tank in such a way that the tank cover could not be removed for plumbing repairs.
>
> 2. A double medicine chest, which is centered by a mirror with a cabinet on either side, was hung with one door on upside down and the other so badly warped that it wouldn't close properly. The mirror was cracked at the bottom, leading the contractor to suggest cutting off the cracked portion, ignoring the obvious fact that it would then be out of proportion to the cabinets.
>
> 3. An exhaust fan was put in place without being properly grounded.
>
> 4. The tub was installed without a complete check of the existing plumbing. It drained so slowly that three months after the bathroom was finished, we had to have an access door cut into a wall of the adjoining bedroom and copper pipe installed. (This cost an extra $375.)
>
> 5. Insulation around the tub was not replaced and the pipes froze on the first cold day.
>
> 6. The sink was installed with a crack in it.
>
> 7. The swag lighting fixture was so shoddy and inferior that we refused to have it hung and brought another one of much better quality at half the price.

Substantial Completion and Acceptance

Contractor _____   Owner _____

Address _____   Job Address _____

City _____ State _____ Zip _____   City _____ State _____ Zip _____

Phone _____   Job Number _____

Were the craftsmen courteous and cooperative?   Yes   No

Remarks _____

_____

Work completed:

☐ Vanity   ☐ Walls
☐ Tops   ☐ Ceiling
☐ Lavatories   ☐ Accessories
☐ Medicine cabinet   ☐ General Construction
☐ Toilet   ☐ Plumbing
☐ Tub   ☐ Heating
☐ Fixtures and Fittings   ☐ Electric
☐ Ventilating   ☐ Lighting
☐ Enclosures   ☐ Demolition
☐ Storage   ☐ Repairs
☐ Floor   ☐ Written Changes

Work yet to be completed:

_____
_____
_____
_____
_____
_____
_____
_____
_____
_____
_____

Any other deficiencies or requests for service should be made in writing to the address above. Please be as specific as possible about the defect. Emergency requests may be phoned in to the number above. Any claim concerning an applicance or fixture under warranty should be made directly to the manufacturer.

I (we) have inspected the work performed and find the job completed to our satisfaction with the exception of the items listed above as yet to be completed.

Date _____   Signature _____

Figure 6-47

This renovation was supposed to take eight working days. The job was started on December 15th and finished on January 20th, inconveniencing family and guests during the Christmas holidays.

Our contractor appeared only on the first day. After that, the work was left in the hands of a very inept young man, who, we discovered after paying the bill, was not a union member, although we were charged union wages for his work.

We made certain that all of the errors were corrected before we paid the bill, but it was an unhappy experience which could have been avoided.

Remember that your client has to live with your mistakes. Make as few mistakes as possible, recognize that some will occur, and do what is necessary to leave a remodeled bathroom that is defect free and a source of pleasure to your client. Have the man who made the sale call on the clients when the job is finished. He should check to be sure that everything has been completed and the homeowners are satisfied. If work still must be done, a "Substantial Completion and Acceptance" form such as in Figure 6-47 should be made up to indicate work yet to be done. Have the job specification sheet (Figure 6-43) and contract (Figure 6-44) available as you complete this form with your client to settle any dispute about what you agreed to do. Most of the discrepancies you find will be "touch-up" type problems that a painter can remedy. Give your clients a copy of the list and advise them that you will work off these defects in a few days. Most homeowners will be satisfied with this and will make the final payment or authorize their lender to release the final payment for the job. Plan to have these final defects corrected in a week or less. If there is a major item that has not yet been completed, such as a shower door that was defective on arrival and the replacement has not yet arrived, go back to the Job Specification sheet (Figure 6-43) and request that full payment be made except for the amount listed for the incomplete item. Again, this is a reasonable request and most homeowners will be willing to comply.

No matter what guarantee you make on the job, treat "call backs" as important sales leads. Perhaps 95 percent of all requests are legitimate and 100 percent of your return service calls offer an opportunity to reestablish contact with someone who can put you in touch with potential customers. A man trained to handle "call backs" can satisfy your customer while he collects the names of people who admired your customer's new bathroom and might be interested in doing something about their own home.

Plaster, Tile, Marble

Most bathroom jobs involve some plastering, tile setting or marble setting. Quite often remodelers handle plaster, tile or marble with their own crews since little specialized equipment is necessary and the amount of work involved usually is too small to make getting a sub bid practical.

Plastering bathrooms involves three distinct coats of plaster to ceiling and walls: Scratch, brown and finish. The initial or scratch coat is applied directly and then scratched with a special raking tool before it sets. It is allowed to set a day or more before applying the second layer of plaster (brown coat). The finish (white coat), a smooth relatively thin covering, is applied several days later. There are several different types of lathing systems which can be applied as a base for plaster. Metal lath is simply a metal screen or mesh. Wooden lath consists of thin wood strips nailed to the wall studs and spaced at least 3/8 inch apart. Gypsum wallboard is now being used as lath in most construction. It is strong and stable, but requires gypsum plaster.

Before your plasterer applies the plaster, he must make sure that the walls are squared. This involves one of two basic methods: The right triangle method or the square method.

When the right triangle method is used, the plasterer chooses a wall and measures out two feet from one of the corners. From the same corner, he measures out a distance of 18 inches on the adjacent wall, at which point he scribes an arch with an 18 inch radius. Finally, he measures the distance from the 2 foot mark to the 18 inch mark. The distance should be 2½ feet. If it is greater than 2½ feet, a 2½ foot length is measured from the end of the 2 foot line to the arc. The arc and the 2½ foot line should intersect. The true line of the room is found by drawing a line from the corner to the point of intersection. The process is repeated in the remaining corners.

When using the square method, the plasterer takes an straightedge board and places it in the starting corner against one wall. Then a wooden square is placed in the corner, butted against the straightedge. The adjacent side of the square should run smoothly along the wall. If it does not, mark the floor with a chalkline where the line of squaring should be.

There are a wide variety of plasters, each of which is mixed and used in a slightly different way. Plasters are mixed with specific amounts of water, sand and cement or lime or gypsum (binder). If too much water is used, the plaster will be hard to apply and will tend to run. Too much sand will weaken a plaster wall because there will not be enough cementing material to bind the mixture together. Extra material is sometimes added to retard the rate at which the plaster sets.

When plaster is mixed by hand in a box, it is thoroughly sifted first to ensure that the mixture is evenly spread. Water is added slowly and the mixture is stirred with a hoe until the proper consistency is reached.

Machine mixing requires that the water be added first, followed by half the sand that will be needed, all the binder and, finally the rest of the sand. Often plaster is left to age after mixing. Water is added to the mixture to replace that lost by evaporation.

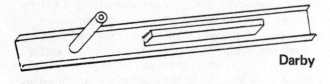

Darby

Darby
Figure 6-48

Applying Plaster

Plaster is applied to the wall with an upward motion, until all the plaster has rolled off the trowel. The angle at which the trowel is held determines the thickness of the plaster. If the trowel is pressed flat to the wall it will leave a thick layer of plaster. If it is pressed to the wall at an angle it will deposit a thin layer. When a fairly large section of wall has been plastered it is smoothed off (darbied). The darby (Figure 6-48) is a long, flat piece of wood or metal with handles. This is held at a slight angle and pushed over the plaster to smooth and even it up. After a wall has been darbied in one direction, it is darbied at right angles to the first direction.

Your plaster crew should make sure that the proper thickness of the plaster coat is maintained as work progresses. To accomplish this, narrow bands of plaster, called screeds, are applied to walls and ceilings, prior to the main coat. They are used as guides to control the thickness of the plaster.

Screeds are plumbed and leveled between small spots of plaster (dots). Several different methods may be used. Sometimes dots are placed on the wall about a straightedge length apart. This is called the rod-and-level-from-a-base ground method. The ground is a length of wood or metal to which moldings can be nailed. Dots of plaster are placed about 6 feet above the floor. The plasterer takes a small block of wood or a piece of paper and embeds it in the dot. Then he takes a straightedge and lays one end of it flat against the base ground and places the other end on the paper or wood on the dot. Then a level is placed vertically along the length of the straightedge. The plasterer presses into the dot until the level shows the straightedge to be exactly upright. Excess plaster is cut away.

A second method for placing dots is called the plumb bob method, employing a plumb bob, a line and two gauges. Gauges transfer a mark from one plane to another. They may be made by cutting wooden strips such as 1 x 2 x 4 inches into identical gauges. Once gauges are cut, dots are placed on the walls as before. The plasterer holds the gauge on the dot with the plumbline running through the cut out. Then the other gauge is held against the base ground. The upper gauge is pressed into the dot until the plumbline just clears the cut out on the lower gauge. Finally, the plasterer smoothes away whatever excess mortar is left around the indentation.

Once dots have been placed, the plasterer can then set up horizontal screeds or lines of plaster exactly leveled. The plasterer is applied slightly thicker than the dots in a 4 inch wide band between the dots. A straightedge, with each end resting on a dot, is smoothed back and forth over the screed until it is exactly level with the dots. A vertical screed is formed by placing mortar between a dot and the ground, and then using the straightedge to smooth and level it.

Dot and screed systems are also placed on celings. First, make up a water level. Water levels are about 50 inches long and made of light plastic or rubber hose. Most levels have an inside diameter of 3/8 or 1/2 inch. Twelve inch long glass tubes are placed in each end of the level. To use the level, place one end in a pail of clean water and suck water through the tube until siphoning starts. Once the level is filled, it must be checked to make sure no air is trapped inside it. This is done by holding both ends pointed upward so that the level looks like a large "U". If no air is trapped inside, the water levels should be exactly the same height. Within the glasses the water forms a ring around the sides of the tube and then slumps in the middle.

Choose either point to mark by, but always use the same point. The level is used by marking the desired height. Then one tube is held on the mark and the other tube is moved around the room and the level is carefully marked off. Usually, marks are made at 2 foot intervals around the room. Then a ceiling dot and screed system can be installed. First, the plasterer places a dot on the ceiling above each level mark. Then he takes a gauge and places one leg on the mark and the other on the dot. He presses into the dot until the gauge lines up with the mark. Finally, he cuts excess material away from the indentation. Once the dots have dried, screeds are set up between them as on a wall. When the screeds have dried the room is laid out and is ready to be plastered.

As note earlier, three coats of plaster are usually applied to the walls of a room: Scratch, brown and finish coat. All these coats are applied in much the same way using the same tools.

The scratch coat stiffens the lath to form a good

foundation for the rest of the plaster. Work begins at the bottom of the wall and moves toward the top. Plaster should be applied with few overlapping strokes. Before the scratch coat sets, the surface is scratched in two directions with the plasterers' rake. For metal lath, the mortar should be rich and thick and applied with a very light pressure. More pressure can be used on paperback gypsum and wooden lathing. Concrete block must be kept damp while the plaster is being applied. After the scratch coat has set hard, the walls are ready for the second (brown) coat.

The brown coat is used to build up the plaster even with the screeds. A ½ to ¾ inch thick brown coat is used. Brown coat mortar is made thinner than scratch coat mortar. A mixture of 2½ to 1 is usually used with brown coat mortar. The brown coat is applied just as the scratch coat is applied. The surface is rodded level using the screeds as guidelines. Low spots will need to be filled in, then the area should be rodded again. Next, it is darbied horizontally and vertically so that the ridges left by the rod are smoothed out. Finally, the corners of the room are trimmed with a clean trowel. The forward edge of the trowel (toe), cuts away excess plaster on the angle. Some plasterers prefer to use a special tool called an angle float. Once the brown coat has set for about 24 hours, the surface is ready for the finish coat.

Usually the finish coat is applied in two parts. The first coat is a very thin layer of finishing plaster applied directly to the brown coat. Over this coat, a second layer is applied and troweled smooth. Putty coat is the most common finishing plaster used now, and it is applied in seven steps. First, apply a coat of plaster to each side of each corner in the room. Second, smooth it out with a straightedge. This is called featheredging. Then the thin first layer of the finish coat is applied to all the walls in the room. The second step is to apply the second coat of the finish coat immediately. This coat is worked with the forward edge (toe) of the trowel. Then the plasterer floats the angles by squaring them up and fills in using an aluminum or plastic float. After this, the entire wall is worked again to remove small surface irregularities (cat faces). The wall is next water troweled. The work is dashed with water from a brush and worked in long up and down strokes holding the trowel at a very sharp angle. Then the plasterer goes over the angles with a trowel or with a small paddle. Last, water is brushed over the entire surface. Once the surface has been brushed with water the job is finished.

Sand finishing is also often used. The procedures used in sand finishing are the same as those used in the putty coat, except that it can be applied over a wet brown coat. Other finish coat materials include Keene's cement and Portland cement.

Often plaster is colored before it is applied. Before a coloring material is mixed in, it should be checked to be sure that it is resistant to lime and that it is waterproof. The problems with colored plasters are that they tend to streak and make it difficult to obtain uniform coloring throughout a room. Colored plasters have advantages in that they are cheaper and coloring makes the plaster easier to work. Colored plasters must be applied to an entire wall at one time, or the boundaries between applications will be visible.

A textured surface is sometimes preferred over a smooth one. A stippling brush may be daubed or swirled through the finished plaster to give two different effects. A rubber sponge dabbed on the plaster will give still another effect. Travertine effects are produced by brushing the surface with a wire brush and then troweling it lightly. A trowel finish is formed by moving the trowel back and forth in arcs across the wall. This creates alternating ridges and smooth spots. Sometimes an old fashioned effect is produced by applying plaster with the hands instead of with a trowel. A Keene's cement finish can be impressed with a pattern to resemble ceramic tile.

Setting Marble
Marble and similar materials should be cut to size and polished before they are delivered. Some minor cutting to make the materials fit exactly may be necessary. In setting marble, the work is laid out, anchor holes are drilled in the marble for wall-work, nonferrous anchors (usually brass or copper) are fastened to the marble, a special plaster mixture is applied to the backing material and the marble is set by pieces in place. When necessary, the marble is braced until the setting plaster has hardened. Special grout is packed into the joints between the marble pieces, and the joints are slightly indented (pointed up) with a pointing trowel or wooden paddle. Bolt holes need to be drilled if attachments to the marble are necessary.

The setting of marble on floors involves the preparation of Portland cement mortar, applying sufficient mortar for one piece of marble, placing the marble on the mortar and tamping it to the proper elevation. The marble setter then removes the marble piece, brushes or trowels a coat of neat cement to the back surface, resets the piece of marble on the setting bed and retamps it to the proper line and elevation.

Setting Tile
The first thing your tile setter needs to know is how to cut and fit tile to a desired size. The tile setter marks off the length of tile to be cut with a ruler and a pencil. Then with a pencil and a straightedge, a line is marked off on the tile from edge to edge on both sides. Using a glass cutter,

the tile setter scratches a line along the mark made on the tile. Then the tile is laid over an 8d (penny) finishing nail with the scratched surface directly on the nail. The setter presses the tile with his hands on both sides of the tile to break it. The broken edge of the tile should be polished with a carborundum stone or electric polishing wheel. Sometimes a tile needs to be cut in a curve rather than in a straight line to fit around a pipe. First, the centerline of the pipe should be scratched on the tiles with a glass cutter. Then the outline of the pipe should be marked off with a pencil on the tile. Using the penciled outline as a guide, the outline of the pipe should be scratched with a glass cutter. Then the area outlined should be nibbled away with glass pliers or glass nippers until a hole outlining the pipe appears. Finally, the edge of the cut is smoothed off with carborundum or an electric grinding wheel.

The first step in applying tile with mortar is laying it out properly. On a vertical wall surface, the tile setter should first figure how many tile will be needed in each course. Then the course is centered so that the fewest cuts possible are made. Tiles should always be cut in pairs, so that the first tile and last tile in a row are of equal size. The tile setter should lay out the courses before doing anything else. Once the courses are laid out, the tile setter is ready to prepare the backing to receive the tile. Loose paint and plaster on the wall should be scraped away before the scratch coat is applied. Then strips of 15 pound building felt should be attached to the wall. Finally, the lath should be placed over this and fastened to the wall with nails or staples. A scratch coat should be applied, which consists of one part Portland cement to 1½ parts of fine sand. The scratch coat should be troweled on smoothly and evenly, and checked with a spirit level after drying. If the scratch coat varies more than ¼ inch per 8 feet, a leveling coat should be applied. The scratch coat is mixed and applied with long, even trowel strokes. Finally, whatever excess mortar is left on the wall is removed with a 4 to 6 foot straightedge.

The float coat is allowed to dry and is then ready for tile application. The tile should be soaked for about ½ hour, and the float coat should be dashed with water. Pure Portland cement should be mixed in a clean container with clean water and allowed to sit for 20 minutes, then remixed to a creamy consistency. The cement should be troweled on 1/32 to 1/16 inch thick with long, even trowel strokes. The tiles should be applied in a level row and tapped into place with a flat rubber-covered block. The tile should be set four rows at a time and the process repeated. Once the tile dries, it should be grouted.

When tile is laid on a floor, the setting bed is tamped and then screeded, or leveled with a straightedge to the right elevation. Once the proper elevation is reached, a layer of pure Portland cement is brushed onto the setting bed and dampened. The tile is then set and tapped into place. After the tile is set, the joints are filled with grout, a Portland cement mortar and dampened. Finally, the tile faces are wiped with a damp cloth or sponge to remove excess grout.

When tile is set in adhesive, a primer should be used over the setting bed before applying the adhesive. The particular type of primer to be used will be recommended by the tile manufacturer. Then with a spatula or trowel, the tile setter applies adhesive to the back of the tile and presses it into place with its top exactly along the guideline. Each successive tile is butted up against the spacer lugs on its predecessor. Once the first course is laid it should be checked with a level. Then with a spatula or notched trowel the tile setter applies adhesive to the wall for the next course. Once tile has been set on a wall, the tile setter then grouts them. The grout is brushed or troweled into the spaces between tiles and binds them together.

There are several relatively simple steps used in applying grout. First, the tile is allowed to set for 24 to 36 hours before grouting. Next the joints or spaces between the tiles are dampened with a small sponge and clean water. The tile setter takes a clean bucket containing a small quantity of clean water, and adds Portland cement until the mixture is thick as cream. A brush or a sponge is dipped into the mixture, wiping it across the face of the tiles. After the grout has set for a few minutes, the tile setter takes a toothbrush handle and begins working the grout into the joints by drawing the toothbrush handle down the joint. The grout begins to set and the face of the tile is wiped with a clean damp sponge. Finally, the joint is sprayed down with a fine water mist and the tile is dried.

If tile has been set above or around a bath tub or sink, the joint between the tile and the butt is caulked with a sealing compound. There is a wide variety of such sealing compounds on the market, each of which has a different set of directions for its use.

Occasionally you may be called on to apply ceramic tile on a floor. There are two ways to lay a floor and both methods start with the longest straight wall in a room. In one method the tile setter snaps a chalkline and lays a row of tiles absolute straight along the master wall. Then, with a straightedge as a guide, the tile setter marks off the location of each tile to be set with a set of dividers or a marked stick the exact length of one tile plus one joint. This course is laid out perpendicular to the master wall. Finally, the tile setter mortars and sets the tile in the row (leg) perpendicular to the master wall. The other method of laying tile floors does not use legs. Tile setting without legs is done with a pole and two

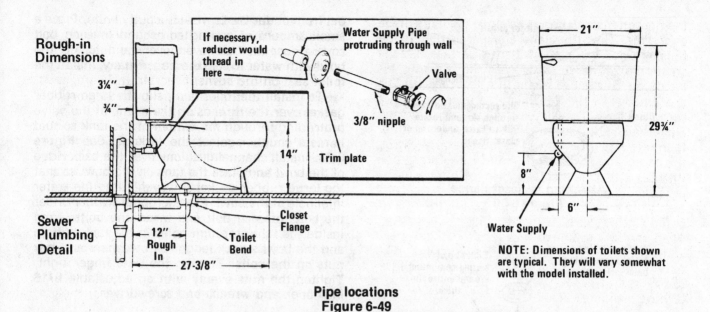

Rough-in Dimensions

If necessary, reducer would thread in here

Water Supply Pipe protruding through wall

Valve

3/8" nipple

Trim plate

3¼"

¾"

14"

21"

29¾"

8"

6"

Water Supply

Sewer Plumbing Detail

12" Rough In

Toilet Bend

Closet Flange

27-3/8"

NOTE: Dimensions of toilets shown are typical. They will vary somewhat with the model installed.

**Pipe locations
Figure 6-49**

layout sticks exactly 3 tiles plus 3 joints long and marked off in tile and joint lengths. A line should be stretched out parallel to the master wall and exactly one tile length away from it. The straightedge should be placed perpendicular to the master wall. Where it goes under the line, a nail should be driven. A row of tile should be laid against the straightedge. Then, with the two layout sticks, the pole should be moved up 3 rows and another course of tile laid. The area between the two courses should then be filled in with courses of tile and the process should continue. As each course is completed, it should be tapped into place using a rubber covered wooden block. When the tile setter must put weight on the newly laid tile, a kneeling board should be used so that weight will be distributed evenly and will not put a dent in one part of the floor. Kneeling boards shoud be no less than 12 inches wide and weight should not be placed on the edge of them. Once the floor is laid, it should be grouted. The process for mixing and applying grout to floors is the same as that used for applying grout to vertical walls.

Replacing Toilets

There are differences in installation requirements for various toilet models. Some toilets are shipped with good instructions and some arrive with little or no guidance for installation. The following instructions will apply in most situations. Note that many building codes regulate the types of toilets that may be installed and piping requirements.

First, check the waste and supply pipe location. Note Figure 6-49. Measure the distance from the finished wall to the center of the rear closet flange bolt. This distance must be a minimum of 12 inches to allow for installation of the tank. The water supply pipe must be located at least 6 inches from the center line of the bowl for two piece toilets and 8½ inches for one piece toilets. A wall installation must be approximately 8 inches above the floor for two piece toilets and 2½ inches for one piece toilets and protrude a minimum of 3¼ inches from the wall.

To remove the old toilet, turn off the water supply at the main source and open any cold water faucet to relieve pressure in the pipes. Flush the toilet to drain the tank. Remove the remaining water with a sponge. Disconnect the water supply pipe from the underside of the tank; also disconnect the pipe at the wall or floor. Remove the old toilet by unscrewing the closet flange bolts. Temporarily stuff a rag into the sewer outlet to prevent sewer gas escaping. Clean the old bowl sealing compound from the floor with a putty knife. Remove the old wax gasket or compound from the closet flange or from around the waste outlet opening in the floor. If flooring is rotted, remove and replace it.

If the old bowl was anchored by bolts screwed into the floor rather than with a closet flange, check the position of the bolt holes and waste outlet opening of the new bowl to assure proper fit with the existing bolts. If not suitable, they will have to be removed and new bolts installed—or you will have to install a closet flange.

If you are going to replace the toilet supply valve, connect the threaded 3/8 inch water supply pipe nipple of the wall or floor type toilet supply pipe to plumbing in the floor or wall. (If the 3/8 inch nipple does not fit, install a reducer of proper size.) Slip the chrome trim plate on the nipple. Thread the valve on the nipple and tighten. Turn the handle on the valve clockwise to close the valve. At the main source, turn the water on so it is available to the rest of the house. Close the faucet which had been opened to relieve pressure.

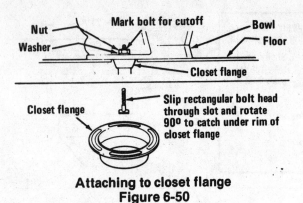

**Attaching to closet flange
Figure 6-50**

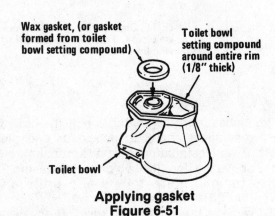

**Applying gasket
Figure 6-51**

Before preparing the bowl for installation, set the bowl in place over the waste outlet opening (Figure 6-50) and mounting bolts. Place washers on the bolts and thread on nuts, tightening them finger tight. If a bolt projects above the nut, mark the bolt for cutting and remove the nuts, washers and bowl. Thread the nuts on the bolts to below the cut mark and cut off the excess length with a hacksaw. Reinstall the bolts in the closet flange. Knead some toilet bowl setting compound to a doughy consistency. Place a uniform layer, about 1/8 inch thick, around the bowl pedestal and place the wax gasket around the bowl waste pipe opening. See Figure 6-51. Where local codes do not permit a wax gasket, toilet bowl setting compound must be used.

Next, remove the rag that has been temporarily stuffed into the waste opening in the floor. Carefully set the bowl on the closet flange, making sure that the bowl mounting bolts pass through the bolt holes in the bowl's base and that the waste opening is lined up correctly with the waste pipe.

Press down on each side at the center of the bowl, as shown in Figure 6-52, using your full weight and a very slight twisting motion to imbed the bowl evenly in the wax gasket (or compound). Place washers and nuts on the bowl mounting bolts. Tighten the nuts finger tight then tighten ½-turn with an adjustable or open end wrench. If the bowl does not "rock," it is tight.

Trim away excess setting compound squeezed

out from under the bowl with a putty knife. Place a small amount of setting compound in each bolt cover. Place the bolt covers over the nuts. Fill the bowl with water to seal the trap. Finally, install the toilet seat on the bowl.

To install the toilet tank, slip the large rubber gasket over the threads on the shank of the valve protruding through the bottom of the tank so that it rests snugly against the locknut. See Figure 6-53. Install channel cushions over the back ridge of the bowl and place the tank on the bowl so that the large rubber gasket fits evenly into the water inlet hole of the bowl. Place a rubber washer under the head of each bolt and insert the bolts, from inside the tank, through holes in the tank bottom and the bowl's back ledge. Slip washers and hex nuts on the bolts. Tighten the nuts finger tight. Tighten the nuts evenly with an adjustable 9/16 inch open end wrench and screwdriver.

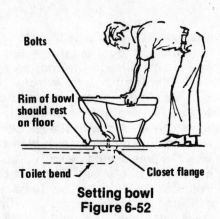

**Setting bowl
Figure 6-52**

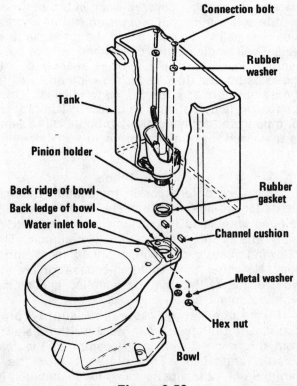

Figure 6-53

Finally, make the water supply connection. Slip the coupling nut (supplied with the toilet), valve nut and valve collar (supplied with valve) on the water supply pipe as shown in Figure 6-54. Insert the straight section of the pipe squarely into the valve until it seats firmly. Slip the valve collar and nut into place on the valve and tighten securely with an adjustable wrench. Holding the pipe securely where it protrudes from the valve (so as not to loosen the seal just made), bend the pipe as necessary to have its nose-piece seat squarely into the shank of the riser tube assembly. Slip the coupling nut into position and tighten securely. Turn on the valve. The tank should start filling with water. Check for leaks at pipe connections.

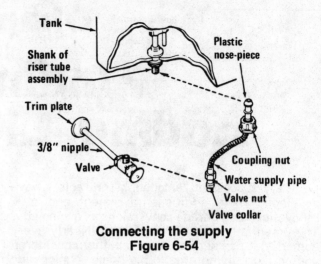

Connecting the supply
Figure 6-54

Chapter 7
Finding and Adding Living Area

Many successful remodeling contractors have developed room additions, basement and attic renovations and garage conversions as a specialty. The potential market is very broad. Nearly every homeowner has at one time or another considered adding more living area to his home. Nationwide more money is spent on creating more space (family rooms, bedrooms, bathrooms) and converting storage space (basements, attics, garages) to living area than any other type of residential remodeling. It has been estimated that adding and converting residential space was a 2 billion dollar market in 1975 and will grow to over a 3 billion dollar market by 1980. This is a very substantial portion of the total residential construction market and will probably continue to grow faster than residential construction in general.

Finding or adding space makes sense in many situations. Families tend to outgrow their homes and homeowners often discover that it is less expensive to add space to their existing home than buy another home with the floor area they require.

Homeowners often are not aware of the possibilities or the limitations in their home. An experienced remodeling contractor will usually be able to suggest ways to gain living space by converting storage or wasted area. Your ingenuity here will be a valuable sales tool. Suggest possibilities such as adding a shed dormer, loft, or second story, or removing a partition. Naturally, you should suggest the possibility of using buildable space on the lot when a room addition is appropriate. Your knowledge of the local code requirements and what can be approved will be essential. Consider setback requirements, minimum ceiling heights, and other code requirements.

Using The Attic

Making an attic usable may be a simple matter of installing finish ceiling, wall, and floor covering; however, it will often require adding a shed or gable dormer for more space or for natural light and ventilation. Furthermore, the existing joists may not be adequate to support a floor load. Usually building codes require a floor design live load of 30 pounds per square inch for attics used as sleeping quarters. Some ceiling joists are designed only for 10 or 20 pounds per square inch live loads. If the joists are inadequate, the best solution is usually to double existing joists.

Where light and ventilation are the main requirements rather than additional space, gable dormers are often used. They are more attractive in exterior appearance than shed dormers but, due to the roof slope, they are usually limited to a small size. They are also more complicated to build then shed dormers.

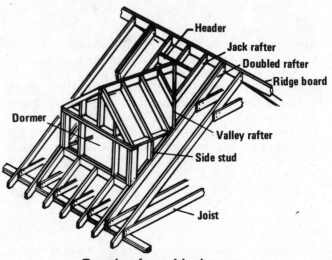

Framing for gable dormer
Figure 7-1

The roof of the gable dormer usually has the same pitch as the main roof of the house. The dormer should be located so that both sides are adjacent to an existing rafter (Figure 7-1). The rafters are then doubled to provide support for the side studs and short valley rafters. Tie the valley rafter to the roof framing at the roof by a header. Frame the window and apply interior and exterior covering material as appropriate.

One of the most critical items in dormer construction is proper flashing where the dormer walls intersect the roof of the house (Figure 7-2). When roofing felt is used under the shingles, it should be turned up the wall at least 2 inches. Shingle flashing should be used at this junction. This consists of tin or galvanized metal shingles bent at a 90° angle to extend up the side of the wall over the sheathing a minimum of 4 inches. Use one piece of flashing at each shingle course, lapping successive pieces in the same manner as shingles. Apply siding over the flashing, allowing about a 2 inch space between the bottom edge of

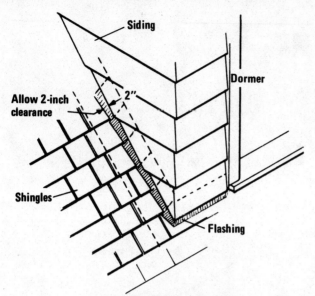

Flashing at dormer walls
Figure 7-2

the siding and the roof. Cut ends of siding should be treated with water repellent preservative.

Shed dormers can be made any width, and are sometimes made to extend across the entire length of the house. They are less attractive than gable dormers and, for this reason, are usually placed at the back of the house. Sides of the shed dormers are framed in the same manner as the gable dormer, so the sides should coincide with existing rafters. The low slope roof has rafters framing directly into the ridgepole (Figure 7-3). Ceiling joists bear on the outer wall of the dormer, with the opposite ends of the joists nailed to the main roof rafters. The low slope of the dormer roof means that requirements of roofing application will be different from those of the main roof.

Sides of the attic rooms are provided by nailing 2 by 4 or 2 by 3 studs to each rafter at a point where the stud will be at least 5 feet long (Figure

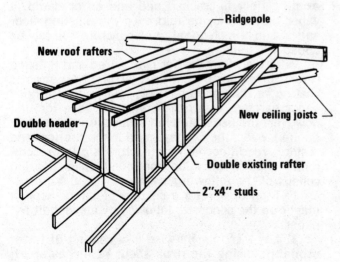

Framing for shed dormer
Figure 7-3

7-4). Studs should rest on a soleplate in the same manner as other partitions. Nail blocking between studs and rafters at the top of the knee wall to provide a nailing surface for the wall finish.

Nail collar beams between opposite rafters to serve as ceiling framing (Figure 7-4). These should be at least 7½ feet above the floor. Nail blocking between collar beams and between rafters at their junction to provide a nailing surface for the finish wall and ceiling materials.

An alternate method of installing the ceiling is to eliminate the collar beams and apply the ceiling finish directly to the rafters (Figure 7-5). This results in what is commonly called a cathedral ceiling. If the dormer is wide, cross partitions or some other type of bracing is required for stability.

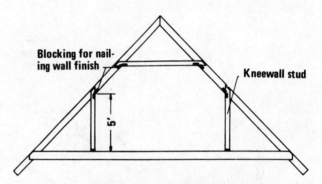

Installation of knee walls and blocking
Figure 7-4

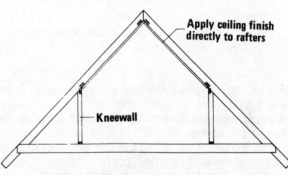

Attic finished with a cathedral ceiling
Figure 7-5

If a chimney passes through the attic, it must either be hidden or worked into the decor. Never frame into the chimney. Keep all framing at least 2 inches from the chimney. Where framing is placed completely around the chimney, fill the space between the chimney and framing with noncombustible insulation.

Basements

Basements can be finished to any desired quality, depending on the investment required and the use to be made of the finished space. This may vary from insulated walls with quality paneling, wood floors, and acoustical ceiling, to merely painting the existing concrete walls and

103

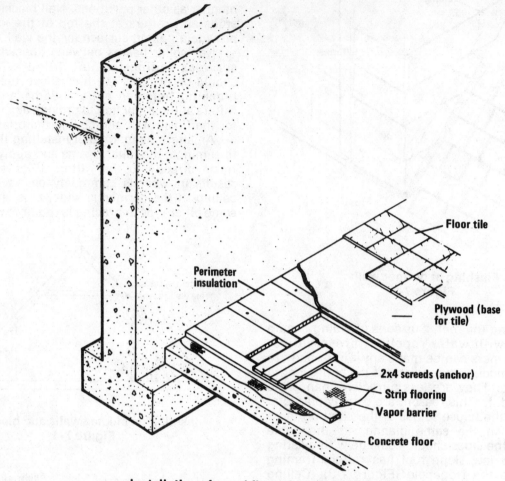

Installation of wood floors in a basement
Figure 7-6

floors. Keep in mind that basement areas, even where the grade is low, tend toward dampness thus requiring dehumidification, and tend to be cooler so the rooms may be uncomfortably cool in periods of light heating. A small amount of auxiliary electric heat can partially correct this. Insulation of below grade concrete or masonry walls is very important.

If a concrete floor is dry most of the year, it indicates that a vapor barrier was probably applied under the slab at the time it was constructed. In such a basement, resilient tile or indoor-outdoor carpeting could be applied directly to the smooth slab. Any protrusions from the slab should be chipped off. Extensive unevenness in the slab indicates that an underlayment over sleepers should be used as a base for the tile or carpet. Install the tile following the manufacturer's recommendations on adhesives and installation practices. Carpeting is usually just cut to size and laid flat without special installation except for double-faced tape at edges and seams.

A very low cost floor can be achieved by merely applying a deck paint. It should be a latex paint to avoid chipping or peeling. Although painting may not give the finished appearance provided by a floor covering, it does brighten the basement and produces a smooth surface that is easily cleaned. It is particularly suited to shops, utility rooms, and playrooms.

Where a concrete floor is desired, but the existing floor is cracked and uneven or damp, a vapor barrier can be laid over the existing floor and a 2 to 3 inch topping of concrete fill can be added.

Where a vapor barrier is required and finished flooring is planned, apply an asphalt mastic coating to the concrete floor, followed by a good vapor barrier. This can serve as a base for tile; however, the use of furring strips with finish floor applied over them may produce a better end result. Wood flooring manufacturers often recommend that preparation for wood strip flooring consist of the following steps:

1. Mop or spread a coating of tar or asphalt mastic on the concrete, followed by an asphalt felt paper.

2. Lay 2 by 4's flatwise in a coating of tar or asphalt, spacing the rows about 12 inches apart; start at one wall and end at the opposite wall.

3. Place a 2 foot width of insulation around the perimeter, between 2 by 4's, where the outside

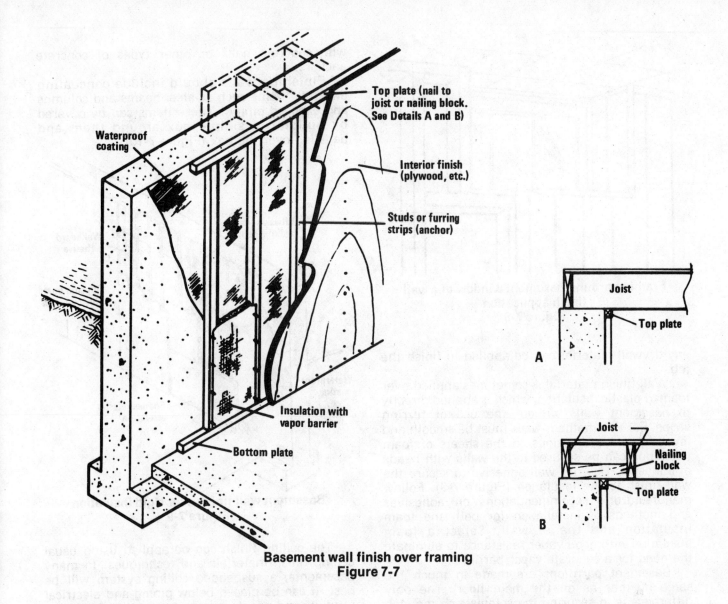

Waterproof coating

Top plate (nail to joist or nailing block. See Details A and B)

Interior finish (plywood, etc.)

Studs or furring strips (anchor)

Insulation with vapor barrier

Bottom plate

Joist

Top plate

A

Joist

Nailing block

Top plate

B

Basement wall finish over framing
Figure 7-7

ground level is near the basement floor elevation.

4. Install wood strip flooring across the 2 by 4's.

A variation of this preparation for flooring consists of laying on a good quality vapor barrier directly over the slab and anchoring the furring strips to the slab with concrete nails (Figure 7-6). Insulation and strip flooring are then applied in the same manner described above. Plywood, 1/2 and 5/8 inch thick, can also be applied over furring strips to provide a base for resilient tile or carpet (Figure 7-6).

Where basement walls were waterproofed on the outside, there should be no problem in applying most any kind of wall finish. However, if there is any possibility of water entry, it is important to apply a waterproof coating to the inner surface. Numerous coatings of this type are available commercially. These coatings usually cannot be applied over a wall that has previously been painted. Walls can simply be painted for a bright, clean appearance where a more finished appearance is not as essential. If a better interior

finish is desired, it is applied over furring strips or rigid insulation.

Furring strips, 2 by 2 inches or larger, are used at the walls in preparation for the interior finish (Figure 7-7). Strips should be pressure treated for decay resistance, especially if the walls are not waterproofed. Anchor a 2 by 2 inch bottom plate to the floor at the junction of the wall and floor. Fasten a 2 by 2 inch or larger top plate to the bottom of the joists above (Figure 7-7A), to nailing blocks (Figure 7-7B), or to the wall. Then fasten 2 by 2 inch or larger furring strips, at 16 or 24 inch spacing, vertically between top and bottom plates. Concrete nails are sometimes used to anchor the center of the furring strips to the basement wall.

Before proceeding any further with the finish wall, install all required electrical conduit and outlet boxes between furring strips. Place blanket-type insulation with vapor barrier on the inside face in each space between furring strips. Frame around windows, as shown in Figure 7-8. The wall is then ready to receive the interior finish. Almost

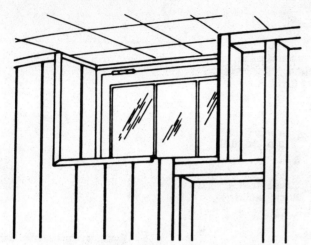

Framing around basement window of a wall finish application
Figure 7-8

any drywall material can be applied to finish the job.

Wall finish material is sometimes applied over foamed plastic insulation, which is applied directly to basement walls without the use of furring strips. For this method, walls must be smooth and level without protrusions so the sheets of foam insulation can be secured to the walls with beads of adhesive. Use dry wall adhesive to secure the wall finish to the insulation (Figure 7-9). Follow manufacturer's recommendations on adhesives and methods of installation for both the foam insulation and the drywall. Select a foam insulation with good vapor resistance to eliminate the need for a separate vapor barrier.

Basement partitions are made in much the same manner as on the main floor. The only difference is in securing the soleplate to the slab

with concrete nails or other types of concrete anchors.

Finishing walls should include concealing unsightly items such as steel beams and columns and exposed piping. These items can be covered by building a simple box around them and paneling over the box (Figure 7-10).

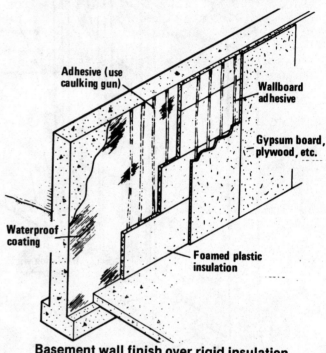

Basement wall finish over rigid insulation
Figure 7-9

The ceiling finish can be applied using usual ceiling finish materials and techniques. In many basements, a suspended ceiling system will be best. It can be placed below piping and electrical conduit, and panels are easily removed for repair

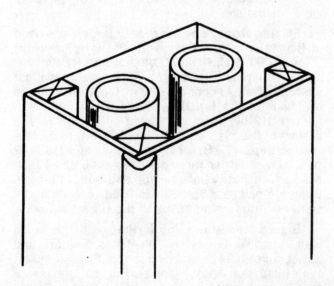

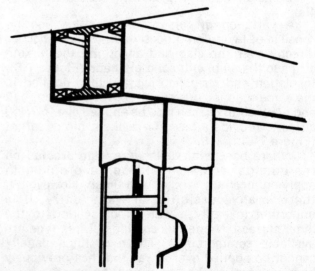

Application of paneling to cover columns, beams, and pipes
Figure 7-10

or changes in utilities. In considering the use of a suspended ceiling, make certain that at least 7 feet will still be available from floor to ceiling. It may be necessary to apply the ceiling directly to the bottom of floor joists and box around or paint to match the ceiling.

Before installing any ceiling, insulate cold water pipes. When the basement has a high relative humidity, water condenses on the cold pipes and this condensation drips down on the ceiling where it is under the pipes. Molded insulation that can be fitted over the pipe is commercially available. Commercial insulations are made specifically for wrapping around pipe.

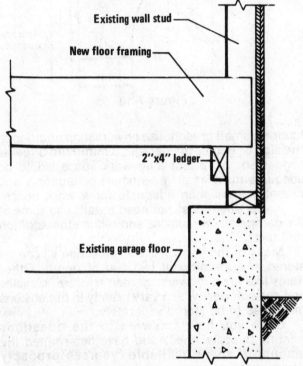

Framing to bring new garage floor to the level of the house floor
Figure 7-11

Garage Conversions

The garage that was built integrally with the house can be used as living space simply by insulating it and adding floor, wall, and ceiling finish. There may be some additional requirement for windows and for heating.

The new finish floor is applied directly over the existing concrete slab, or, where headroom is sufficient, over new floor framing above the slab. Where the ceiling of the finished garage is at the same elevation as the house ceiling, new floor framing can be installed to place the new floor at the same level as the house floor. The framing may rest on the foundation wall or be supported on ledgers nailed to the wall studs (Figure 7-11). Consult joist tables to determine the correct size for the required span. The floor is installed as a conventional floor and insulated in the same

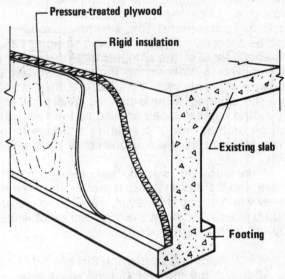

Application of insulation to outer face of garage foundation wall
Figure 7-12

manner as the floor in a crawl space house.

The garage roof is often lower than the house roof, so that the floor must be placed directly on the concrete slab one or two steps below the house floor. Finish floor can be installed using the same materials and techniques as for a basement slab. If the garage floor is near ground level, insulation is required. If the finished floor is on sleepers over the concrete, place insulation between sleepers.

Another method of insulating is to dig the soil away from the foundation and apply rigid insulation to the outer face of the foundation wall to the depth of the footing (Figure 7-12). It must be a moisture proof insulation, such as polystyrene or polyurethane, and should be attached with a mastic recommended by the insulation manufacturer. The insulation should be covered with a material suitable for underground use, such as asbestos-cement board or preservative-treated plywood. Where subterranean termites are a threat, soil can be poisoned during the backfill.

The garage walls are not usually insulated, so blanket insulation with a vapor barrier on the inside face should be installed in the space between studs. Then apply any type of drywall. It may be convenient to use the existing garage door opening for a sliding glass door, or a window wall. If it is not to be used for either purpose, fill it in, using conventional 2 by 4 framing, and install covering materials. Be sure to insulate the ceiling well and ventilate the attic space.

Additions

Additions to existing houses are made by employing new construction techniques, and building to match or blend with the existing structure. The main item peculiar to additions is making the connection between the addition and the existing structure.

Where the addition is to be made by extending

107

the length of building, the structure as well as the siding and roofing must match the existing portion. To accomplish this, a complete roofing job may be necessary. Where existing shingles can be matched, some of the shingles near the end must be removed in order to lap the saturated felt underlayment. The new shingles are then worked into the existing shingle pattern. Short pieces of lap siding must also be removed so the new siding can be toothed into the existing siding with end joints offset rather than as one continuous vertical joint.

If the addition is to be perpendicular to the house, the siding can either match or contrast with the existing siding. However, the roofing material should match. Siding at the intersection of walls is applied with a corner strip.

Adding Garages and Storage

Adding a garage and storage space is a very popular home improvement in many regions. A scale drawing of the lot showing the location of the house, property lines, the setback and clearances required by local zoning ordinances, trees, walkways, etc., will be an invaluable source of information concerning existing conditions. If you have the information needed to locate your property lines exactly, you may wish to make your own drawing. If you don't have the information, you may want to recommend that the property be surveyed. In this case, an engineer can provide a scale drawing of the lot, locating any additional features on it that are required. These features would include anything that would prevent you from locating the garage in any particular position. Besides those noted previously, you would need to know the location of the septic tank, leaching field and well if the property has them. Also, it would be wise to note the location of any moderate or steep slopes.

Next, ask your client important planning questions such as the following:

1. Do you need space for one____or two____ cars?

2. Do you need space for play or recreation? For a laundry? A greenhouse? A workshop?

3. Do you need to store any of the following items: garden tools, window screens, freezer, storm sash, building materials, lawn furniture, snowblower, firewood, bicycles, snowmobiles, boats, travel trailer, carpentry tools, and motor bikes.

4. Do you need complete protection from the wind, rain, snow and sun?

5. Do you need partial protection from the weather for living space, play area recreation?

6. Is there an absolute limit to the money you can spend on the project?

For storing an average car, a 12 foot by 22 foot structure will be adequate. A 24 by 22 foot structure is needed for two cars.

For those considering a separate area for a

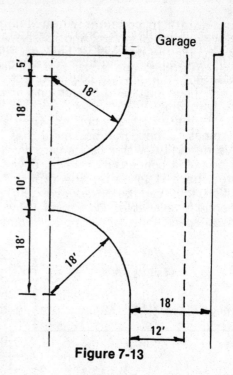

Figure 7-13

laundry, small greenhouse or workshop addition at the side or end of the garage, a minimum 5 feet is suggested. A 12 foot long work space would be adequate for a small greenhouse or laundry, and probably more than adequate for a work bench.

If your client does not need a wall, use some of the garage for work space and reduce the addition to 3 feet.

Many of the items listed in Question #3 can be stored in an additional two feet of depth. Other items like lawn mowers, garden tractors, bicycles and snowblowers which vary greatly in dimensions may need to be stored separately.

Once you have answers to the questions relating to space needs and have determined the amount of space available, you can probably decide whether a carport or garage would be most suitable. If your client is trying to save money and needs only partial protection from the weather for activities and storage, a carport will probably provide the best solution.

Where you will locate the garage or carport may be determined initially by the amount of space available between the house and the edge of the property which is unrestricted by zoning or other existing conditions on the lot.

One item not previously discussed is the orientation of the garage. If the garage will block light and air from entering the house where needed, you should consider either changing its location or building a carport instead. Another factor is the location of the driveway and turning space. For a straight driveway with a "Y" for turning and backing you will need a minimum of 15 feet and a maximum of 20 feet between the edge of the garage and the property line. The 18 feet shown on the driveway in Figure 7-13 is for backing and turning a car of average length.

Other points to consider when designing driveways are ease of access and safety. A safe driveway avoids the dangers of backing into the street, steep grades, and areas that will collect water or be covered with ice in winter.

Where to locate the garage or carport in the plan is frequently easier than designing the exterior so that it looks as if it belongs there instead of being an afterthought. This can be done by adding a single car garage directly to the end of a building the same width, continuing the same roof and eave lines, and using the same exterior material and finish. Such a solution is particularly suitable when making additions to small homes 32 feet to 48 feet in length.

Where steep grades are encountered it is usually wise to detach the garage or carport from the house. It may not be as convenient but it avoids difficult roofing, foundation and design problems. Frequently the family is grateful for the semi-secluded space created when this is done.

Grading for good drainage is absolutely essential to satisfactory performance. Roofs collect water. This must be considered. So must the runoff onto the surrounding land. Structures added to existing buildings can cause problems if connections are poorly planned. The bottom of all foundation footings should be placed below the prevailing frost penetration. Detached structures can, however, be satisfactorily built on floating slab foundations if there is good drainage under and away from this slab.

Masonry is not the only alternative for use as a permanent foundation material. Do not overlook pressure-treated wooden posts or poles. Wood treated with penetrating preservatives under pressure can be very durable. In fact, treated wood pilings driven deep into the ground support many large buildings and some major highways.

Carport and garage floor drains seem to be a perennial problem. A simple but effective way to drain a floor is to slope the whole floor toward the driveway opening. Three inches of slope in 24 feet is often recommended. But constructing this slope is easier said than done, and mistakes made laying concrete are hard to correct.

Roof and wall design varies in different parts of the country according to what is known as live load. Live load is usually determined by the prevailing winds and snow. Unfortunately, many small buildings are structurally unsound and require much maintenance or they may be subject to total failure.

Doors in new, single-car garages should be at least 9 feet wide. Two-car garages might have two 9 feet doors or a single large door 16 to 18 feet wide. Wide doors are helpful when the approach to a garage cannot be straight. Interior posts should be avoided whenever possible. Garages used for storage should also have a pedestrian door.

Chapter 8
Flooring

Traditionally people have used linoleum in areas such as the kitchen, ceramic tile in the bathroom, and wood floors throughout the rest of the house. With the variety of flooring materials available today—approximately 30 types and hundreds of varieties—there is little reason to be bound by tradition.

In making a choice, livability, aesthetics, durability, maintenance, and cost should be considered. Remember, no one material is ideally suited to the requirements of every room in the home.

Resilient surfaces refer to a number of different types of water resistant materials that range from the traditional linoleum available in rolls to thin sheets or tile materials differentiated according to their ingredients: asphalt, vinyl, vinyl asbestos, rubber, and cork. Resilient surfaces are dense and have non-absorbent surfaces. Their resilience aids in sound control and provides resistance to indentation. Density of the material usually provides long life and ease of maintenance. The most expensive material will usually give the most beauty and highest wear resistance. The lowest cost materials give the least wear and should be used only as a short term covering.

The sheet materials are more difficult to install than tile; however, in comparable materials, sheet usually costs less than tile. Most resilient surfaces are secured to the subfloor with an adhesive: linoleum paste, asphalt emulsion, latex, or epoxy. What you use is dictated by the flooring manufacturer's specifications.

Linoleum is a blend of linseed oil, pigments, fillers, and resin binders bonded to a backing of asphalt-saturated felt. It is available in solid colors or with inlaid, embossed, or textured patterns simulating stone, wood, or tile. It is available in rolls eight feet or wider and in tiles either 9 inches square or 12 inches square with thicknesses of 1/16 inch, .090 inch, and 1/8 inch. Linoleum provides fair wear resistance, and its color extends completely through to the backing material. Inlaid linoleum has a hard durable surface, is grease-proof, and easy to clean; however it is damaged by cleaning products containing alkali solutions. It should not be installed on a concrete slab on grade since moisture permeating through the concrete from below will cause the material to rot.

Asphalt tile is a combination of asbestos fibers, ground limestone, and mineral pigments with an asphalt binder. It is the least expensive and most commonly used tile. Its price depends on the color; dark colors are the least expensive, light colors and special patterns are the most expensive. Asphalt tile are manufactured with the pattern through the total thickness. Some tile patterns simulate other materials. In this case, the pattern does not penetrate its thickness and it will wear rapidly under heavy use. Normal tile size is 1/8 inch thick and 9 inches square.

Asphalt tile will stain and break down if it contacts animal fats and mineral oils. It is brittle and breaks easily. Its recovering from indentation is negligible. It can, however, be used on concrete slabs on grade and where there may be a moisture problem.

Vinyl floor covering's chief ingredient is polyvinyl chloride (PVC). It also contains resin binders, with mineral filters, stabilizers, plasticizers, and pigments. The vinyl may be filled or clear.

Clear vinyl consists of a layer of opaque particles or pigments covered with a wearing surface of clear vinyl bonded to a vinyl or polymer-impregnaed asbestos fiber or resin-saturated felt. The clear vinyl surface provides high resistance to wear. Filled vinyl is made of chips of vinyl of varied color and shape immersed in a clear vinyl base and bonded by heat and pressure. When used in a basement, a vapor barrier or an epoxy adhesive should be used to install it.

Vinyl tile is the most costly, but also the most wear resistant and easily maintained of the various tiles. It is produced in standard size squares, 9 inches square or 12 inches square, in standard thicknesses of 1/16 inch, .080 inch, 3/32 inch, and 1/8 inch.

Sheet vinyl may be produced with a layer of vinyl foam bonded to the backing or between the finish surface and the backing. The result is a resilient flooring with good walking comfort and an effective sound absorbent quality. The vinyl is produced in rolls eight feet wide or wider and can be installed over most subsurfaces. While the material has high resistance to grease, stains, and alkali, its surface is easily damaged by abrasion and indentation since it is, generally, a soft product.

Vinyl-asbestos tile consists of blended compositions of asbestos fibers, vinyls, plasticizers, color pigments, and fillers. The tiles, without backing, are 9 inches square or 12 inches square and 1/16 inch, 3/32 inch and 1/8 inch thick. It may be obtained in marbleized patterns, or textured to simulate stone, marble, travertine, and wood. It is semiflexible and requires a rigid subfloor for support. The tile has high resistance to grease, oils, alkaline substances, and some acids. It is quiet underfoot and many forms can go without waxing for extended periods of time. It can be used almost anywhere, and can be obtained with a peel-and-stick backing.

Rubber tile is based on natural or synthetic rubber. Mineral fillers and nonfading organic pigments are used to produce a narrow range of colors and patterns. The standard sizes are 9 inches square or 12 inches square. Larger sizes are available at higher cost. The thicknesses are 0.080 inch, 1/8 inch, and 3/16 inch.

Rubber tile is resilient and has high resistance to indentation. The material is softened by petroleum products and its resistance to grease and kitchen oils depends on its method of manufacture. Waxing and buffing are necessary to maintain a high gloss. The surface becomes slippery when wet. Use of a vapor barrier or epoxy adhesive for slab on grade installation is required.

Cork tile consists of granulated cork bark combined with a synthetic resin as a binder. The best tile has a clear film of vinyl applied to improve its durability, water resistance, and ease of maintenance. Tile sizes are 6 inches square and 12 inches square with a range of thicknesses from 1/8 inch to 1/2 inch. Cork floors are great for foot comfort and sound control. They wear rapidly and do not resist impact loads well. Maintenance is difficult since the material is broken down by grease and alkalies.

Wood floorings. Many varieties of both hard and soft woods are available for flooring. Certain hardwoods, because of their high resistance to wear, are more often used than others. Two are oak and maple. Wood flooring is finished with a combination of coatings such as a sealer and varnish, or a liquid plastic. Wood flooring may be simply nailed to the subfloor or, when used over a concrete slab, nailed to wood "sleepers" fastened to the slab. In either case, the floor is sanded smooth, and finished with stain and sealer.

The most commonly used hardwood flooring is oak because of its beauty, warmth, and durability. Maple flooring is produced from the sugar, or rock, maple. It is smooth, strong, and hard. The grain of maple does not have as much contrast as oak; however, where a smooth polished surface is necessary, maple makes a superior floor. Beech, birch, hickory, and several other hardwoods are also used.

Hardwood strip flooring is hollowed or has "V" slots cut into its back surface to minimize warping. It is produced in thicknesses of 3/8 inch, 1/2 inch, or 25/32" inch and widths varying from 1½ inch to 3¼ inches, with the most popular width being 2¼ inches, and is tongue and grooved to provide tight joints.

Hardwood flooring is graded on its appearance according to the number of defects, variations of color, and surface characteristics. Strength and wear are not dependent on grading since all grades are comparable in these respects.

Strip flooring is available prefinished. The finish is applied at the factory and the floor can be used right after installation. It comes as imitation peg style, random width, and simulated plank.

Hardwood squares 9 inches square or 12 inches square by 5/16 inch or 1/2 inch thick can be purchased to produce a parquet floor. These squares are available in several types of wood such as oak, maple, mahogany, cherry, and teak. The blocks may be nailed to the subfloor or secured with a mastic. Although hardwood squares cost more than strip flooring, they are normally produced in prefinished form and are therefore competitive in completed cost.

Softwood flooring. The softwood most used is southern yellow pine, Douglas fir is next, with western hemlock and larch following. Some woods such as redwood, cedar, cypress, and eastern white pine are used in areas where they are common and available. Softwood flooring is available in several sizes and thicknesses; the most common is 25/32 inch thick and 4½ inches wide. The long edges of the flooring are tongued and grooved or side matched in order to give tight joints. As with hardwood, the underside is hollowed or V-grooved to minimize warping.

Non-resilient flooring. These include brick "pavers", ceramic and clay tile, stone, and terrazzo. These materials are more difficult to install than other flooring materials and usually are the most expensive. However, they have a long life. They may be installed using a special "thin-set" cement, or in the traditional 3/8 inch bed of mortar. They require a "grout" (cement fill) between the tiles.

Glazed ceramic tile and terra cotta are relatively non-porous and as a result resist staining. These glazed tiles are, however, susceptible to scratching and crazing (formation of minute cracks) with age. Ceramic tiles range in size from what is called "mosaic" tile 3/8 inch square to a large 16 inch by 18 inch size. Mosaic tiles commonly are sold on a backing sheet, making possible the installation of larger areas at one time. It is necessary to grout the joints between each tile after they are set in place.

Unglazed ceramic tile, slate, and flagstone are porous unless treated with special stain-resistant sealants. Clay or quarry tile, usually unglazed, is produced from clays that result in a strong, long-wearing surface. It is relatively easy to maintain and withstands impact well. The color range is reds, buffs, blacks, browns, greys, and gold. A semiglazed type is produced in greys, browns, and greens. The product is available with a variety of surface patterns. The tiles come in several thicknesses from ¼ inch, ½ inch, and up to 1½ inch depending on their width and length. They may be square, rectangular, or some geometric shape.

Terrazzo is made of marble chips in combina-

Species	Grain Orientation	Size		First Grade	Second Grade	Third Grade
		Thickness Inch	Face Width Inch			
Softwoods						
Douglas-fir and hemlock	Edge grain	25/32	2-3/8=5-3/16	B and better	C	D
	Flat grain	25/32	2-3/8=5-3/16	C and better	D	--
Southern pine	Edge grain	5/16-	1-3/4=5-7/16	B and better	C and better	D (& No. 2)
	Flat grain	1-5/16	--	--	--	--
Hardwoods						
Oak	Edge grain	25/32	1-1/2=3-1/4	Clear	Select	--
	Flat grain	3/8	1-1/2, 2	Clear	Select	No. 1 common
		1/2	1-1/2, 2	Clear	Select	No. 1 common
Beech, birch, maple, and pecan[1]	--	25/32	1-1/2=3-1/4	First grade	2nd grade	--
	--	3/8	1-1/2, 2	First grade	2nd grade	--
	--	1/2	1-1/2, 2	First grade	2nd grade	--

[1]Special grades are available in which uniformity of color is a requirement.

Grade and description of strip flooring of several species and grain orientation
Table 8-1

tion with portland cement mortar and is ground and polished to a smooth finish. It is very resistant to moisture and therefore relatively easy to maintain. It is very noisy and is a tiring walking and work surface.

Most non-resilient flooring is installed using a masonry mortar. This demands a higher degree of skill than other types of flooring and adds to the installed cost.

Preparing The Floor

Before any floor is laid, a suitable base must be prepared. Unless existing wood flooring is exceptionally smooth, it should receive a light sanding to remove irregularities before any covering is put over it. If a thin underlayment or no underlayment is being used, wide joints between floor boards should be filled to avoid showthrough on the less rigid types of finish floor. An underlayment of plywood or wood-base panel material installed over the old floor is required when linoleum or resilient tile is used for the new finish floor.

Where underlayment is required, it should be in 4 by 4 foot or larger sheets of untempered hardboard, plywood, or particleboard 1/4 or 3/8 inch thick. Some floor coverings are not guaranteed over all types of underlayments, so check the manufacturers' recommendations before choosing an underlayment. Underlayment grade of plywood has a sanded, C-plugged or Better face ply and a C-ply or Better immediately under the face. It is available in interior types, exterior types, and interior types with exterior glue. The interior type is generally adequate but one of the other two types should be used where there is possible exposure to moisture. Underlayment should be laid with 1/32 inch edge and end spacing to allow for expansion. Nail the underlayment to the subfloor using the type of nail and spacing recommended by the underlayment manufacturer.

Wood flooring, sheet vinyl with resilient backing, seamless flooring, and carpeting can all be installed directly over the old flooring after major voids have been filled and it has been sanded relatively smooth. These coverings can also be installed over old resilient tile which is still firmly cemented.

Wood Flooring

Wood flooring may be hardwood or softwood. Grades and descriptions are listed in Table 8-1. Types are illustrated in Figure 8-2.

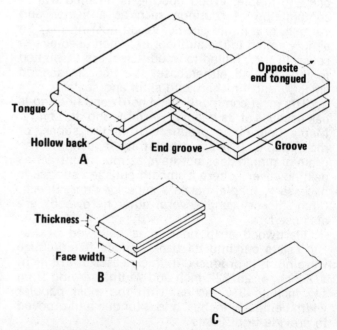

Strip flooring, A, side and end matched; B, side matched; C, square edged
Figure 8-2

Hardwood flooring is available in strip or block and is usually tongued and grooved and end matched, but it may be square-edged in thinner patterns. The most widely used pattern of hardwood strip flooring is 25/32 by 2¼ inches with hollow back. Strips are random lengths varying from 2 to 16 feet long. The face is slightly wider than the bottom so that tight joints result.

Softwood flooring is also available in strip or block. Strip flooring has tongued and grooved edges, and some types are also end matched. Softwood flooring costs less than most hardwood species, but is less wear resistant and shows surface abrasions more readily. However, it can be used in light traffic areas.

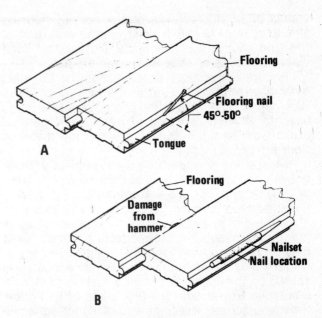

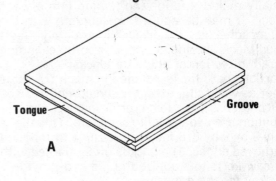

Nailing of flooring; A, angle of nailing; B, setting the nail without damage to the flooring
Figure 8-4

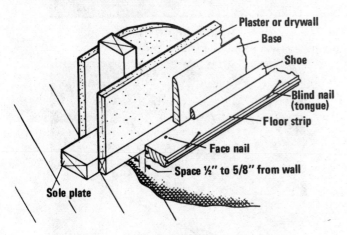

Installation of first strip of flooring
Figure 8-3

Bundles of flooring should be broken and kept in a heated space until the moisture content common to the interior finish in the locale is achieved.

Strip flooring is normally laid crosswise to the floor joists; however, when laid over old strip flooring, it should be laid crosswise to the existing flooring. Nail sizes and types vary with the thickness of the flooring. For 25/32 inch flooring use eightpenny flooring nails; use sixpenny flooring nails for ½ inch flooring; and use fourpenny casing nails for 3/8 inch flooring. Other nails, such as the ring-shank and screw-shank types, can be used, but it is well to check the flooring manufacturer's recommendations on size and diameter for specific uses. Flooring brads with blunted points which prevent splitting of the tongue are also available.

Begin installing matched flooring by placing the first strip 1/2 to 5/8 inch away from the wall to allow for expansion when the moisture content increases. Nail straight down through the board near the grooved edge (Figure 8-3). The nail should be close enough to the wall to be covered by the base or shoe molding and should be driven into a joist when the flooring is laid crosswise to the joists. The tongue should also be nailed, and

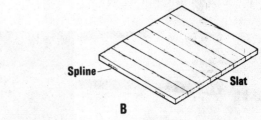

Two types of wood block flooring
Figure 8-5

consecutive flooring boards should be nailed through the tongue only. Nails are driven into the tongue at an angle of 45° to 50° and are not driven quite flush, to prevent damaging the edge by the hammer head (Figure 8-4). The nail is then set with the end of a large nail set or by laying the nail set flatwise against the flooring. Contractors use nailing devices designed especially for nailing flooring. These drive and set the nail in one operation. Select lengths of flooring boards so that butts will be well separated in adjacent courses. Drive each board tightly against the one previously installed. Crooked boards should be forced into alignment or cut off and used at the ends of a course or in closets. The last course of flooring should be left 1/2 to 5/8 inch from the wall, just as the first course was. Face-nail it near the edge where the base or shoe will cover the nail.

Square-edged strip flooring must be installed over a substantial subfloor and can only be face-nailed. The installation procedures relative to spacing at walls, spacing of joints, and general attachment are the same as those for matched flooring.

Most wood or wood-base tile is applied with an adhesive to a smooth base such as underlayment or a finished concrete floor with a properly installed vapor barrier. Wood tile may be made up of a number of narrow slats held together by a membrane, cleats, or tape to form a square, or it may be plywood with tongued and grooved edges (Figure 8-5). To install wood tile, an adhesive is spread on the concrete slab or underlayment with a notched trowel, and the tile is laid in it. Follow the manufacturer's recommendations for adhesive and method of application.

Wood block flooring usually has tongues on two edges and grooves in the other two edges and is usually nailed through the tongue into a wood subfloor. It may be applied on concrete with the use of an adhesive. Shrinkage and swelling effects of wood block flooring are minimized by changing the grain direction of alternate blocks.

Particleboard tile is installed in much the same manner as wood tile, except it should not be used over concrete. Manufacturer's instructions for installation are usually quite complete. This tile is usually 9 by 9 by 3/8 inches in size, with tongued and grooved edges. The back is often marked with small saw kerfs to stabilize the tile and provide a better key for the adhesive.

**Approximate fitting
Figure 8-7**

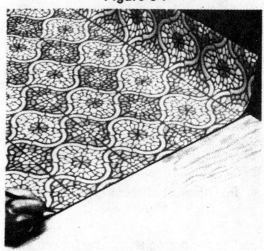

**Drawing the seam line
Figure 8-8**

**Unrolling the vinyl
Figure 8-6**

Sheet Vinyl Flooring

Sheet vinyl with resilient backing smooths out minor surface imperfections. Most vinyl will lay flat, so no adhesive is required. Double-faced tape is used at joints and around the edge to keep the covering from moving. Most sheet vinyls are available in widths of 6, 9, 12, and 15 feet, so that complete rooms can be covered with a minimum of splicing. This permits a fast, easy installation. The material is merely cut to room size, using scissors, and is then taped down. Sheet vinyl also has the ability to compensate for mistakes in application. If the price is cut a bit too loosely, bulges or wrinkles will gradually disappear as the product gently contracts. If slightly undercut, the vinyl floor can be stretched. After cleaning the original floor and applying latex fill over uneven areas, unroll the material in the room and cut the flooring 2 inches longer than required (Figure 8-6 and 8-7). If one sheet will not cover the entire room, draw a pencil line along the straight, factory-cut edge of the first piece from one wall to the other. This marks the position where the two pieces of flooring will be seamed (Figure 8-8). Fold back and stick 1½ inch double-faced carpet tape to the subfloor along the inside of the pencil line, stopping 5

**Carpet tape at seam line
Figure 8-9**

**Fitting second piece
Figure 8-10**

force to cut through the tape, too. After removing both strips of scrap created by the pattern-matching cut, fold back a section at a time and, in crisscross fashion (Figure 8-14) peel off the paper backing of the carpet tape. As you proceed, press flooring onto the tape. This holds the two pieces of floor together until the seam is cemented and the cement dries.

**Matching the pieces
Figure 8-11**

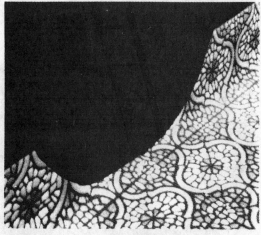

**Cutout at seam
Figure 8-12**

inches from each wall. Note Figure 8-9. Do not remove the tape's paper backing yet. After the tape is down, return the folded-back portion to its former position. Next, unroll the second piece and maneuver it into position as in Figure 8-10. Take care not to move the first piece as this is being done. Overlap the two pieces where they are to be seamed. Match up the pattern along the entire width. If possible, use one of the simulated grout lines as the match point (Figure 8-11). Then, when the seam is cut, the two pieces of flooring will join at a grout line.

With a utility knife, cut and remove a wide U-shaped section from the excess floor where the seam area extends up the wall at each side of the room. The bottom of the U should be at the juncture of the wall and subfloor (plywood, in Figure 8-12). This partial fitting allows the seam area to lie flat. Next, as illustrated in Figure 8-13, slice through both pieces where they match and overlap, guiding the utility knife with a metal straightedge or carpenter's square. Apply enough

Next, apply a bead of special seam sealing cement to the surface of the seam (Figure 8-15). Move the bottle from one end of the seam to the other as the cement is squeezed out. When the clear cement dries, it will match the gloss level of the floor surface. Another strip of carpet tape (Figure 8-16) placed across the two pieces of floor at both sides of the room will further help prevent the fresh seam from tearing until the cement ''sets up.'' When the cement hardens, it forms a very

115

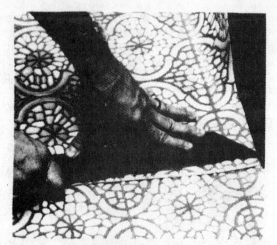

Cutting both pieces
Figure 8-13

Exposing tape adhesive
Figure 8-14

Sealing the seam
Figure 8-15

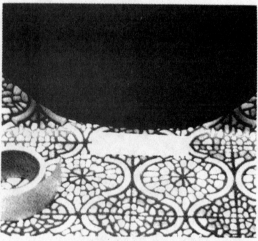

Taping the sealer
Figure 8-16

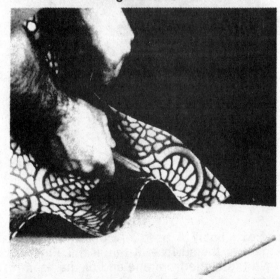

Cutting the corner
Figure 8-17

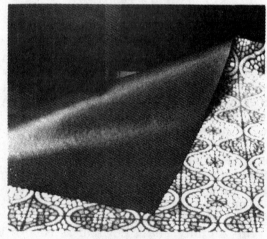

Folding back at corner
Figure 8-18

tight bond — and a very strong seam. The cement, not the tape, is what gives the seam a permanent weld.

The next steps are the same for a one-piece installation as for a two-piece installation requiring seaming. Trimming and fitting the floor start with cutting the inside corners. The cut is made upward from the base of the corner. See Figure 8-17. This allows the excess materials to flash

Cutting corner
Figure 8-19

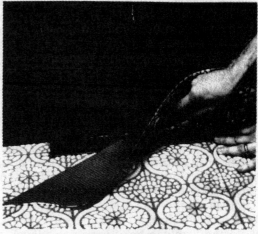

Trimming edge
Figure 8-20

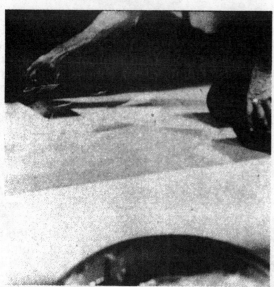

Applying full floor adhesive
Figure 8-21

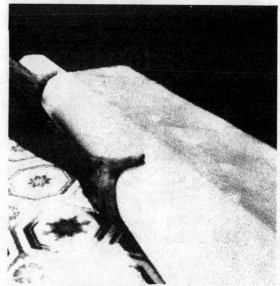

Rolling back over adhesive
Figure 8-22

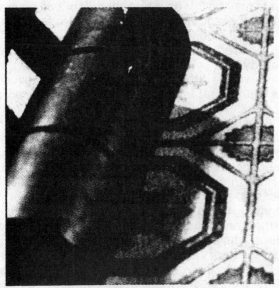

Bonding vinyl to adhesive
Figure 8-23

Cementing an edge
Figure 8-24

117

neatly up the walls. Outside corners are fitted by first folding back the floor diagonally at the base of the corner (Figure 8-18) and then cutting from the base upward as in Figure 8-19, awaiting the final trim. Pressing a straightedge or square into the joint where the wall and floor meet, cut away the excess material with the utility knife. Note Figure 8-20. Follow this procedure around the perimeter of the room until the flooring lies flat and snug and the entire floor has been trimmed to size. If full floor adhesive is recommended for the vinyl you are using, roll back half of one sheet and apply the adhesive as in Figure 8-21. Then carefully roll the material back into place being careful to work out air pockets (Figure 8-22). Finally, use a 3-section, 100 pound roller in both directions to eliminate air pockets and firmly imbed flooring to adhesive. See Figure 8-23.

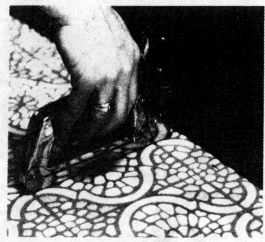

Stapling an edge
Figure 8-25

Some types of vinyl are laid loose and only stapled or cemented around the edges. Use staples where possible and cement only in inaccessible areas. The edges that are to be cemented are put down first. Fold back the vinyl and apply a narrow band of cement to the subfloor as shown in Figure 8-24. Firmly press the flooring down into the fresh cement. Fasten the other edges with a heavy duty staple gun loaded with staples no smaller than 3/8 inch or larger than 9/16 inch. See Figure 8-25. Space the staples at 3 inch intervals. This completes the installation. The staples will be hidden by the molding.

Liquid Seamless Flooring
Seamless flooring consisting of resin chips combined with a urethane binder can be applied over any stable base, including old floor tile. This is applied as a liquid in several coats with drying between coats. Complete application may take from ½ to 2 days depending on the brand used. Manufacturer's instructions for application are quite complete. This floor covering is easily renewed by additional coatings, and damaged

spots are easily patched by adding more chips and binder.

Carpeting
Carpeting also lends itself well to rehabilitation. It can be installed over almost any flooring that is level, relatively smooth, and free from major surface imperfections. Carpeting is now available for all rooms in the house, including the kitchen. A very close weave is used for kitchen carpeting so that spills stay on the surface and are easily wiped up. The cost of carpeting may be two or three times that of a finished wood floor, and the life of the carpeting before replacement would be much less than that of a wood floor. However, carpeting requires less maintenance and has the advantages of sound absorption and resistance to impact.

Linoleum and Tile
Linoleum and resilient tile both require a smooth underlayment or a smooth concrete slab to which they are bonded with an adhesive. Regardless of the type of floor cover, the surface must be prepared carefully to keep the flooring from pealing up and cracking. If the flooring is to be laid over old concrete, the surface should be prepared carefully. Cracks, expansion joints, uneven and rough areas require the application of a good quality latex underlayment to level. The underfloor must be firm and free from moisture, dust, solvent, scaly paints, wax, oil, grease, asphalt, sealing compounds, and other extraneous foreign materials.

Paints should be removed by sanding the floor until clean. Use coarse Number 4 or Number 5 open grit sandpaper. A strong solution of tri sodium phosphate or lye may be required in difficult cases. If alkaline solutions are used, the floor must be neutralized with acid as described below. Good quality chlorinated rubber base paints may be left on suspended floors or on grade or below grade floors if the paint film is found to be securely bonded to the subfloor. To test for chlorinated rubber base paint, apply a small quantity of a solution of 2 tablespoons of lye to a cup of water. If the paint has not been removed by this solution after four hours, it may be left on the floor.

Concrete floors which have been treated with alkali must be neutralized before installation of tile. One part of muriatic acid and nine parts of water makes a satisfactory neutralizing solution. During this process the use of rubber gloves, galoshes and protection for the eyes is recommended. Flood the floor with the neutralizing solution and allow it to remain at least one hour before rinsing off with clear water. Allow the concrete to dry thoroughly.

Concrete slabs in contact with the earth at any point, or those which do not have at least 18 inches of cross ventilated air space underneath require special attention. Unless such construction in-

118

corporates a continuously effective permanent barrier, bonding failures may result. If a permanent moisture barrier (such as six mil polyethylene film or equivalent) is used which is sufficiently effective to restrict transmission of moisture through the floor, no special flooring adhesive is needed. If such a barrier is not used, an epoxy adhesive should be used or the flooring should be laid over a special asbestos felt material designed for this purpose. Linoleum should not be used to cover concrete slabs below grade or in contact with the ground. The requirements which apply to suspended or on grade concrete floors should be observed with floors over radiant heating systems.

If a wood subfloor is used, it must be dry, smooth, free of vertical movement or horizontal expansion, and contain no moisture, oil, grease, dirt, or waxes to impair the adhesive bond. Wood floors should be of sufficient strength to carry the intended loads without deflection. They should also have at least 18 inches of cross ventilated air space between the underside of the joists and the ground. Under no circumstances should sheet materials or tile be laid over wood subfloors which are subject to conditions which might cause buckling or rotting of the wood. This condition occurs generally on wood floors that are below grade, on grade or above the ground without heat or adequate ventilation underneath the floors.

Where an underlayment is required, it is important to use the correct plywood grade and surface. The following grades have a solid surface with the special inner-ply construction needed for direct application of resilient floorings: Underlayment grade ¼ inch hardboard (Masonite or equal), UNDERLAYMENT INT DFPA, UNDERLAYMENT INT DFPA with Exterior Glue, UNDERLAYMENT EXT DFPA and 2.4.1 both INT DFPA and with Exterior glue. Panels marked as UNDERLAYMENT are made with a touch sanded or fully sanded surface. When thin resilient surfaces are to be used, only fully sanded panels should be installed. Where floors may be subject to unusual moisture conditions as in bathrooms, closets, and under counters, sinks, etc. only panels with exterior glue are recommended.

Any leveling or filling on wood subfloors should be done with a good quality latex underlayment. Plaster patching compounds are not suitable, since they disintegrate under traffic.

Loose or broken boards should be renailed or replaced. Fill all cracks with latex underlayment and sand the floor level. Use 15 pound asphalt saturated felt paper or asbestos felt cemented in place with linoleum paste. Lay the felt paper across the boards, butt the edges (do not overlap the felt) and roll with a 150 pound roller from the center to the edges to insure a good bond and to eliminate air bubbles. Some types of vinyl tile need special lining felt underlayments. The tile manufacturer will make a specific recommendation.

Locating room center
Figure 8-26

To lay tile, locate the center of the room by finding the center of each wall. Then pull a chalked string taut to the opposite wall across the center of the room, and snap a straight line on the subfloor. See Figure 8-26. This allows the room to be "squared off," so that the tile is laid parallel to the walls. Next, lay a row of loose tiles along the chalkline from the center point to one side wall and one end wall. Measure the distance between the wall and the last full tile. If this space is less than a half a tile wide, snap a new chalkline and move half a width of tile closer to the opposite wall. Check the right angles. Now, do the same with the other row. Note Figure 8-27. This will improve the appearance of the floor and eliminate the need to fit small pieces of tile next to the walls.

Adhesive should be brushed, troweled, or rolled on thinly, (Figure 8-28) so that when the tile is laid, the adhesive will not push up between the tiles or cause them to slip underfoot. If a "shelf-stick" or adhesive-backed tile is used, this step is not necessary.

Resetting chalk lines
Figure 8-27

One quarter section of the room is laid at a time, (Figure 8-29) starting at the center point, which has been marked with the chalk string, and moving toward the walls. Each tile should be set down firmly and tightly to the adjoining tile, so that there are no "joints" between the tiles. Do not slide the tiles into place, as this may cause the adhesive to come up between the tiles. To cut and fit the tile next to the wall, place a loose tile (A,

Rolling on adhesive
Figure 8-28

Laying each quarter
Figure 8-29

Figure 8-30), squarely on top of the last full tile closest to the wall. On top of this, place a third tile (B) and slide it until it butts against the wall. Using the edge of the top tile as a guide, mark the tile under it with a pencil. With a pair of household shears, cut tile (A) along the pencil line. To fit around pipes as in Figure 8-31, make a paper pattern to fit the space exactly. Then trace the outline onto the tile and cut with the shears. Insert tile into the border space with the rough edge against the wall.

Many types of resilient flooring can be installed over an existing resilient floor if the

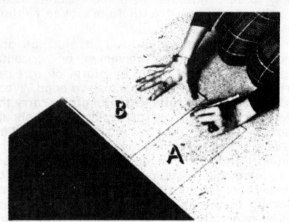

Fitting border tile
Figure 8-30

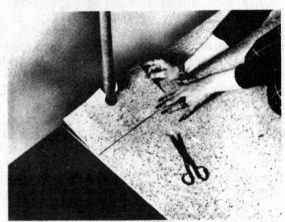

Fitting around an obstruction
Figure 8-31

surface is reasonably flat, smooth and completely bonded to the subfloor. Consult a flooring material dealer about the type of adhesive to use. Generally, flooring will not bond to asphalt tile, resilient flooring installed below grade, cushioned sheet vinyl or urethane coated floors. Before installing the new floor, be certain that all wax and other floor finishes are completely removed by sanding or stripping. Open seams or other indentations must be repaired prior to installation.

Chapter 9
Siding and Soffits

Installing new siding can do more for the exterior of a house than nearly any other home improvement project. The exterior wall cover says something about the owner of a house and many wise homeowners recognize the importance of keeping the exterior of their home in good condition. Most important from a sales standpoint, replacing the exterior cover of a home is usually a much smaller expense than adding a bathroom or remodeling a kitchen.

Many different types of exterior wall covering are popular for home improvement. Wood siding is probably the most common. Wood siding includes cedar, plywood, redwood, lap siding, shingles and shakes. Advantages are that wood is a natural material and an excellent insulator. It is easy to work with and can be installed without specialized tools. Wood siding is available in most parts of the country and its in-place cost is generally low. Disadvantages lie primarily in the need for painting or staining every few years, higher fire insurance rates, and susceptibility to termites and weather.

Advantages of hardboard siding lie in its generally reasonable cost and availability. It is relatively easy to work with and, like wood siding, can be installed without specialized tools. Disadvantages closely follow those of wood siding. They include the need for paint, and susceptibility to termites and weather. Hardboard siding may require more maintenance through the years than most other materials. Some homeowners feel this is offset by the modest initial cost.

Aluminum siding is often less expensive than wood sidings, and aluminum is fire resistant, rot-proof, termite-proof and has a number of textures and colors readily available to builders. It is often sold and erected in packages which include aluminum gutters, downspouts and soffits. Disadvantages include the possibility of fading, denting, and the need for ground in case of an electrical storm. Also, some people feel that aluminum lacks the quality image of good quality wood siding.

Advantages of vinyl siding are somewhat similar to aluminum. However, vinyl siding won't fade or rust and tends to resist dents. Disadvantages are the possibility of cracking in low temperatures, and difficulty of installation. Also, the fire resistance of vinyl siding is less than that of aluminum or masonry. Nevertheless, this is a relatively new material which is constantly being improved. At a distance, it is a dead ringer for aluminum siding.

Builders and remodelers have a choice of a number of other finished sidings including steel, asphalt, and asbestos-cement. They are not nearly as widely used as wood and the major finished sidings. Their advantages primarily lie in the area of durability, while disadvantages include selection and aesthetics.

Stucco is a cement-lime material that may be applied over masonry (concrete block) or over a metal lath. Advantages are that the material is readily available, since it is generally prepared on site, and has a unique appearance. Stucco can have a rugged Mediterranean textured look or appear to be smooth as glass. And it can be prepared in a wide range of colors. Disadvantages of stucco are sensitivity to climatic changes, tendency to crack, and need for skilled craftsmen.

Wood Siding

The solution for wood siding problems often involves corrective measures in other components of the house. Failure of paint is frequently not the fault of the siding but can be attributed to moisture moving out through the wall or to water washing down the face of the wall. Corrective measures are discussed in the chapters on roofing and insulation. After adopting these corrective measures, siding may need only refinishing as discussed in the chapter on painting and finishing. Some of the "permanent" sidings that require no painting may cause serious difficulty in time by trapping moisture in the wall, thus creating a decay hazard. It is desirable, therefore, to have a siding that will let water vapor escape from inside the wall.

If new horizontal wood or nonwood siding is used, it will probably be best to remove old siding. Vertical board and panel-type siding may be successfully applied over old siding.

The main difficulty in applying new siding over existing siding is in adjusting the window and door trim to compensate for the added wall thickness. The window sills on most houses extend far enough beyond the siding so that new siding should not affect them; however, the casing may be nearly flush with the siding and require some type of extension. One method of extending the casing is by adding an additional trim member over the existing casing (Figure 9-1). When this is done, a wider drip cap may also be required. The drip cap could be replaced, or it could be reused with blocking behind it to hold it out from the wall a distance equal to the new siding thickness (Figure 9-2). Another method of extending the casing would be to add a trim member to the edge of the existing casing, perpendicular to the casing (Figure 9-3). A wider drip cap will also be required. Exterior door trim can be extended by the same technique used for the window trim.

Any of the conventional siding materials can be used for remodeling, but some may be better suited to this application than others. Panel-type siding is probably one of the simplest to install and one of the most versatile. It can be applied over most surfaces, and will help to smooth out unevenness in the existing walls.

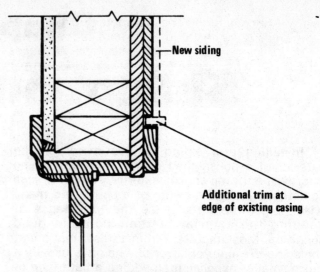

Top view of window casing extended for new siding by adding trim at the edge of existing casing
Figure 9-3

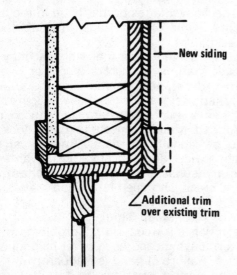

Top view of window casing extended by adding trim over existing trim
Figure 9-1

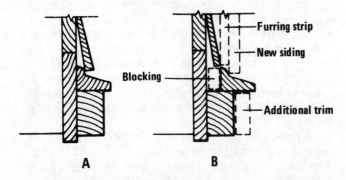

Change in drip cap with new siding: A, existing drip cap and trim; B, drip cap blocked out to extend beyond new siding and added trim
Figure 9-2

Panel Siding

Panel-type siding is available in plywood, hardboard, and particleboard, as well as numerous nonwood materials. The most popular of these are probably plywood and hardboard. Always specify exterior type for both, and the hardboard must be tempered. The grade of plywood depends on the quality of finished surface desired.

Plywood panel siding is available in a variety of textures and patterns. Sheets are 4 feet wide and are often available in lengths of 8, 9, and 10 feet. Rough textured plywood is particularly suited to finishing with water repellent preservative stains. Smooth surfaced plywood can be stained, but it will not absorb as much stain as rough textured plywood, and, therefore, the finish will not be as long lasting. Paper-overlaid plywood (called medium density overlaid or M.D.O. plywood) is particularly good for a paint finish. The paper overlay not only provides a very smooth surface, but also minimizes expansion and contraction due to moisture changes. Most textures can be purchased with vertical grooves. The most popular spacings of grooves are 2, 4, and 8 inches. Battens are often used with plain panels. They are nailed over each joint between panels and can be nailed over each stud to produce a board-and-batten effect. No building paper is required for plywood panel siding if joints are shiplap covered with battens. Building paper is required for unbattened square butt joints where panels are applied directly over the studs. No building paper is required when siding is applied over sheathing. In new construction, plywood applied directly over framing should be at least 3/8 inch thick for 16 inch stud spacing and ½ inch thick for 24 inch stud spacing. Grooved plywood is normally 5/8 inch thick with 3/8 by 1/4 inch deep grooves.

For installation over existing siding or sheathing, thinner plywood can be used; however, most of the available sidings will be in the thicknesses listed above. Use hot-dip galvanized or aluminum nails to prevent staining of the siding from nail weathering and rusting. 6d box, siding, or casing nails are recommended for 3/8 inch or 1/2 inch panels. Use 8d nails for thicker panels or when applying panels over existing siding. Drive nails every 6 inches along the perimeter of the panel and every 12 inches at intermediate supports. See Figure 9-4 for nailing shiplapped panels. All edges of panel siding must be backed by solid lumber framing or blocking. Unfinished panels with shiplap joints should be treated along the joint with water repellent preservative before installation. Square-edge butt joints between plywood panels should be caulked with a sealant with the plywood nailed at each side of the joint. Caulk butt joints at all inside and outside wall corners, using any of the various high performance polyurethane, thiokol, or silicone caulks. Caulk around all windows and doors.

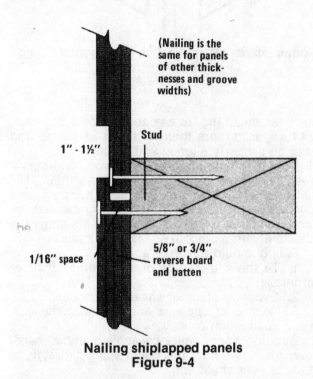

Nailing shiplapped panels
Figure 9-4

Where battens are used over the joint and at intermediate studs, nail them with eightpenny galvanized nails spaced 12 inches apart. Longer nails may be required where thick existing siding or sheathing must be penetrated. Nominal 1 by 2 inch battens are commonly used.

Lap the corner trim strip over the siding joint as shown in Figure 9-5, so that there is no continuous joint through the corner trim and siding. Where siding is applied with a face grain across the studs, apply battens to conceal the vertical butt joints at the panel ends. Block behind horizontal joints. Nails must penetrate through the battens and at least one inch into the studs.

Remember to leave a 1/16 inch space between all ends and edges of panels. This is necessary to ensure that panels will stay flat under all weather conditions.

If existing siding on gable ends is flush with the siding below the gable, some adjustment will be required in applying panel siding in order to have a new siding at the gable extend over the

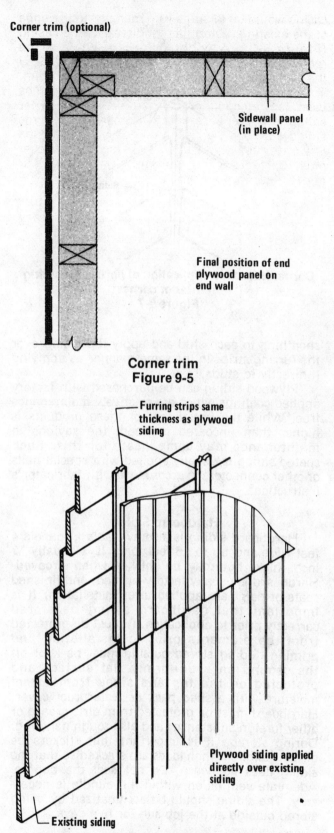

Corner trim
Figure 9-5

Application of plywood siding at gable end
Figure 9-6

siding below. This is accomplished by using furring strips on the gable (Figure 9-6). Furring must be the same thickness as the new siding applied below. Nail a furring strip over siding or

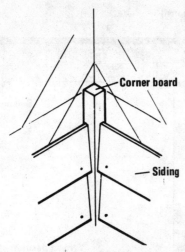

Corner board for application of horizontal siding at interior corner
Figure 9-7

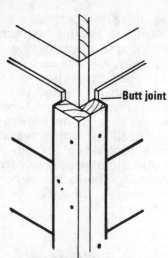

Corner boards for application of horizontal siding at exterior corner
Figure 9-8

sheathing to each stud and apply the siding over the furring strips in the same manner as applying it directly to studs.

Plywood siding can be purchased with factory applied coatings which are relatively maintenance free. While the initial cost of these products is higher than uncoated plywood, the savings in maintenance may compensate for this. Such coated siding is usually applied with special nails or other connectors in accordance with fabricator's instructions.

Hardboard Siding

Hardboard siding is also available in panels 4 feet wide and up to 16 feet long. It is usually ¼ inch thick, but may be thicker when grooved. Hardboard is usually factory-primed, and finished coats of paint are applied after installation. It is important that hardboard siding be stored correctly prior to application. It must be protected from the elements prior to installation and priming. Siding stored outside must be kept off the ground on a uniformly flat surface and protected on the top and sides from direct moisture with a shed pack or waterproof cover. Hardboard must be protected from dirt, grease or other foreign substances and also rough handling. During storage it is important that stickers be directly aligned when loads are stacked so that the siding does not warp. Do not seal the bundle. Adequate ventilation within the bundle is necessary. The siding should be acclimatized by being stored outside at the job site for five days.

Hardboard Lap Siding

For the soundest installation hardboard lap siding should be applied over standard sheathing material. Where the local building codes permit, the siding may be applied directly over studs. Some waviness may appear in the installed siding but such waviness will be minimized by installa-

tion over sheathing. In any application studs must not be spaced more than 16 inches on center and must be properly aligned and braced.

Primed aluminum "H" expansion moldings or plastic spacers are available for butt joints. "H" molding will hide the natural expansion and contraction of hardboard lap siding caused by normal changes in the weather. "H" moldings are available with a smooth or textured surface.

Siding should never be applied:

1. To the 4 inch side of studs other than at corners.

2. Over wet studs or sheathing.

3. Next to or opposite an uncured cement or stucco foundation or wall.

Building paper must be applied under hardboard siding whether it is being applied directly to studs or over sheathing.

Corners are finished by butting the panel siding against corner boards as shown for horizontal sidings (Figures 9-7 and 9-8).

Use a 1-1/8 by 1-1/8 inch corner board at interior corners and 1-1/8 inch by 1-1/2 and 2-1/2 inch boards at outside corners. Apply caulking wherever siding butts against corner boards, window or door casings, and trim boards at gable ends. See Figure 9-9.

Use aluminum or galvanized siding or box nails. Use 8d (2½ inch) nails when siding is applied directly to the studs. Use 10d (3 inch) nails when applied over sheathing. Panels should be nailed 4 inches along the perimeter and 3/8 inch from the outside edge. Nails should be spaced 8 inches on center along intermediate support studs. Nails must penetrate at least 1 inch into studs which must be spaced not more than 16 inches on center.

Hardboard Panel Siding Over Old Siding

Hardboard panel siding is available in a number of designs, textures and finishes and is an

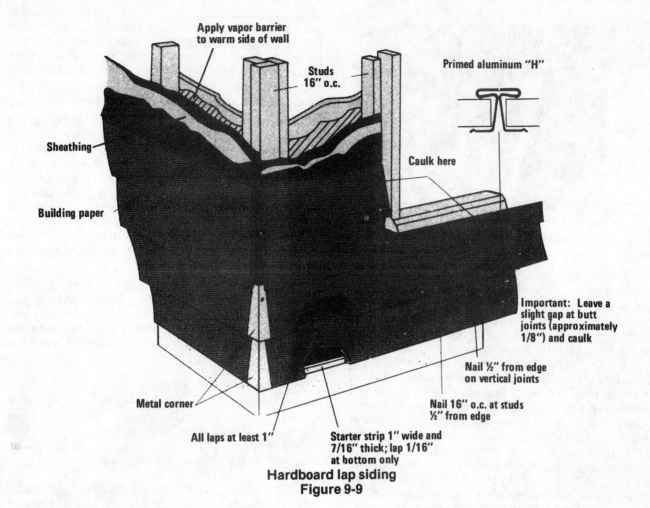

Apply vapor barrier
to warm side of wall

Studs
16" o.c.

Primed aluminum "H"

Sheathing

Caulk here

Building paper

Important: Leave a
slight gap at butt
joints (approximately
1/8") and caulk

Nail ½" from edge
on vertical joints

Nail 16" o.c. at studs
½" from edge

Metal corner

All laps at least 1"

Starter strip 1" wide and
7/16" thick; lap 1/16"
at bottom only

Hardboard lap siding
Figure 9-9

excellent material to use over existing outdated and worn siding.

As much of the work as possible should be completed at ground level. For instance, the panels should receive an edge coat of paint, applied by roller, while still stacked. Then they can be separated and each face can be rolled with an initial coat of paint. To avoid constant repositioning of ladders and scaffolding, the work should be divided into two or three panel sections, and each section should be completed before the next is started.

A work station consisting of a pair of saw horses with a couple of 2x4's laid across from one to the other can be set up at one corner of the home to provide a convenient work table (Figure 9-10).

All the original siding which is split or warped should be nailed back into position. See Figure 9-11. Then building paper should be applied over the old siding. Place the top of the ladder against the top of the paper to hold it while fastening the lower portion with a staple gun. See Figure 9-12.

The ground course is applied first. To assure

Setting up work table
Figure 9-10

Refastening loose siding
Figure 9-11

Holding and nailing paper
Figure 9-12

perfect alignment and avoid repetitive measuring, tack a long 1x4 temporarily along the house as a guide where the top of each first-floor panel would come. Siding is then butted up against this stop to insure an even line. As each panel is prepared for installation, several nails should be started in it while it is on the ground. Then the panel is butted against the 1x4 and fastened with 8 penny galvanized nails, placed approximately 6 inches on center around all edges and 12 inches on center across the face. A short piece of 2x4, with a rock as a fulcrum, can serve as a foot lever to hold the panel up against the 1x4 while it is nailed into position. Wherever possible, all nails should be placed through the textured face rather than in battens. The rough texture will hide any stray hammer marks and small nail heads will disappear in the texture. When trimming and cutting to size, place the siding face down on the 2x4's between the saw horses. Mark accurate measurements on the smooth back of the panel. Cut up through the back of the face. To simplify fitting the new siding over pipes and meter boxes, exterior hardware such as water faucet handles should be removed, chalk applied to protruding edges and the siding panel fitted and pressed against the chalked area, then removed. This will transfer the outline to the

Installing new window trim
Figure 9-13

back of the panel, which can then be drilled or cut out with a sabre saw along the accurately positioned guidelines. Brackets holding downspouts and TV lead-in wires should be loosened, then the siding slid behind the items, fastened, and the brackets re-applied to the new surface.

Metal "Z" flashing should be applied to the top of the first course to provide a waterproof joint between the top and bottom panels. Metal flashing also should be installed above windows and doors. New window trim will usually be required after the siding is installed because the new siding will project beyond the original shallow trim. See Figure 9-13. Rabbeted stock, 1¼ x 2 inches can serve as the new trim. Nail heads should be set in the trim and all joints caulked before painting.

For the upper course of siding panels, temporary wooden scaffolding with sturdy diagonal bracing should be erected. The scaffolding can extend along the full side of the house, with the planking moved as necessary. Note Figure 9-14.

Scaffold at second story level
Figure 9-14

Vinyl Siding

Most vinyl siding manufacturers make horizontal siding in two styles, one with an 8 inch exposure, and one with two 4 inch exposures per panel of siding, called "double-four." A panel of siding is typically 12 feet 6 inches long. Vinyl siding comes in both horizontal and vertical styles. When re-siding a two or three story home, particularly when it is boxy, use the wider clapboard because it gives fewer shadow-lines to the house, lessening its height visually. For a ranch style home, the double-four clap-board or vertical style is more attractive because they double the number of shadow-lines, making the house look taller and larger. Vertical siding is also good for gables or accent areas on horizontally sided homes. Vinyl, like most sidings, comes in a variety of colors, some having woodgrains. Figures 9-15 to 9-36 show the application procedure over existing siding.

126

Make sure the sidewall is as straight as possible
Figure 9-15

Install furring to shim out the wall at low spots, achieving an even base for nailing
Figure 9-16

Install furring strips if the surface is very uneven
Figure 9-17

Begin by making a chalk line for the starter strip, parallel to foundation at lowest point of house
Figure 9-18

Install the special starter strip, fastening eight inches on center
Figure 9-19

Nail corner post 16'' on center. Do not nail too tightly
Figure 9-20

A joint is necessary where the building height requires an additional length of corner post. Nail through the vinyl side flange within two inches of the joint, above and below. Nail the remainder of the post 16'' on center
Figure 9-21

Slip the lock of first panel into the starter strip and snap into place
Figure 9-22

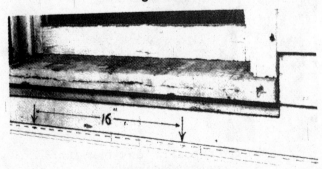

Nail every 16'' on center
Figure 9-23

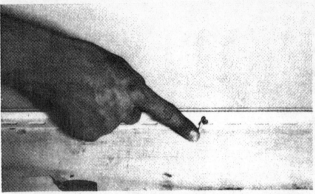

Do not nail tightly. You should be able to slide a paper match between the nailhead and siding
Figure 9-24

At vertical joints, nail six to eight inches away from the edge of the underlap panel
Figure 9-25

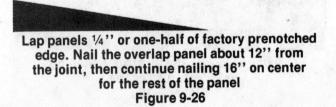

Lap panels ¼'' or one-half of factory prenotched edge. Nail the overlap panel about 12'' from the joint, then continue nailing 16'' on center for the rest of the panel
Figure 9-26

Laps should be staggered for best appearance
Figure 9-27

Fit vinyl window trim under window sill
Figure 9-28

Cut panel under window to fit around sill
and into window trim
Figure 9-29

Trim the sides of the window with
J-channel trim
Figure 9-30

Install drip cap at the window tops
Figure 9-31

Panels can be bowed to simplify fitting them into
place between corner post and window trim. Cut
siding ¼'' short to allow for expansion
Figure 9-32

Before hanging last course, install finishing trim
Figure 9-33

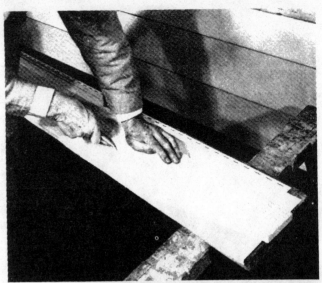

Trim last course to size, cutting off nailing strip
Figure 9-34

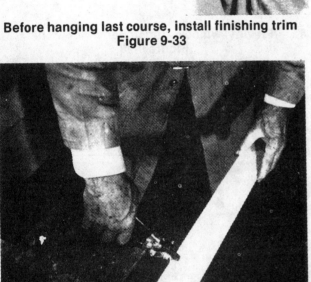

Crimp top of last panel with snap-lock tool
Figure 9-35

Hook last panel into lower course, and insert
crimped-edge panel top into finishing trim
Figure 9-36

Horizontal Board Siding

Bevel siding has been one of the most popular sidings for many years. It is available in 4 to 12 inch widths. The sawn face is exposed where a rough texture is desired and a stain finish is planned. The smooth face can be exposed for either paint or stain. Siding boards should have a minimum of 1 inch horizontal lap. In application, the exposed face should be adjusted so that the butt edges coincide with the bottom of the sill and the top of the drip cap of window frames (Figure 9-37).

Horizontal siding must be applied over a smooth surface. If the old siding is left on, it should either be covered with panel sheathing or have furring strips nailed over each stud. Nail siding at each stud with a galvanized siding nail or other corrosion resistant nail. Use sixpenny nails for siding less than ½ inch thick and eightpenny nails for thicker siding. Locate the nail to clear the top edge of the siding course below. Butt joints should be made over a stud. Interior corners are finished by butting the siding against a corner board 1-1/8 inches square or larger depending on thickness of siding (Figure 9-7). Exterior corners can be mitered, butted against corner boards 1-1/8 inches or thicker and 1½ and 2½ inches wide (Figure 9-8), or covered with metal corners.

Vertical Wood Siding

Vertical siding is available in a variety of patterns. Probably the most popular is matched (tongued and grooved) boards. Vertical siding can be nailed to 1 inch sheathing boards or to 5/8 and

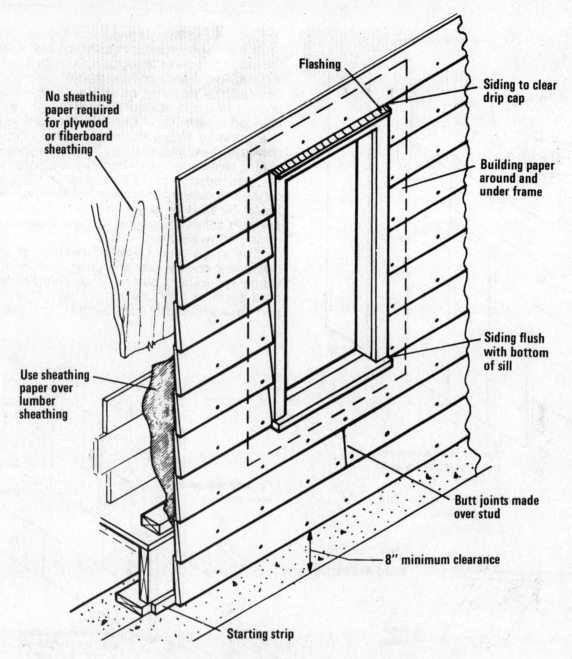

Flashing

Siding to clear drip cap

No sheathing paper required for plywood or fiberboard sheathing

Building paper around and under frame

Use sheathing paper over lumber sheathing

Siding flush with bottom of sill

Butt joints made over stud

8" minimum clearance

Starting strip

Application of bevel siding to coincide with window sill and drip cap
Figure 9-37

3/4 inch plywood. Furring strips must be used over thinner plywood because the plywood itself will not have sufficient nail holding capacity. When the existing sheathing is thinner than 5/8 inch, apply 1 by 4 inch nailers horizontally, spaced 16 to 24 inches apart vertically. Then nail the vertical siding to the nailers. Blind-nail through the tongue at each nailer with galvanized sevenpenny finish nails. When boards are nominal 6 inches or wider, also face-nail at midwidth with an eightpenny galvanized nail (Figure 9-38). Vertical siding can be applied over existing siding by nailing through the siding into the sheathing.

Another popular vertical siding is comprised of various combinations of boards and battens. It also must be nailed to a thick sheathing or to horizontal nailers. The first board or batten should be nailed with one galvanized eightpenny nail at center or, for wide boards, two nails spaced 1 inch each side of center. Close spacing is important to prevent splitting if the boards shrink. The top board or batten is then nailed with twelvepenny nails, being careful to miss the underboard and nail only through the space between adjacent boards (Figure 9-39). Use only corrosion resistant nails. Galvanized nails are not recommended for some materials, so be sure to follow siding manufacturer's instructions.

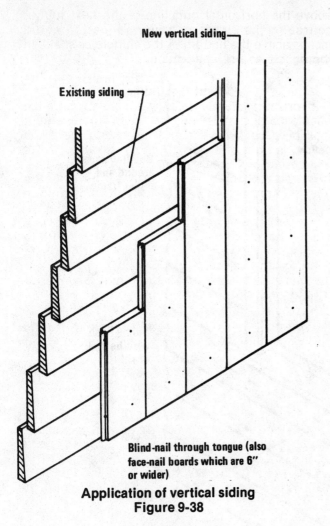

Application of vertical siding
Figure 9-38

Blind-nail through tongue (also
face-nail boards which are 6"
or wider)

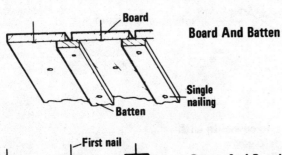

Board And Batten

Board

Single
nailing

Batten

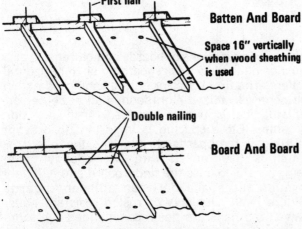

First nail

Batten And Board

Space 16" vertically
when wood sheathing
is used

Double nailing

Board And Board

Application of vertical wood siding
Figure 9-39

Wood Shingle and Shake Siding

Some architectural styles may be well suited to the use of shakes or shingles for siding. They give a rustic appearance and can be left unfinished, if desired, to weather naturally. They may be applied in single or double courses over wood or plywood sheathing. Where shingles are applied over existing siding which is uneven or over a nonwood sheathing, use 1 by 3 or 1 by 4 inch wood nailing strips applied horizontally as a base for the shingles. Spacing of the nailing strips will depend on the length and exposure of the shingles. Apply the shingles with about 1/8 to 1/4 inch space between adjacent shingles to allow for expansion during rainy weather.

The single-course method consists of simply laying one course over the other, similar to lap siding application. Second-grade shingles can be used because only one half or less of the butt portion is exposed (Figure 9-40).

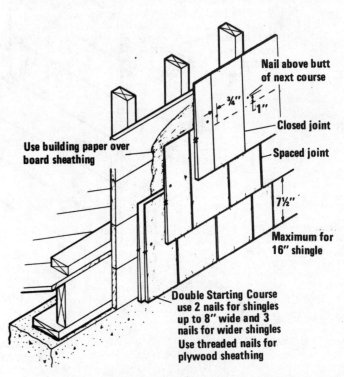

Nail above butt
of next course

¾"
1"

Closed joint

Use building paper over
board sheathing

Spaced joint

7½"

Maximum for
16" shingle

Double Starting Course
use 2 nails for shingles
up to 8" wide and 3
nails for wider shingles
Use threaded nails for
plywood sheathing

Single-course application of shingle siding
Figure 9-40

The double-course method of laying shingles consists of applying an undercourse and nailing a top course directly over it with a ¼ to ½ inch projection of the butt over the lower course shingle (Figure 9-41). With this system, less lap is used between courses. The undercourse shingles can be lower quality, such as third grade or the undercourse grade. The top course should be first grade because of the shingle length exposed. Recommended exposure distances for shingles and shakes are given in Table 9-42. Regardless of the method of applying the shingles, all joints must be broken so that the vertical butt joints of the upper shingles are at least 1½ inches from the undershingle joint.

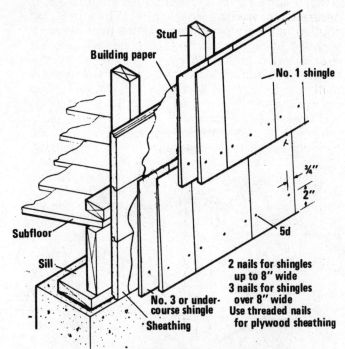

Double-course application of shingle siding
Figure 9-41

Material	Length Inch	Maximum Exposure		
		Single Coursing Inch	Double Coursing	
			No. 1 Grade Inch	No. 2 Grade Inch
Shingles	16	7½	12	10
	18	8½	14	11
	24	11½	16	14
Shakes (handsplit and resawn)	18	8½	13	--
	24	11½	20	--
	32	15	--	--

Exposure distances for wood shingles and
shakes on sidewalls
Table 9-42

Use rust resistant nails for all shingle applications. Shingles up to 8 inches wide should be nailed with two nails. Wider shingles should be secured with three nails. Threepenny or fourpenny zinc-coated ''shingle'' nails are commonly used in single-coursing. Zinc-coated nails with small flat heads are commonly used for double-coursing where nails are exposed. Use fivepenny for the top course and threepenny or fourpenny for the undercourse. When plywood sheathing less than ¾ inch thick is used, threaded nails are required to obtain sufficient holding power. Nails should be ¾ inch from the edge. They should be 1 inch

above the horizontal butt line of the next higher course in the single-course application and 2 inches above the bottom of the shingle or shake in the double-course application.

Aluminum Siding

Horizontal aluminum siding is one of the most popular siding materials for home improvement jobs because it covers existing siding materials easily, is light and easy to apply, and has a life expectancy of 20 years or more in most situations. It is practically maintenance free and never needs paint. You can cover all side walls with aluminum siding, but you may want to recommend that an attractive entrance with brick veneer or wood siding be preserved as is, with only a little refinishing or repainting. Several designs and many colors are available, so select colors and designs that go well with the shutters, exterior trim, or roof of the home to be resided.

Aluminum siding and accessories are shown in Figure 9-43. The most common siding in use is the horizontal with an 8 inch exposure. Double 4 and double 5 inch are also available. Vertical siding is similar in appearance to board and batten or V-groove siding and is used for emphasis, for gable ends, and for breaking up the long horizontal sweep that is so common to homes. Some styles of homes are sided entirely in vertical siding. It is available in 8 inch, 12 inch and 16 inch widths. The inside corner post is a siding accessory used to finish an inside corner. The post accepts siding from two directions and locks it neatly together. The outside corner post performs the same function on an outside corner. The starter strip engages the first panel of siding. Window and door casing effectively covers weathered and damaged wood and gives a uniform, low maintenance finished job. Various shaped moldings can be bought or field formed to gain a finished look at building breaks other than corners. Outside corner cap completes the junction of panels of siding. They are applied one at a time rather than an entire wall at a time as in the use of outside corner posts. Outside corner caps are used with 8 inch horizontal siding. Back-up strip or backer is used under ends of 8 inch horizontal siding to provide stiffening of each panel. The use of back-up strip helps give the siding job an even, smooth appearance.

The quality of the finished job depends on good preparation of the work surface. Use your chalk line, the edge of a piece of siding or any straight edge to locate low places in the plane of the wall. Shim these out to provide as straight a wall as possible. Nail all boards and repair all loose edges. Back up and build out all bad places to provide good nailing surfaces. The appearance is not important, only the ability to accept nailing. Furr out around bellys and low places to get a final even surface. Furring is simply building out from the wall surface to provide an even base for nailing the

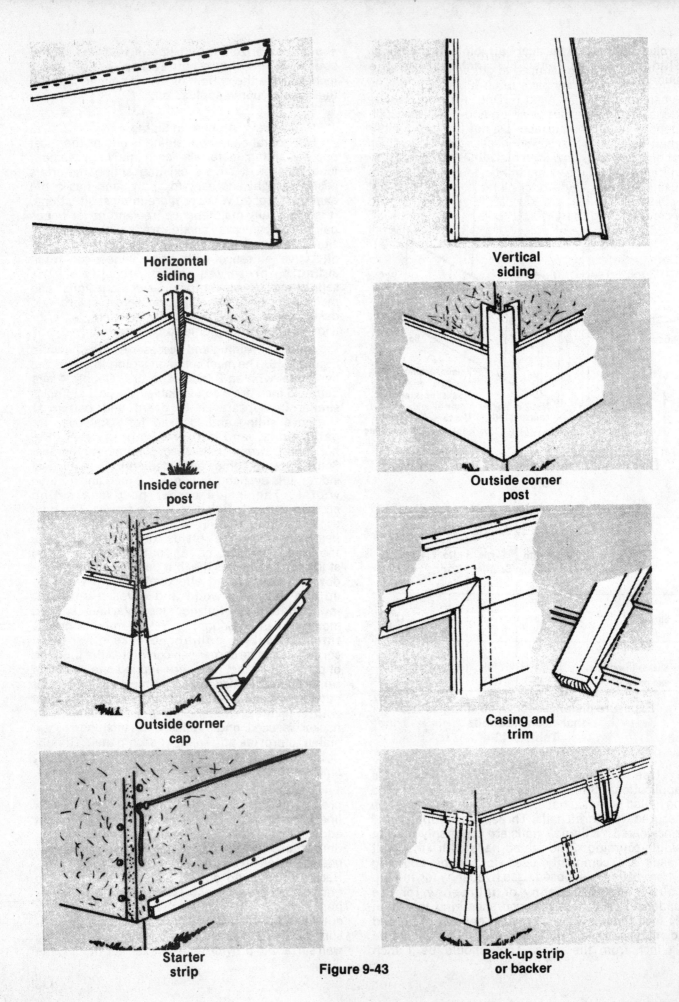

Horizontal
siding

Vertical
siding

Inside corner
post

Outside corner
post

Outside corner
cap

Casing and
trim

Starter
strip

Back-up strip
or backer

Figure 9-43

134

siding. Furring strips are narrow strips of wood lath or other material used to get an even surface. The entire wall may have to be furred out or only small sections may need furring. Furring may be done in a left-to-right or an up-and-down direction depending on the surface. Do not furr out further than absolutely necessary because complications at doors and windows can make the siding job very difficult and time consuming. Many applicators use insulation board or backing board to furr out walls and build up low places. Rotten wood should be cut out and replaced by solid or strip material.

Scrape lumps of paint, blisters and hardened putty off if they will interfere with trim or siding. Care in surface preparation will save you time and work as the job progresses.

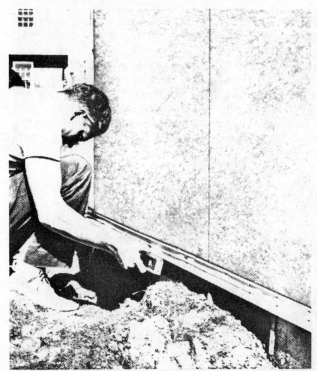

Installing the starter strip
Figure 9-45

Setting the chalk line
Figure 9-44

To establish the line necessary for aligning the starter strip, drive a nail part way in, far enough up the wall to clear a full siding panel. For example, if 8 inch siding is used, set the nail 9 to 10 inches above the foundation. See Figure 9-44. Stretch the string line between the two nails. Check the line with your level and adjust until the string is level.

Using the chalk line or string as a guide, install the starter strip along the bottom of the building. See Figure 9-45. Allow room at the building ends for corners and other accessories. Nail through the starter strip every 8 inches. Be sure to set the strip securely. Maintain a straight line, being careful not to bend the strip. Check with your level or chalk line to be sure the starter strip is in the right place. Occasionally a starter strip will be needed above the garage doors or in locations other than at the foundation. See Figure 9-46. These unusual situations must be planned on an individual basis as they occur. Other accessory items such as channels or trim may serve better in these cases. In some cases it may be necessary to build out by nailing furring strips under the starter strip to permit the siding to be hung flat against the

Starter above windows
Figure 9-46

building. This is most often needed when insulated siding is used.

A plumb line should be used when applying a starter strip for vertical siding.

Inside corner posts are used to receive siding at insets, porches and other wall meeting points.

The post is set into the corner and nailed every 12 inches. The siding is later butted to the corner and nailed into place. Allow 1/16 inch of space between the corner post and the siding for expansion. The critical part of inside corner work is the initial placement of the post. If it is true and even all work that follows will be neat and attractive. The corner is completed by caulking the junction of the siding and the corner post. Note Figure 9-47.

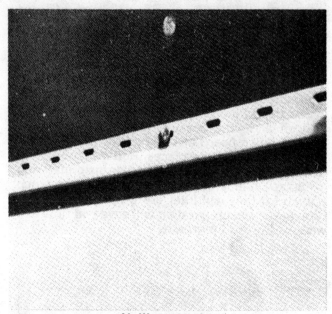

**Nailing panels
Figure 9-49**

**Inside corner post
Figure 9-47**

**Locking the first course
Figure 9-48**

Take special care when applying the first course of siding. The first course establishes a base upon which the entire wall of siding is built. Be sure the first course is straight and is securely locked into the starter strip (Figure 9-48).

Aluminum siding is hung on the nails, not nailed to the wall. When nailing, drive the nail through the slotted hole to within 1/32 inch of the nailing flange. See Figure 9-49. Do not drive it tight, as locked siding cannot move as temperature changes and takes on a wavy look. Do not drive the nail head into the siding edge. The forcing of the nail into the siding will bend the nailing flange and give the job a distorted look. In remodeling, when siding is installed over old wood, be sure that the nails are holding securely. Be certain that rotted or broken boards are not the nailing base. When nailing siding or siding accessories, always use aluminum nails. Nails should be driven directly into the wall, never up or down. Always nail through the nailing slots in the siding. Do not make holes in the siding between nailing slots. The slots are elongated to permit the siding to contract and expand freely as temperature changes. Nail in the center of the slot. Be careful not to bend or distort the edges of the siding as bent edges make it difficult to install the next panel. The panel must be hung with aluminum nails through the slotted holes every 16 inches along its entire length. Skipping of nails will cause the siding to bow away from the wall and result in a wavy appearance. Keep checking to be sure that the panel is locked into the starter strip along the entire length of the panel.

As you install the first course, use your chalk line to be sure the course is straight and level. Exert upward pressure on the siding from the bottom as you work. Do not pull hard or the siding will become distorted. A check with your level at the top of the first course will guarantee the correctness of the chalk line, starter strip and first course work.

Two methods of completing outside corners are used in siding work. The first uses outside corner posts. Refer again to Figure 9-43. Their use is similar to the use of inside corner posts and the same care that is used on inside corners applies here. Outside corner posts are installed before the siding itself is applied, so getting it straight is very important. The second method of completing

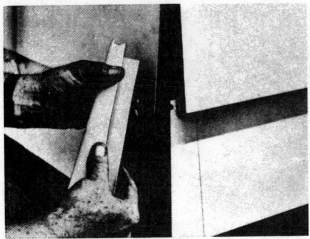

Squeezing the corner cap
Figure 9-50

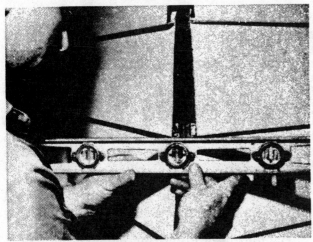

Checking corner caps
Figure 9-51

outside corners involves the use of outside corner caps or individual building corners (also in Figure 9-43). These are installed after the siding itself is in place. Keep siding back from the corner to let the cap lock in. Where siding meets from two directions, the individual corner caps are installed working from the bottom to the top of the building. Before installing the corner cap, squeeze it gently in the palm of your hand. Note Figure 9-50. The bottom lip of the corner cap is pressed over the siding panels and nailed into place at the corner. Use 2 or 2½ inch aluminum nails here as the nailing is done at an angle and the greater length is desirable for assuring a permanent installation. The use of outside corner caps requires proper alignment of the panels on the two sides of the building. A mismatch will prevent the holding lip from locking securely into the siding. Make sure the bottom edge of each two panels that meet is horizontal. Note Figure 9-51.

Usually the horizontal panels are not as long as the sidewalls to be covered. At least one joint is necessary in each course on most jobs. Try to join the panels so they look as much as possible as if a single panel had been run from one end of the building to the other. To get this effect, put the cut end under the first panel when the first panel has been installed beginning at a doorway or at the driveway side of the building. Put the new end over the panel when your first panel was started on the edge of the building away from the entrance, the back or the side away from the driveway. Overlap at least ½ inch when locking the panel into position. To keep a straight look to the lap, backers or back-up strip should be used at all laps. Backers or back-up strip are not always needed when insulated siding is applied. Try to leave the factory finished end exposed and the cut end lapped under the edge of the panel. This requires some planning before starting to hang siding, but the finished result is well worth the extra effort.

Cutting siding, like lapping away from traffic and entrances, demands careful planning. Avoid extremely short siding pieces as they lack strength and detract from the appearance of the job. As a general rule, never use a piece of siding less than 20 inches long. If it is not possible to hang a short length on two or more nails, it should not be hung unless it will fit by itself between openings. Arrange cutting to stagger the joints or positions of lapping. As a general rule, every lap should be passed both above and below by a solid panel. If the first is a full panel, the second course should start with a shorter one, possibly six or seven feet. Staggering of the joints gives the siding greater strength.

Furring around windows
Figure 9-52

Cutting is done either with a tin snips or with a power saw. If you are working with insulated siding, first cut the insulation away. In cutting with the tin snips, score the siding with a knife along a straight edge and cut in the score line being careful to get a clean cut where the metal is folded and formed. In cutting with a power saw,

137

Cutting for a window
Figure 9-53

Fitting around a window
Figure 9-54

merely line up the siding, mark the cutting point and saw away from you across the siding. The use of an ordinary candle or a bar of soap on the blade will greatly reduce burring and increase the blade life.

You will probably need some furring around windows and doors on most jobs. See Figure 9-52. Furr out to the proper distance so the siding lays smooth and even. If it isn't, the furring is too thick or too thin and must be adjusted. When it is right, you may want to cover the channel with trim for a neatly finished look. If spacing worked out so that the entire panel fits under the window, just nail a channel below the window frame and work the siding into the channel. If the spacing works out so that the siding must be cut, measure the siding, cut the sides, score and then bend the piece that you want to remove back and forth until it breaks

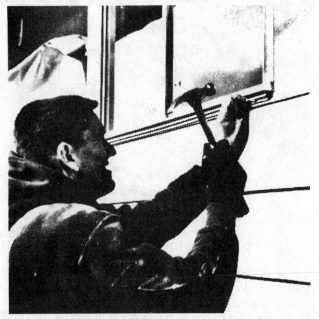

Installing "L" moulding
Figure 9-55

off. See Figures 9-53 and 9-54. Then lock the siding into the channel and lock the channel against the siding. Install siding window and door trim the same way as inside corner posts except use the L-molding or window and door trim instead of the corner posts. Note Figure 9-55. Over the window or door the material cut away is from the bottom of the panel rather than the top. The trim used has a long lip on it which serves as a drip cap. An L-channel should be installed on the sides to accept the siding and reduce the amount of caulking needed. Before installing the window or door trim, scrape the old frame smooth and level the surface so the new trim fits snugly against the entire frame without bending.

You may have to furr out all along the top of the sidewall so the last panel is nailed in place at the proper angle. See Figure 9-56. If so, a channel is nailed onto the furring. On gable ends where short lengths of siding must be used, extra care must be taken to lock in all edges and to avoid bending or deforming these pieces.

If the full panel is not required, then the top panel must be cut in a manner similar to that used below windows, except the entire length is cut rather than the small section required for clearing the window. General purpose trim, channel or closure strip should be used to accept the cut piece of siding. A molding trim piece is applied to finish the job.

Caulking is done to get a better looking finished job and to seal areas where water can get under the siding. Caulking is available in bulk for filling guns, and in cartridges. The cartridge feed gun is most popular because it is easier and quicker to refill than the bulk unit. In addition, there is little chance of it getting lumpy or dirty in the cartridge. Exposed caulking can be held to a minimum if proper siding accessories are used. Caulking is available in colors to match any siding used. For best results, apply a single bead of caulking to the seam or line moving the gun evenly and with a steady even pressure. Note Figure

Fitting the top panel
Figure 9-56

Caulking around a window
Figure 9-57

9-57. Caulking is done around doors, windows and gables where metal meets wood and when metal meets metal except where the accessories are used to make caulking unnecessary. Caulking is also needed where metal meets brick or stone around chimneys and at brick walls. If old wood soffits and fascia are retained, it may be necessary to caulk these junctions, also.

If normal care has been taken in installing the siding, the clean-up job is very simple. Fingerprints and smudges will usually wipe right off. If they are greasy, a little mild soap and water will almost always do the job. If you need to use a solvent of any type, get the one that the manufacturer of the siding tells you is safe. Harsh or abrasive type cleaning agents should not be used. Scrap ends can be dangerous to children and should be hauled away from the job. Leaving a job site neat and clean is the sign of the good siding mechanic. Nails, scrap and packing materials (boxes, wrapping paper and the like) should all be taken away at the end of the job.

Many varied and attractive effects can be obtained by combining horizontal and vertical aluminum siding on a single building. Note Figure 9-59. Use of vertical siding in porch areas or along the entrance will help break up the long look of a house.

Insulation siding is made by building into each panel either shiny reflective aluminum foil that increases the insulation value of air space, or trapped air insulation that reduces the passage of heat or cold. The second type comes with either foam material fastened to the siding during the manufacture or a fiberboard material that can be nailed to the wall or simply dropped in place behind the siding. Insulation use changes the job only slightly. In some cases different trim may be needed. Cutting away of insulation at the ends and junctions or greater build out of windows and doors may be needed.

Repairing Masonry Veneer

Where brick or stone veneer is used as siding, mortar may become loose and crumble, or uneven settlement may cause cracks. In either case, new mortar should be applied both to keep out moisture and to improve appearance. Repair is accomplished in much the same manner as for masonry foundations, except that more attention to appearance is required. After removing all loose mortar and brushing the joint to remove dust and loose particles, dampen the surface. Then apply mortar and tamp it well into the joint for a good bond. Pointing of joints should conform to existing joints. Particular care should be excercised in keeping mortar off the face of the brick or stone unless the veneer is to be painted.

Many older houses used soft bricks and porous stone trim. After repair of the cracks and mortar joints, the entire surface may require treatment with transparent waterproofing. Painted, stained, or dirty brick and stone can be restored to its original appearance by sandblasting. It can then be repainted or waterproofed. Caution should be used in sandblasting soft stone and brick.

Combining vertical and horizontal siding
Figure 9-58

Adding Roof Overhang

The addition of a roof overhang, where there is none, will soon pay for itself in reduced

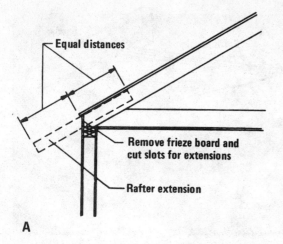

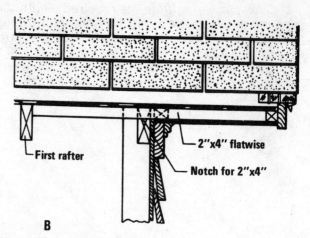

Extension of roof overhang: A, rafter extension at eaves; B, extension at gable end
Figure 9-59

maintenance on siding and exterior trim. Without the overhang, water washes down the face of the wall, creating moisture problems in the siding and trim and, consequently, more frequent painting is required. Additional roof overhang also does much to improve the appearance of the house.

Where new sheathing is being added, the sheathing can be extended beyond the edge of the existing roof to provide some overhang. This is a minimum solution and the extension should not be more than 12 inches where ½ inch plywood sheathing is used. Any greater extension would require some type of framing. Framing can usually be extended at the eave by adding to each rafter. First, remove the frieze board, or, in the case of a closed cornice, remove the fascia. Nail a 2 by 4 to the side of each rafter, letting it extend beyond the wall the amount of the desired overhang. The 2 by 4 should extend inside the wall a distance equal to the overhang (Figure 9-59,A). Framing for an overhang at the gable ends can be accomplished by adding a box frame. Extensions of the ridge beam and eave fascia are required to support this boxed framing. An alternate extension is possible with a plank placed flat, cut into gable framing, and extending back to the first rafter (Figure 9-59,B).

Installing Manufactured Soffit

A new soffit should generally be recommended to finish off a new siding job. An attractive aluminum soffit will usually improve attic ventilation while providing better protection from insects and birds. Figure 9-60 shows a manufactured soffit installed above brick veneer. Hanging a soffit system is comparatively easy and fast. Remove any protrusions and existing molding first and touch up any exposed wood surfaces. Most manufactured soffit comes in 12 inch widths. For a typical rafter overhang 40 feet long and 36 inches wide (Figure 9-61) you need 120 running feet. Of this, about 25 percent (or 30 feet) should be perforated soffit which allows the house to

"breathe." You will also require 40 feet of molding or frieze board, 40 feet of fascia cap and fascia board and 1 pound each of 1½ inch siding nails and white trim nails.

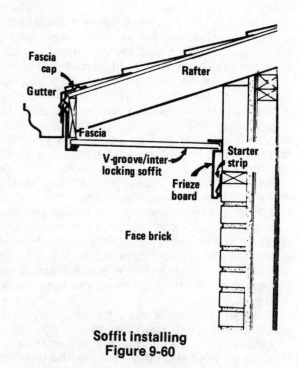

Soffit installing
Figure 9-60

First, remove any old trim. To insure a level soffit, chalk a line on the wall exactly level with the fascia bottom. The existing fascia should be level but check it to be sure. Select your molding which will form the rear support for the soffit. Secure the molding by nailing it to the joist or stud. If possible, cut a tab as shown and nail the tab to a stud or joist. See Figure 9-62.

Next, measure in several places (about 12 inches apart) from the molding to the outside of the fascia board to determine the correct lengths of soffit pieces to be cut. Note Figure 9-63.

Cut a piece of soffit to the required length. Slip one end into the frieze molding and nail through the nailing flange into the bottom of the fascia board. See Figure 9-64. The exposed edge will be covered neatly with a fascia board when complete. No center nailing is required in soffit lengths up to 48 inches.

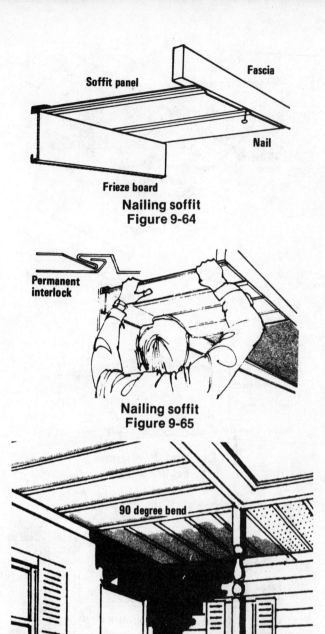

Nailing soffit
Figure 9-64

Nailing soffit
Figure 9-65

Butt corner
Figure 9-66

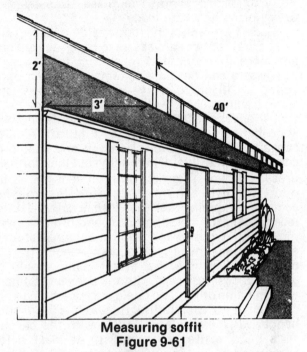

Measuring soffit
Figure 9-61

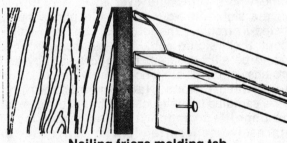

Nailing frieze molding tab
Figure 9-62

Measuring soffit lengths
Figure 9-63

Cut the remaining soffit to size and slide it into place as in Figure 9-65. Most systems will have interlocking grooves which are secured by pulling the soffit toward you. Nail the outside edge to the bottom of the fascia as before. Repeat this procedure until the entire run is in place. Each fourth panel should be perforated.

If your overhang goes around the corner, you can mount your runs square, as in Figure 9-66, or at 45 degree angles as in Figure 9-67. Where soffits butt, nail two pieces of "J" channel back-to-back (Figure 9-68) to the overhang and fascia as shown in Figure 9-66. In 45 degree turns you must cut soffit pieces at a 45 degree angle, (shorter as required) and insert them into the "J" channel.

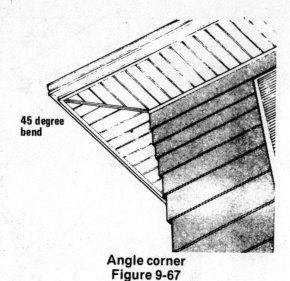

45 degree bend

Angle corner
Figure 9-67

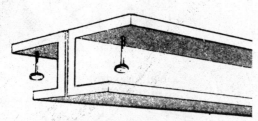

Double "J" channel
Figure 9-68

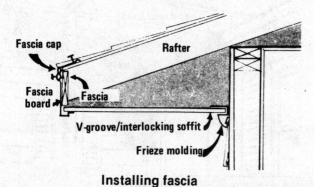

Fascia cap

Rafter

Fascia board

Fascia

V-groove/interlocking soffit

Frieze molding

Installing fascia
Figure 9-69

Cover the exposed soffit ends with fascia board, making sure the bottom fascia flange is snug up against the soffits. See Figure 9-69. Nail securely at 12 inch intervals to the fascia. Install the fascia cap under shingles and over fascia board. Nail the fascia cap under the shingles.

Selling Siding Jobs

Firms that specialize in siding work have discovered that siding jobs can be sold most easily through a system of jobsite convassing. Unlike kitchen or bathroom remodeling, it will be obvious from the curb when a home needs improved siding. Generally, where one house in a community needs new siding, you can bet that many homes in the immediate vicinity can use the same. And best of all, there is no better demonstration of your service than the dramatic facelift your crew can give a house in one or two eight hour days. You can take advantage of these facts by usng every job to generate leads.

Each time you send out a siding crew, also send a crew of two canvassers to cover every home not more than one-half mile from the job. The canvassers can be male or female and can be as young as 18 or 20. A very modest salary plus commission is usually all the incentive they need. Give them about 100 of your calling cards and door knob hangers as illustrated in Figure 9-70. The canvasser's job is strictly soft sell and should go something like this: The neatly but not formally dressed convasser greets each homeowner at the door, identifies himself or herself and your company, explains that your firm is doing a siding job today in the neighborhood, and offers to walk the homeowner over to the job to demonstrate how easily and inexpensively modern residing can be done. Homeowners that won't come should be offered a business card with the convasser's hand written name on it. If no one answers the door, the canvasser is instructed to leave a doorknob hanger, at the residence, again with the canvasser's hand written name on it. At least a few homeowners will come over to look at the job. Have a sign with your name and phone number posted in front of the job so anyone can call you about their siding job. The job superintendent should be alert to greet neighbors as they arrive and answer questions about technique, colors and siding descriptions. Naturally he should have some samples in his truck and be ready to explain the benefits of modern siding materials. An aggressive superintendent or canvasser may occasionally be able to close sales on the spot. If so, they should get the full sales commission (5% to 10%). Any canvasser who develops a lead that becomes a buyer should receive a bonus equivalent to about 1 or 2% of the job. The canvasser's main job is to take names and phone numbers of people who would like to have your salesman call for an appointment to discuss residing their home.

Any remodeling contractor who is careful to canvass on every residing jobsite should develop at least 2 or 3 good leads for every job completed. If your crew does good work, leaves the owner satisfied with the job he bought and follows up on the leads developed at each job, at least 50 percent of your new work should come from prospects developed by canvassing.

Sorry!

Please excuse the noise as we finish re-siding

The_____'s

house at_____

Stop over and see your neighbor's new remodeling job, it's a beauty!

We would be glad to give you a free estimate on any home improvement remodeling that you may have been thinking of. Just fill out the card and mail it in today.

Remodeler

Phone

Address

City,   State,   Zip

Name_____

Address_____

City_____

Sorry!

Please excuse the noise as we finish re-siding

The_____'s

house at_____

Stop over and see your neighbor's new remodeling job, it's a beauty!

We would be glad to give you a free estimate on any home improvement remodeling that you may have been thinking of. Just fill out the card and mail it in today.

Remodeler

Phone

Address

City,   State,   Zip

Name_____

Address_____

City_____

Figure 9-70

Chapter 10
Roofing and Gutters

A wide variety of roof coverings is available, and most can be used in rehabilitation in the same manner as in new construction. First cost usually influences the choice. In most houses the roof is a major design element and the covering material must fit the house design. Heavy materials such as tile or slate should not be used unless they replace the same material or unless the roof framing is strengthened to support the additional load. The most popular covering materials for pitched roofs are wood, fiberglass, asphalt, and asbestos shingles. These can be applied directly over old shingles or over sheathing under the old surface; however, if two layers of shingles exist from previous reroofing, it may be well to remove the old roofing before proceeding. Roll roofing is sometimes used for particularly low cost applications or over porches with relatively low pitched roofs. The most common covering for flat or low pitched roofs is built-up roof with a gravel topping.

An underlay of 15 or 30 pound asphalt saturated felt and roof edging should be used on asphalt shingle roofing. Note Figure 10-1. Felt and edging are not commonly used under wood shingles or shakes. A 45 pound or heavier smooth surface roll roofing should be used as a flashing along the eave line in areas where moderate to severe snowfalls occur. The flashing should extend to a point 36 inches inside the warm wall. If two strips are required, use mastic to seal the joint. Also use mastic to seal end joints. This flashing gives protection from ice dams (Figure 10-2). Ice dams are formed when melting snow runs down the roof and freezes at the colder cornice area. The ice gradually forms a dam that backs up water under the shingles. The wide flashing at the eave will minimize the chances of this water entering the ceiling or the wall. Good attic ventilation and sufficient ceiling insulation are also important in eliminating ice dams. Roll roofing 36 inches wide is also required at all valleys.

Where shingle application is over old wood or asphalt shingles, industry recommendations include certain preparations. First remove about 6 inch wide strips of old shingles along the eaves and gables, and apply nominal 1 inch boards at these locations. Thinner boards may be necessary where application is over old asphalt shingles. Remove the old covering from ridges or hips and replace with bevel siding with butt edges up. Place a strip of lumber over each valley to separate old metal flashing from new. Double the first shingle course.

Wood Shingles

Wood shingles used for house roofs should be Number 1 grade, which is all heartwood, all edge grain, and tapered. Principal species used commercially are western red cedar and redwood,

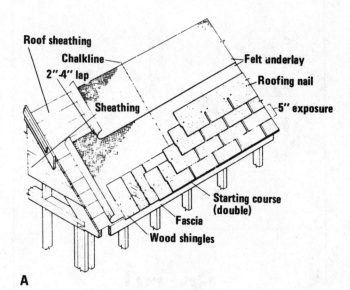

Application of asphalt-shingle roofing over plywood: A, with strip shingles; B, metal edging at gable end
Figure 10-1

144

which have heartwood with high decay resistance and low shrinkage. Widths of shingles vary, and the narrower shingles are most often found in the lower grades. Recommended exposures for common shingle sizes are shown in Table 10-3.

General rules to follow in applying wood shingles are illustrated in Figure 10-4:

1. Extend shingles 1½ to 2 inches beyond the eave line and about ¾ inch beyond the rake (gable) edge.

2. Nail each shingle with two rust resistant nails spaced about ¾ inch from the edge and 1½ inches above the butt line of the next course. Use threepenny nails for 16 and 18 inch shingles and fourpenny nails for 24 inch shingles. Where shingles are applied over old wood shingles, use longer nails to penetrate through the old roofing and into the sheathing. A ring-shank nail (threaded) is recommended where the plywood roof sheathing is less than ½ inch thick.

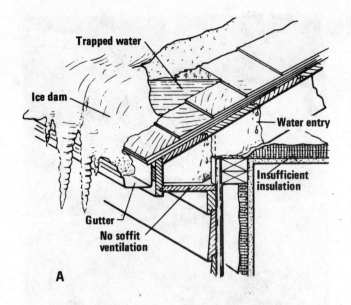

A

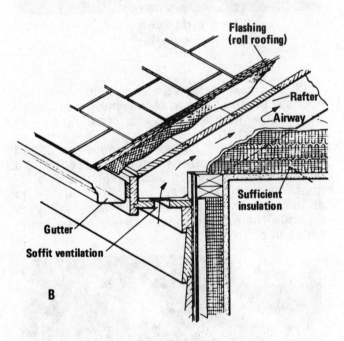

B

Flashing and ventilation to prevent damage from ice dams
Figure 10-2

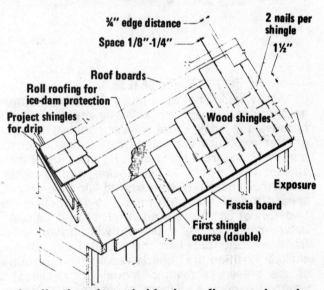

Application of wood-shingle roofing over boards
Figure 10-4

3. Allow a 1/8 to 1/4 inch space between each shingle for expansion when wet. Lap vertical joints at least 1½ inches by the shingles in the course above. Space the joints in succeeding courses so that the joint in one course is not in line with the joint in the second course above it.

4. Shingle away from valleys, selecting and precutting wide valley shingles. The valley should be 4 inches wide at the top and increase in width at the rate of 1/8 inch per foot from the top. Use valley flashing with a standard seam. Do not nail through the metal. Valley flashing should be a minimum of 24 inches wide for roof slopes under 4 in 12; 18 inches wide for roof slopes of 4 in 12 to 7 in 12; and 12 inches wide for roof slopes of 7 in 12 and over.

5. Place a metal edging along the gable end of the roof to aid in guiding the water away from the end walls.

Shingle Length Inch	Shingle Thickness (Green)	Maximum Exposure	
		Slope less[1] Than 4 in 12 Inch	Slope 4 in 12 and over Inch
16	5 butts in 2 in.	3¾	5
18	5 butts in 2¼ in.	4¼	5½
24	4 butts in 2 in.	5¾	7½

[1]Minimum slope for main roofs—4 in 12. Minimum slope for porch roofs—3 in 12.

Exposures for wood shingle roofing
Table 10-3

Starter course
Figure 10-5

First Course
Figure 10-6

Wood Shakes

Apply wood shakes in much the same manner as shingles, except longer nails must be used because the shakes are thicker. Shakes have a greater exposure than shingles because of their length. Exposure distances are 8 inches for 18 inch shakes, 10 inches for 24 inch shakes, and 13 inches for 32 inch shakes. Butts are often laid unevenly to create a rustic appearance. An 18 inch wide underlay of 30 pound asphalt felt should be used between each course to prevent wind driven snow from entering between the rough faces of the shakes. Position the underlay above the butt edge of the shakes a distance equal to double the weather exposure. Where exposure distance is less than one-third the total length, underlay is not usually required.

When reroofing over the old roof cover, follow the following procedure:

1. A 36 inch wide strip of 15 pound minimum asphalt felt is applied over the old shingles at the eave line. 18 inch wide strips of felt are then applied shingle style, at 10 inch intervals up the roof. Next, a starter course is applied at the eave line extending approximately 1½ inch over the eave edge. See Figure 10-5.

2. The first course of shakes is then applied directly over the starter course. The tip ends of each course of shakes are tucked under the felt strips. With a 24 inch long shake laid at 10 inch exposure, the top 4 inches of the shake will be covered by felt. The shakes should be spaced about ½ inch apart to allow for possible expansion. These joints or spaces between the shakes should be broken or offset at least 1½ inch in adjacent courses. See Figure 10-6.

3. Felt covers the old shingles, and each felt overlaps the top 4 inches of the shake. See Figure 10-7. Longer nails are used in overroofing than in new construction.

4. Metal valley, at least 26 gauge galvanized iron, painted for anti-rust protection, center and edge crimped, is laid in place. When new valley metal may come in contact with old metal, a strip of lumber should be placed in each valley to separate them. Figure 10-8 shows valley being nailed in place.

Felt overlapping between shakes
Figure 10-7

Applying valley
Figure 10-8

5. Shakes are cut parallel with valleys as courses are applied over the metal valley as in Figure 10-9. Valley gutters should be approximately 6 inches in width. Metal valley sheets should be at least 20 inches wide with a 4 to 6 inch head lap.

Shakes in valley
Figure 10-9

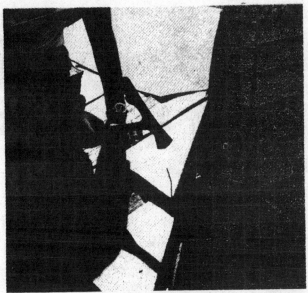

Shakes at ridge
Figure 10-10

Factory ridge
Figure 10-11

6. Shakes are cut parallel to hip line. (Figure 10-10).

7. Factory assembled "hip-and-ridge" units "finish" the hips and ridges. Hip-and-ridge units should be applied at the same exposure as the rest of the roof, and longer nails should be used to ensure approximately ¾ inch sheathing penetration. See Figure 10-11.

Asphalt and Fiberglass Shingles

The most common type of asphalt or fiberglass shingles is the square-butt strip shingle, which is 12 by 36 inches, has three tabs, and is usually laid with 5 inches exposed to the weather. Bundles should be piled flat so that strips will not curl when the bundles are opened for use. An underlayment of 15 pound saturated felt is often used. Table 10-12 shows the requirements for applying underlayment.

Underlayment[1]	Minimum Roof Slope	
	Double Coverage[2] Shingles	Triple Coverage[2] Shingles
Not required	7 in 12	[3]4 in 12
Single	[3]4 in 12	[4]3 in 12
Double	2 in 12	2 in 12

[1] Headlap for single coverage of underlayment should be 2 in. and for double coverage 19 in.
[2] Double coverage for a 12-by-36-in. shingle is usually an exposure of about 5 in. and about 4 in. for triple coverage.
[3] May be 3 in 12 for porch roofs.
[4] May be 2 in 12 for porch roofs.

**Underlayment requirements for asphalt shingles
Table 10-12**

Begin application of the roofing by preparing the existing roof deck. Deck preparation includes cutting out old wood shingles at the rake with the help of a power saw. See Figure 10-13. Removal of the old shingles wil enable the roofer to apply a new wood rake strip to insure a smooth rake under the new shingles. Next, all loose or badly curled wood shingles should be securely nailed down. This will insure that the old shingles do not distort the appearance of the new roof. A mineral surfaced starter strip, 18 inches wide, is installed, overhanging the eave about ½ to ¾ inch. See Figure 10-14. The starter strip will provide protection for the cut-outs of the first course of asphalt shingles. Next, apply the first course even with the bottom edge of the starter strip. On the second and following courses use a gage hatchet to insure the proper exposure and correct alignment of courses. Note Figure 10-15. A five inch exposure is usually recommended.

**Trimming rake
Figure 10-13**

**Applying starter strip
Figure 10-14**

Care should be taken, particularly in flashing and valley details, to insure proper application where leaks are more likely to occur. Because drainage runoffs concentrate at the valleys, these are very vulnerable to leakage and contractors must provide adequate protection under the shingles for any water that may flow back under the tabs by wind-driven rain. Closed valley techniques are recommended when reroofing with heavyweight asphalt shingles. A closed or laced valley offers a smooth, unobstructed drainage path constructed from the same material used on the remainder of the roof. Because of this, the

148

Gaging with a hatchet
Figure 10-15

Roll roofing in valley
Figure 10-16

center of the valley at the bottom (Figure 10-18). Shingles on the overlying side are than trimmed according to the chalk line. Note in Figure 10-19 that all nails near the valley are kept at least 6 inches from the center line of the valley so as not to penetrate the underlying sheet in this critical area. The finished closed valley construction, shown in Figure 10-20, assures the homeowner the same quality of material in the valley as on the

Shingling the valley
Figure 10-17

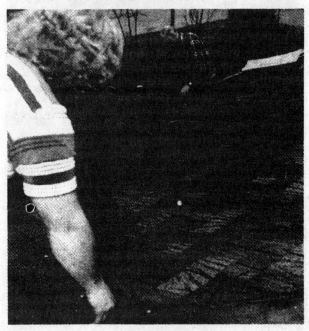
Snapping a chalk line
Figure 10-18

valley will not wear out before other roof sections.

In addition to valleys and flashings, hips and ridges require special care. In these areas, proper placement of the nails are important. When applying hip and ridge shingles, none of the nails should be exposed since they have a tendency to work loose and leak.

Begin a closed valley by applying 36 inch wide rolled roofing, either smooth or mineral surfaced, centered in the valley and secured with only enough nails to hold it in place until the shingles are applied. Note Figure 10-16. Extend the shingles of the underlying side at least 12 inches beyond the valley center as in Figure 10-17. Note that the overlay strips from the portions of the shingle which extend across the valley have been removed. After shingles of the overlying side have been applied, snap a chalk line 1 inch from the center of the valley at the top and 2 inches from

Trimming valley
Figure 10-19

Completed valley
Figure 10-20

not under it. Flange corners should then be cemented down.

When installing shingles along any large expanse of roof, as in Figure 10-22, apply the shingles in a diagonal method up the roof for proper aesthetic appeal. Shingles should never be

Proper vent placement
Figure 10-21

Diagonal application
Figure 10-22

remainder of the roof. Unlike an open valley, a closed valley will not wear out before other sections of the roof.

Proper flashing procedures are crucial to a roof's performance. Figure 10-21 shows correct installation around a vent, with upper courses covering the metal flange of the vent while metal covers the lower courses of shingles so that the natural flow of water will run over the flashing and applied in continuous horizontal or vertical courses. Hips and ridges also require extra care in application. Projecting portions of the shingles on the hip in Figure 10-23 have already been trimmed flush with the center line of the hip. To attain a straight application and proper alignment of the hip shingles, snap a chalk line along one side of the hip as is being done in Figure 10-24. Start the

Applying ridge shingles
Figure 10-25

Trimming at the ridge
Figure 10-23

Hip and ridge joint
Figure 10-26

Aligning the ridge
Figure 10-24

hip at the bottom, along the chalk line. The hip and ridge shingles are applied in double thickness, one directly upon another, to enhance the distinctive appearance of a heavyweight asphalt roof. Note Figure 10-25. To insure good weather protection, it is important to cover the joint where hips meet ridges. Note in Figure 10-26, for example, how the completed ridge totally covers the junction of the completed hips.

Applying the starter course
Figure 10-27

Applying the first course
Figure 10-28

Applying the second course
Figure 10-29

Using The ''Butt-Up'' Method Over
An Asphalt Shingle Roof

On reroofing jobs, your crews should apply the starter course shingles so that they are even with the existing roof at the eaves and do not overlap the course above. The shingles should be installed with the factory-applied adhesive strip located near the eaves. Note Figure 10-27. Remove three inches from the end of the starter course shingle at the rake to insure that all cutouts will be covered when the first course is applied. Apply the first course of shingles over the starter strip with two inches or more removed from the tops of the new shingles so they do not bridge the butts of the old shingles in the course above. See Figure 10-28. Start application at a rake with a full length shingle, locating roofing nails 5/8 of an inch above the cutouts and one inch and 12 inches in from the edges. Use four nails for each shingle, and place each nail below, not in or above, the factory-applied adhesive strip.

Use full width shingles for the second and succeeding course (Figure 10-29), aligning the top edge with the butt edges of the old shingles in the next course. This reduces first course exposure to three inches, and provides an automatic five inch exposure for each succeeding course. Roofers say this method gives a smoother, more uniform and faster application, with a new horizontal nailing pattern two inches below the old.

Avoiding Problems

Correct application includes two procedures that are often overlooked by field crews: correct nailing and across-and-up application of shingles.

Common nailing errors include use of nails of incorrect size for reroofing, incorrect placement of nails, and use of too few nails. Nailing mistakes such as these can lead to many difficulties. On reroofing jobs, use galvanized or aluminum nails long enough to penetrate through the roofing material and at least ¾ inch into the lumber of a wood deck or through plywood decks. For reroofing over old asphalt roofing, 1¼ to 1½ inch nails should be used; for reroofing over wood shingles and for tearoffs, use 1¾ inch nails, the same as recommended for new construction. With regard to the location of nails on reroofing jobs, the most common error occurs with the improper placement of nails in or above the factory-applied adhesive strip. Place nails below the adhesive strip and 5/8 of an inch above the tops of the cutouts. Correct placement affords better wind protection until the sealing action of the factory adhesive takes effect. Use four nails for each three-tab, two-tab, and no-cutout square butt strip shingle. Nails should be driven straight so the edge of the nail head will not cut into the shingle, with heads flush and not sunk into the surface. Inadequate and improper nailing practices decrease a roof's integrity and resistance to high winds. Correcting nailing errors at the time they are made improves the overall quality of an application. It also saves time and money on callbacks and repairs. In the long run, careful roofing application techniques will mean a lasting impression of quality on many home improvement and remodeling projects.

Across-and-up shingle application minimizes color variations by blending together shingles from different bundles as it also avoids improper nailing. "Racking," or straight-up application requires an applicator to slip shingles under adjoining shingles in every other course. All too often, the applicator who does this fails to use four nails in each shingle as he should.

New shingles should be applied across and up (Figure 10-30) with the cutouts breaking joints on thirds or halves, or with random spacing. For instance, when shingle joints are broken on thirds as in Figure 10-31, the second course is started with a strip from which four inches has been cut, and the third course with a shingle from which eight inches has been cut. For the fourth course, repeat with a full shingle.

Roofing applicators should also be reminded to begin nailing from the shingle nearest to the one just applied, and then proceed across. Otherwise, buckling of shingles may result.

Breaking joints at thirds
Figure 10-31

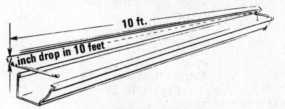

Locating the gutter top
Figure 10-32

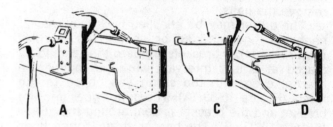

Attaching bracket hangers
Figure 10-33

Across and up application
Figure 10-30

Other Types Of Roofing

Built-up roof coverings are limited to flat or low pitched roofs and are installed by contractors who specialize in this work. The roof consists of three, four, or five layers of roofers' felt, with each layer mopped down with tar or asphalt. The final surface is then coated with asphalt and usually covered with gravel embedded in asphalt or tar.

Other roof coverings, such as asbestos, slate, tile, and metal, require specialized applicators, so their application is not described in detail. They are generally more expensive and are much less widely used than wood or asphalt shingles and built-up roofs.

Gutters And Downspouts

New gutters are often installed when a new roof cover is required. Though wood and fiberglass gutters are used occasionally, most gutters installed on existing residential structures are the O.G. or box type, 4 inches or 5 inches wide and made of either 26 gauge or 28 gauge steel or aluminum. Recent developments in forming equipment have made it more economical for contractors who install large amounts of gutters to form both gutters and downspouts on the job.

In replacement guttering, a fascia installation is nearly always used and the gutter is sloped about ½ inch for each 20 feet of run. Begin by tacking a string on the house to mark the location of the top edge of the gutter. See Figure 10-32. Take measurements and do the necessary cutting using tinsnips. Often you can save time by assembling sections on the ground, especially where sections and accessories are attached by rivets. Some assembly is done most easily from the roof. If you are working alone, you can temporarily support the gutter in place by tying a wire around it and hooking the wire over two nails

Spikes and ferrules
Figure 10-34

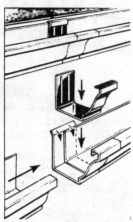

Connection at a joint
Figure 10-35

about 8 inches apart. After the section is fastened, remove the nails.

The gutter may be attached by fascia bracket hangers (Figure 10-33) or spikes and ferrules (Figure 10-34). If brackets are used they should be placed at 2 foot intervals with the top of the bracket even with the string you nailed to the fascia (Figure 10-33, A). The gutter goes into the bracket and the bracket is bent around the gutter (Figure 10-33, B). The hanger strap (dotted lines in Figure 10-33, C) supports the front lip of the gutter and the bracket and strap are bent together (Figure 10-33, D) to form a solid unit. Some manufacturers provide a one piece bracket and strap or use bar hangers which interlock with both edges of the gutter. Spikes and ferrules have to be drilled or punched through the gutter every three feet or every second rafter.

Install succeeding sections in the same manner. Snap connectors are used to join sections and fittings. Note Figure 10-35. Apply gutter mastic to the inside of the gutter, joint, and edges of snap connectors. Slide the outside connector over the joint. Hook the inside connector under the front lip of the gutter and snap into place. Bend the tab of the outside connector down over the inside part. Be sure to lap the sections at least 1½ inches. If you are installing a run of 40 feet or more or on any job where expansion and contraction are restricted, use an expansion joint available from many manufacturers. The joints of gutter sections made by some manufacturers are not sealed or bent together and have about 3/8 inch play to allow for expansion and contraction.

Locate drop outlets so that the downspout can be installed adjacent to a corner of the building. Join the outlet sections to the gutter with a snap connector and attach an end cap to close the end of the run. Note Figure 10-36. If the gable end of the building has an overhang, install the drop outlet at the downspout location and add a short section of gutter. Some downspout systems use an eave tube attachment which is cut into the bottom of the gutter and then riveted into place.

Some gutter systems have inside and outside corner sections that create a one piece corner as in Figure 10-37. Other types of gutter systems must be cut at a 45 degree angle. The corner is then reinforced with a miter section. Regardless of the method, the unit must be supported on both sides of the corner.

Drop outlet
Figure 10-36

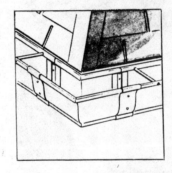

Corner installation
Figure 10-37

Each downspout pipe section and elbow has one end slightly smaller than the other. Parts are joined by fitting the large end of one over the small end of the one above and forcing them together tightly. Downspouts should fit snugly against the wall. If the house has a roof overhang, start your downspout installation by fitting an elbow on the gutter outlet and securing it with a sheet metal screw. Add a second elbow to direct the downspout downward. Note Figure 10-38. If necessary, add a short section of pipe between the elbows. Install the connector pipe by fitting its large end over the end of the elbow above and secure with a cadmium plated sheet metal screw.

Fasten the pipe against the wall with pipe straps—2 for each 10 foot section. Attach with 2 nails, automatically tightening around the pipe as nails are driven home. Finish your installation by

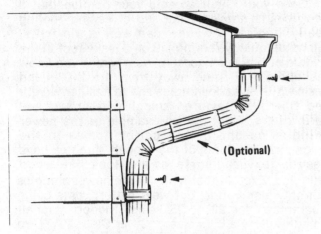

**Downspout connection
Figure 10-38**

attaching an elbow at the bottom of each conductor pipe to direct water away from the house.

Decorative painting of unpainted aluminum gutters may proceed after cleaning the surface with a dust brush only if the metal has weathered for several weeks after installation. Such exposure is usually enough to remove oil film from the surface. Spots of grease or dirt must be removed before painting. Either mineral spirits or painters' naphtha is a suitable solvent.

If it is necessary to paint before the system has has time to weather, a simple chemical treatment may be used. First, heavy accumulations of dirt and grease are removed. Then, an acid solvent solution (sold commercially under such names as Deoxidine, Alumi-Prep, Sol-Clean, Rusticide or Oakite 35) is diluted with water according to instructions provided and brushed or swabbed over the metal. The surface should be kept wet for five to ten minutes, then rinsed thoroughly with clean water. When completely dry, the aluminum is ready for painting.

Common house paints with a titanium base are entirely satisfactory for priming and finish coats. For priming, lead free paints are recommended and may be used with little or no thinning.

Factory painted aluminum or steel guttering do not need repainting. Galvanized guttering should be painted to prevent rust, but you should let it weather for at least three months before painting. Use any good gutter primer as a base coat, then an exterior paint or anti-rust finish. Whenever aluminum contacts dissimilar metals, masonry, or bare wood, or where water may wash over copper or other dissimilar metal before entering the gutter, coat the contacting surface or the dissimilar metal with aluminum asphalt paint, aluminum metal-and-masonry paint, or mastic. An impervious coating must be maintained for continued protection. Under no circumstances use nails, screws or rivets made with anything other than aluminum.

Selling Roofing Jobs

Remodeling contractors who are looking for more and better profit opportunities should pursue more reroofing jobs, especially the application of heavyweight asphalt or fiberglass shingles over old wood shingles or wood shake roofs.

Remodeling time is a particularly good time for reroofing. Since the roof is the largest visual expanse of a house, a successful remodeling project that includes reroofing can greatly improve its aesthetic appearance. The wide variety of shingle colors, textures and designs offer a good deal of flexibility for exterior design coordination.

Your prospect probably never thought about his roof very much. Very likely, he doesn't really want to have much to do with the roof. Its a source of annoyance rather than pleasure and pride. As a salesman your job is half done if you can convince him that his roof is an important decorative element in the design of the house and that roof color, style, and texture as well as durability are important.

Let your prospect explain any trouble he has been having with the roof. Then, explain that most shingle and built-up roofs are designed to last about 15 years. Ask him when his roof was installed and about the problems he has had with it. Your obvious concern for his problem and his willingness to discuss it establish a professional to client relationship that makes your opinion more valuable when you finally give it. Go with your prospect to inspect any leaks from the interior of the house. This is your prospect's real problem and he will want to show you what happened and emphasize how serious it is.

Next, go up on the roof to make an inspection. Take your prospect with you if he is willing to follow you and the pitch is not too steep. Take a pencil, tape measure and note pad or checklist (see Figure 10-39) with you. When you reach the roof, emphasize two things at once. Walking on a roof is both dangerous, because of the possibility of a fall, and hazardous to the roof. Foot traffic on a roof will only aggravate any problem the roof already has. Your prospect will get the point: A roof is no place for an unskilled amateur. Now begin your written evaluation of the roof. Make notes about each of the seven problem areas listed below. Be thorough. List the good points as well as the bad. It establishes your credibility. Be honest. You're sure to loose the job if your prospect suspects your truthfullness. Be professional. Your prospect wants to believe that you are a real ''pro'' in the roofing business and that you know what you are talking about. Don't disappoint him.

General roof condition. Look for missing, blistered, split or curled shingles. Explain that shingles can be replaced inexpensively in a new roof but in an older roof it is easier to replace the entire roof rather than make piecemeal repairs.

Flashing. Look for rust spots and loose areas. Explain that rusted areas should be cleaned and painted. Metal that is rusted through should be replaced. Check for dried out caulking and emphasize that dry, stiff caulking should be removed and replaced.

Loose and missing nails. Look for nail heads that have torn the shingle or are no longer tight against the shingle. If you see poorly nailed shingles, comment on it but don't dwell on it. Your prospect will note that you recognize substandard practice but you won't gain stature in his eyes by criticizing whoever installed the existing roof, especially if he installed the roof himself.

Worn spots. Check the roof for patches of dark gray or black where granules have worn off. Also, check gutters and the area around downspouts for concentrations of loose granules. Explain that large amounts of loose granules are the best indication that the roof has lost its ability to withstand further sun, wind and moisture.

Valleys. Be especially aware of these critical areas. Look for torn and worn spots. Explain that leaks caused by a faulty valley can appear anywhere and should be corrected at once.

Gutters. Explain the importance of good gutters and that water from one slope should never be allowed to drain onto a different lower slope. Explain that leaves and debris in a gutter can cause water to back up and soak under shingles.

Roof attachments. Look for signs that anything has been attached to or nailed to the roof. Explain that a T.V. antenna anchored to a vent without proper guy wires will vibrate and loosen caulking around the flashing.

Now give a well considered evaluation of what is needed. Don't recommend a new roof if one isn't needed. Go ahead and recommend minor repairs if that is what is called for. Explain that temporary repairs will be just that, temporary. Moreover, you can't guarantee that a similar problem won't develop next season. If you recommend repair rather than replacement and you do a good job of making the repair, you can be sure of getting a call when replacement is necessary.

If replacement is necessary, measure the roof while you are there and make a sketch to use in preparing the estimate. Explain why replacement is necessary and use your notes to make sure you cover every point. Make a sketch, if you want, to show how water can enter a roof at one point and run under shingles, felt and sheathing for some distance before leaking into the house. Explain that it may be possible to add a new roof without removing the old roof and that this will both save money and provide better roof insulation when completed. Don't make any promises about the sheathing if the roof is badly deteriorated. If a tear-off is required, make it clear that you cannot evaluate the cost of replacing sheathing until it is exposed. If the flashing is in good condition, explain that it can be cleaned, painted and reused.

Once your client accepts your opinion that a new roof is needed, your job of selling roofing really begins. Anyone can sell a no-frills, minimum quality roofing job on a leaky roof in the middle of the rainy season. As a roofer you know that there is a trend away from the 240 pound basic white or dark shingles and to heavier weight and fiberglass textured shingles in brown and earth colors. Point out an example of the newer shingles on one of the better homes in the neighborhood. Some of the asphalt shingles have heavily textured surfaces and look like wood shingles. Explain that many of the newer shingles not only look better, but wear better, some being guaranteed for 20 or 25 years. Modern asphalt shingles carry a U.L. class "C" or Class "A" fire rating and are wind resistant. In short, in the last 15 or 20 years there have been many improvements in roofing shngles that make it sensible to modernize the roof. Reroofing with the newer shingles will improve both the appearance and resale value of the property. Recommend one particular high quality shingle and offer an opinion on what color goes well with his house.

Now is the time to bring out some samples and pictures. Nearly all manufacturers of roofing products offer full color pictures and samples of their products. Many general circulation magazines carry full page color advertisements placed by roofing manufacturers featuring large, neatly landscaped homes with attractive shingle roofs. Have some of these and some color samples available so your prospect can visualize the new roof on his home. Stand back and let your prospect compare shades and textures against the house. Remember, your prospect will buy the admiration of family, friends and neighbors. He will buy security and piece of mind. He is **not** going to buy just a set of shingles.

Your first samples and pictures should be of top of the line shingles you recommend. Point out the rich texture and deep shadow lines. Have your prospect pick up a sample and feel the heavy weight. Emphasize that these better shingles carry a 20 or 25 year guarantee rather than the usual 15 year guarantee. Then offer to figure an estimate for a reroof job using these better shingles. As you are figuring the cost, make these important points about top of the line shingles. A good 25 year roof costs only about 1/3 more than the least expensive asphalt shingle roof but is guaranteed to last nearly twice as long. More important, many buildings will support only one additional roof before the entire roof surface has to be removed. The first reroof should be done with a very good shingle. Finally, emphasize that the better shingles just look better and will continue to look better for many years.

You know that selling better quality shingles increases your volume and profits. Your client knows this too. As a salesman you want to convince him that the better shngles offers the benefit of better appearance and lower cost per

Roof Inspection Checklist

Address ———————————— Name Of Owner ———————————— Date ————————————

1. **Approximate pitch** ————————————————————————————

2. **Existing roof cover** ————————————————————————————

3. **General roof condition**
   - ☐ Missing shingles ————————————————————————
   - ☐ Blistered area ————————————————————————————
   - ☐ Split or curled shingles ——————————————————————
   - ☐ Worn areas ————————————————————————————
   - ☐ Other ————————————————————————————————

   ——

   ——

4. **Flashing**
   - ☐ Loose areas ————————————————————————————
   - ☐ Rusted ————————————————————————————————
   - ☐ Caulking deteriorated ————————————————————————
   - ☐ Other ————————————————————————————————

   ——

5. **Nailing**
   - ☐ Loose heads ————————————————————————————
   - ☐ Torn shingles ————————————————————————————
   - ☐ Rusted out heads ————————————————————————
   - ☐ Nails missing————————————————————————————
   - ☐ Other ————————————————————————————————

   ——

6. **Worn spots**
   - ☐ Granules worn off————————————————————————
   - ☐ Discolored areas ————————————————————————

   ——

7. **Valleys**
   - ☐ Wear evident————————————————————————————
   - ☐ Tears ————————————————————————————————
   - ☐ Faulty design ————————————————————————————
   - ☐ Poorly constructed ——————————————————————
   - ☐ Other ————————————————————————————————

   ——

8. **Gutters**
   - ☐ Faulty design ————————————————————————————
   - ☐ Poorly built ————————————————————————————
   - ☐ Filled with debris ————————————————————————
   - ☐ Deteriorated ————————————————————————————
   - ☐ Other ————————————————————————————————

   ——

9. **Roof attachments**
   - ☐ Type ————————————————————————————————

   ——
   - ☐ Damage apparent————————————————————————

10. **Recommendation** ————————————————————————

   ——

   ——

11. **Roof dimensions**————————————————————————

Roofing inspection checklist
Figure 10-39

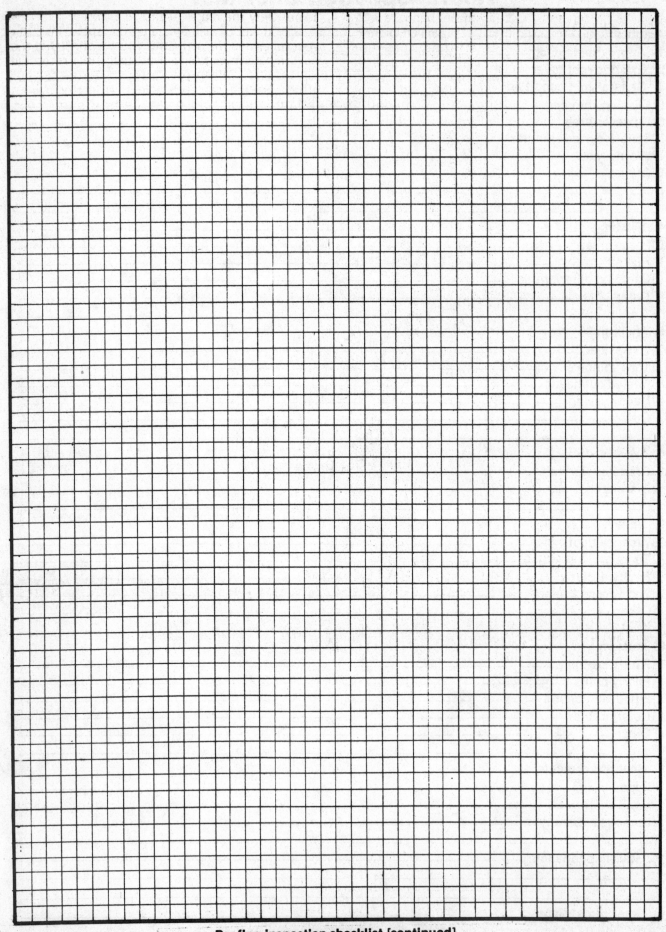

Roofing inspection checklist [continued]
Figure 10-39

year of useful life and that these benefits outweigh the higher initial cost.

If your prospect balks at accepting your suggestions, offer to refigure the job based on lower cost, standard quality shingles. Be honest in explaining the difference in durability and weight between the better quality and lower quality shingle. Explain that the labor cost is nearly the same. Only the material cost is different.

If you can't make a sale on the spot, file away your notes and the estimate. You may find that the next rainstorm makes your prospect very receptive to your offer. When your prospect agrees to your offer, use the contract below. It protects you, is fair, and should make getting paid much easier.

Proposal and Contract

Date _____ 19 _____

To _____

Job Location: _____

Dear Sir:
   We propose to furnish all materials and perform all labor necessary to complete the following:
_____
_____
_____
_____

   All of the above work to be completed in a substantial and workmanlike manner for the sum of
_____ Dollars ($ _____)

   Payments to be made as follows: _____
_____

   Except as noted above, the price quoted does not include removing or replacing fascia, trim, sheathing, rafters, structural members, siding, masonry, vents, roofing, caulking, metal edging or flashing of any type. If, during the course of work, it should become apparent that any portions of the structure should be repaired or replaced, Owner may authorize Contractor to do additional work and charge Owner for the additional labor and materials required plus a reasonable profit.

   The price quoted is for prompt acceptance. Delay in acceptance will require a verification of prevailing labor and material costs.

By _____   Address _____

_____   License No. _____

   You are hereby authorized to furnish all materials and labor required to complete the work according to the terms of this proposal, for which we agree to pay as described above.

Owner _____

Date _____   Owner _____

Chapter 11
Windows and Doors

A little imagination goes a long way to create entrances and windows of distinction in home remodeling.

Over 5 million American homes will have some form of exterior remodeling done on them each year. The opportunities to up-sell each job are spelled out by consumer research which says, for example, that decorative panel doors are preferred 3 to 1 over flush doors in the residential entry door market. Of all the exterior products on the market today, nothing has a stronger impact on a home's "curb appeal" than entry doors and windows. To complete a siding job without suggesting a new entry door and windows deprives you of profit and the homeowner of a chance to do what he wanted done in the first place—beautify the exterior of his home.

In some cases it is as easy as replacing the doors and windows. A number of suppliers have elegant single and double door entrance systems that are elegant to behold and which can be installed in a matter of hours. More than your own profits are involved when you are selling any form of exterior remodeling. That job becomes your neighborhood hallmark. Every salesman should be encouraged to sell entrance doors and windows on every exterior job.

Windows

Windows may need repair, replacement, or relocation. Until recent years, windows were not generally treated with a preservative, so moisture may have gotten into some joints and resulted in decay. Also, older windows may allow more air infiltration than newer types. It will often be desirable to replace the windows. This is not difficult where the same size window can be used; however, where the window size is no longer produced, replacement of the window will also mean reframing it.

Since window openings can account for as much as 40 percent of the exterior wall area in an older building, selection of the proper windows for a specific remodeling job can have an important effect on the building's facade and function. The simple job of changing old windows for new can give a tired building a fresh personality and even influence its architectural styling without making major structural changes

Modifying Opening Sizes

Once the old window is removed there remains a rough opening to fill—just as in new construc-

tion. And, as in new construction, a wide range of window styles and sizes is available to carry out the architectural theme, bring light into the building, or filter it out where excessive sunlight might be a problem. Most important, double pane insulating glass in quality windows reduces heat loss and heat gain for maximum conservation of energy and low long range operating costs. The selection of windows that require only minimum maintenance over many years of service also can help to increase the return on a remodeling investment.

Ease of installation in existing openings is a factor that can substantially affect the cost of such a remodeling job. Some stock size window will nearly always fit, or can be made to fit into a rough opening in an older building. Installation of new windows is expedited by factory accessories such as jamb extenders, which fit windows to different wall thicknesses. Several jamb extenders of different widths are available to adapt stock windows readily to a wide range of standard wall thicknesses. Where wall thickness is not standard, as in many old frame and masonry homes, windows can be fitted by cutting off enough material from the edges of an oversize set of jamb extenders to bring them flush with the existing interior wall surface. Alignment of these jamb extenders is easy and fast; the tongued edge of the extender fits snugly into a groove in the jamb and is secured by nailing.

Windows set in off-size framed walls and in some masonry walls do not always require jamb extenders; recessing the window so that the inside of the jamb is flush with the interior wall may be sufficient. Trimming out with interior casing around the jamb and a stool and apron where applicable completes the window installation on the inside.

Rigid vinyl auxiliary exterior casing further simplifies installation of windows and seals out the weather. Where the exterior wall material outlines the rough opening exactly, as in masonry walls where new windows fit without modifying the opening, metal jamb-fastening clips and auxiliary vinyl casing are available to expedite installation. The auxiliary casing is applied to the sides and head of the jamb to overlap the exterior wall when the window is set in place. Then the metal clips are nailed to the jambs and secured to masonry walls with masonry fasteners to hold the window securely in position. In frame walls, the clips are simply bent over the inside edges of the

WINDOW TYPE	HOW DOES IT OPERATE?	IS IT EASY TO CLEAN?	HOW IS IT FOR VIEWING?	HOW IS IT FOR VENTILATION?
FIXED	Does not open, so requires no screens or hardware	Outside job for exterior of window	No obstruction to views or light	No ventilation Minimum air leakage
SLIDING				
Double-Hung	Sash pushes up and down. Easy to operate except over sink or counter	Inside job if sash is removable	Horizontal divisions can cut view	Only half can be open
Horizontal-Sliding	Sash pushes sideways in metal or plastic tracks	Inside job if sash is removable	Vertical divisions cut view less than horizontal divisions	Only half can be open
SWINGING				
Casement (Side Hinged)	Swings out with push-bar or crank Latch locks sash tightly	Inside job if there is arm space on hinged side	Vertical divisions cut view less than horizontal	Opens fully. Can scoop air into house
Awning (Top Hinged)	Usually swings out with push-bar or crank. May swing inward when used high in wall	Usually an inside job unless hinges prevent access to outside of glass	Single units offer clear view. Stacked units have horizontal divisions which cut view	Open fully. Upward airflow if open outward; downward flow when open inward
(Bottom Hinged)	Swings inward, operated by a lock handle at top of sash	Easily cleaned from inside	Not a viewing window, usually set low in wall	Airflow is directed upward
Jalousie	A series of horizontal glass slats open outward with crank	Inside job, but many small sections to clean	Multiple glass divisions cutting horizontally across view	Airflow can be adjusted in amount and direction

Guide to window selection
Table 11-1

framing studs and nailed after setting and aligning the window.

Blocking with 2x4s brings odd-shaped openings to the required size. Usually, these tall windows were set low in the walls of older buildings, with the head at door height and the sill close to the floor. Just nailing new sill framing across the bottom of the opening at a height appropriate for the new window often will bring the rough opening to the exact size. If the opening still is a bit wide for the stock-size window, nailing trimmer studs at one or both sides will reduce the width. A plywood panel, or any low-maintenance material, trimmed with molding if desired for a decorative effect, can be nailed beneath the window to fill the space formerly occupied by the tall window. Also, a veneer of used brick can be laid up to the new sill to retain the continuity of a masonry wall.

Many window distributors employ architectural representatives who have the technical background and product "know-how" to assist remodeling contractors in designing the arrangement and proportioning of windows for older buildings in the planning stage of remodeling. These field men will also help work out any structural or installation problems that might arise in fitting stock-size windows to non-standard openings, or grouping windows together vertically or horizontally to create large glass areas or entire walls of windows.

Selecting The Window

Today's windows are picked more for the view and the light they let in than for the amount of ventilation they give. Ventilation, of course, has taken a back seat due to the more sophisticated heating and cooling systems of today's homes. But with the concerns for energy conservation, ventilation may come back as a substitute for air conditioning.

With all windows, you should consider four important selling points: What kind of viewing will it give? How easy is it to clean? Will it be easy to open and close where it's located? And, if important, will it give good ventilation? There are three basic types of windows to look at: 1) fixed, 2) sliding, and 3) swinging. Of course, there are many variations in design and style under these categories. See Table 11-1.

Choosing the type of window is only one part of window selection. You also need to pay attention to materials used in the makeup of that window. What's the glass area made of? How about the frame around it? And what about the quality of hardware? Note Table 11-2. Of course, cost may be a major factor in making the decision. To get a true picture, be sure you're comparing total costs. Examine the cost of the window unit itself. Then add to it the extras such as hardware and screens, plus installation and finishing costs. In general, it's usually wise to recommend the best quality windows.

GLASS AREA

Single-strength glass	Suitable for small glass panes. Longest dimension—about 40 inches
Double-strength glass	Thicker, stronger glass suitable for larger panes. Longest dimension—about 60 inches
Plate glass	Thicker and stronger for still larger panes. Also more free of distortion. Longest dimension—about 10 feet
Insulating glass	Two layers of glass separated by a dead air space and sealed at edges. Desirable for all windows in cold climates to reduce heating costs. Noise transmission is also reduced
Safety glass	Acrylic or plexiglass panels eliminate the hazard of accidental breakage. Panels scratch more easily than glass. Laminated glass as used in automobiles also reduces breakage hazard
Wood	Preferable in cold climates as there is less problem with moisture condensation. Should be treated to resist decay and moisture absorption. Painting needed on outside unless frame is covered with factory-applied vinyl shield or other good coating.
Aluminum	Painting not needed unless color change is desired. Condensation a problem in cold climates unless frame is specially constructed to reduce heat transfer. Often less tight than wood frames.
Steel	Painting necessary to prevent rusting unless it is *stainless* steel. Condensation a problem in cold climates
Plastic	Lightweight and corrosion-free. Painting not needed except to change color
HARDWARE	
	Best handles, hinges, latches, locks, etc. usually are steel or brass. Aluminum satisfactory for some items but often less durable. Some plastics and pot metal are often disappointing.

**Guide for window materials
Table 11-2**

Repair of Existing Windows

Where the wood in windows is showing some signs of deterioration but the window is still in good operating condition, a water-repellent preservative may arrest further decay. First, remove existing paint; then brush on the perservative, let it dry, and repaint the window. Paint cannot be used over some preservatives, so make sure a paintable preservative is used.

Double-hung sash may bind against the stops, jambs, or the parting strip. Before doing any repair, try waxing the parts in contact. If this doesn't eliminate the problem, try to determine where the sash is binding. Excessive paint buildup is a common cause of sticking and can be corrected by removing paint from stops and parting strips. Nailed stops (Figure 11-3) can be moved away from the sash slightly; if stops are fastened by screws, it will probably be easier to remove the stop and plane it lightly on the face contacting the sash. Loosening the contact between sash and stop

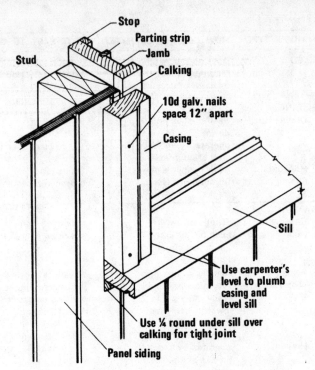

**Installation of double-hung window frame
Figure 11-3**

too much will result in excessive air infiltration at the window. If the sash is binding against the jamb, remove the sash and plane the vertical edges slightly.

It may be desirable to add full width weatherstrip and spring balance units to provide a good airtight window that will not bind. These are easily installed, requiring only removal of parting strip and stops. Install the units in accordance with manufacturer's instructions and replace the stops.

Replacement of Existing Windows

The sequence of window replacement will depend on the type of siding used. Where new panel siding is being applied, the window is installed after the siding. Where horizontal siding is used, the window must be installed before the siding.

If windows require extensive repairs, it will probably be more economical to replace them. New windows are usually purchased as a complete unit, including sash, frame, and exterior trim. These are easily installed where a window of the same size and type is removed or where the opening is modified to a standard size as explained previously. Most window manufacturers list rough opening sizes for each of their windows. Some general rules to follow for rough opening size are:

A. Double-hung window (single unit).
Rough opening width = glass width **plus** 6 inches.
Rough opening height = total glass height **plus** 10 inches.
B. Casement window (two sash)

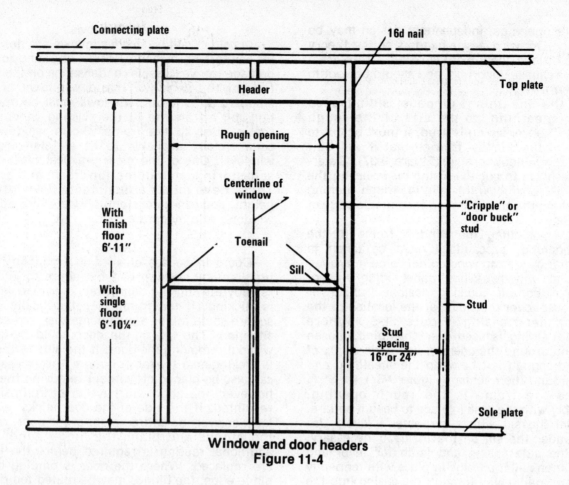

Window and door headers
Figure 11-4

Labels in figure: Connecting plate; 16d nail; Header; Top plate; Rough opening; Centerline of window; "Cripple" or "door buck" stud; With finish floor 6'-11"; With single floor 6'-10¼"; Toenail; Sill; Stud; Stud spacing 16" or 24"; Sole plate

Rough opening width = total glass width **plus** 11¼ inches.

Rough opening height = total glass height **plus** 6-3/8 inches.

After the existing window is removed, take off the interior wall covering to the rough opening width for the new window. If a larger window must be centered in the same location as the old one, half the necessary additional width must be cut from each side; otherwise, the entire additional width may be cut from one side. For windows 3½ feet or less in width, no temporary support of ceiling and roof should be required. Where windows more than 3½ feet wide are to be installed, provide some temporary support for the ceiling and roof before removing existing framing in bearing walls. Remove framing to the width of the new window and frame the window as shown in Figure 11-4. The header must be supported at both ends by cripple studs. Headers are made up of two 2 inch thick members, usually spaced with lath or wood strips to produce the same width as the 2 by 4 stud space. The following sizes might be used as a guide for headers:

Maximum Span	Header Size
Feet	Inches
3½	Two 2 by 6
5	Two 2 by 8
6½	Two 2 by 10
8	Two 2 by 12

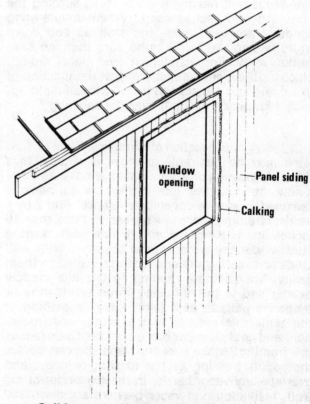

Calking around window opening before installing frame
Figure 11-5

Labels in figure: Window opening; Panel siding; Calking

For wider openings, independent design may be necessary. Do not oversize headers on the theory that, if a little is good, more is better. Cross-grain shrinkage causes distortion, and should be kept to a minimum.

Cut the sheathing, or panel siding used without sheathing, to the size of the rough opening. If bevel siding is used, it must be cut to the size of the window trim so that it will butt against the window casing (Figure 9-37). Determine the place to cut the siding by inserting the preassembled window frame in the rough opening and marking the siding around the outside edge of the casing.

Before installing the window frame in the rough opening, precautions must be taken to insure that water and wind do not come in around the finished window. Where panel siding is used, place a ribbon of caulking sealant (rubber or similar base) over the siding at the location of the side and head casing (Figure 11-5). Where horizontal siding is used over sheathing, loosen the siding around the opening and slide strips of 15 pound asphalt felt between the sheathing and siding around the opening (Figure 9-37)

Place the frame in the rough opening, preferably with the sash in place to keep it square, and level the sill with a carpenter's level. Use shims under the sill on the inside if necessary. Check the side casing and jamb for level and square; then nail the frame in place with tenpenny galvanized nails. Nail through the casing into the side studs and header (Figure 11-3), spacing the nails about 12 inches apart. When double-hung windows are used, slide the sash up and down while nailing the frame to be sure that the sash works freely. For installation over panel siding, place a ribbon of caulking sealer at the junction of the siding and the sill and install a small molding, such as a quarter-round, over the caulking.

Relocation of Windows

It may be desirable to move a window to a different location or to eliminate a window and add a new one at another location. Where a window is removed, close the opening as follows: Add 2 by 4 vertical framing members spaced no more than 16 inches apart. Keep framing in line with existing studs under the window or in sequence with wall studs so covering materials can be nailed to them easily. Toenail new framing to the old window header and to the sill using three eightpenny or tenpenny nails at each joint. Install sheathing of the same thickness as that existing, add insulation, and apply a vapor barrier on the inside face of the framing. Make sure the vapor barrier covers the rough framing of the existing window and overlaps any vapor barrier in the remainder of the wall. Insulation and vapor barriers are discussed more fully in Chapters 13 and 14. Apply interior and exterior wall covering to match the existing coverings on the house.

Storm Windows

In cold climates, storm windows are necessary for comfort, for economy of heating, and to avoid damage from excessive condensation on the inside face of the window. If old windows are not standard sizes, storm windows must be made by building a frame to fit the existing window and fitting glass to the frame. Storm windows are commercially available to fit all standard size windows. One of the most practical types is the self-storing or combination storm and screen. These have minor adjustments for width and height, and can be custom fabricated for odd size windows at a moderate cost.

Doors

Doors in houses of all ages frequently cause problems by sticking and by failure to latch. To remedy the sticking door, first determine where it is sticking. If the frame is not critically out of square some minor adjustments may remedy the situation. The top of the door could be planed without removing the door. If the side of the door is sticking near the top or bottom, the excess width can also be planed off without removing the door; however, the edge will have to be refinished or repainted. If the side of the door sticks near the latch or over the entire height of the door, remove the hinges and plane the hinge edge. Then additional routing is required before the hinges are replaced. Where the door is binding on the hinge edge, the hinges may be routed too deeply. This can be corrected by loosening the hinge leaf and adding a filler under it to bring it out slightly. If the latch does not close, remove the strike plate and shim it out slightly. Replace the strike plate by first placing a filler, such as a matchstick, in the screw hole and reinserting the screw so that the strike plate is relocated slightly away from the stop.

Exterior Doors

If exterior doors are badly weathered, it may be desirable to replace them rather than attempt a repair. Doors can be purchased separately or with frames, including exterior side and head casing with rabbeted jamb and a sill (see Figure 11-6).

Exterior doors should be either panel or solid-core flush type. Several styles are available, most of them featuring some type of glazing (Figure 11-7). Hollow-core flush doors should be limited to interior use, except in warm climates, because they warp excessively during the heating season when used as exterior doors. Standard height for exterior doors is 6 feet 8 inches. Standard thickness is 1¾ inch. The main door should be 3 feet wide, and the service or rear door should be at least 2 feet 6 inches and preferably up to 3 feet wide.

Where rough framing is required either for a new door location or because old framing is not square, provide header and cripple studs as shown in Figure 11-4. Rough opening height should be

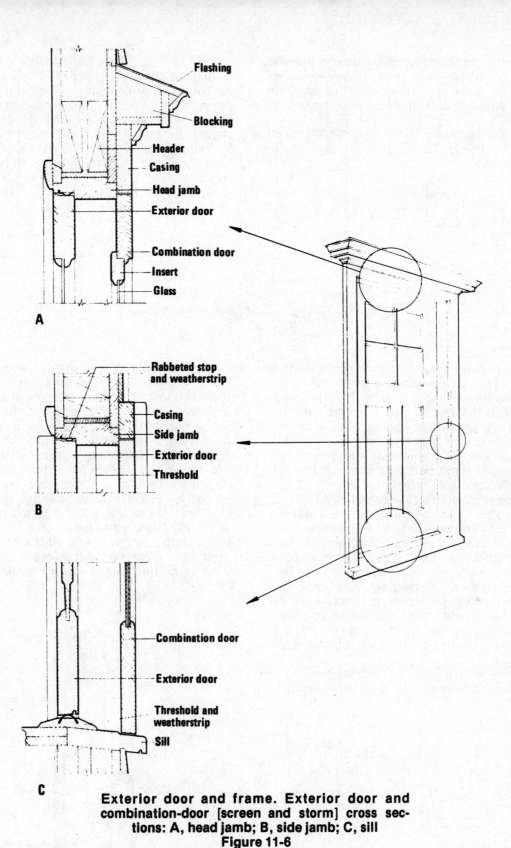

Flashing

Blocking

Header

Casing

Head jamb

Exterior door

Combination door

Insert

Glass

A

Rabbeted stop and weatherstrip

Casing

Side jamb

Exterior door

Threshold

B

Combination door

Exterior door

Threshold and weatherstrip

Sill

C

Exterior door and frame. Exterior door and combination-door [screen and storm] cross sections: A, head jamb; B, side jamb; C, sill
Figure 11-6

the height of the door plus 2¼ inches above the finished floor; width should be the width of the door plus 2½ inches. Use doubled 2 by 6's for headers and fasten them in place with two sixteenpenny nails through the stud into each member. If the stud space on each side of the door is not accessible, toenail the header to studs. Nail cripple (door buck) studs, supporting the header, on each side of the opening to the full stud with twelvepenny nails spaced about 16 inches apart and staggered.

After sheathing or panel siding is placed over

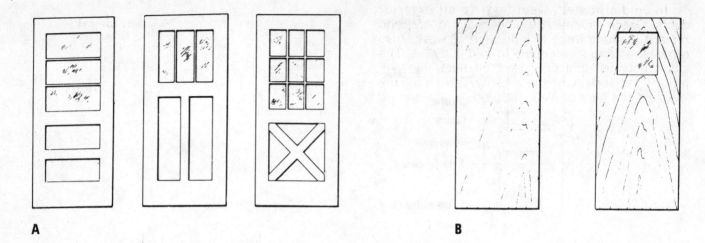

A　　　　　　　　　　　　　　　　　**B**

Exterior doors: A, panel type, B, flush type
Figure 11-7

the framing, leaving only the rough opening, the door frame can be installed. Apply a ribbon of caulking sealer on each side and above the opening where the casing will fit right over it. Place the door frame in the opening and secure it by nailing through the side and head casing. Nail the hinge side first. In a new installation the floor joists and header must be trimmed to receive the sill before the frame can be installed (Figure 11-8). The top of the sill should be the same height as the finish floor so that the threshold can be installed over the joint. Shim the sill when necessary so that it will have full bearing on the floor framing. When joists are parallel to the sill, headers and a short support member are necessary at the edge of the sill. Use a quarter-round molding in combination with caulking under the door sill when a panel siding or other single exterior covering is used. Install the threshold over the junction with the finish floor by nailing it to the floor and sill with finishing nails.

Exterior doors are usually purchased with an entry lock set which is easily installed. Any

trimming to reduce the width of the door is done on the hinge edge. Hinges are routed or mortised into the edge of the door with about 3/16 or 1/4 inch back spacing (Figure 11-9). In-swininging exterior doors require 3½ by 3½ inch loose-pin hinges. Nonremovable pins are used on out-swinging doors. Use three hinges to minimize warping. Bevel the edges slightly toward the side that will fit against the stops. Clearances are shown in Figure 11-10. Carefully measure the opening width and plane the edge for the proper side clearances. Next, square and trim the top of the door for proper fit; then saw off the bottom for the proper floor clearance. All edges should then be sealed to minimize entrance of moisture.

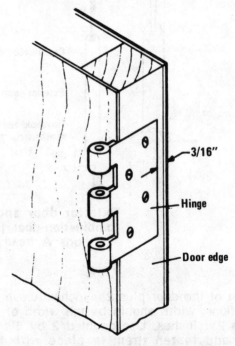

Installation of door hinges
Figure 11-9

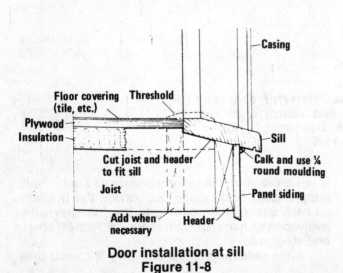

Door installation at sill
Figure 11-8

In cold climates, weatherstrip all exterior doors. Check weatherstripping on old doors, and replace it where there is indication of wear. Also consider adding storm doors in cold climates. The storm door will not only save on heat, but will protect the surface of the exterior door from the weather and help prevent warping.

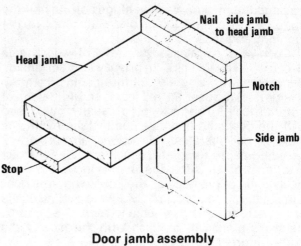

Door jamb assembly
Figure 11-12

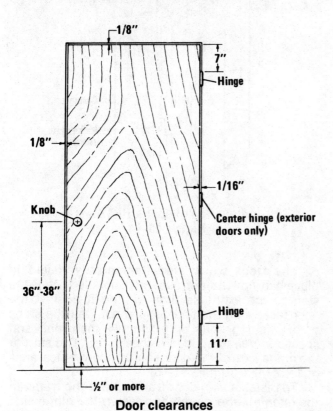

Door clearances
Figure 11-10

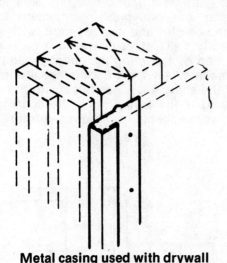

Metal casing used with drywall
Figure 11-11

Interior Doors

If a new interior door is added or the framing is replaced, the opening should be rough framed in a manner similar to that for exterior doors. Rough

framing width is 2½ inches plus the door width; height is 2 inches plus the door height above the finished door. The head jamb and two side jambs are the same width as the overall wall thickness where wood casing is used. Where metal casing is used with drywall (Figure 11-11), the jamb width is the same as the stud depth. Jambs are often purchased in precut sets, and can even be purchased complete with stops and with the door prehung in the frame. Jambs can also easily be made in a small shop with a table or radial-arm saw (Figure 11-12). The prehung door is by far the simplest to install, and is usually the most economical because of the labor savings. Even where the door and jambs are purchased separately, the installation is simplified by prehanging the door in the frame at the building site. The door then serves as a jig to set the frame in place quite easily. Before installing the door, temporarily put in place the narrow wood strips used as stops. Stops are usually 7/16 inch thick and may be 1½ to 2¼ inches wide. Install them with a mitered joint at the junction of the head and side jambs. A 45 degree bevel cut 1 to 1½ inch above the finish floor will eliminate a dirt pocket and make cleaning easier (Figure 11-13). This is called a sanitary stop.

Fit the door to the frame, using the clearances shown in Figure 11-10. Bevel the edges slightly toward the side that will fit against the stops. Route or mortise the hinges into the edge of the door with about a 3/16 or 1/4 inch back spacing. Make adjustments, if necessary, to provide sufficient edge distance so that screws have good penetration in the wood. For interior doors, use two 3 by 3 inch loose-pin hinges. If a router is not available, mark the hinge outline and depth of cut, and remove the wood with a wood chisel. The surface of the hinge should be flush with the wood surface. After attaching the hinge to the door with screws, place the door in the opening, block it for proper clearances, and mark the location of door hinges on the jamb. Remove the door and route

167

the jamb the thickness of the hinge half. Install the hinge halves on the jamb, place the door in the opening, and insert the pins.

Lock sets are classed as: (a) Entry lock sets (decorative keyed lock); (b) privacy lock set (inside lock control with a safety slot for opening from the outside); (c) lock set (keyed lock); and (d) latch set (without lock). The lock set is usually purchased with the door and may even be installed with the door. If not installed, directions are provided, including paper templates which provide for exact location of holes. After the latch is installed, mark the location of the latch on the jamb when the door is in a near-closed position. Mark the outline of the strike plate for this position and route the jamb so the strike plate will be flush with the face of the jamb (Figure 11-14).

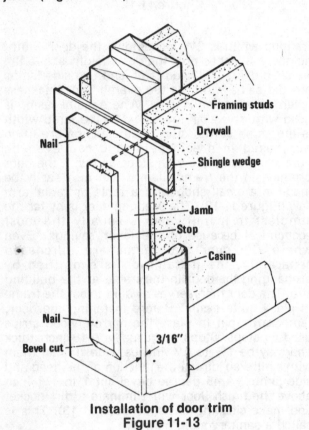

Installation of door trim
Figure 11-13

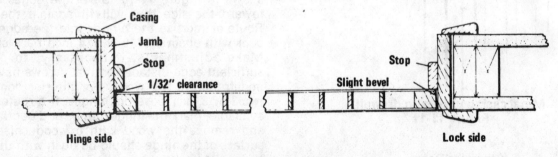

Installation of door strike plate
Figure 11-14

The stops which were temporarily nailed in place can now be permanently installed. Nail the stop on the lock side first, setting it against the door face when the door is latched. Nail the stops with finishing nails or brads 1½ inch long and spaced in pairs about 16 inches apart. The stop at the hinge side of the door should allow a clearance of 1/32 inch (Figure 11-15).

To install a new door frame, place the frame in the opening and plumb and fasten the hinge side of the frame first. Use shingle wedges between the side jamb and the rough door buck to plumb the jamb (Figure 11-13). Place wedge sets at hinge and latch locations plus intermediate locations along the height, and nail the jamb with pairs of eightpenny nails at each wedge area. Continue installation by fastening the opposite jamb in the same manner. After the door jambs are installed, cut off the shingle wedges flush with the wall.

Casing is the trim around the door opening. Shapes are available in thicknesses from ½ to ¾

Plan View

Door stop clearances [plan view]
Figure 11-15

inch and widths varying from 2¼ to 3½ inches. A number of styles are available (Figure 11-16). Metal casing used at the edge of drywall eliminates the need for wood casing (Figure 11-11).

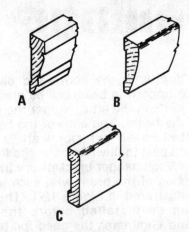

Styles of door casings: A, colonial; B, ranch; C, plain
Figure 11-16

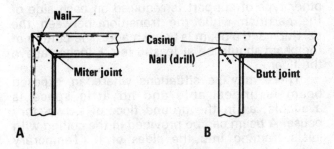

Installation of door trim: A, molded casing; B, rectangular casing; C, metal casing
Figure 11-17

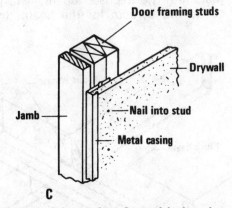

Position the casing with about a 3/16 inch edge distance from the face of the jamb (Figure 11-13). Nail with sixpenny or sevenpenny casing or finishing nails, depending on the thickness of the

casing. Casing with one thin edge should be nailed with 1½ inch brads along the edge. Space nails in pairs about 16 inches apart. Casings with molded forms must have a mitered joint where the head and side casings join (Figure 11-17A), but rectangular casings are butt-joined (Figure 11-17B).

Metal casing can be installed by either of two methods. In one method, the casing is nailed to the door buck around the opening; then the dry wall is inserted into the groove and nailed to the studs in the usual fashion. The other method consists of first fitting the casing over the edge of the drywall, positioning the sheet properly, and then nailing through the drywall and casing into the stud behind (Figure 11-17C). Use the same type of nails and spacing as for drywall alone.

Interior doors are either panel or flush type (Figure 11-18). Flush doors (Figure 11-18 A) are usually hollow core. Moldings are sometimes included on one or both faces. Such moldings can also be applied to existing doors where added decoration is desired. The panel-type doors are available in a variety of patterns. Two popular patterns are the five-cross-panel and the colonial.

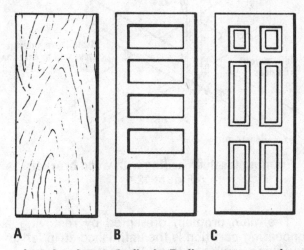

Interior doors: A, flush; B, five-cross-panel; C, colonial panel type
Figure 11-18

Standard door height is 6 feet 8 inches; however, a height of 6 feet 6 inches is sometimes used with low ceilings, such as in the upstairs of a story-and-a-half house or in a basement. Framing is limited in hollow-core doors so, if the framing is removed in major cutting to length, that framing must be replaced. Door widths vary, depending on use and personal taste; however, minimums may be governed by building regulations. Usual widths are: (a) bedroom and other rooms, 2 feet 6 inches; (b) bathroom, 2 feet 4 inches; (c) small closet and linen closet, 2 feet.

Chapter 12
Walls and Ceilings

One of the most common occurrences in remodeling is the moving of partitions for a more convenient room layout. It may mean eliminating a partition or adding a partition in another location.

Removing a Partition

The nonbearing partition is easily removed because none of the structure depends upon it. If the covering material is plaster or gypsum board, it cannot be salvaged, so remove it from the framing in any manner. The framing can probably be reused if it is removed carefully.

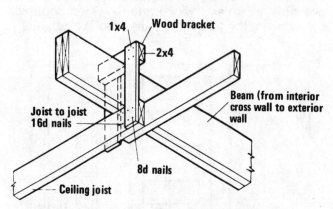

Framing for flush ceiling with wood brackets
Figure 12-1

The main problem presented by removing a nonbearing partition is the unfinished strip left in the ceiling, wall, and floor. This unfinished strip in the ceiling and wall is easily finished by plastering to the same thickness as the existing plaster or by cutting strips of gypsum board to fit snugly into the unfinished strip and finishing the joints with joint compound and tape. Wood flooring can also be patched by inserting a wood strip of the same thickness and species as the existing floor. If the existing flooring runs parallel to the wall, patching is fairly effective; but where the flooring runs perpendicular to the wall, the patch will always be obvious unless a new floor covering is added. In making the patch, cut the flooring to fit as snugly as possible. Even where the flooring is well fitted and of the same species, it may not be exactly the same color as the existing flooring.

Removing a load-bearing partition involves the same patching of walls, ceiling, and floor as the nonbearing partition but, in addition, some other means of supporting the ceiling joists must be provided. If attic space above the partition is available, a supporting beam can be placed above the ceiling joists in the attic, so that the joists can hang from the beam. The ends of the beam must be supported on an exterior wall, a bearing partition, or a post that will transfer the load to the foundation. Wood hanger brackets are installed at the intersection of the beam with each joist. One method is illustrated in Figure 12-1. This type of support can be installed before the wall is removed, and eliminates the need for temporary support.

Where an exposed beam is not objectionable, it can be installed after the partition is removed. A series of jacks with adequate blocking or some other type of support is required on each side of the partition while the transition between the partition and a beam is being made. The bottom of the beam should be at least 6 feet 8 inches above the floor.

There may be situations where an exposed beam is undesirable and no attic space is available, as in the ground floor of a two-story house. A beam can be provided in the ceiling with joists framing into the sides of it. Temporary support for the joists is required similar to that used for installing the exposed beam. The joists must be cut to make room for the beam. Install

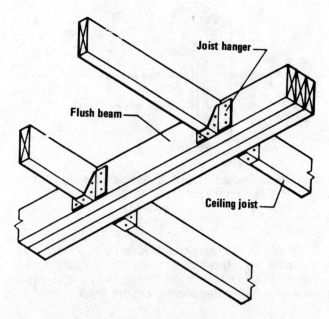

Flush beam with joist hangers
Figure 12-2

joist hangers on the beam where each joist will frame into it (Figure 12-2). Put the beam in place and repair the damaged ceiling.

The size beam required will vary greatly, depending on beam span, span of joists framing into it, and material used for the beam. Determination of beam size should be made by an engineer or someone experienced in construction.

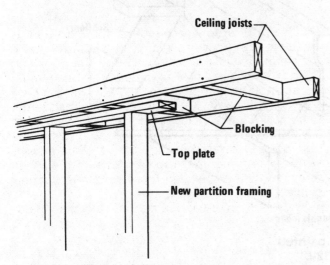

Blocking between joists to which the top plate of a new partition is nailed
Figure 12-3

Adding a Partition

A partition is added by simply framing it in, much as in new construction. Framing is usually done with 2 by 4's, although 2 by 3 inch framing is also considered adequate for partitions. The first step in framing should be to install the top plate. If ceiling joists are perpendicular to the partition, nail the top plate to each joist using sixteenpenny nails. If ceiling joists are parallel to the partition and the partition is not directly under a joist, install solid blocking between joists at no more than a 2 foot spacing (Figure 12-3) and nail the top plate to the blocking. To assure a plumb partition, hold a plumb bob along the side of the top plate at several points and mark these points on the floor. Nail the sole plate to the floor joists or to solid blocking between joists in the same manner as the top plate was nailed to the ceiling joists. The next step is to install studs to fit firmly between the plates at a spacing of 16 inches. Check the required stud length at several points. There may be some variation. Toenail the studs to the plates, using eightpenny nails. If conditions permit, it may be easier to partially assemble the wall on the floor and tilt it into place. First, the top plate is nailed to the studs and the frame tilted into place, after which studs are toenailed to the bottom plate as above. Frame in doors where desired in the manner described in Chapter 11. The partition is then ready for wall finish and trim, which are discussed later in this chapter.

Framing For Utilities

Certain construction practices are required to accommodate plumbing, heating and electrical service. Because heating ducts, plumbing stacks and drains, water piping, and electrical conduit must be run throughout the house, it is usually difficult to avoid cutting into the structure to accommodate them. Mechanical trades should be cautioned to avoid cutting if possible, and instructed in the manner of cutting where it is unavoidable. At critical points the structure can be altered, but only by or at the direction of someone qualified. Cut members can often be reinforced and any new walls or other framing can be built to accommodate mechanical items.

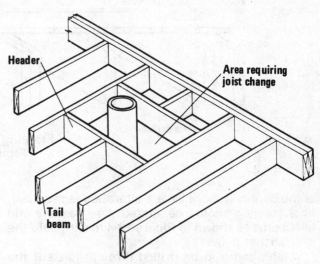

Headers for joists to eliminate cutting
Figure 12-4

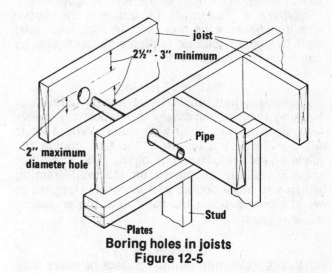

Boring holes in joists
Figure 12-5

Cutting Floor Joists

Floor joists must often be cut to accommodate pipe for water supply lines or electrical conduit. This may be in the form of notches at top or bottom of the joist, or as holes drilled through the joist. Notching of joists should be done only in the end quarter of the span and to not more than one-sixth

171

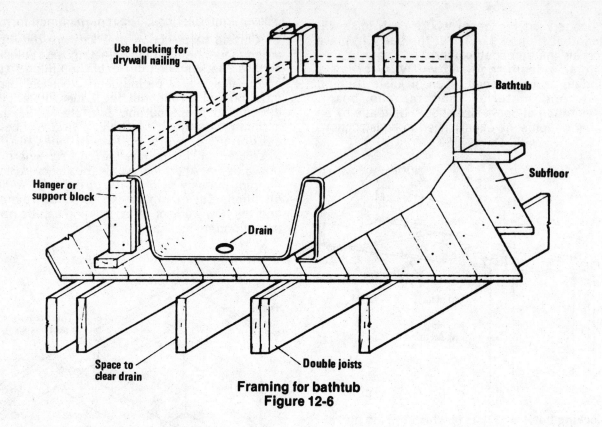

Framing for bathtub
Figure 12-6

of the depth. If more severe alteration is required, floor framing should be altered. Use headers and tail beams as shown in Figure 12-4 to eliminate the joist at that point.

Holes can also be drilled through joists if the size of hole is limited to 2 inches in diameter and the edges of the hole are not less than 2½ inches from the top or bottom of the joist (Figure 12-5).

Where a joist must be cut and the above conditions cannot be met, add an additional joist next to the cut joist or reinforce the cut joist by nailing scabs to each side.

Bathtub Framing

Where a bathtub is to be added, additional framing may be necessary to support the heavy weight of the tub filled with water. Where joists are parallel to the length of the tub, double the joists under the outer edge of the tub. The other edge is usually supported on the wall framing which also has a double joist under it. Hangers or wood blocks support the bathtub at the enclosing walls (Figure 12-6).

Utility Walls

Walls containing plumbing stack or vents may require special framing. Four inch soil stacks will not fit in a standard 2 by 4 inch stud wall. Where a thicker wall is needed, it is usually constructed with 2 by 6 inch top and bottom plates and 2 by 4 inch studs placed flatwise at the edge of the plates (Figure 12-7, A). This leaves the center of the wall open for running both supply and drain pipes through the wall.

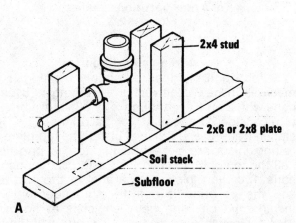

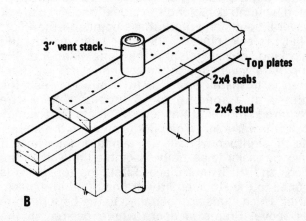

Framing for vent stack: A, 4-inch soil pipe;
B, 3-inch stack vent
Figure 12-7

172

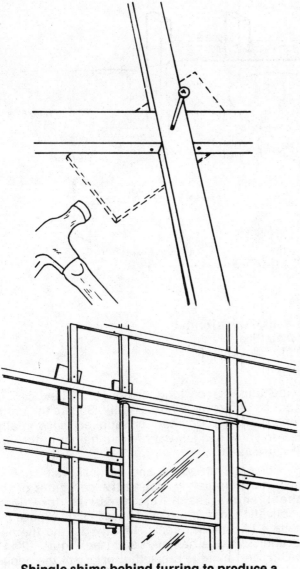

**Shingle shims behind furring to produce a
smooth vertical surface
Figure 12-8**

Three inch vent stacks do fit into 2 by 4 inch stud walls; however, the hole for the vent requires cutting away most of the top plate. Scabs cut from 2 by 4's are then nailed to the plate on each side of the vent to reinforce the top plate (Figure 12-7, B).

Interior Wall Finish

Minor cracks in plaster or drywall can be easily patched by filling the crack with a plaster-patching mix and sanding after the plaster dries. A fiberglass fabric applied over the crack helps prevent recurrence. This works well when the cracks are limited in number, but if plaster is generally cracked or pulled loose from its backing, a new covering material should be used.

In houses that require a new wall finish, some type of drywall sheet material is usually the most practical. The application of most drywall requires few special tools, is inexpensive, and it can be applied in a manner to smooth out unevenness and to cover imperfections. The most common forms of drywall are gypsum board, plywood, hardboard, fiberboard, and wood paneling. Drywall is usually applied to framing or to furring strips over the framing or existing wall finish. If the existing wall finish is smooth, new wall finish can sometimes be glued or nailed directly to the existing wall. In this direct application there is no thickness requirement for the new covering because it is continuously supported. For drywall applied over framing or furring, the recommended thicknesses for 16 and 24 inch spacing of fastening members are listed in the following tabulation:

| Finish | Minimum material thickness (inches) when framing is spaced | |
	16 Inches	24 inches
Gypsum board	3/8	1/2
Plywood	1/4	3/8
Hardboard	1/4	—
Fiberboard	1/2	3/4
Wood paneling	3/8	1/2

The ¼ inch plywood or hardboard may be slightly wavy unless applied over 3/8 inch gypsum board.

In order to prepare a room for a new wall finish, first locate each stud. They are usually spaced 16 inches apart and at doors and windows. The easiest way to find them is to look for nailheads on drywall or baseboard. These nails have been driven into studs. Where there is no evidence of nailheads, tap the wall finish with a hammer. At the stud, the sound will be solid; whereas, the space between studs will sound hollow. Commercial stud finders are also available at hardware and building supply stores. These operate by the use of a magnet that points to nail heads. Mark the stud locations in order to attach horizontal furring strips or to nail on paneling applied without furring strips.

Check walls for flatness by holding a straight 2 by 4 against the surface. Mark locations that are quite uneven. Also check for true vertical alignment by holding a large carpenter's level on

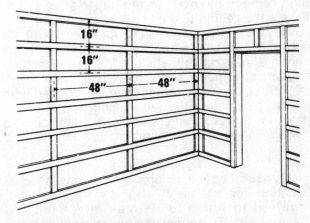

**Application of horizontal furring to interior wall
Figure 12-9**

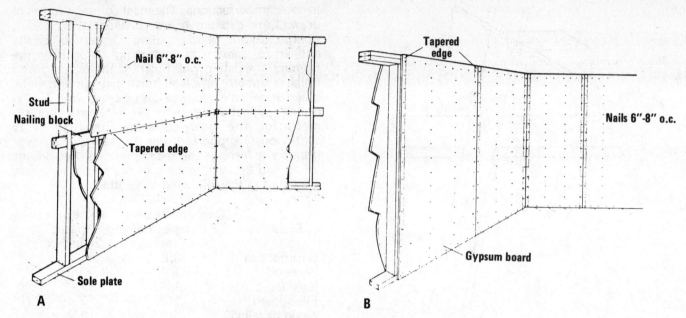

Installing gypsum board on walls: A, horizontal application; B, vertical application
Figure 12-10

the straight 2 by 4 against the wall. As furring is applied, use shingles as shims behind the furring where needed to produce a smooth vertical surface (Figure 12-8).

Apply standard 1 by 2 inch furring horizontally at 16 or 24 inch spacing, depending on the cover material to be used (Figure 12-9). Nail the furring at each stud. Remove existing base trim and window and door casings and apply furring around all openings. Also use vertical furring strips where vertical joints will occur in the drywall.

After this preparation, any of the usual drywall materials can be applied.

Ceilings can be finished with gypsum board or other sheet materials in much the same manner as interior walls. A variety of ceiling tiles can also be used, including the type for use with suspended metal or wood hangers. The suspended ceiling is particularly useful in rehabilitation of houses with high ceilings. It covers many imperfections, and lowers the ceiling to a more practical height. The space above the new ceiling may be useful for electrical wiring, plumbing, or heating ducts added during rehabilitation, and this mechanical equipment remains easily accessible.

Cracks in plaster ceilings can be repaired with plaster patching in the same manner as walls are patched; however, where cracks are extensive, a new ceiling is the only cure.

Gypsum Board

Gypsum board is one of the lowest cost materials for interior finish; however, the labor required to finish joints may offset the low material cost. This sheet material is composed of a gypsum filler faced with paper. Recessed edges accommodate tape for joints. Sheets are 4 feet

wide and 8 feet long or longer. They can be applied vertically or horizontally. Sheets the entire length of a room can be applied horizontally, leaving only one joint at the mid height of the wall (Figure 12-10).

For both horizontal and vertical wall application, nail completely around the perimeter of the sheet and at each furring strip; for direct application to framing, nail at each stud. Space the nails 6 to 8 inches apart. Lightly dimple the nail location with the hammerhead (see Figure 12-11), being careful not to break the surface of the paper. Minimum edge nailing distance is 3/8 inch. Screws or adhesive with nails are sometimes used instead of nails alone.

Gypsum board can be applied directly to ceiling joists by first removing existing ceiling material. It may also be applied directly over plaster or to furring strips nailed over the existing ceiling where plaster is uneven. Use 2 by 2 inch or 2 by 3 inch furring strips oriented perpendicular to the joists and spaced 16 inches on center for 3/8 inch gypsum board or 24 inches on center for ½ inch gypsum board. Nail the furring strips with two tenpenny nails at each joist.

A little thought and planning before you start can result in a better job and savings in materials and time. Make a sketch of the areas to be surfaced and lay out the board panels. Install the boards across (perpendicular to) the joists or studs whenever possible. Use as long a board as can be handled to eliminate or reduce end joints. For example, in a 12 foot by 13 foot room where the ceiling joists run parallel to the 13 foot dimension, it is desirable to use boards 12 feet long. If they are 8 feet long, an end joint would be necessary in

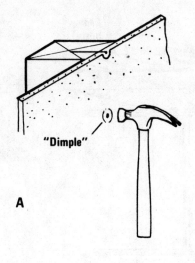

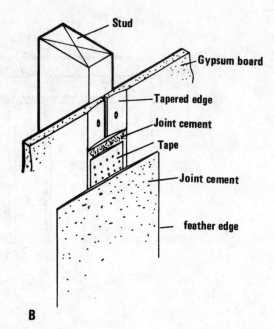

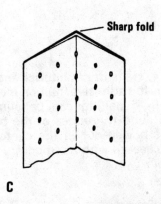

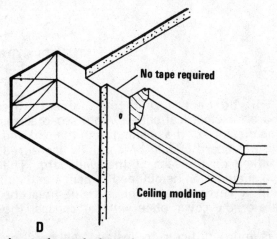

Preparing gypsum drywall sheets for painting: A, drive nails in "dimple" fashion; B, detail of joint treatment; C, corner tape; D, ceiling molding
Figure 12-11

each course. Where end joints cannot be avoided, they should be staggered. It is usually better to apply the board on the ceiling first, then the sidewalls.

It is often easier to use the adhesive/nail-on method of application. This method results in fewer nails to drive and conceal and makes a higher quality installation. Drywall adhesive is applied to the joists and studs before each piece of wallboard is positioned and nailed. The adhesive is applied to the framing member from a caulking gun in about a 3/8 inch diameter bead. For each 1,000 square feet of wallboard use eight quart size tubes of adhesive.

Using your sketch, determine the lengths and number of boards required. Nails can be estimated from Table 12-12. About 50 percent fewer nails are required if the adhesive/nail-on method is used.

Use a T-square and wallboard knife to cut the board to size. With the knife at right angles to the board, score completely through the face paper. Then apply firm even pressure to snap the board. Fold back the partially separated portion of the board and use the knife again to cut the back paper. Rough edges should be smoothed. Panels can be cut with a saw if desired.

Wallboard Thickness	Nail Type	Approx. lbs. per 1000 S.F. of gypsum wallboard
3/8″, 1/2″	1-5/8″ coated type drywall nail	5¼ lbs.
5/8″	1-1/8″ coated type drywall nail	5¼ lbs.

Estimating nails
Table 12-12

175

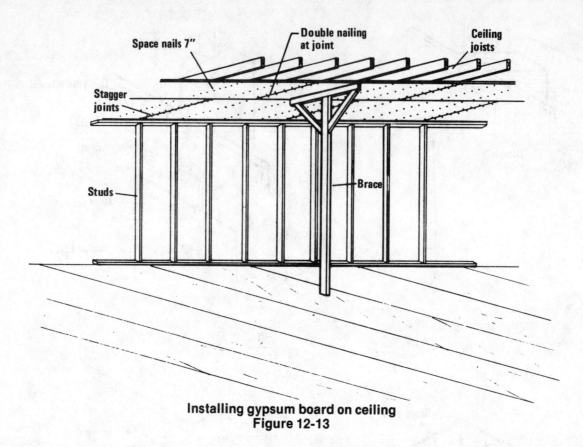

Installing gypsum board on ceiling
Figure 12-13

It will be necessary to cut holes in the wallboard for electrical outlets, light receptacles, etc. The distance of the opening from the end and edge of the board should be carefully measured and marked on the face of the wallboard. The opening should then be outlined in pencil and cut out with a keyhole saw. The cut out must be accurate or the cover plate will not conceal the hole.

It is more difficult to install wallboard on ceilings because of the overhead positioning. It is desirable to have T-braces to hold the board in place while it is being nailed. A satisfactory T-brace consists of a 2 foot piece of 1 x 4 nailed onto the end of a 2 x 4. See Figure 12-13. The length should be about an inch longer than the floor to ceiling height. Nails should be driven into the ceiling joists 7 inches apart. When the adhesive/nail-on method is used, the edges should be nailed, but only 1 nail per joist in the field of the board. All edges should be supported on framing. The nails should be driven to bring the board tight to the framing, then another blow struck to dimple the nail, being careful not to break the face paper. Apply the gypsum boards with end joints staggered and centered on a joist or furring strip. Place the sheets so there is only light contact at joints.

In horizontal application on sidewalls, install the top board first. Push the board up firmly against the ceiling and nail, placing nails 7 inches apart. One exception, however, is to keep all nails back 7 inches from interior ceiling angles. See Figure 12-14. Nails in the interior angles are quite apt to pop. If the adhesive/nail-on method is used, all of the field nailing can be eliminated. The nailing is around the edges of the board only. If the board is bowed out in the center, it may be advisable to secure with a temporary nail until the adhesive sets.

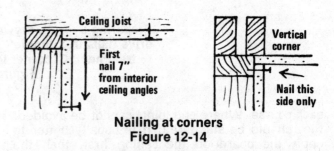

Nailing at corners
Figure 12-14

A vertical application places the long edges of the wallboard parallel to the framing members. This is more desirable if the ceiling height of your wall is greater than 8 feet 2 inches or the wall is 4 feet wide or less. Nailing recommendations are the same as for horizontal application.

To protect corners from edge damage, install metal corner bead after you have installed the wallboard. Nail the metal corner bead every 5 inches through the gypsum board into the wood frame. See Figure 12-15A.

The conventional method of finishing gypsum sheets includes the use of a joint cement and perforated joint tape. Some gypsum board is

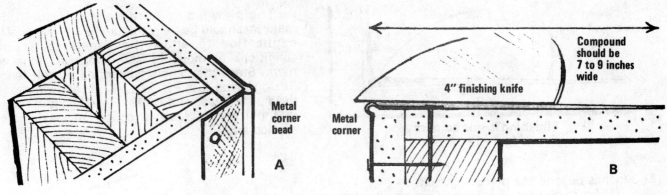

Corner bead: A, application; B, finishing
Figure 12-15

supplied with a strip of joint paper along one edge, which is used in place of the tape. After the gypsum board has been installed and each nail driven in a "dimple" fashion (Figure 12-11A), the walls are ready for treatment. Joint cement comes in powder or ready-mix form and should have a soft putty consistency so that it can be easily spread with a trowel or wide putty knife. The gypsum board edges are usually tapered so that, where two sheets are joined, there is a recessed strip to receive joint cement and tape. If a sheet has been cut, the edge will not be tapered. A square edge is taped in much the same manner as the beveled edge except the joint cement will raise the surface slightly at the seam, and edges have to be feathered out further for a smooth finish. A minimum of three coats of joint compound is recommended for all taped joints. This includes an embedded coat to bond the tape and two finishing coats over the tape. Each coat should dry thoroughly, usually 24 hours, so that the surface can be easily sanded. When sanding, sand the surface evenly. Do not over-sand or sand the paper surface. This may outline the joint or nail head through the paint.

The junction of the wall and ceiling can also be finished with a wood molding in any desired shape, which will eliminate the need for joint treatment (Figure 12-11, D). Use eightpenny finishing nails spaced 12 to 16 inches apart and nail into the top wallplate.

Treatment around window and door openings depends on the type of casing used. When a casing head and trim are used instead of a wood casing, the jambs and the beads may be installed before or during application of the gypsum wall finish.

Applying Tape and Cement

After the wallboard is installed, the flat joints and inside corners need to be reinforced with a paper tape and joint compound. See Figure 12-11, B. With a joint finishing knife, apply the joint compound fully and evenly into the slight recess created by the adjoining tapered edges of the

board. Next, center the wallboard tape over the joint and press the tape firmly into the bedding compound with your wallboard knife held at a 45 degree angle. The pressure should squeeze some compound from under the tape, but enough must be left for a good bond. When thoroughly dry, after at least 24 hours, apply a full coat extending a few inches beyond the edge of the tape and feather the edges of the compound. When the first finishing coat is thoroughly dry, use a 10 inch joint finishing knife to apply a second coat. Feather the edges about 1½ inch beyond the first coat. When this coat is dry, sand lightly to a smooth even surface. Wipe off the dust in preparation for the final decoration. Total width should be 12 to 14 inches.

Use the same procedure with end or butt joints as you do with tapered edges. The end joints are not tapered so care must be taken not to build up the compound in the center of the joint. This encourages ridging and shadowed areas. Feather the compound well out on each side of the joint. The final application of joint compound should be 14 to 18 inches wide.

To finish nail holes, draw a 4 inch joint finishing knife across the nails to be sure they are below the surface of the board. Apply the first coat of joint compound with even pressure to smooth the compound level with the surface of the board. Do not bow the knife blade with excess pressure as this tends to scoop the compound from the dimpled area. When dry, apply a second coat. Let dry, sand lightly and apply a third coat. Sand lightly before applying your decoration. An additional coat may be needed depending on temperature and humidity.

Before finishing the corner bead, make sure the corner is attached firmly. With a 4 inch finishing knife spread the joint compound 3 to 4 inches wide from the nose of the bead, covering the edges. Note Figure 12-15, B. When completely dry, sand lightly and apply a second coat, feathering edges 2 to 3 inches beyond the first coat. A third coat may be needed depending on your coverage. Feather the edges of each coat 2 or

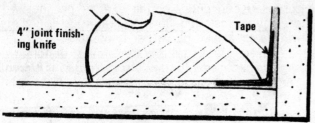

4" joint finishing knife

Tape

Finishing inside corners
Figure 12-16

3 inches beyond the preceding coat.

To finish corners, cut the tape the length of the corner angle you are going to finish. Apply the joint compound with a 4 inch knife. Spread the compound evenly about 1½ inch on each side of the angle. Use sufficient compound to embed the tape. Fold the tape along the center crease (Figure 12-11, C) and firmly press it into the corner. Use enough pressure to squeeze some compound under the edges. Feather the compound 2 inches from the edge of the tape. When the first coat is dry, apply a second coat. Note Figure 12-16. Feather the edge of the compound 1½ inch beyond the first coat. Apply a third coat if necessary, let dry and sand to a smooth surface. Use as little compound as possible at the apex of the angle to prevent hairline cracking.

After the joint treatment is thoroughly dry, all surfaces should be sealed or primed with a vinyl or oil base primer/sealer. This equalizes the absorption difference between the exposed surface and the joint compound surface. You will then have a uniform texture and suction over the entire wall or ceiling.

Special Wallboards

There are several wallboards that are made for specific uses. Predecorated vinyl faced wallboards are available in various designs. They are installed vertically on walls and joints are left unfinished.

Tile backer board is a water resistant wallboard which serves as a backer for ceramic tile in bath or shower areas. Fire rated board is also available where building codes require fire resistant wall covering.

Plywood and Hardboard

Plywood and hardboard are usually in 4 by 8 foot sheets for vertical application. However, 7 foot long panels can sometimes be purchased for use in basements or other low ceiling areas. Plywood can be purchased in a number of species and finishes with wide variation in cost. Hardboard imprinted with a wood grain pattern is generally less expensive. The better hardboard paneling uses a photograph of wood to provide the woodgrain effect, which produces a very realistic pattern. Both plywood and hardboard can be purchased with a hard, plastic finish that is easily wiped clean. Hardboard is also available with vinyl coatings in many patterns and colors.

The plywood or hardboard interior finish material should be delivered to the site well before application to allow the panels to assume conditions of moisture and temperature in the room. Stack the panels, separated by full length furring strips as in Figure 12-17, to allow air to get to all panel faces and backs. Panels should remain in the room at least a couple of days before application.

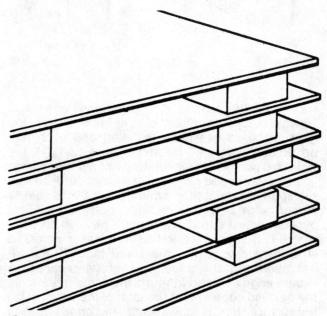

Stacking panels for conditioning to room environment prior to use
Figure 12-17

Generally with sheet paneling there's very little surface preparation needed. However, if there's loose plaster, tear it out and build out the wall with furring or plywood. If wallboard isn't exactly flat and tight, nail it tight to the studs. You can also rip it out and build out with plywood to match the rest of the wall. Paneling will hide many minor wall defects, but be certain that the defect won't get worse and spoil the paneling in the future.

It is best to remove moldings around doors and windows. Do it carefully to avoid splitting, with a wedge or carpenter's pry-bar. You can also drive the nails through moldings with hammer and nailset. Should you discover a void in the wall, or if you've removed an electric outlet or are filling in a larger opening (such as to change a window into a smaller pass-through) build out with studding or shims to match the vertical plane of the rest of the wall. Build a simple box frame around pipes that you don't want to relocate. If the plan includes built in closets, wall shelving, or cabinets, frame them out before you start to panel.

Some panels may vary somewhat in color and texture. Before putting up the first panel, arrange all panels into the sequence that you want to follow. Line them against the walls in their approximate positions. Number the panel backs with crayon to correspond with the order you want them installed.

Paneling can be installed directly to smooth studs without furring. Use a wood plane to smooth the high spots or shim out low spots if studs are not perfectly straight. Building paper, plastic sheeting, or other vapor barrier installed against studs will protect paneling from moisture and provide the insulating benefits of dead air space.

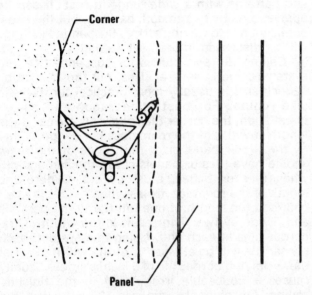

Scribing of cut at panel edge to provide exact fit in a corner or at ceiling
Figure 12-18

In applying the panels to the wall, the start depends partly on whether or not the wall corners are truly vertical. If the starting corner is straight, the first panel is merely butted to the corner and subsequent panels located so they lap on studs. If the panel corner or other surface is not truly vertical, place the panel edge on a vertical line 50 inches from the surface, plumb that edge with a carpenter's level, and use an art compass to scribe the outline of the cut to be made on the other edge (Figure 12-18). Cut the panel and move it against the corner. It might not be necessary to scribe and fit to compensate for slight irregularities or for corners only slightly out of plumb, especially if you are planning to use inside corner molding.

After the first panel is placed, install successive panels by butting edges against the previous panel, being careful to maintain a true vertical line. Any misalignment is less noticeable, of course, if all walls are paneled. A similar procedure can be used for fitting panels against the ceiling.

Cut panels slightly oversize at the edge that will go into the room corner, leaving an inch or more excess to permit scribing or accurate measuring to compensate for an out-of-plumb

corner. Keep panel edges exactly parallel with previously installed panels. (Slight irregularities will be concealed by corner molding.)

If you are installing hardboard paneling, a slight space should be left at joints. To avoid butting, use a thin coin or paper clip to separate panels temporarily. These panels are 1/32 inch scant in width to accommodate this spacing and permit modular installation. To utilize the storage feature of perforated paneling, this material should not be applied directly to a flat surface.

Panels can be fastened with nails or adhesive. Adhesive is sometimes preferable because there are no nailheads to mar the finish. Most adhesives include instructions for application, and these instructions should be followed carefully. Use an adhesive that allows enough open assembly time to adjust the panel for a good fit. Where panels are nailed, use small finishing nails (brads). Use 1½ inch long nails for 1/4 or 3/8 inch thick materials and space 8 to 10 inches apart on edges and at intermediate supports. Most panels are grooved and nails can be driven in these grooves. Set nails slightly with a nail set. Many prefinished materials are furnished with small nails having heads that match the color of the finish; thus no setting is required.

When cutting panels, use a hand or power saw, whichever you prefer. Putting masking tape along the line to be cut helps prevent edge splinters and chips. Use a crosscut hand saw. A rip saw will surely chip the face of the paneling. Have the panel face up so the saw cuts into the face on the downstroke. Start cuts carefully at the edge of the paneling. Support cut-off material during the final saw strokes. A combination hollow ground blade is recommended for power saws. The blade should cut into the face of the panel.

When making cutouts for doors, windows outlets, radiators, etc., it is always more satisfactory to measure and mark when you can visualize the cutout to be made. Even if it should require considerable moving and jockeying of full-size panels back and forth, it's worth the trouble to have the panel fit snugly. For any of these shapes or cutouts, wrapping paper or the divider sheet in the package may be used for easy, accurate marking. Tape it in position as if it were a panel. Mark off or cut out the opening or shape. Use the sheets as a pattern to transfer the marks onto the panel. Recheck and cut.

Measure outlet boxes from the edge of the adjacent panel. Mark the panel to be cut. Drill pilot holes at the corners of the cutout area for cutting with a jig saw or keyhole saw. You can also drill holes close together within the cutout area, and press out the scrap piece. Pull the outlet flush with the panel face. Another way would be to cover the outlet with chalk, powder or shaving cream. Then place the panel in position and tap the face at the approximate outlet location. The panel back will pick up the chalk or other marking in the spot to be cut out.

For a door or window, stand the panel next to the opening. Plumb the panel into position. Measure for the cutout and transfer the measurements to the panel with china-marking grease pencil or anything that will be readily seen and will wipe off clean. If walls are furred out, door jambs must be extended to compensate for the thickness of the furring. Clear pine is recommended.

Masonry walls, especially when below ground level, pose a moisture problem. Be certain that dampness released by concrete, brick or stone will not harm the panels. Do what you can to cure any problems of dampness. Be sure the paneling you select is appropriate for the location you have in mind. Some types of paneling are less likely to be affected by moisture. A vinyl plastic vapor barrier and insulation should be used according to the manufacturers' instructions. You can also attach furring strips to the wall with masonry nails, as previously described, or the preferred method— build a suitable frame of 2x3 lumber over the entire wall. Shim the frame out as necessary. Fasten the frame directly to the wall with nails or screws into plugs or expansion shields, or fasten with hardened cut nails, bolt anchors, adhesive anchors or adhesive. Or (easiest of all) add spacers between the frame and the masonry wall and wedge and nail the frame tight against the floor and ceiling. Install paneling to the frame just as you would over furring. Use the same nailing schedule or adhesive method as with furring on a dry wall.

Wood and Fiberboard Paneling

Wood and fiberboard paneling elements are tongued-and-grooved and are available in various widths. Wood is usually limited to no more than 8 inches in nominal width. Fiberboard paneling is often 12 or 16 inches wide. Paneling should also be stacked in the room to be paneled, as recommended for plywood and hardboard, to stabilize at the temperature and moisture conditions of the room.

Paneling is usually applied vertically, but at times is applied horizontally for special effects. When applying paneling vertically, it is desirable to install 1x4 nailing strips or 2x4 cross blocking at 24 inch intervals for nailing. One by four strips are also applied where paneling is to go over concrete or plaster. Unfinished wood paneling should be cut to size, sanded and stained before installation.

Paneling over masonry first requires nailing strips attached to the masonry surface. The nailing strips should be treated with a commercial toxic water repellent. Nailing strips may be attached with concrete nails or fiber plugs in prebored holes. Nailing strips then can be screwed to the plugs. Paneling, if applied horizontally over framing, can be attached to the studs. If applied vertically, 1x4 nailing strips over the studs will also facilitate insulation installation.

Vertically applied paneling is nailed to horizontal furring strips or to nailing blocks between studs. Nail with 1½ to 2 inch finishing or casing nails. Blind nail through the tongue, and for 8 inch boards, face nail near the opposite edge. Where 12 or 16 inch wide fiberboard is used, two face nails may be required. Color-matched nails are sometimes supplied with the fiberboard. Staples may also be used in the tongue of fiberboard instead of nails. Where adhesive is used, the only nailing is the blind nail in the tongue.

Ceiling Tile

Ceiling tile is available in a variety of materials and patterns with a wide range in cost. It can be applied directly to a smooth backing, but the usual application is to furring strips. Suspended ceilings will be discussed in the next section.

Ceiling tile and panels for suspended ceiling systems come in several sizes, but the 12 inch by 12 inch size is usually most practical for average size rooms. For best appearance and easier installation, the border tiles or panels are cut if the length or width of the room does not divide evenly by the tile or panel size. Both ends of the room should have tiles or panels cut to the same width when tiles or panels must be cut. Where the length of the room leaves less than a full tile at an end, add the length of one full tile to the part tile and divide by two. The result is the width of the border tiles at each end. Using this system, each border tile will be at least ½ of the width of a full tile. Applying border tile of 6 inches or less usually causes a noticeable irregularity in the finished ceiling. For example, suppose 12 by 12 inch tile are being applied to a ceiling 18 feet 2 inches long. Eighteen 12 by 12 tiles will reach to within 2 inches of one end. Rather than cut tiles down to a 2 inch width, use the rule given above. Add 12 inches to the 2 inches (14 inches) and divide by 2. The result, 7 inches, is the width of the border tile at each end of the room. The length of the room will be covered by 17 full tile and one 7 inch tile at each end. Do the same for the width of the room to equalize the border tile at each side.

Tile should be cut with a sharp handsaw or with a power saw. Before cutting, the exposed side should be scored with a sharp utility knife on the finished side. Saw the tile with the finished side up. Because there may be slight variation in color and texture of the tile from one box to the next, it is a good practice to take tile alternately from several boxes as the work progresses. The tile may expand or contract slightly as they absorb moisture in the air or give up moisture to the air. Avoid expansion or contraction after application by exposing the tile to room conditions 24 hours before they are to be applied. Remember that lighter color tile shows smudges or stains and may be difficult to clean after application. Some craftsmen apply talcum powder or corn meal to their hands before handling the tile. When using adhesive, be especially careful to keep the adhesive off the exposed side of the tile. When nailing tile to unexposed ceiling joists you will save time

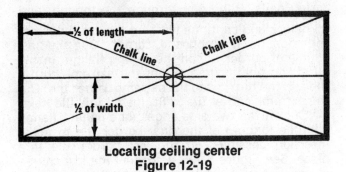

Locating ceiling center
Figure 12-19

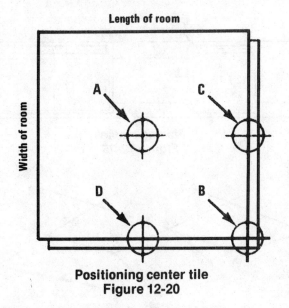

Length of room

Width of room

A   C

D   B

Positioning center tile
Figure 12-20

placed over the room center.

Apply a thin prime coat of tile adhesive to each corner of the tile about 2 inches from the corner. Then place a walnut-size daub over each prime coat. See Figure 12-21. Tile 12 by 24 inches should have adhesive applied as illustrated in Figure 12-21. Each gallon will cover 50 to 60 square feet of tile. Press the tile firmly to the ceiling, moving it back and forth slightly. Press succeeding tile to the ceiling, carefully sliding them into position to engage the tongue and groove. Cut all border tile with a reverse bevel so that the exposed face fits flush against the wall. Clean up excess adhesive with mineral spirits or naphtha.

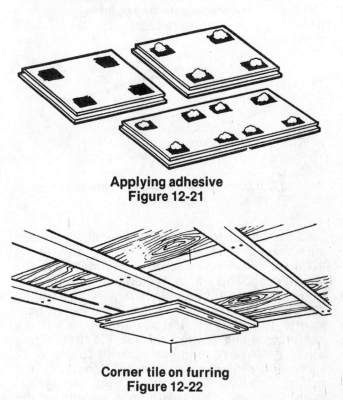

Applying adhesive
Figure 12-21

Corner tile on furring
Figure 12-22

and nails by snapping a chalk line on the plaster or wallboard under each joist once the joists have been located.

If the existing ceiling has a flat surface, tile can be fastened with adhesive. Use an adhesive recommended by the tile manufacturer and follow directions carefully. When applying tile with adhesive, work is begun from the center of the room. Snap a chalk line between opposite corners. Where the lines cross is the center point of the room (Figure 12-19). Placement of the first tile will determine the size of the border tile on all four walls. The size of the border tile will never be less than 6 inches if the following rule is observed. Figure 12-20 represents one 12 by 12 inch tongue and groove ceiling tile (as seen from below), oriented so that the top edge is parallel to the length of the room and the side edge is parallel to the width of the room. Select the correct center point on the tile that will cover the center point of the room:

1. When both the width and length of the room in feet are even numbers, the point A should be placed over the room center.

2. When both the width and length are odd, place point B over the room center.

3. When only the room length is odd, point C is placed over the room center.

4. When only the width is odd, point D is

A more common method of installing ceiling tile is through fastening them to furring strips. Nominal 1 by 3 or 1 by 4 inch furring strips are used where ceiling joists are spaced no more than 24 inches apart. Nail the strips with two sevenpenny or eightpenny nails to each joist. Where trusses or ceiling joists are spaced up to 48 inches apart, use nominal 2 by 2 or 2 by 3 inch furring nailed with two tenpenny nails to each joist. The furring should be a low density wood, such as the softer pines, if tiles is to be stapled to the furring. Furring strips can be used even where only exposed joints are present (Figure 12-22). Accurately measure each of the four walls where they join the ceiling. Divide each wall dimension in half and place a mark on the ceiling near the walls. Snap a chalk line between these marks to locate the center point. See Figure 12-23. Determine the number of full tiles that will be required: Subtract the border tile widths to determine if an even or an

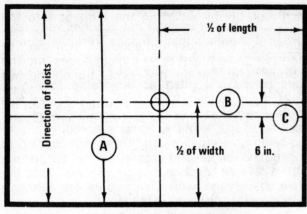

Locating starting point
Figure 12-23

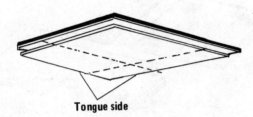

Tongue side

Cut corner board tile on tongue side
Figure 12-24

grooves of the tile already applied. Nail or staple as before and continue working diagonally across the room to the opposite corner. Cut the last course of border tile on the nailing flange. Install all of the final course except the extreme corner tile and the tile next to the extreme corner tile. Cut the extreme corner tile to fit. Install this tile next.

Cut off the two side edges (one nailing flange and one tongue) of the final border tile to make insertion possible. Fit the last border tile into place. See Figure 12-26. Face nail the last course of border tile.

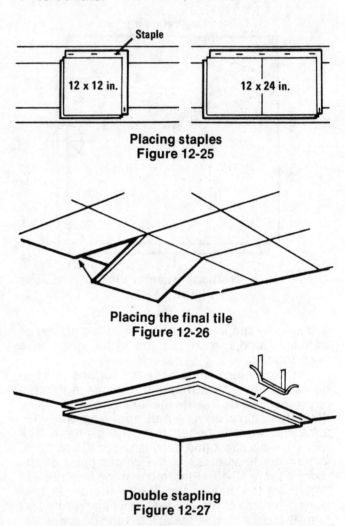

Staple

12 x 12 in.   12 x 24 in.

Placing staples
Figure 12-25

Placing the final tile
Figure 12-26

Double stapling
Figure 12-27

odd number of full size tiles is required in each direction. If an even number of full size tile is required for ceiling direction A (Figure 12-23), the first furring strip will be centered on line B. If an odd number of full size tile is required for ceiling direction A, snap a new chalk line C 6 inches from center line B. The first furring strip will be centered on line C. Working from the first furring strip, install remaining furring strips spaced at 12 inches on center across the full length of the ceiling. One furring strip should be placed against each side wall. A chalk line snapped on the existing ceiling under the center of each joist will save time and nails.

When all furring is in place, check the level of the strips with a 6 foot straight edge or with a true length of lumber. If necessary to bring the furring into a level plane, insert wood filler shims between the strips and the joists.

As previously described, calculate the width of each border tile. Border tile at each end of the room should be the same width and border tile at each side should be the same width. Cut a corner border tile to the size of the two adjacent borders. Cut off only the tongue sides. See Figure 12-24. Face nail or staple the extreme corner tile in place (Figure 12-22). Use 1 1/8 inch blue nails or 9/16 inch staples at four points on the flanges. When stapling, hold the gun firmly against the tile flange so the staple is driven straight and to the full depth. Staples should be placed as shown in Figure 12-25.

Insert the tongue of the next tile into the

If a sound gypsum wallboard ceiling exists it may not be necessary to install furring strips. Double action staples, one staple driven in over the staple below (Figure 12-27), will usually be enough to hold the tile firmly. Except for installing furring strips, each step above remains the same.

Metal furring strips can be used over an existing ceiling or over exposed joists. Installation begins at one corner but, unlike wood furring work, the entire first border is installed first. Succeeding courses are installed one row at a time. Unlike tile applications over wood furring, the tile are not nailed or stapled. Instead, a molding strip holds the tile in place at the walls

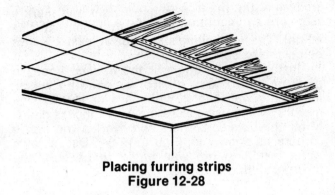

Placing furring strips
Figure 12-28

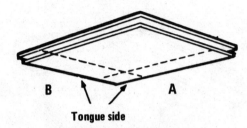

B   A

Tongue side

Starting corner border tile
Figure 12-29

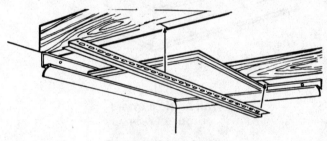

Corner tile on furring
Figure 12-30

each four tile by inserting the tongues into the grooves of the preceding course. For the last course, cut the border tile about ¼ inch short of the space to be filled. Slide the cut edges into the molding strip. Cut the last corner tile on the two nailing flange sides (Figure 12-31). Install the last corner tile before the adjacent border tile. Cut off sides D, Figure 12-31 and insert the last border tile. See Figure 12-32.

Cut off side protection

A   B

D   D

Nailing flange

Cutting last border tile on nailing flange sided
Figure 12-31

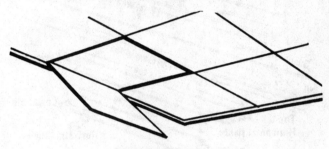

Placing final tile
Figure 12-32

and the metal furring strip interlocks with the other side of the tile to hold each tile in place. See Figure 12-28. Using metal furring strips, some unevenness can be allowed in the existing ceiling or the joists without producing a noticeably inferior finished ceiling.

Place the molding strips on the wall against the existing ceiling or exposed ceiling joists around the perimeter of the room. Nail the molding at 12 inch intervals. Determine the size of the border tile and cut the tile to size. Cut the first corner tile on both tongue sides (Figure 12-29, A and B). Cut the remaining tile for the first border on side B only. Snap a chalk line on the ceiling or on the ceiling joists where the first metal furring strip is to be placed. This distance from the wall should be the width of the border tile plus one inch. Insert the first corner tile and the next three border tile into the molding. Nail or staple the furring strip in place so that it supports the first four tile. Use 1-1/8 inch blue nails or 9/16 inch staples. See Figure 12-30. Install the entire border strip, inserting four border tile into the molding and attaching the metal furring strip. For the next course, install

Often 24 by 12 inch center scored tile are used with metal furring strips. The 24 inch dimension should then be installed parallel to the metal furring strip.

Suspended Ceiling

Suspended ceilings consist of a grid of small metal or wood hangers, which are hung from the ceiling framing with wire or strap, and drop-in panels sized to fit the grid system. This type of ceiling can be adjusted to any desired height. Where the existing ceiling is normal height, the hangers can be supported only 2 or 3 inches below the ceiling and still cover any bulging plaster or other unevenness. Where existing ceilings are high, adjust the hangers to the desired ceiling height. A suspended ceiling system reduces the sound transfer from the floor above and increases the insulating value of the ceiling. It provides access for ceiling lights, heat supply and return ducts, and eliminates the need for any other ceiling finish.

Determine the room size in even number of feet. If the room length or width is not divisible by 2 feet, increase the dimension to the next unit divisible by 2 feet. For example, a room 11 feet 3 inches by 14 feet 2 inches should be considered a 12 foot by 16 foot room. Next, multiply the length times the width to arrive at the material required. In the example, 12 feet times 16 feet equals 192 square feet.

Figure 12-33 illustrates typical grid components for a suspended ceiling system. Main tees must run perpendicular to joists. For 2 foot by 2 foot grid, cross and main tees are 2 feet apart. For 2 foot by 4 foot grid, main tees are 4 feet apart and 4 foot cross tees connect the main tees. Then 2 foot cross tees can be used to connect the four foot cross tees. Plan to position the main tees so that border panels at the room edges are as equal and as large as possible. Try several layouts to see which looks best for the main tees. Small rooms should use 2 by 2 foot panels.

**Grid components
Figure 12-33**

Calculate the Materials Required:

1. The wall angle is installed around the perimeter of the room. Wall angle usually comes in 12 foot sections.

2. Main tees come in 12 foot sections and should run at right angles to the joists across the length of the room. If the room length exceeds 12 feet, two or more main tees may be spliced together.

3. Make a sketch of the layout and count the number of 2 foot or 4 foot cross tees. No more than 2 cross tees for border areas can be cut from one full length cross tee.

4. Count the number of ceiling panels from the layout sketch.

5. The quantity of wire depends on the "drop" distance of the ceiling. Figure one suspension wire and one hook or suspension screw for each four feet of main tee.

Determine the new ceiling height and mark a line around the perimeter of the room at this height. The new ceiling should be at least 2 inches below the existing ceiling.

Install the wall angle at the marked line. On gypsum wallboard, plaster or paneled walls, install the wall angles with nails or screws. On masonry walls use concrete nails 24 inches apart. Miter the outside corners and overlap the inside corners as shown in Figure 12-34. Use a level while installing the wall angle to be sure the finished ceiling is level.

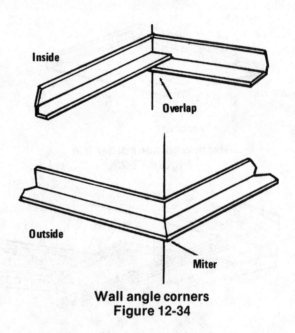

**Wall angle corners
Figure 12-34**

Suspension wires are required every 4 feet along main tees and on each side of splices. Cut the suspension wires to the proper length - at least 2 inches longer than the distance between the old ceiling and the new ceiling. Before attaching the first suspension wire, extend a length of wire between the wall angles at right angles to the main tees under the point where the first row of suspension wires will be attached to the main tees. This wire serves as a guide for leveling the suspension wires.

Position one suspension wire for the first main tee. Measure out from the walls the distance of the two border courses to position the screw eyelet or hook. Pull the wire taught to remove kinks and make a 90 degree bend in the wire at the point where the suspension wire crosses the guide wire. If suspension wires are attached to the ceiling joists, the nail or hook can be attached at either the side or bottom edge of the joist. Position the first suspension wire for the next main tee. Use the same guide wire to find the correct length for the 90 degree bend.

Continue placing suspension wires over the guide wire for each main tee. When the first row of suspension wires is complete, move the guide wire four feet toward the other end of the room and continue placing suspension wires. The last row of suspension wires should be no more than 4 feet from the wall. Remember to place suspension wires on both sides of spliced main tees.

Determine the length of the main tees and cut all the tees to the correct length. Aluminum tees should be cut with a hacksaw. Steel tees are cut most easily with tin snips. If the length of the room is over 12 feet, two main tees will be spliced together to span the room. If spliced tees are necessary, 6 inches should be cut from one end of each main tee so that cross tees do not end at the splice joints.

Install the main tees. For rooms less than 12 feet long, rest both ends of a main tee on the wall angles and attach one suspension wire in the middle. Use a level to adjust the length of the wire. For main tees longer than 12 feet, rest one end on the wall angle and attach the suspension wire closest to the other end. Use a level to adjust the length of the remaining suspension wires. Measure the remaining length of main tee needed, resting one end on the opposite wall angle. Connect the main tees with a splice plate (Figure 12-35). If the main tee splice falls at a cross tee, it may be necessary to cut 12 inches off the first main tee. Start the next main tee with the length cut off the first main tee. Be sure to line up the cross tee slots in the main tees, however.

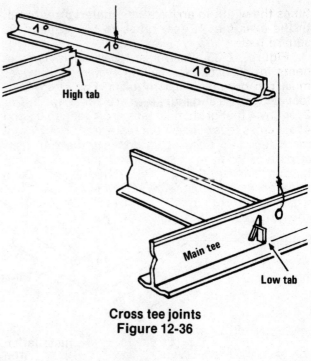

Cross tee joints
Figure 12-36

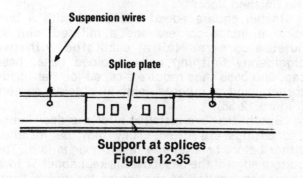

Support at splices
Figure 12-35

Next, install the cross tees. Simply insert the tab of the cross tee into the slot of the main tee. In some aluminum systems, cross tees have one "high" and one "low" tab as in Figure 12-36. Install all the border cross tees on the first main tee by cutting off the "high" tab and resting the cut end in the main slot. On the opposite side of the room, cut off the "low" tab end and rest the cut end on the wall angle. If the border edge is less than half the length of the cross tee, save the balance for use on the opposite side of the room.

Install the ceiling panels. Cut the border panels to size with a sharp knife and a straight edge. Lay the panel on a flat surface with the white side up and cut the panel with the straight edge protecting the panel you intend to use rather than the scrap end. Most ceiling panels have a random pattern and do not require orientation. Some panels do require orientation and usually are marked with arrows on the back side. Keep the

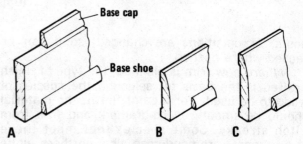

Baseboard: A, two-piece; B, narrow; C, medium width
Figure 12-37

arrows pointing in the same direction.

If translucent plastic lighting panels are to be installed, the panels should be cut by scoring repeatly with a sharp tool until the panel is completely cut through. Lay in the panel with the glossy side up. If recessed light panels are to be dropped into the grid, each corner of the light fixture must be supported by a suspension wire.

If fluorescent lighting fixtures are used above the grid, the light tubes should be centered over the translucent panel. The lighting efficiency will be improved if the sides and ends of the fixtures are enclosed with foil or wood painted a highly reflective white.

Interior Trim

Interior trim consists of window and door casings and various moldings. Such trim in existing houses varies considerably, depending on age, style, and quality of the house. The trim found in many older houses is probably no longer on the market, so matching it requires expensive custom fabrication. If the plan is to use existing trim, remove pieces carefully where windows,

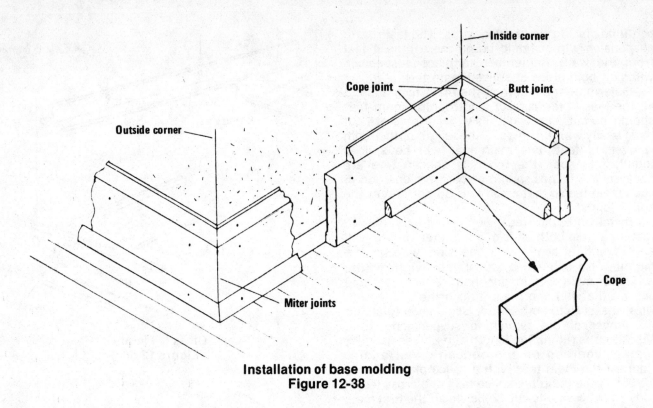

Installation of base molding
Figure 12-38

doors, or partitions are changed, so it can be reused where needed.

Where new trim is planned, the type of finish desired is the basis for selecting the species of wood to be used. For a paint finish, the material should be smooth, close-grained, and free from pitch streaks. Some species that meet these requirements are ponderosa pine, northern white pine, redwood, mahogany, and spruce. The additional qualities of hardness and resistance to hard usage are provided by such species as birch, gum, and yellow poplar. For a natural finish, the wood should have a pleasing figure, hardness, and uniform color. These requirements are satisfied by species such as ash, birch, cherry, maple, oak, and walnut.

Casing is the interior edge trim for door and window openings. New casing patterns vary in width from 2¼ to 3½ inches and in thickness from ½ to ¾ inch. Place the casing with about a 3/16 inch edge distance from door and window jambs. Nail with sixpenny or sevenpenny casing or finishing nails, depending on the thickness of the casing. Space nails in pairs about 16 inches apart, nailing to both jambs and framing. Rectangular casings can be butt-joined at corners (Figure 11-17B), but molded forms must have a mitered joint (Figure 11-17A).

Baseboard

Baseboard, the finish between the finished wall and floor, is also available in several sizes and forms (Figure 12-37). It may be either one or two piece. The two piece base consists of a baseboard topped with a small base cap which conforms to any irregularities in the wall finish. Most baseboards are finished with a base shoe, except where carpet is installed. The base shoe is nailed into the subfloor, so it conforms to irregularities in the finished floor.

Install square edged baseboard with a butt joint at inside corners and a mitered joint at outside corners. Nail at each stud with two eightpenny finishing nails. Molded base, base cap, and base shoe require a coped joint at inside corners and a mitered joint at outside corners (Figure 12-38).

Baseboards are installed after the door jambs and casings are fitted. They may be installed either before or after the finish floor is laid. The bottom edge of the baseboard is kept about ¼ inch above the location of the top of the finish floor. This space will later be covered up by the base shoe. If no base shoe is to be used, the finish floor is laid before the baseboard which is then scribed to the floor. Select clear baseboard stock free from twists and imperfections. If it is the molded type, inspect the molded edges for imperfections such as chipped edges or poorly machined surfaces. Assume you were installing base in the room illustrated in Figure 12-39. Measure the distance from wall to wall along the end of the room marked A. Use an extension rule for this purpose. Transfer this length to a length of baseboard. Mark these points on the top edge of the baseboard and square from these points. Taper the cut at each end so that the length of the base at the bottom edge will be about ¼ inch shorter than the top edge.

Place the baseboard in position, being careful not to injure the wall covering or the baseboard. Use a block of wood to tap the baseboard in place.

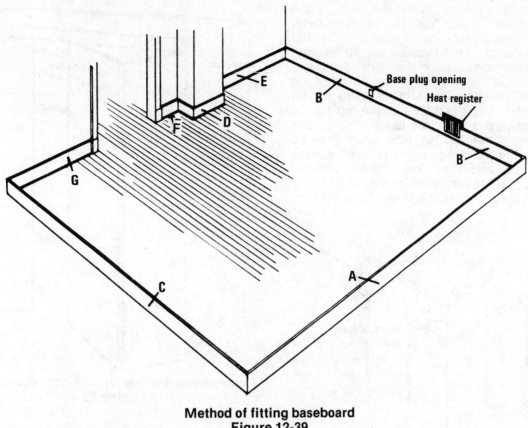

Method of fitting baseboard
Figure 12-39

If the finish floor has not been laid, use a block of finish flooring to keep the baseboard the correct distance above the subfloor. Nail the baseboard in place with two 8d finishing nails in each stud.

Measure and cut the baseboard marked B, Figure 12-39. This measurement should be taken from the face of base A to the opposite wall. Allow enough stock to form a coped joint to fit against base A. In most cases this coped joint will have to be scribed to the surface of base A and there should be an additional ½ inch allowed for this purpose.

Scribe the straight part of the baseboard B, and from the top end of this line, mark a 45 degree line. Cut this miter and cope the molded section. Then, cut the straight section to the scribed line.

Measure, cut and nail the baseboard around the pilaster (Figure 12-39). The measurement is taken from the wall, allowing enough stock for the joint at the outside corner of the pilaster. The application of the baseboard around the pilaster before the base E is placed is only necessary when the pieces around the pilaster are short and apt to split when nailed. Applying the baseboard around the pilaster before applying baseboard E permits these short pieces to be longer and to be held in place by the baseboard E which would be coped on both ends. If the pilaster is large, piece E needs to be scribed only to piece B. This same principle holds true in fitting piece F.

Cut the piece F to the plinth block or casing and scribe it to the surface of the baseboard on the pilaster. Be careful not to cut these pieces so long that they will force the jamb out of line when the pieces are forced and nailed in place. Set the nails in all the baseboards.

Cut and place piece C. Cope the end that fits against piece A and butt the other end against the wall. Fit piece G in place. One end should be coped to fit against piece C and the other end should be fitted against the door casing or plinth block.

Figure 12-39 shows a base plug opening in the baseboard. To locate and cut the opening in the baseboard, proceed as follows: Fit the baseboard in its proper place on the wall. Be sure to put the finish floor blocks under the bottom of the base if the finish floor has not been laid. Mark the outer edges of the outlet box with blue chalk. Replace the baseboard in position and tap it lightly against the edges of the plug box. Remove the baseboard and the outline of the box should show on the back. Bore a hole at each corner of the box outline until the spur of the bit shows through. Remove the bit and complete the hole from the other side. Cut around the chalk lines with a reciprocating saw, allowing about 1/8 inch clearance on all sides. Replace and nail the baseboard.

Figure 12-39 shows a hot air register that projects above the top of the baseboard. The baseboard is fitted against and around the register. Cut the baseboard so that it butts against each side of the register. Keep the ends of the baseboard about ½ inch from the register metal lining. Rip pieces about 2 inches wide from the top edge of base stock. These pieces should be long enough to be mitered around the top and sides of the register. Butt the two side pieces onto the top edges of the baseboard and miter them to the piece that extends across the top of the register. If molded baseboard is used, the same procedure is followed, except that the two side pieces may be mitered into the molded base. Next, cut the base shoe to length and cope the internal corners. Temporarily nail the shoe in place. It may be permanently nailed after the finish floor is laid.

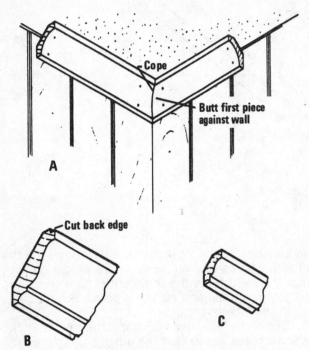

Ceiling mouldings; A, installation [inside corner]; B, crown; C, small crown
Figure 12-40

Ceiling Moldings [Figure 12-40]

Ceiling moldings are used at the junction of wall and ceiling. They may be strictly decorative where there is a finished joint or they may be specifically for the purpose of hiding a poorly fitted joint. They are particularly useful with wood paneling or other dry wall that is difficult to fit exactly, and where plaster patching cannot be used to finish the joint. Attach the molding with finishing nails driven into the upper wallplates. Large moldings should also be nailed to the ceiling joists. For gypsum drywall construction, a small simple molding might be desirable (Figure 12-40, C). Finish nails should be driven into the upper wall plates and also into the ceiling joints for large moldings when possible. As in the base moldings, inside corners should also be cope-jointed. This insures a tight joint and retains a good fit if there are minor moisture changes.

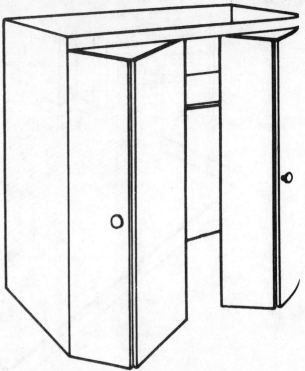

Double-hung door set for full-width opening closet
Figure 12-41

Miscellaneous Decorative Moldings

Many decorative moldings can be used in a variety of ways. They can be applied to walls or doors to give the effect of relief paneling or carved doors. They can also be used to add interest to existing cabinetwork. Check with a local building supply dealer for available types and ideas on how to use them.

Closets

Many older homes are lacking in closet space, and often the closets provided are not well arranged for good usage. Remodeling may involve altering existing closets or adding new ones.

One thing that can do much to improve closets is to provide fullfront openings to replace small doors. Doors are available in a great variety of widths, particularly accordian doors. Remove wall finish and studs to the width required for a standard accordian-fold or double-hinged door set (Figure 12-41). Standard double doors may also be suitable. Where the closet wall is load-bearing, use the header sizes listed for window openings. The header can be eliminated for a nonload-bearing wall, and a full ceiling-height accordian-fold door can be used. Frame the opening in the same manner as other door openings, with the rough framed opening 2½ inches wider than the door or set of doors. Install closet doors in a manner similar to other interior doors. Special types of doors, such as the double-hinged doors, are usually supplied with installation instructions. Closets can sometimes be made more useful by

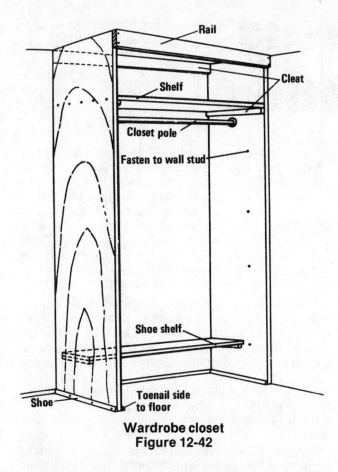

Wardrobe closet
Figure 12-42

clothes rods to the cleats that support the shelf.

New closets can be built in a conventional manner, or wardrobe closets of plywood or particleboard can be built at a lower cost.

Conventional closets are constructed by adding a partition around the closet area, using 2 by 3 inch or 2 by 4 inch framing and gypsum board or other cover material. Provide a cased opening for the closet door.

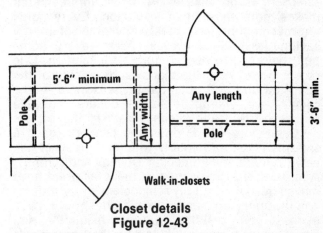

Closet details
Figure 12-43

Wardrobe closets require less space because the wall is a single material without framing. Build wardrobe closets as shown in Figure 12-42. Use 5/8 or 3/4 inch plywood or particleboard supported on cleats.

Use a 1 by 4 inch top rail and back cleat. Fasten the cleat to the wall, and in a corner, fasten the sidewall to a wall stud. Toenail base shoe moldings to the floor to hold the bottom of the sidewalls in place. Add shelves and closet poles where desired. Similar units can be built with shelves for linens or other items. Any size or combination of these units can be built. Add plywood doors or folding doors as desired.

Walk-in clothes closets can be added when space permits. Figure 12-43 shows a plan view of two walk-in closets. They should be provided with the necessary drawers, shelves, hanging rods and hook strips for the storage of wearing apparel.

additions or alterations in shelves and clothes rods. The usual closet has one rod with a single shelf over it. Where hanging space is quite limited, install a second clothes rod about half way between the existing rod and the floor. This type of space can be used for children's clothing or for short items of adult clothing. Shelves can also be added in any manner to fit a particular need. To add either a shelf or a pole, support them with 1 by 4 inch cleats nailed to the end walls of the closet. Nail these cleats with three sixpenny nails at each end of the cleat and at the intermediate stud. Shelf ends can rest directly on these cleats. Attach

Chapter 13
Insulation

There is no area where home improvement makes more sense than insulation. Every home improvement project can take advantage of energy saving possibilities because nearly every home built before 1975 needs improved insulation. One recent survey reported that over 80 percent of the home owners and potential home owners in the country are willing to invest dollars today to reduce heating and cooling costs in the future.

Common sense is often the best approach a salesman has, particularly when his prospect can easily understand immediate savings and comfort. A number of exterior product manufacturers have moved quickly to supply remodelers with products and supporting research which should help increase your residing, window, insulation and insulated door business. While the research is necessary to document savings, common sense tells a remodeling salesman that his best selling tool is that the cost of fuel is going to increase for the forseeable future.

Using Insulation

Proper use of insulation is the most positive and efficient means of conserving home conditioning energy. This chapter is designed to give you information which will help you evaluate insulation requirements and install the right insulation materials.

Heat energy always flows from a warm material, body, or space to a relatively cooler material, body, or space. In winter, heat energy flows from the warm interior of a house to the cooler exterior. This is referred to as "heat loss" when discussing heating systems. In summer, heat flows from the warmer exterior to the relatively cooler interior of a house. This is referred to as "heat gain" when considering air conditioning.

Heat is transferred in three basic ways. By Conduction, heat passes through a dense material such as an iron skillet sitting on a hot stove or a steel rod held in a hot fire. By Convection, heat is transferred by movement of air over a heated panel or radiator surface. By Radiation, heat is radiated in the form of infra-red rays from one place to another, such as from the sun to the earth.

The term insulation used in this chapter refers to thermal insulation, not electrical insulation. Some very good electrical insulators, such as glass, are poor thermal insulators.

Thermal insulation materials are those developed specifically to reduce heat transfer materially. Insulation materials are generally very light in weight and are produced in four common forms. These are batts or blankets, loose fill or granulated, rigid boards, and reflective.

Batt and blanket insulation is made of matted mineral, glass, or cellulose fibers. Insulation batts or blankets are usually encased in paper, one face of which is made to serve as a vapor barrier. This may be an asphalt paper or a paper with a reflective metal foil backing. Proper installation of this vapor barrier is very important.

Blankets and batts range in thickness from 1 inch to 6 inches and in widths to fit between joists, rafters, and studs spaced 12, 16 and 24 inches on center. Blanket insulation may come in lengths of 50 feet or more. Batts generally are between 2 feet and 8 feet in length. The batt and blanket type insulation is normally used during initial construction when it can be easily placed between structural members of the sidewall, ceiling and floor.

Loose fill or granulated insulation generally consists of mineral wool, vermiculite, treated cellulose fiber, granulated polyurethane and other material. It is usually sold in bags for easy storing and handling. Depending on the particular job, it can either be poured directly from the bag or blown in place by mechanical means. Loose fill insulation is especially convenient for insulating the ceilings of existing houses where there is access to the attic. Granulated forms such as vermiculite are well adapted for placing in the cores of masonry blocks, thereby reducing heat loss through masonry walls considerably.

Placing loose fill insulation into sidewalls and ceilings of existing houses by special blowers is about the only method of insulating when the spaces are inaccessible otherwise, but this can create moisture problems from a lack of proper vapor barriers.

Rigid type insulation, commonly referred to as insulation board, has many special uses. Its rigidity and strength give it advantages which other types of insulation do not have. Rigid insulation boards are used on the outside of wall studs as sheathing or inside as a wall finish. Insulation board is relatively dense, and therefore less effective as an insulator than batt or blanket types of the same thickness.

Rigid insulation boards 1 to 2 feet wide and 1 to 4 inches thick are used as perimeter insulation around the outside edges of houses built with concrete slab floors. This insulation must be moisture resistant so it is generally made of foamed glass, or foamed plastic such as polyurethane.

Reflective insulation is made from reflective foils such as aluminum, or polished metallic flake adhered to a reinforced paper. It retards the flow of infra-red heat rays passing across an air space. To be effective, the foil surface must face an air space of ¾ inch or more. Reflective insulation is available in single sheets, strips formed to create 3 or 4 separated air spaces of ¾ inch each and as a combination vapor barrier and reflective surface attached to batts or blankets.

The critical factor to insure efficiency of reflective insulation is installing it so the reflective surface always faces an air space of ¾ inch or more. Once the reflective surface touches surrounding materials it is ineffective as an insulator because it does not retard conducted heat or heat of convection.

Many types and forms of materials are used in manufacturing building insulation. Also, there are many building materials which are mistakenly assumed by the uninformed to have much more insulation value than they really have.

Because of these variations and misunderstandings, it is unrealistic to compare insulating values of materials by thickness in inches. For example many people are shocked when told that 1 inch of mineral- wool insulation has more insulating value than an 18 inch masonry wall of common brick.

Fortunately, reliable standard rating methods have been developed to help you rate various insulations. Although the rating method is quite reliable, some of the terms used to explain it may be new and somewhat confusing to you at first.

Effectiveness of insulation is specified in two ways. One is stated as the resistance offered by a material or materials to the flow of heat under known conditions. This "Resistance" is generally designated by the letter "R". The second specification is stated as the amount of heat that will pass through a material or materials under known conditions. This is designated by the letters "C" or "U".

Most manufacturers of insulation have adopted the "R" rating and stamp it on their product for easy evaluation by the purchaser. For example, a 3 inch thick batt of one insulation might have an R10 rating, whereas a 3 inch batt of another type might have an R12 rating. On the other hand a 2 inch thick rigid foam material might have an R10 rating stamped on it. In all cases when you are given the "R" rating it is a simple matter to evaluate and make comparisons.

When you are building a wall containing several materials and have the "R" values for each, you can determine the total "R" value or insulating value by simply adding them directly. For example: R4 + R10 + R5 = R19 total. Another example of how the rating can be used is in determining values of various thicknesses of materials. If 2 inches of insulation "A" has a rating of R8 then 4 inches would have a rating of R16, for all practical purposes.

You cannot add the "C" or "U" values directly. When you can only find the "C" or the "U" rating you change them to "R" before adding. Simply stated the procedure is this: when you know the "C" or "U" value, you divide one (1) by this value to get "R". Example: a "C" or "U" value is .5, then "R" = 1 ÷ .5 = 2.

These simple rating designations also make it easy to compare costs of insulations. If one insulation rated R13 meets your job requirements and costs 10 cents per square foot, and another is identical in every way except cost, your choice is clear. Also, you can easily determine the amount of insulation required for a desired "R" value for any type of construction.

Some characteristics or features of insulation materials besides "R" values and cost could influence your choice of materials used for particular conditions. A few of these are:

1. Structural requirements, which must be considered when you want to use rigid insulation for such things as sheathing, plaster base, interior ceiling or wall finish, and roof decks.

2. Fire resistance of any insulating material should be considered regardless of where it is used in the house.

3. Effects of moisture on an insulation material sometimes determine whether or not it can be used for some jobs such as perimeter insulation under concrete slabs. When any insulation material becomes saturated, the insulating value decreases to practically nothing.

4. Vermin resistance is an important consideration, but you must make relative evaluations. No material is absolutely vermin proof.

Windows and Doors

When planning any insulation work or remodeling, consider all major causes of heat loss. Heat loss per unit area through windows and exterior doors is much greater than through most wall materials. Another major heat loss is from air infiltration around doors and windows. Storm doors and storm windows will reduce air infiltration, and reduce heat transfer by 50 percent or more through the exposed surface. Properly fitted windows can be the biggest single energy saver. A poor fitting window without weatherstripping, for example, can allow 5½ times as much air infiltration as an average fitting window that is weatherstripped. Weatherstripping can significantly cut down on air infiltration even with a poorly fitted window, but the place to begin is still with a quality window, properly fitted.

Thermal break type windows reduce heat loss. Note Figure 13-1. The window frames are basically metal with some less conductive material (like vinyl) sandwiched in between the outside and inside metal frame units to cut down on the amount of cold transmitted through the metal. Wood window frames conduct less heat than metal, and are now made with vinyl facings which reduce the likelihood of warpage over a period of time. Authorities disagree on the amount of heat conducted through the metal sash compared to wood sash, but in any event the thermal break type does reduce heat loss and condensation.

Sliding glass doors with large glass areas should be avoided. They are thermally inefficient and have large areas subject to possible air infiltration. In those cases where sliding glass doors must be used, make sure that a thermal break type, double glazed door (Figure 13-2) is used. Properly weatherstripped and installed, they are much more energy efficient than standard sliding glass doors.

Storm windows and doors vary widely in basic design, durability, and cost. Storm windows range from single glass panels, that must be put in place each fall and removed each spring, to permanent triple-track assemblies, which include sliding upper and lower windows and a screen. Triple-track windows can be used both during heating and cooling periods and can be opened for natural ventilation at other times. Because they are left in place permanently, wear and tear and the chance of breakage is minimized.

You should be aware that it is not the storm window itself that keeps the warmth inside in

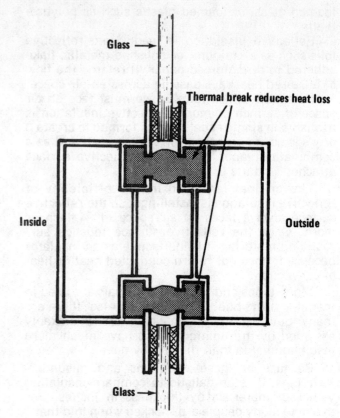

Thermal break type metal windows
Figure 13-1

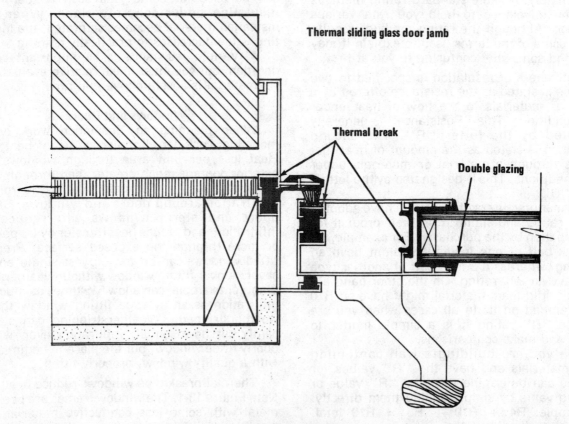

Sliding glass doors
Figure 13-2

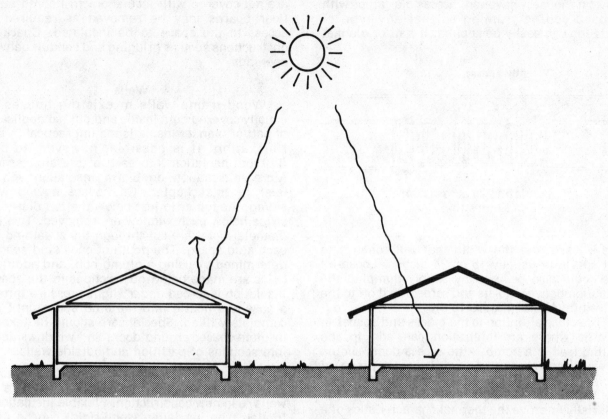

Lighter roof reflects heat   Darker roof absorbs heat

**Roof shingle color
Figure 13-3**

winter and outside in summer. It is actually the dead air space—at least ¾ inch—between the two windows that saves energy.

The size of the storm window is determined by the size of the window frame and not by the glass area. Storm windows should be properly installed and fit tightly to do the most good. To assure a tight fit, permanent storm windows should be sealed to the outer window frame with caulking compound or other sealing material. Storm window frames should have a tiny opening at the bottom to allow water vapor to escape. Storm windows are generally more economical than double pane windows in existing houses, because they usually cost less to install and they reduce infiltration of air around the window sash.

Storm doors may not always be economical when considered for winter heating savings alone unless the doorways are frequently used. However, they may still be a good investment, since they can be used as a screen door in the summer months if they have interchangeable glass and screen inserts. If the house already has a screen door, it generally will not pay to replace it with a storm door. Storm doors over doors with inset glass areas save more energy than those over solid doors.

Roofs and Ceilings

Dark colors tend to absorb heat while light colors reflect heat. Note Figure 13-3. So even with a well insulated ceiling, the color of the roof does make some difference where heat gain is concerned. Recommend lighter color shingles where cooling is a major expense in warm weather.

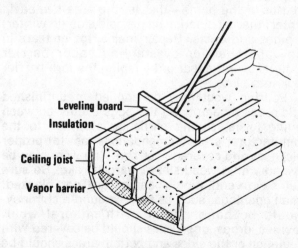

Leveling board
Insulation
Ceiling joist
Vapor barrier

**Installation of loose-fill ceiling insulation
Figure 13-4**

Most houses have an accessible attic with exposed ceiling framing so that any type of insulation can easily be applied. If batt or blanket

Attic opening

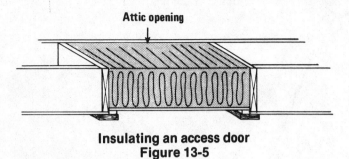

Insulating an access door
Figure 13-5

type is used, get the width that will conform to joist spacing—usually 16 or 24 inches. Loose-fill type could also be used by simply dumping the insulation between joists and screeding it off to the desired thickness (Figure 13-4).

Give close attention to the places and spaces in an attic where air infiltration can add to the heating load of a home. Attic access doors should be weatherstripped. In addition, a piece of insulation board cut to the size of the attic door and tacked to the attic side of the door can materially improve the thermal characteristics of a panel or hollow core door. Note Figure 13-5. If there is an attic scuttle hole, weatherstrip it and insulate the back of the scuttle closure panel. Stuff insulation around pipes, flues, or chimneys penetrating into the attic space, especially in cold climates.

When insulating a ceiling, extend the insulation over the top of the top plate as in Figure 13-6. Mineral fiber blankets are installed in ceilings by stapling vapor barrier flanges from below, installing unfaced pressure-fit blankets, or laying blankets in from above after the ceiling is in place. Unfaced (no vapor barrier) pressure-fit batts are held between joists without fastening. Install insulation so the vapor barrier side faces the interior of the home—that is, the area heated in winter. Place insulation on the cold side (in winter) of pipes and ducts. Repair major rips or tears in the vapor barrier by stapling vapor barrier material over the tear or by taping the torn barrier back into place.

Pouring wool may be applied in unfinished attic areas by emptying the bags evenly between ceiling joists, paying particular attention to the manufacturer's recommendations as to proper thickness and coverage per bag. The wool may be leveled with a wood slat or garden rake. Be sure that eave ventilation openings are not blocked. Small openings, such as those around a chimney, should be hand packed with mineral wool. Cavities, drops, or scuttles should be covered with insulation or the sides and bottom areas should be insulated. Be sure that recessed lighting fixtures and exhaust fan motors protruding into the ceiling

are **not** covered with insulation. In floored attics, floor boards may be removed as required for access to the space to be insulated. Check for obstructions such as bridging and conduit between openings.

Walls

Wood-frame walls in existing houses are usually covered both inside and out, so application of batt or blanket insulation is impractical in such renovating. It is possible, however, to blow fill-type insulation into each of the stud spaces. More information on blown insulation appears later in this chapter. On houses having wood siding, the top strip just below the top plates and strips below each window are removed. Two inch diameter holes are cut through the sheathing into each stud space. The depth of each stud space is determined by using a plumb bob, and additional holes are made below obstructions in the spaces. Insulation is forced under slight pressure through a hose and nozzle into the stud space until it is completely filled. Special care should be taken to insulate spaces around doors and windows and at intersections of partition and outside walls.

When the studs are exposed during remodeling or when building an addition, insulate the walls as in new construction. Flexible insulation or friction type insulation (semi-rigid) can be used. Flexible insulation in blanket or batt form is normally manufactured with a vapor barrier. These vapor barriers contain tabs at each side, which are stapled to the frame members. To minimize vapor loss and possible condensation problems, the best method of attaching consists of stapling the tabs over the edge of the studs (Figure 13-7A). However, many contractors do not follow this procedure because it is more difficult and may cause some problems in nailing of the rock lath or drywall to the studs. Consequently, in many cases, the tabs are fastened to the inner faces of the studs. This usually results in some openings along the edge of the vapor barrier and, of course, a chance for vapor to escape and cause problems. When insulation is placed in this manner, it is well to use a vapor barrier over the entire wall. This method is described in the next section.

Insulate non-standard width spaces by cutting the insulation and the vapor barrier an inch or so wider than the space to be filled. Staple the uncut flange as usual. Pull the vapor barrier on the cut side to the other stud, compressing the insulation behind it, and staple through the vapor barrier to the stud. Unfaced blankets are cut slightly oversize and wedged into place.

Another factor in the use of flexible insulation having an integral vapor barrier is the protection required around window and door openings. Where the vapor barrier on the insulation does not cover doubled studs and header areas, additional vapor barrier materials should be used for

194

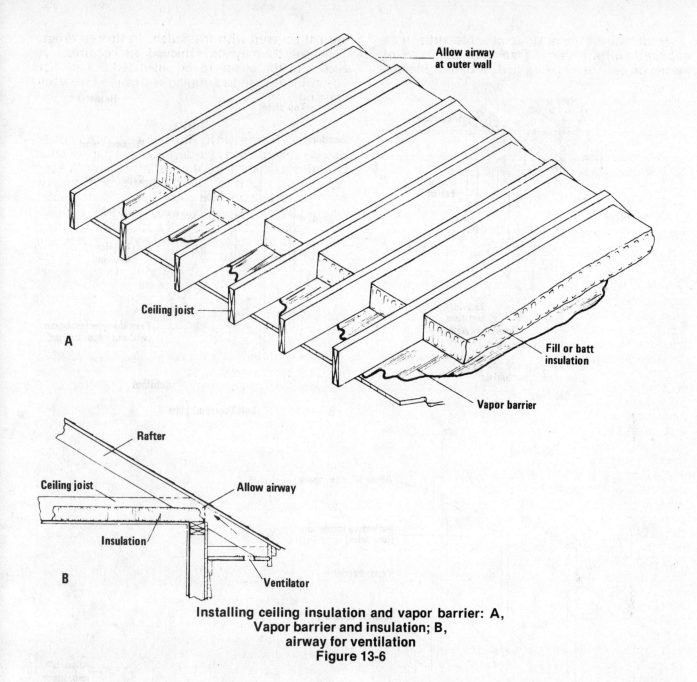

**Installing ceiling insulation and vapor barrier: A,
Vapor barrier and insulation; B,
airway for ventilation
Figure 13-6**

protection (Figure 13-7A). Most well informed contractors include such details in the application of their insulation.

At junctions of interior partitions with exterior walls, care should be taken to cover this intersection with some type of vapor barrier. For best protection, insulating the space between the doubled exterior wall studs and the application of a vapor barrier should be done before the corner post is assembled (Figure 13-7A). However, the vapor barrier should at least cover the stud intersections at each side of the partition wall.

Some of the newer insulation forms, such as the friction-type without covers, have resulted in the development of a new process of installing insulation and vapor barriers so as to practically eliminate condensation problems in the walls. An unfaced friction-type insulation batt is ordinarily supplied without a vapor barrier, is semi-rigid, and is made to fit tightly between frame members spaced 16 or 24 inches on center. ''Enveloping'' is a process of installing a vapor barrier over the entire wall (Figure 13-7B). This type vapor barrier often consists of 4-mil or thicker polyethylene or similar material used in 8 foot wide rolls. After insulation has been placed, rough wiring or duct work finished, and window frames installed, the vapor barrier is placed over the entire wall, stapling when necessary to hold it in place. Window and door headers, top and bottom plates, and other framing are completely covered, (Figure 13-7B). After rock lath plaster base or drywall finish is installed, the vapor barrier can be trimmed around window openings.

Reflective insulations ordinarily consist of either a kraft sheet faced on two sides with

195

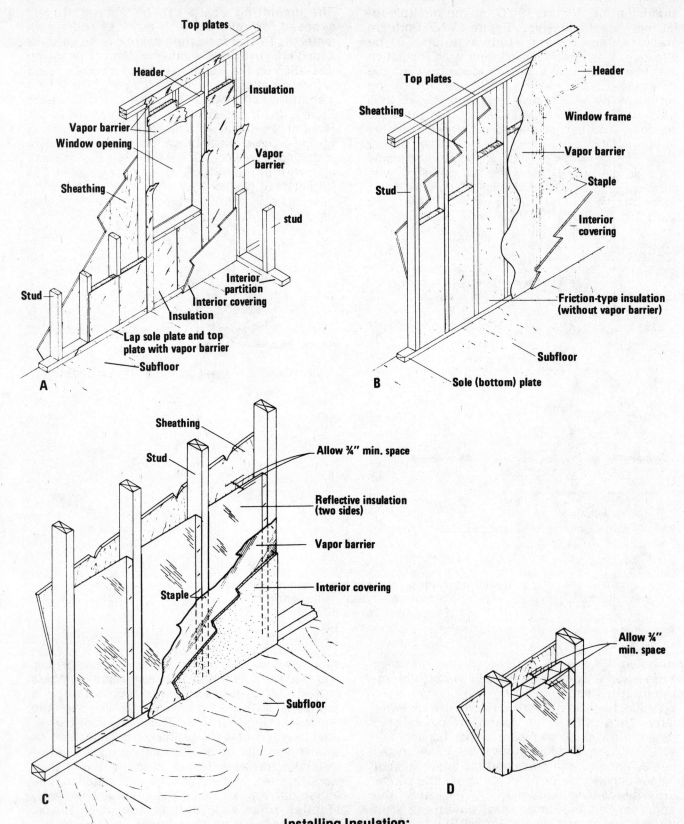

Installing Insulation:
A, Blanket insulation and vapor barriers in exterior wall. B, Vapor barrier over friction-type insulation (enveloping). C, Single sheet two side reflective insulation. D, Multiple sheet reflective insulation.
Figure 13-7

aluminum foil, Figure 13-7C, or the multiple-reflective "accordion" type, Figure 13-7D. Both are made to use between studs or joists. To be effective, it is important in using such insulation that there is at least a ¾ inch space between the reflective surface and the wall, floor, or ceiling surface. When a reflective insulation is used, it is good practice to use a vapor barrier over the studs or joists. The barrier should be placed over the frame members just under the drywall or plaster base (Figure 13-7C). Gypsum board commonly used as a drywall finish can be obtained with aluminum foil on the inside face which serves as a vapor barrier. When such material is used, the need for a separate vapor barrier is eliminated.

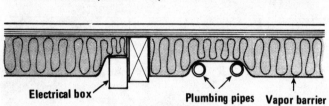

Insulate behind pipes and electrical boxes
Figure 13-8

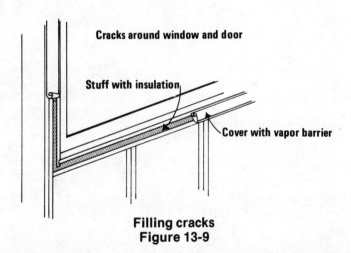

Filling cracks
Figure 13-9

Push insulation behind pipes, ducts, and electrical boxes. Note Figure 13-8. The space may be packed with loose insulation or a piece of insulation of the proper size can be cut and fitted into place. Pack small spaces between rough faming and door and window headers, jambs, and sills with pieces of insulation. Note Figure 13-9. Staple insulation vapor barrier paper or polyethylene over these small spaces.

Regardless of type and location of insulation, vapor barriers are required on the warm side of the insulation. Vapor barriers will be discussed in the next chapter.

Solid masonry walls, such as brick stone, and concrete, can be insulated only by applying insulation to the interior surface. Remember that in doing this some space is being lost. One method of installing such insulation is to adhesively bond insulating board directly to the interior surface.

The insulating board can be plastered, left exposed, or covered with any desired finish material. Thicker insulating board can be used for added effectiveness. Another method of installing insulation on the inside surface of masonry walls is through attachment of 2 by 2 , 1 by 2, or 2 by 4 inch furring strips 16 or 24 inches on center. Note Figure 13-10. Staple vapor barrier flanges to the faces of the furring strips. With 1 by 2 furring, wedge unfaced masonry wall blankets (normally 1 inch thick) between the furring strips; apply a separate polyethylene vapor barrier over the insulation or use foil-backed gypsum board as the interior finish. With the 2 by 4 furring frequently used in colder climates, insulation blankets are stapled to the sides of the furring strips.

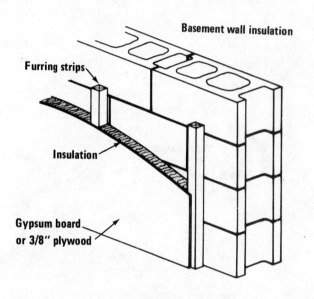

Masonry wall insulation
Figure 13-10

Floors
An unheated crawl space in cold climates offers insufficient protection to supply and disposal pipes during winter months. It is common practice to use a large vitrified or similar tile to enclose the water and sewer lines in the crawl space. Insulation is then placed within the tile to the floor level.

Insulating batts, with an attached vapor barrier, are normally located between the floor joists. They can be fastened by placing the tabs over the edge of the joists before the subfloor is installed when the cover (vapor barrier) is strong enough to support the insulation batt. However, there is often a hazard of the insulation becoming wet before the subfloor is installed and the house enclosed. Thus, it is advisable to use one of the following alternate methods:

Friction type batt insulation is made to fit tightly between joists and may be installed from the crawl space as shown in Figure 13-11. It is good practice to use small "dabs" of mastic

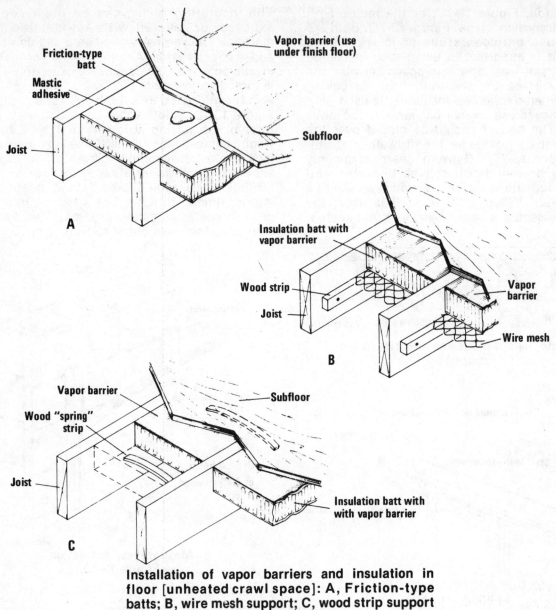

Installation of vapor barriers and insulation in floor [unheated crawl space]: A, Friction-type batts; B, wire mesh support; C, wood strip support
Figure 13-11

adhesive to insure that it remains in place against the subfloor. When the vapor barrier is not a part of the insulation, a separate film should be placed between the subfloor and the finish floor.

One method of heating which is sometimes used for crawl space houses, utilizes the crawl space as a plenum chamber. Warm air is forced into the crawl space, which is somewhat shallower than those normally used without heat, and through wall-floor registers, around the outer walls, into the rooms above. When such a system is used, insulation is placed along the perimeter walls as shown in Figure 13-12. Flexible insulation, with the vapor barrier facing the interior, is used between joists, at the top of the foundation wall. A rigid insulation such as expanded polystyrene is placed along the inside of the wall extending below the groundline to reduce heat

loss. Insulation may be held in place with an approved mastic adhesive. To protect the insulation from moisture and to prevent moisture entry into the crawl space from the soil, a vapor barrier is used over the insulation below the groundline (Figure 13-12). Seams of the ground cover should be lapped and held in place with bricks or other bits of masonry. Some builders pour a thin concrete slab over the vapor barrier. The crawl space of such construction is seldom ventilated.

Second Stories
One of the areas of a two-story house where the requirement of a vapor barrier and insulation is often overlooked is at the perimeter area of the second floor joists. The space between the joists at the header and along the stringer joists should be protected by sections of batt insulation which

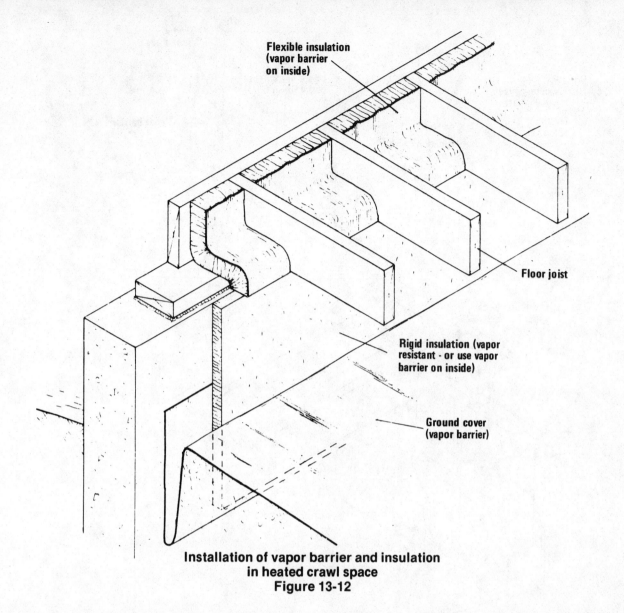

Flexible insulation (vapor barrier on inside)

Floor joist

Rigid insulation (vapor resistant - or use vapor barrier on inside)

Ground cover (vapor barrier)

Installation of vapor barrier and insulation in heated crawl space
Figure 13-12

contain a vapor barrier (Figure 13-13). The sections should fit tightly so that both the vapor barriers and the insulation fill the joist spaces. Insulation and vapor barriers in exposed second floor walls (Figure 13-13) should be installed in the same manner as for walls of single-story houses.

A two-story house is sometimes designed so that part of the second floor projects beyond the first. This projection varies but is often about 12 inches. In such designs, the projections should be insulated and vapor barriers installed as shown in Figure 13-14.

In 1½-story houses containing bedrooms and other occupied rooms on the second floor, it is common practice to include knee walls. These are partial walls which extend from the floor to the rafters (Figure 13-15). Their height usually varies between 4 and 6 feet. Such areas must normally contain vapor barriers and insulation in the following areas: (a) In the first floor ceiling area, (b) at the knee wall, and (c) between the rafters. Insulation batts with the vapor barrier facing down

should be placed between joists from the outside wall plate to the knee wall. The insulation should also fill the entire joist space directly under the knee wall (Figure 13-15). Care should be taken when placing the insulating batt to allow an airway for attic ventilation at the junction of the rafter and exterior wall.

Insulation in the knee wall can consist of blanket or batt type insulation with integral vapor barrier or with separately applied vapor barriers, as described for first and second floor walls.

Batt or blanket insulation is commonly used between the rafters at the sloping portion of the heated room, Figure 13-15. As in the application of all insulations, the vapor barrier should face the inner or warm side of the roof or wall. An airway should always be allowed between the top of the insulation and the roof sheathing at each rafter space. This should be at least a 1 inch clear space without obstructions such as might occur with solid blocking. This will allow movement of air in the area behind the knee wall to the attic area above the second floor rooms.

199

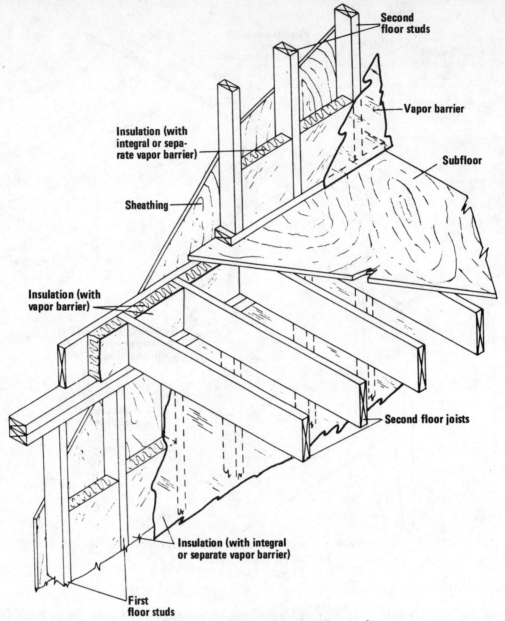

**Insulation in walls and joist space of
two-story house
Figure 13-13**

Labels on figure:
- Second floor studs
- Vapor barrier
- Subfloor
- Insulation (with integral or separate vapor barrier)
- Sheathing
- Insulation (with vapor barrier)
- Second floor joists
- Insulation (with integral or separate vapor barrier)
- First floor studs

Finished Basement Rooms

Finished rooms in basement areas with fully or partly exposed walls should be treated much the same as a framed wall with respect to the use of vapor barriers and insulation (Figure 13-16). When a full masonry wall is involved, several factors should be considered: (a) When drainage in the area is poor and soil is wet or when the basement has a history of dampness, drain tile should be installed on the outside of the footing for removing excess water; (b) in addition to an exterior wall coating, a waterproof coating should also be applied to the interior surface of the masonry to insure a dry wall.

Furring strips (2 by 2 or 2 by 3 inch members) used on the wall provide (a) space for the blanket insulation with the attached vapor barrier and (b) nailing surfaces for interior finish, (Figure 13-16).

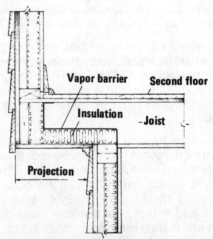

**Insulation and vapor barrier at second
floor projection
Figure 13-14**

Labels on figure:
- Vapor barrier
- Second floor
- Insulation
- Joist
- Projection

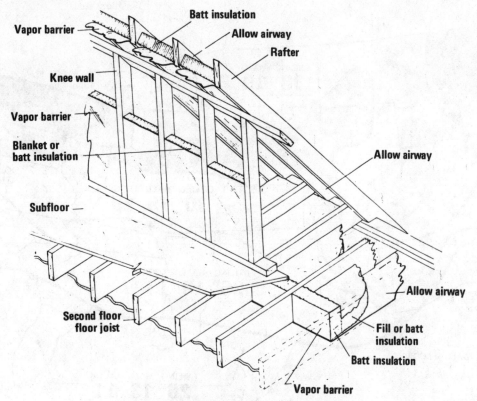

Installing vapor barrier and insulation in
knee-wall areas of 1½ story house
Figure 13-15

One or 1½ inch thicknesses of friction-type insulation with a vapor barrier of plastic film such as 4-mil polyethylene or other materials might also be used for the walls.

Other materials which are used over masonry walls consist of rigid insulation such as expanded polystyrene. These are installed with a thin slurry of cement mortar and the wall completed with a plaster finish. The expanded plastic insulations normally have moderate resistance to vapor movement and require no other vapor barrier.

When a vapor barrier has not been used under the concrete slab, it is good practice to place some type over the slab itself before applying the sleepers. One such system for unprotected in-place slabs involves the use of treated 1 by 4 inch sleepers fastened to the slab with a mastic. This is followed by the vapor barrier and further by second sets of 1 by 4 inch sleepers placed over and nailed to the first set. Subfloor and finish floor are then applied over the sleepers.

To prevent heat loss and minimize escape of water vapor, blanket or batt insulation with attached vapor barriers should be used around the perimeter of the floor framing above the foundation walls (Figure 13-16). Place the insulation between the joists or along stringer joists with the vapor barrier facing the basement side. The vapor barrier should fit tightly against the joists and subfloor.

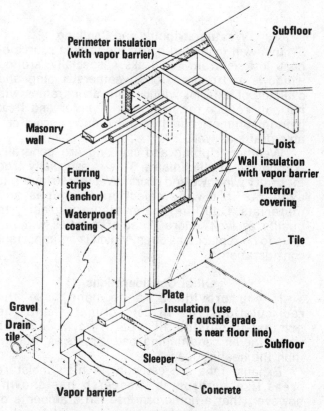

Installing vapor barrier in floor and wall of
finished basement
Figure 13-16

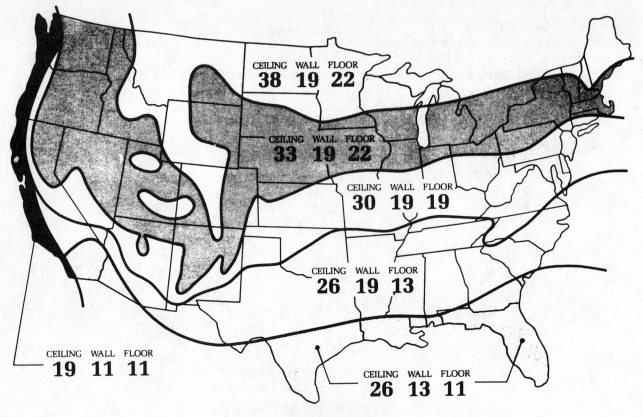

Recommended "R" values
Figure 13-17

CEILING WALL FLOOR
38 19 22

CEILING WALL FLOOR
33 19 22

CEILING WALL FLOOR
30 19 19

CEILING WALL FLOOR
26 19 13

CEILING WALL FLOOR
19 11 11

CEILING WALL FLOOR
26 13 11

Weatherstripping and Caulking

In a well insulated house the largest source of heat loss is from air leaks, especially around windows and doors. Good weatherstripping and caulking of exterior window and door frames will not only reduce the heat loss in winter and heat gain in summer, but they will reduce uncomfortable drafts as well.

Weatherstripping and caulking are generally economical in all climates. This is especially true for drafty windows and doors. Weatherstripping is available in a wide selection of shapes and materials. Caulking materials vary greatly in quality as well. More durable varieties will not have to be replaced as often, which is an important consideration.

Other Considerations

If you provide a range hood, use the recirculation type in cold climates and the exhaust-to-outside air type in warm climates where the air conditioning load is more significant than the heating load.

Exhaust fans are essential in high moisture areas like bathrooms. Research has shown, however, that a home can lose large amounts of heated air through them, so use a model with a positive shutter closure.

If a fireplace is used, be sure to install a damper and instruct the homeowner to close it whenever the fireplace is not in use. A properly constructed fireplace draws air up through the chimney to expel smoke from the fire, but without a damper, it will continually draw out heated room air in the winter when the fireplace is not in use.

Try to avoid running ducts through non-conditioned spaces like attics. But if this is not possible, insulate the ducts well. Insulating ducts or using pre-formed fiberglas ducts can significantly reduce heat loss. Even if a sufficient amount of insulation exists, you might want to loosen it temporarily at the joints to check the condition of the ducts. Escaping air indicates the need for retaping of the duct joints. Most houses have no more than 1 or 2 inches of insulation wrapped around ducts in unheated areas. More is often desirable. Duct wrap in layers of 2 inch thicknesses may be used to insulate ductwork. As an alternative, ordinary mineral fiber batts can be placed around the ducts. Avoid crushing the insulation because this lowers its resistance to heat flow. Where ductwork is used for heating purposes only, a vapor barrier is not usually needed. However, if ducts are used for air conditioning as well as heating, a vapor barrier is required around the outside of the insulation to prevent condensation on the ducts. In either case, some protective covering should be used. Check with insulation manufacturers for the correct procedure for your climate.

Ceilings, double layers of batts
R-38, Two layers of R-19 (6'') mineral fiber
R-33, One layer of R-22 (6½'') and one layer of R-11 (3½'') mineral fiber
R-30, One layer of R-19 (6'') and one layer of R-11 (3½'') mineral fiber
R-26, Two layers of R-13 (3-5/8'') mineral fiber

Ceilings, loose fill mineral wool and batts
R-38, R-19 (6'') mineral fiber and 20 bags of wool per 1,000 S.F. (8¾'')
R-33, R-22 (6½'') mineral fiber and 11 bags of wool per 1,000 S.F. (5'')
R-30, R-19 (6'') mineral fiber and 11 bags of wool per 1,000 S.F. (5'')
R-26, R-19 (6'') mineral fiber and 8 bags of wool per 1,000 S.F. (3¼'')

Walls, using 2''x6'' framing
R-19, R-19 (6'') mineral fiber batts

Walls, using 2''x4'' framing
R-19, R-19 (6'') mineral fiber batts and 1'' polystyrene foam sheathing
R-13, R-13 (3-5/8'') mineral fiber batts
R-11, R-11 (3½'') mineral fiber batts

Floors
R-22, R-22 (6½'') mineral fiber
R-19, R-19 (6'') mineral fiber
R-13, R-13 (3-5/8'') mineral fiber
R-11, R-11 (3½'') mineral fiber

Insulation recommendations
Table 13-18

How Much Insulation Is Needed?

Figure 13-17 summarizes the optimum insulation values for residential heating and cooling as calculated for each climate zone. The figures are the recommendation of Owens-Corning Fiberglas and take into account national weather data, energy costs and projected increases, and insulation costs. Table 13-18 shows how much insulation is required to reach the recommended ''R'' values.

Figure 13-17 shows that most areas in the United States require R19 walls insulation. Filling the cavity in a 2 by 4 inch frame wall will produce no more than a R14 rating. A nominal 6 inch cavity will carry R19 insulation. Six inch studs spaced 24 inches on center (Figure 13-19) will give the same strength as 4 inch studs 16 inches on center but use fewer studs, less lumber, and provide better insulation because the studs transmit heat better than the fully insulated spaces between the studs.

Blown Insulation

Loose-fill, blown insulation for reinsulation of ceilings and outside walls can be very effective in reducing heating and cooling energy requirements. Three or 4 inches of insulation properly placed in wall air space can reduce the heat transfer through the walls by as much as two thirds. Blown wall insulation in a 3½ inch cavity wall is usually given a R14 rating. Typically, blown wall and ceiling insulation alone will reduce heating costs by about 35 percent in houses with little or no insulation. Savings and improved

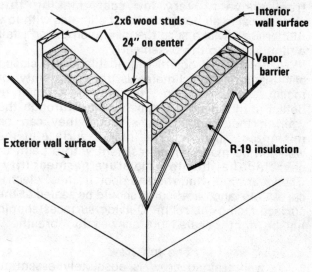

Modern 2''x6'' framing
Figure 13-19

comfort begin immediately when a good insulation job is done and continues throughout all seasons.

The advantage of blown insulation is that you are able to reinsulate exterior walls without tearing up the interior surface. The lightweight material is fed by a machine, located outside the building, through a flexible hose to the point of installation. The equipment is obtained from the insulation manufacturer. Holes are bored through sidings and later covered with an inconspicuous plug, or a section of the exterior facing material can be removed and replaced. The fibers can be blown over existing insulation, filling voids which might have occurred with older insulation. It can be put into a ceiling, for example, where extra weight might be a problem and where long spans might otherwise require reduced insulation depth or special trussing.

Loose-fill materials, usually mineral fiber or cellulose, are the insulation forms best suited for this job. Cellulose is remarkably well suited for reinsulation jobs. The pneumatic blowing of short fibers results in a completely filled cavity without "hang-up" on wiring, nails, rough plaster or splinters. The monolith of millions of wood pulp particles retains its non-settling density over years of service. Moisture dissipation properties also are significant. When moisture vapor enters the wall it is held in suspension within the natural cellular fibers preventing it from collecting at the base of the wall. The moisture dissipates naturally, during temperature changes, similar to the evaporation of perspiration from the skin. It is essential that proper insulation be installed in all building areas exposed to the outside. Blown insulation generally can be applied to all ceiling and sidewall areas as well as to other exposed areas such as overhanging floors. For those areas where it is not practical to install loose fill insulation, such as under floors, batts or blankets should be installed.

In some older houses, access can be gained to the wall space from the attic. In this case, loosefill insulation material can be dropped into the space from above at very low cost, making this economical in all but the mildest climates with low fuel prices. (Make sure the insulation doesn't fall all the way into the basement!)

A potential problem with insulation in closed cavities in some climates is the possibility of moisture accumulation. This may be difficult to detect until moisture begins to show through the wall. If moisture problems occur, they can be minimized, however. The interior surface of the wall can be made vapor resistant with a paint or covering that has low moisture permeability. Cracks around windows and door frames, electrical outlets, and baseboards should be sealed at the surface facing the room. Outside surfaces should not be tightly sealed but allowed to "breathe."

The Job Crew

A well trained crew is absolutely essential, both to insure that the insulation work is installed correctly and to see that a profit is made on the job. The size of the crew will vary, depending upon the type of blowing equipment and the particular job. A minimum of two men is recommended even on ceiling jobs with automatic machines. Where both ceilings and sidewalls are to be insulated, it may be advisable to use as many as four men, one of whom should be a competent carpenter.

Before starting any work, the foreman should make a thorough survey of the job to note any pre-existing damage to the property. Any damage which could be blamed upon the insulation crew should be shown to the homeowner. The foreman should request permission to enter the house to check the condition of interior walls and to suggest removal of dishes on shelves, mirrors, and pictures which might be dislodged during the application process. At job completion, the foreman should supervise the careful cleaning of buildings and grounds. It is generally an excellent idea for the foreman to offer the homeowner the opportunity of seeing those parts of the completed insulation job which are accessible.

Frequently the foreman can pick up leads to other prospective insulation purchasers. Although it is not his specific responsibility, a good foreman never misses the opportunity to get his company's name to an interested prospect or to turn the prospect's name over to his boss. Make sure your foreman has a supply of your business cards and receives a commission on jobs he sells.

Attics

Since the attic area is generally the most accessible part of the house to be insulated, most crews will start in this area. For practical scheduling, the attic area should always be saved for inclement weather.

Be sure there is sufficient light by which to work. Every crew should carry a long extension cord with extra bulbs for use whenever the attic is not well lighted. The hoseman should also have a flashlight available.

Before starting to blow insulation in the attic, anything which might interfere with the movement of the hoseman or proper application should be placed in an area where it will give the least amount of trouble. Items stored in the attic should be protected as much as possible. Clothes and other items which could be damaged by the insulation should be removed from the attic.

Before beginning the blowing, a batt or other baffle should be placed around the perimeter of houses having eave ventilation. This batt or baffle prevents the loss of insulation into the eave vents and, more important, prevents air moving through the eave vents from moving the insulation back away from the plate. Apply batts over the tops of wells and drop ceilings to assure that there is a continuous insulating blanket over the top floor ceiling.

In an unfloored attic, keep the hose parallel to the floor with insulation falling 6 to 8 feet in front. Where possible, back away from the work to prevent packing. Where work space is tight, prevent insulation from packing by allowing it to blow off your hand.

Blow three or four joists from one position by aiming the hose to the right or left. Always blow in the direction of the joists, not across them. Keep the hose close to the floor where the insulation must go underneath obstructions such as cross-bracing and wiring. Insulation must be blown on both sides of this kind of obstruction. Where an obstruction may cause a low spot to occur, move around, check this spot and, if necessary, fill in the low area. Be sure that the insulation is installed on both sides of obstructions such as solid cross bracing. (If an area is not properly blown, dust marks will appear on the ceiling below after a short time due to the difference in temperature of the surface.)

If a batt or blanket is not used to block off the ends of joists, be sure that the full depth of insulation is applied all the way to the end of the plate. When roof construction does not allow the application to go to the ends of the joists to assure full depth, it is possible to bounce the insulation off the underside of the roof in order to build up the required depth at the plate. Care should be taken, however, not to get the eave vents blocked.

There are two areas in the attic over which insulation must not be placed. One is on top of exhaust fans and the other on top of recessed light fixtures commonly used in kitchens and bathrooms. A recessed fixture must dissipate the heat upward and it is absolutely essential that no insulation be on top of its box.

Water pipes run in the attic area must be given protection, since the attic temperature during cold weather will now be very close to that of the outdoors. The severity of the winters and the location of the pipes will determine what additional protection must be given to prevent freezing. As a minimum, however, apply the same depth of insulation on top of pipes as there is between pipes and the warm surface.

After the attic is blown, smooth down the job as much as possible to make it look neater and to even out any of the high and low spots. Do not remove the hose from the attic until the foreman has inspected and determined that no areas have been accidently missed. Install a batt or blanket on top of areas where loose fill has not been applied, such as access panels, stair wells, and fan covers. The completed job should provide a continuous blanket of insulation over the ceiling area.

Many attic areas contain some floored areas which present no real problem. It is not advisable to attempt to blow more than six feet under flooring, so a floor board should be removed approximately every twelve feet. If some flooring is difficult to remove because of its nailing pattern, take care during removal to prevent any possible damage to the ceiling surface below. Since it is difficult to see bracing underneath the flooring, take particular care to assume that the flow of insulation under the floor is not being blocked. When there is a large amount of bracing under the floor, it may be necessary to take off several boards in a small area. Insert the hose approximately six feet under the floor and gradually pull it out as the joist area fills. Twist and turn the hose as it is removed to assure complete coverage of the area under the floor.

Knee Walls

Although it is possible to put up a retainer and blow knee walls, the easiest method is to use a batt or blanket for the height of the knee wall. Slopes, if not too long, can be insulated with batts, or the bottom of the slopes may be blocked off and the area blown full of insulation. When blowing slopes, be sure that trapped air does not leave void spots. When insulating story and a half houses or houses with attic rooms, be sure not to miss any flat areas beyond the knee walls. Many times these flat areas are accessible by cutting a hole through a closet wall or by making a roof opening which should be patched immediately. These flat areas are very easy to miss if the job is not thoroughly checked.

Attic Stairways

Many older homes have an attic stairway. A ceiling insulation job is not complete until the stairway is insulated, including the soffit area, the walls, and the door. Many times the soffit area can be insulated by either removing the treads or drilling holes and filling with insulation. Close holes by means of plugs. Plugs can be finished and restained if necessary. Stairway walls can be insulated in the same manner as any other wall. Finish the openings as required to match the existing stairway finish. An alternate to insulating the entire stairway assembly is to install a trap door, operated by a counter-balance, over the top of the stairway, and insulate it by means of batts or blankets.

Sidewalls

Insulation of sidewalls is not a difficult job, providing you understand how sidewalls are framed out. A crew that has not previously blown a sidewall of an existing house should spend some time studying the general framing principles in a house which is under construction. By studying a wall section before the interior finish is put up, a man has a much better idea of how such things as fire stops, junction boxes, romex cables, and bracing will affect his job.

Regardless of the outside finish, all sidewalls are insulated in a similar manner. Some of the outside finish is removed and openings are made in the sheathing so that insulation can be blown into the empty stud area.

Some blown insulation principles apply to all

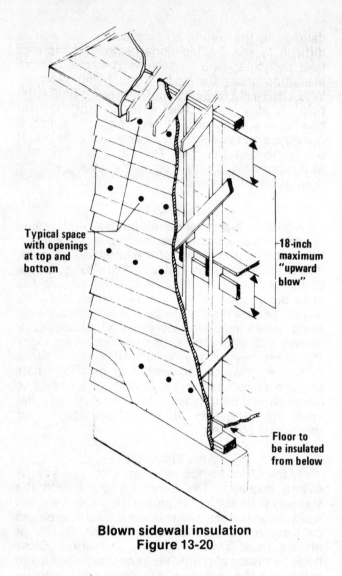

Typical space with openings at top and bottom

18-inch maximum "upward blow"

Floor to be insulated from below

**Blown sidewall insulation
Figure 13-20**

types of wall construction:

1. The "double blow" method (Figure 13-20), with two openings, for sidewalls is recommended, although some reliable applicators can fill the spaces with one opening. Some stud sections may even require three or more openings because of construction features. Openings should be made into the stud area for each eight to ten feet of height. This is essential to assure that the stud area is completely full. Never try to blow more than 4 feet down or 18 inches up unless you have made two openings. Blowing through a single opening in an 8 foot wall could leave the bottom of the stud space with no insulation.

2. Remove the trim whenever possible. Generally this will expose studding.

3. Many homes have eaves which are actually below the level of the plate. Frequently access to the stud area can be gained by removing the eave panels. As with trim, the removal of eave sections will generally open a large amount of stud area with very little work.

4. Plumb bob all openings to determine the amount of area that can be filled through that opening. A plumb bob should be of sufficient size

to readily reveal objects which would stop the flow of insulation. Areas under windows and below fire stops and bracing must be opened to completely fill the area.

Siding, shingles and brick veneer can be opened and replaced in many cases. Plywood panel siding and stucco are usually bored and slugged to match the previous condition as much as possible.

After the sidewall has been properly opened, insulation is blown into all areas. It is strongly recommended that a sidewall nozzle be used in blowing sidewalls so that the applicator can best control the direction in which his insulation is being blown. Your material manufacturer or equipment supplier can recommend the best nozzle for your needs.

Different applicators have different methods for filling up sidewall panels. It is generally recommended that the lower holes be filled first, particularly if the wall is partially insulated. If not, some lower holes will appear to be filled from the opening above. The hose should always be inserted in such holes to make sure that the insulation in the lower stud space is at the correct density.

Machine Pressure

The amount of pressure at which a machine should operate will vary with the job. The blowing machine should be equipped with a properly operating pop-off valve so that when the wall section is filled, pressure will bleed off at the machine rather than into the sidewall section, eliminating the danger of blowing out the inside surface. Considerably less pressure should be used on a sidewall in which the inner surface is drywall construction than on one in which the inner surface is metal lath and plaster. As the insulation is being blown into the sidewall, continually move the blowing nozzle from one side to the other so that the entire stud area will be filled.

The hoseman should be able to judge the approximate length of time it will take to fill any opening. If an opening is filled too quickly, it probably indicates that the insulation is hung up on some type of obstruction. When this occurs, it may be possible, by adjusting the nozzle, to break the insulation away from whatever it is hung on. However, it may also be necessary to make another opening below the stoppage. The applicator should remember that any sidewall section which is not properly insulated will be very evident to the homeowner as it will appear as a cold surface on his inside wall.

Houses Without Sheathing

On some houses, siding is nailed directly to the face of the studs without any sheathing. With this type of construction, removal of the outside facing will result in an opening into the stud area which is too large for proper use of the sidewall nozzle. In such cases, use a length of lumber approximately 24 inches long and the appropriate width as a substitute for sheathing, drilling a hole in it for inserting the sidewall nozzle.

Opening Procedures For Sidewalls

The information contained in the following paragraphs covers the procedure for the removal and replacement of different types of sidewall materials. There can be many variations of the procedures shown, and any method which gives efficient access to the sidewall area can be used.

Since the removal of the outside finishes on any existing home largely involves the principles of good carpentry, it is strongly recommended that one of the crew be a proficient carpenter. If one of the crew is not a carpenter, it is suggested that a carpenter be hired to work and train your men on the first few sidewall jobs. For brick or stone sidewalls it is recommended that an experienced mason be used until your crew is familiar with the procedure for entering these walls.

Bevel Siding

Tools usually required are a wide-faced chisel, curved siding chisel, claw hammer, nail set, nail saw, and back saws. The siding to be removed is shown as board Number 1 in Figure 13-21. Removal and preparatory steps are as follows:

1. Using the back saw, free the vertical end joints of the board on either side, or cut new vertical joints if only a portion of the board has to be removed.

2. Using the chisel, pry up the bottom of the board leaving a space at least the width of the chisel between the edge of board Number 1 and the one below. With the board in this position, cut with the nail saw nail "A" holding the bottom of the board.

3. Pry board Number 2 with the chisel in the same manner as board Number 1 and, at the same time, kink the nail holding the Number 2 board to the sheathing.

4. Tap the Number 2 board back gently, thus forcing out nail "B" previously kinked.

5. Carefully remove the Number 1 board, using the chisel if the board is not free enough to fall out by itself.

6. After the board is removed, the building paper, usually found between siding and sheathing, should not be removed or damaged any more than necessary. Start by cutting the paper along the horizontal edge about 1 inch above the board still in place.

7. Cut the paper vertically on either side of the board and at least 6 inches on either side of the last hole at the end of the row.

8. Bend up the paper and tack to the board above, allowing free access to the sheathing.

9. Drill the hole and blow the insulation.

After the wall section is blown, replace the siding in a workmanlike manner, filling any nail holes or splits with putty, and touch up. If

possible, leave a very slight opening between the replaced board and the ones already on the house in order to permit breathing in the wall.

A chisel may be used instead of a nail saw for cutting nails. In isolated cases it may be necessary to set the nails with nail sets through both boards and into the sheathing. This method is not recommended unless absolutely necessary. The bottom of the board may be split and the nail set creates large holes requiring a large amount of puttying.

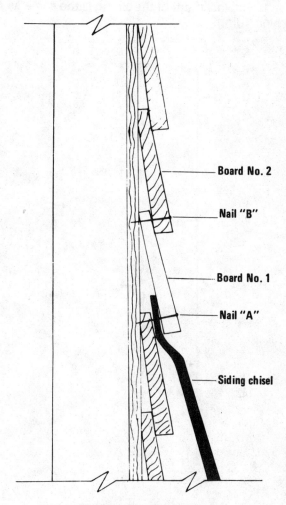

Opening in bevel siding
Figure 13-21

Jointed Siding

Novelty siding (regardless of its exterior appearance) usually has one common characteristic—the boards are joined together by some form of shiplap or tongue and groove joint. In this type of work there is no opportunity to pry the boards loose. Therefore, the following steps are practically the only methods by which the work should be handled. Tools required are a claw hammer, wood chisel, nail set, and a broad curved blade knife with a striking plate on the back side of the blade. The procedure for shiplap siding (Figure 13-22) is as follows:

1. Set the nails directly through the board.

2. After setting the nails, the procedure is practically the same as that for bevel siding. The chisel is inserted under the lower edge of the board to be removed and it is pried outward (Figure 13-22). As the illustration of the shiplap joint shows in Figure 13-22, there is nothing to prevent the removal of the board.

3. Preparations for blowing (treatment of sheathing, etc., and drilling) are the same as for bevel siding.

4. Replacement of the siding is the same as for bevel siding.

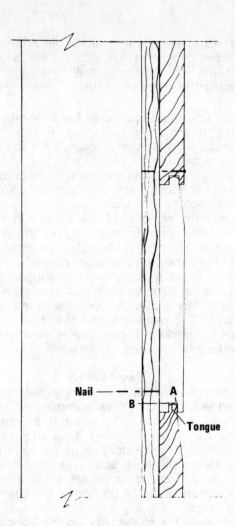

**Opening tongue-and-groove siding
Figure 13-23**

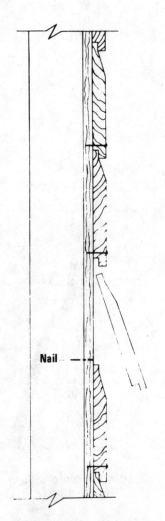

**Opening shiplap siding
Figure 13-22**

Tongue and Groove Siding

With tongue and groove work a slightly different procedure is necessary as the boards cannot be pried out without breaking either the tongue or one of the edges. Two methods are in use at present. Both can be recommended, although where the boards are very tightly fitted, Method Number 1 is likely to split off the top outer edge of the board.

Method 1

1. Set all nails directly through to the sheathing.

2. At the lower edge of the board to be removed, insert an electrician's chisel and pry out and down at the point "A" as shown in Figure 13-23. This will usually cause the board to split off at the rear along the line "B", leaving the tongue of the board below intact.

3. By proper manipulation, board Number 1 will come away from the top groove in the board above and can be removed.

4. Preparations for blowing (treatment of sheathing, etc., and drilling) are the same as for bevel siding.

To replace the siding, set the edge of the tongue into the groove of the upper board and tap the board along the edges until the tongue is started in the groove along the entire length. When the tongue has entered the groove far enough so that bottom edges are level, the bottom edge can be knocked straight in and the board is now ready for face nailing. Finishing nails are used in the same manner as for bevel siding.

Method 2

Where the boards are set in tightly and there is danger that the board cannot be freed at the top tongue, the following alternate method is suggested:

1. Set the nails through the board.

2. Use a broad bladed knife about 12 inches long, with a slight curvature to the blade face and a welded striking plate on the blade. The knife blade is inserted in the groove at both top and bottom of the board to be removed and driven in, in order to cut off the tongue.

3. After the tongues are completely cut away from the board, top and bottom, the board can be removed easily. If the board is toenailed at the top edge through the groove instead of straight-nailed, the procedure is the same except that slightly more prying is required.

4. Preparations for blowing (treatment of sheathing, etc., and drilling) are the same as for bevel siding.

After blowing, cover the sheathing behind the top and bottom edges of the board with mastic the full length of the cut and at least 1 inch either side of the tongue and groove. The mastic should be sufficiently thick so that when the board is pushed back into place, the mastic will work into place between the upper and lower joints, serving in place of the tongue and groove as a protection against the weather. Since the board is absolutely free, it can be easily put in place. Finishing nails are used in the same manner as for bevel siding.

Vertical Siding

Vertical novelty siding is sometimes applied to the face of the house without sheathing. In such cases, there are usually several horizontal stops or nailers in the wall. When this construction is encountered, the only suitable method is removal of all vertical siding or removal of sufficient siding in each panel to provide for blowing or hand packing as the case may be.

The method of removing and replacing vertical siding is exactly the same as for horizontal siding. However, since there is no sheathing behind the siding, the edges must be butted with either white lead or mastic compound to provide a water seal at the joint. If there is an appreciable opening on either side of the board after replacement, the joints should be further sealed by caulking with a fine-pointed caulking gun.

Siding Shingles

Method 1

Where this method is to be used, the owner should be informed of the procedure. Otherwise, the shingles should be pulled as shown under Method 2. Method 1 requires the cutting of the outer course of shingles and involves the use of a claw hammer and a 2½ inch wide blade steel chisel no greater than 8 inches in overall length, the taper of the blade to range from 0 to 3/16 of an inch. The various steps are as follows:

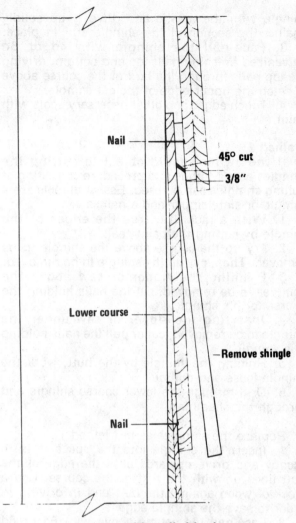

Opening sidewall shingles
Figure 3-24

1. Beginning at a distance of about 3/8 inch up under the butt edge of the shingle course above, start to cut the shingle to be removed on a 45 degree angle upward being careful that each chisel width of cut overlaps the previous cut and that the cutting line is kept straight. (See Figure 13-24). The first cut across should only go half way through the shingle thickness to prevent unnecessary splitting of the shingle.

2. Repeat the cutting across the shingle in the same manner, finishing the cut.

3. Carefully break the paint seal at the bottom and sides of the shingle with a jacknife and remove the shingle.

4. Drill through the lower course of shingles and through the sheathing. Pieces of shingles usually cannot be saved.

After blowing, replace the shingles:

1. If the lower course of shingles is in poor condition, place a piece of asphalt paper over the hole to entirely cover the under surface of the shingle removed and press the paper down into mastic.

2. Lightly daub the cut edge of the removed

shingle with the mastic and drive it up solidly against the section of the shingle still in place.

3. Face-nail the shingle with 4d or 6d galvanized finishing nails top and bottom, driving the top nails through the butt of the course above to catch the upper edge of the cut shingle.

4. Touch edges and other necessary spots with paint.

Method 2

If the owner should object to cutting the shingles as in Method 1, a procedure consisting of pulling shingles can be used. Essential tools are a shingle or slate ripper and a nail saw.

1. With a jacknife, free the edges of the shingle by cutting the paint seal.

2. Pry up the course above the shingle to be removed. Then, pry up the shingle to be removed.

3. Running the ripper or saw above the shingles to be removed, cut the nails holding the course to this shingle.

4. Inserting the ripper or saw under the shingle to be removed, cut or pull the nails holding this shingle.

5. Holding the shingle by the butt, work the shingle loose and pull out.

6. Drill through the lower course shingle and through the sheathing.

Replace the shingles as in Method 1:

1. Insert the shingle into the space it is to occupy and drive upwards until the edge of the butt lines up with the rest of the course. Use a block of wood against the butt end in driving in order to save the shingle edge.

2. Face-nail the course above with 6d or 8d galvanized finishing nails through the butt and shingle below.

3. Face-nail the shingle just replaced at the butt, setting the nails and puttying.

4. Touch edges and other necessary spots with paint.

Brick Veneer

Brick veneer construction consists of a 4 inch brick wall attached to sheathing which is nailed to studs. Where entrance by removal of the trim or eaves is not possible, it will be necessary to remove the brick. It is of utmost importance that the number of openings be minimized. Thus, it is desirable to make openings straddling the studding, as this permits two panels to be serviced through one opening. The removal of one brick will usually give access to two stud spaces. Where it is necessary to remove two or three bricks, the combination may be two bricks beside each other or two beside each other and one above or below. When servicing only one panel, the removal of a single brick will usually suffice.

When blowing brick veneer walls, a 2 inch nozzle rather than a 2½ inch nozzle will make the job easier. When doing brick walls, consider the use of an experienced mason if your crews are not experienced in this type of construction.

There are two methods of removing brick: (1) Removal of the brick itself by removing the mortar around it; or (2) by actually breaking the brick out in pieces. Whenever possible the former method should be used.

To remove the brick in its entirety, it is necessary to remove the mortar around it. This can be done by drilling four holes at the corners of the brick and chiseling out the mortar between them. Power chisels make this a relatively easy operation on some jobs. Another method of removing the mortar joint is by the use of a skill saw and a special cement cutting blade.

After the panels are filled, the bricks should be carefully replaced, matching the old mortar as closely as possible. Mortar may be given an aged appearance with the aid of a blow torch.

How To Sell Insulation

Cold logic is really all you need to sell improved insulation to homeowners. Unlike most home improvement projects that must be sold on the basis of eye appeal and convenience, insulation can be sold on dollars saved alone. Every month homeowners are reminded by their utility company that they are wasting money and energy. Your only task is to translate the service you sell into dollars saved.

Most everyone knows how important insulation is. But when you meet your prospect it wouldn't hurt to emphasize points to your prospect like the following: "Mr. Prospect, you may not have thought about energy conservation as a good investment, but investing in insulation is better than most long term investments you can make. When you invest in energy conservation improvements, you immediately begin to earn dividends in the form of reduced utility bills. These dividends not only pay off your investment, but they pay "interest" as well. And unlike dividends from many other investments, these are not subject to income taxes.

"At current fuel prices, improved insulation will pay for itself many times over during the life of the house. The more poorly insulated the house is to begin with, the shorter the payback period.

"Even though utility bills rise as energy prices increase, the rise will be much less than it would have been without increased insulation. In fact, you might think of energy conservation improvements as a hedge against inflation.

"Even if you don't plan to live in your house long enough to reap the full return on your investment in the form of lowered utility bills, Mr. Prospect, it will probably still pay to invest in energy conservation improvements now. Because of higher energy prices, a well insulated house is likely to sell more quickly and at a higher price than a poorly insulated house that costs a lot to heat and cool. Show your low fuel bills to prospective buyers. They will find the small increase in monthly mortgage payments will be

Energy Conservation Survey

Attic insulation
1. Attic area (sq.ft.) _____
2. Recommended level _____
3. Existing level _____
4. Add _____
5. Cost/sq.ft. _____
6. Total cost (1x5) _____

Wall insulation [blown-in]
1. Wall area (sq. ft.) _____
2. Recommended level _____
3. Existing level _____
4. Add _____
5. Cost/sq.ft. _____
6. Total cost (1x5) _____

Floor insulation
1. Floor area (sq.ft.) _____
2. Recommended level _____
3. Existing level _____
4. Add _____
5. Cost/sq.ft. _____
6. Total cost (1x5) _____

Duct insulation [attic]
1. Length (ft.) _____
2. Perimeter (ft.) _____
3. Area (1x2x1.5) _____
4. Recommended level _____
5. Existing level _____
6. Add _____
7. Cost/sq.ft. _____
8. Total cost (3x7) _____

Duct insulation [other areas]
1. Length (ft.) _____
2. Perimeter (ft.) _____
3. Area (1x2x1.5) _____
4. Recommended level _____
5. Existing level _____
6. Add _____
7. Cost/sq.ft. _____
8. Total cost (3x7) _____

Storm windows

Size (sq.ft.)	Number	Cost each	Sub-total
_____	_____	_____	_____
_____	_____	_____	_____
_____	_____	_____	_____
_____	_____	_____	_____
_____	_____	_____	_____

Total cost _____

Storm doors
1. Doors needed _____
2. Cost per door _____
3. Total cost _____

Weatherstripping
1. Linear feet _____
2. Cost per foot _____
3. Total cost _____

Caulking
1. Location _____
2. Estimated cost _____

Total cost of all improvements _____

Existing material "R" values

Batt or blanket
Wood or cellulose fiber with paper backing and facing, per inch 4.00

Mineral wool
(Rock, slag or glass), per inch 3.80

Loose fill, per inch
Mineral wool (rock, slag or glass) 3.30
Vermiculite, expanded 2.08

Board or rigid per inch
Expanded urethane, foamed in place, sprayed or preformed 5.88
Polystyrene foam, extruded or expanded.... 4.50
Glass fiberboard 4.34

Doors
Solid wood, 1'' 1.56
Solid wood, 2'' 2.33

Construction materials, per inch
Wood fiberboard, laminated sheathing 2.90
Plywoods and softwoods.................. 1.25
Plaster, stucco, brick 0.20

Windows [Glass area only]
Single glazing 0.88
Double glazing with ¼'' air space 1.64
Single glazing with storm window......... 1.89

Insulation "R" values

Ceilings, double layers of batts
R-38, Two layers of R-19 (6'').
R-33, One layer of R;22 (6½'') and one layer of R-11 (3½'').
R-30, One layer of R-19 (6'') and one layer of R-11 (3½'').
R-26, Two layers of R-13 (3-5/8'').

Ceilings, loose fill mineral wool and batts
R-38, R-19 (6'') mineral fiber and 20 bags of wool per 1,000 S.F. (8¾'').
R-33, R-22 (6½'') mineral fiber and 11 bags of wool per 1,000 S.F. (5'').
R-30, R-19 (6'') mineral fiber and 11 bags of wool per 1,000 S.F. (5'').
R-26, R-19 (6'') mineral fiber and 8 bags of wool per 1,000 S.F. (3¼'').

Floors
R-22, R-22 (6½'') insulation.
R-19, R-19 (6'') insulation.
R-13, R-13 (3-5/8'') insulation.
R-11, R-11 (3½'') insulation.

Walls
R-14, (3½'') blow insulation.
R-19, (3½'') blown insulation and 1'' polystyrene foam sheathing.
R-19, (3½'') blown insulation and insulated siding.

Table 13-25

211

more than offset by monthly fuel bill savings, possibly bringing the cost of living in the house within their reach. The increased value of the house alone might cover the cost to you of making the investment in energy conservation improvements."

Explain to your prospect how much insulation he should have in his home. Offer to make an energy conservation survey of his home using Table 13-25. Explain that accepting your proposal should yield savings in fuel bills that will more than equal the cost of improvements in a relatively few years. Have some examples of savings on jobs you have done and offer references to satisfied past customers. If cost is the prime consideration, offer to do the more difficult work (walls, crawl spaces and attics) and give the homeowner assistance in completing the easier work (caulking and weatherstripping). Your approach will make sense and result in a sale to nearly anyone who has an insulation deficient home and is tired of wasting money on heating and cooling.

Chapter 14

Ventilation and Vapor Barriers

Many older homes are badly in need of improvement because excessive moisture and condensation have caused decay and unsightly deterioration. Repairing the structure without solving the moisture problem will be a temporary improvement at best. Water vapor moving through the walls and ceilings creates major maintenance problems, makes frequent repainting necessary and increases heating cost. Properly installed vapor barriers, the proper use of insulation, and adequate ventilation will avoid most of these difficulties. Modern technology and a better understanding of ventilation requirements have made it easy and inexpensive to reduce moisture problems in most homes. This chapter is intended to help you understand the cause of moisture problems and to suggest steps to be taken to remedy excessive moisture conditions in existing structures.

Most remodeling and repair jobs deal in some way with moisture problems. Anytime you remove and replace siding, flooring, roofing or wall covering you have a chance to evaluate the work from a moisture control standpoint. You will have the opportunity to observe condensation problems and will have the chance to recommend an appropriate solution. Any time you add floor space to a house or convert an attic, basement or garage, you will have to deal with moisture. Everyone in the home improvement industry should be aware that professional workmanship demands attention to preventing moisture problems. This chapter will help you become highly competent at preventing and curing a major cause of home deterioration.

Condensation

Condensation can be described as the change in moisture from a vapor to a liquid. In homes not properly protected, condensation caused by high humidities often results in unnecessarily rapid deterioration. Water vapor within the house, when unrestricted can move through the wall or ceiling during the heating season to some cold surface where it condenses, collecting generally in the form of ice or frost. During warm periods the frost melts. When conditions are severe, the water from melting ice in unvented attics may drip to the ceiling below and cause damage to the interior finish. Moisture can also soak into the roof sheathing or rafters and set up conditions which

could lead to decay. In walls, water from melting frost may run out between the siding laps and cause staining or soak into the siding and cause paint blistering and peeling.

Wood and wood-base materials used for sheathing and panel siding may swell from this added moisture and result in bowing, cupping, or buckling. Thermal insulation also becomes wet and provides less resistance to heat loss. Efflorescence may occur on brick or stone of an exterior wall because of such condensation.

The cost of heat losses, painting and redecorating, and excessive maintenance and repair caused by cold weather condensation can be easily reduced or eliminated when proper construction details are used.

Changes in design, materials, and construction methods since the mid-thirties have resulted in houses that are easier to heat and more comfortable, but these changes have accentuated the potential for condensation problems. New types of weatherstripping, storm sash, and sheet material for sheathing in new houses provide tight air-resistant construction which restricts the escape of moisture generated in the house. Newer houses are also generally smaller and have lower ceilings, resulting in less atmosphere to hold moisture.

Estimates have been made that a typical family of four converts 3 gallons of water into water vapor per day. Unless excess water vapor is properly removed in some way (ventilation usually), it will either increase the humidity or condense on cold surfaces such as window glass. More serious, however, it can move in or through the construction, often condensing within the wall, roof, or floor cavities. Heating systems equipped with winter air-conditioning systems also increase the humidity.

Most new houses have from 2 to 3½ inches of insulation in the walls and 6 or more inches in the ceilings. Unfortunately, the more efficient the insulation is in retarding heat transfer, the colder the outer surfaces become and unless moisture is restricted from entering the wall or ceiling, the greater the potential for moisture condensation. Moisture migrates toward cold surfaces and will condense or form as frost or ice on these surfaces.

Inexpensive methods of preventing condensation problems are available. They mainly involve

Darkened areas on roof boards and rafters in attic indicate stain that stemmed from condensation. This usually could be prevented by vapor barrier in the ceiling attic ventilation
Figure 14-1

the proper use of vapor barriers and good ventilating practices. Naturally it is simpler, less expensive, and more effective to employ these during the construction of a house than to add them to existing homes. But most condensation problems can be solved at a reasonable cost and with little extra effort when remodeling or repair becomes necessary.

Condensation will take place anytime the temperature drops below dewpoint (100 percent saturation of the air with water vapor at a given temperature). Commonly, under such conditions some surface accessible to the moisture in the air is cooler than the dewpoint and the moisture condenses on that surface.

Visible Condensation

During cold weather, visible condensation is usually first noticed on window glass but may also be discovered on cold surfaces of closets and unheated bedroom walls and ceilings. Visible surface condensation on the interior glass surfaces of windows can be minimized by the use of storm windows or by replacing single glass with insulated glass. However, when this does not prevent condensation on the surface, the relative humidity in the room must be reduced. Drapes or curtains across the windows hinder rather than help. Not only do they increase surface condensation because of colder glass surfaces, but they also prevent the air movement that would warm the glass surface and aid in dispersing some of the moisture.

Reducing high relative humidities within the house to permissible levels is often necessary to minimize condensation problems. Discontinuing the use of room-size humidifiers or reducing the output of automatic humidifiers until conditions are improved is helpful. The use of exhaust fans and dehumidifiers can also be of value in eliminating high relative humidities within the house. When possible, decreasing the activities which produce excessive moisture, as discussed later in this chapter, is sometimes necessary. This is especially important for homes with electric heat.

Concrete slabs without radiant heat are sometimes subjected to surface condensation in late spring when warm humid air enters the house. Because the temperature of some areas of the concrete slab or its covering is below the dewpoint, surface condensation can occur. Keeping the windows closed during the day, using a dehumidifier, and raising the inside temperature aid in minimizing this problem. When the concrete slab reaches normal room temperatures, this inconvenience is eliminated.

Condensation might also be visible in attic spaces on rafters or roof boards near the cold cornice area (Figure 14-1) or form as frost. Such condensation or melting frost can result in excessive maintenance, such as the need for refinishing of window sash and trim, or even decay. Water from melting frost in the attic can also damage ceilings below. Condensation or frost on protruding nails, on the surfaces of roof boards,

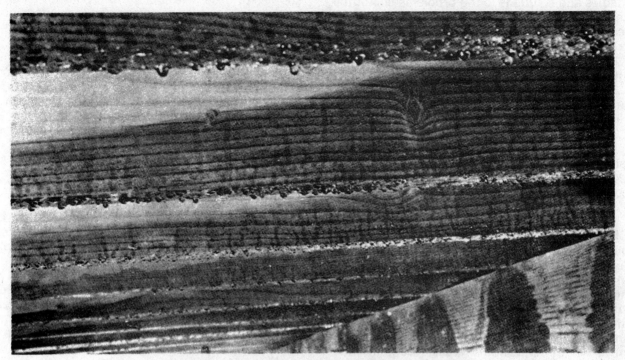

Surface condensation on floor joists in crawl space. A vapor barrier ground cover can prevent this because it restricts water vapor movement from the soil and thus avoids high humidity of crawl space and subsequent surface condensation
Figure 14-2

or other members in attic areas normally indicates the escape of excessive amounts of water vapor from the heated rooms below. If a vapor barrier is not already present, place one between joists under the insulation. Make sure the vapor barrier fits tightly around ceiling lights and exhaust fans, caulking if necessary. In addition, increase both inlet and outlet ventilators to levels recommended later in this chapter. Decreasing the amount of water vapor produced in the living areas is also helpful.

Another area where visible condensation can occur is in crawl spaces under occupied rooms. This area usually differs from those on the interior of the house and in the attic because the source of the moisture is usually from the soil or from warm moisture-laden air which enters through foundation ventilators. Moisture vapor then condenses on the cooler surfaces in the crawl space (Figure 14-2). Such conditions often occur during warm periods in late spring. To eliminate this problem, place a vapor barrier over the soil and use the proper amount of ventilation as recommended later in this chapter.

An increase in relative humidity of the inside atmosphere increases the potential for condensation on inside surfaces. For example, when inside temperature is 70 F., surface condensation will occur on a single glass window when outside temperature falls to -10 F. and inside relative humidity is 10 percent. When inside relative humidity is 20 percent, condensation can occur on the single glass when outside temperature only falls to about +7 F. When a storm window is added or insulated glass is used, surface condensation will not occur until the relative humidity has reached 38 percent when the outdoor temperature is -10 F. The above conditions apply only where storm windows are tight and there is good circulation of air on the inside surface of the window. Where drapes or shades restrict circulation of air, storm windows are not tight, or lower temperatures are maintained in such areas as bedrooms, condensation will occur at a higher outside temperature.

Concealed Condensation

Condensation in concealed areas, such as wall spaces, often is first noticed by stains on the siding or by paint peeling. Water vapor moving through permeable walls and ceilings is normally responsible for such damage. Water vapor also escapes from houses by constant outleakage through cracks and crevices, around doors and windows, and by ventilation, but this moisture-vapor loss is usually insufficient to eliminate condensation problems.

Concealed condensation takes place when a condensing surface is below the dewpoint. In cold weather, condensation often forms as frost. The resulting problems are usually not detected until spring after the heating season has ended. The remedies and solutions to the problems should be taken care of before repainting or residing is attempted. Several methods might be used to correct this problem.

1. Reduce or control the relative humidity within the house.

2. Add a vapor-resistant paint coating such as aluminum paint to the interior of walls and ceilings.

3. Improve the vapor resistance of the ceiling by adding a vapor barrier between the ceiling joists.

4. Improve attic ventilation to the minimums recommended in this chapter.

Moisture Sources

Moisture, which is produced in or which enters a home, changes the relative humidity of the interior atmosphere. Ordinary household functions which generate a good share of the total amount of water vapor include dishwashing, cooking, bathing, and laundry work, to say nothing of human respiration and evaporation from plants. Houses may also be equipped with central winter air conditioners or room humidifiers. Still another source of moisture may be from unvented or poorly vented clothes dryers. Several sources and their effect in adding water vapor to the interior of the house are as listed below in pints of water given off.

Plants = 1.7 for each plant in 24 hours

Showers = 0.5 for each shower and 0.1 for each bath

Floor mopping = 2.9 per 100 square feet, each washing

Kettles and cooking = 5.5 per day

Clothes (washing, steam ironing, drying) = 29.4 per week

Water vapor from the soil of crawl-space houses does not normally affect the occupied areas. However, without good construction practices or proper precautions it can be a factor in causing problems in exterior walls over the area as well as in the crawl space itself. It is another source of moisture that must be considered in providing protection.

People moving into a newly constructed house or room addition in the fall or early winter sometimes experience temporary moisture problems. Surface condensation on windows, damp areas on cold closet walls where air movement is restricted, and even stained siding all indicate an excessive amount of moisture. Such conditions can often be traced to water used in the construction of a house.

There are other sources of moisture, often unsuspected, which could be the cause of condensation problems. One such source can be a gas fired furnace. It is desirable to maintain flue gas temperatures within the recommended limits throughout the appliance, in the flue, the connecting vent, and other areas. Otherwise, excessive condensation problems can result. If all sources of excessive moisture have been exhausted in determining the reasons for a condensation problem, it is best to have the heating unit examined by a competent heating engineer.

There is a distinct relationship in all homes between indoor relative humidity and outdoor temperature. The humidity is generally high indoors when outdoor temperatures are high and decreases as outdoor temperatures drop. In an exceptionally tight modern house where moisture buildup may be a problem, outside air should be introduced into the cold air return ducts to reduce relative humidity.

Vapor Barriers

Many materials used as interior coverings for exposed walls and ceilings, such as plaster, drywall, wood paneling, and plywood, permit water vapor to pass slowly through them during cold weather. Temperatures of the sheathing or siding on the outside of the wall are often low enough to cause condensation of water vapor within the cavities of a framed wall. When the relative humidity within the house at the surface of an unprotected wall is greater than that within the wall, water vapor will migrate through the plaster or other finish into the stud space; there it will condense if it comes in contact with surfaces colder than its dewpoint temperature. Vapor barriers are used to resist this movement of water vapor or moisture in various areas of the house.

The most effective vapor barrier is a continuous membrane which is applied to the inside face of studs and joists in new construction. In rehabilitation, such a membrane can be used only where new interior covering materials are to be applied.

The amount of condensation that can develop within a wall depends upon (a) the resistance of the intervening materials to vapor transfusion, (b) differences in vapor pressure, and (c) time. Plastered walls or ordinary dry walls have little resistance to vapor movement. However, when the surfaces are painted with oil base paint, the resistance is increased. High indoor temperature and relative humidities result in high indoor vapor pressures. Low outdoor vapor pressures always exist at low temperatures. Thus, a combination of high inside temperatures and humidities and low outside temperatures will normally result in vapor movement into the wall if no vapor barrier is present. Long periods of severe weather will result in condensation problems. Though fewer homes are affected by condensation in mild winter weather, many problems have been reported. Where information is available, it appears that the minimum relative humidities in the affected homes are 35 percent or higher.

Vapor barrier requirements are sometimes satisfied by one of the materials used in construction. In addition to integral vapor barriers which are a part of many types of insulation, such materials as plastic-faced hardboard and similar interior coverings may have sufficient resistance when the permeability of the exterior construction is not too low. The permeability of the surface to such vapor movement is usually expressed in "perms," which are grams (438 grams per ounce) of water vapor passing through a square foot of material per hour per inch of mercury difference in vapor pressure. A material with a low perm value (1.0 or less) is a barrier, while one with a high perm value (greater than 1.0) is a "breather."

The perm value of the cold side materials should be several times greater than those on the warm side. A ratio of 1 to 5 or greater from inside to outside is sometimes used as a rule of thumb in selecting materials and finish. When this is not possible because of virtually impermeable outside construction (such as a built-up roof or resistant exterior wall membranes), research has indicated the need to ventilate the space between the insulation and the outer covering.

Vapor barriers are used in three general areas of the house to minimize condensation or moisture problems:

Walls, ceilings, floors.—Vapor barriers used on the warm side of all exposed walls, ceilings, and floors greatly reduce movement of water vapor to colder surfaces where harmful condensation can occur. For such uses it is good practice to select materials with perm values of 0.25 or less. Such vapor barriers can be a part of the insulation or a separate film. Commonly used materials are (a) asphalt coated or laminated papers, (b) kraft backed aluminum foil, and (c) plastic films such as polyethylene, and others. Foil backed gypsum board and various coatings also serve as vapor barriers. Oil base or aluminum paints, or similar coatings are often used in houses which did not have other vapor barriers installed during their construction.

Concrete slabs.—Vapor barriers under concrete slabs resist the movement of moisture through the concrete and into living areas. Such vapor barriers should normally have a maximum perm value of 0.50. But the material must also have adequate resistance to the hazards of pouring concrete. Thus, a satisfactory material must be heavy enough to withstand such damage and at the same time have an adequate perm value. Heavy asphalt laminated papers, papers with laminated films, roll roofing, heavy films such as polyethylene, and other materials are commonly used as vapor barriers under slabs.

Crawl space covers.—Vapor barriers in crawl spaces prevent ground moisture from moving up and condensing on wood members, (Figure 14-2). A perm value of 1.0 or less is considered satisfactory for such use. Asphalt laminated paper, polyethylene, and similar materials are commonly used. Strength and resistance of crawl space covering to mechanical damage can be lower than that for vapor barriers used under concrete slabs.

Exterior Materials

In new constructions where low permeance vapor barriers are properly installed, most commercially available sheathing and siding materials and coatings can be used without creating condensation problems. However, in structures without vapor barriers, a low permeance material or coating on the outside can retard the escape of moisture which has been forced into the wall from the inside. An alternative finish for such situations is a penetrating stain, which does not form a coating on the wood surface and so does not retard the movement of moisture. Penetrating stains are very durable and easily refinished because they do not fail by blistering or peeling. Where the older home has a paint peeling problem due to condensation, the siding should be painted with white latex paint, which is very porous, and then spot painted annually wherever peeling occurs. White paint is recommended because it does not fade and retains a good exterior appearance between yearly touchups.

Thermal Insulation

Thermal insulation has a major influence on the need for vapor barriers. The inner face of the wall sheathing in an insulated wall, for example, is colder than the sheathing face in an uninsulated wall and consequently has a greater attraction to moisture. Thus there is greater need for vapor barrier in an insulated wall than in an uninsulated wall.

Most blanket insulation has a vapor barrier on one side. Place the insulation with the vapor barrier toward the warm surface. Tabs on the blanket must be stapled to the inside face of the stud or joist and adjacent tabs should lap each other. Tabs stapled to the side of studs or joists in the cavity will be ineffective because vapor will move out between the tabs and framing members. In rehabilitation, this type of vapor barrier can be used only where the old interior covering materials are completely removed or where furring strips are added on the inside. Insulation is installed between furring strips in the same manner as between studs.

Vapor Resistant Coating

Where loose fill insulation has been used in walls and ceiling and no new interior covering is planned, a vapor resistant coating should be applied to the inside surface. One method for applying such a coating is to paint the interior surface of all outside walls with two coats of aluminum primer, which are subsequently covered with decorative paint. This does not offer as much resistance to vapor movement as a membrane, so it should be used only where other types cannot be used. If the exterior wall covering is permeable enough to allow moisture to escape from the wall, a vapor resistant coating on the inside should be adequate.

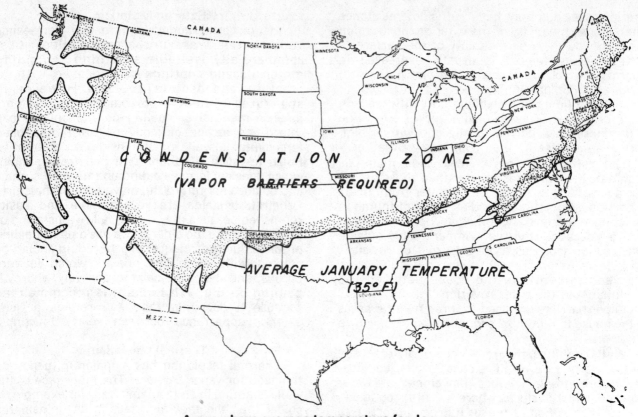

**Areas where average temperature for January
is 35° F. or lower
Figure 14-3**

Good Practice Recommendations

The control of condensation through the use of vapor barriers and ventilation should be practiced regardless of the amount of insulation used. Normally, winter condensation problems occur in those parts of the United States where the average January temperature is 35 F. or lower. Figure 14-3 illustrates this condensation zone. The northern half of the condensation zone has a lower average winter temperature and, of course, more severe conditions than the southern portion. Areas outside this zone, such as the southeast and west coastal areas and the southern states, seldom have condensation problems. Vapor barriers should be installed in all houses built within the condensation zone outlined in Figure 14-3 and proper ventilation procedures should be followed. These will insure control over normal condensation problems.

Location of Vapor Barriers

A good general rule to keep in mind when installing vapor barriers in a house is as follows: Place the vapor barrier as close as possible to the interior or warm surface of all exposed walls, ceilings, and floors. This normally means placing the vapor barrier (separately or as a part of the insulation) on (a) the inside edge of the studs just under the rock lath or drywall finish, (b) on the under side of the ceiling joists of a one-story house or the second floor ceiling joists of a two-story house, and (c) between the subfloor and finish floor (or just under the subfloor of a house with an unheated crawl space in addition to the one placed on the ground). The insulation, of course, is normally placed between studs or other frame members on the outside of the vapor barrier. The exception is the insulation used in concrete floor slabs where a barrier is used under the insulation to protect it from ground moisture.

Placement of vapor barriers and insulation in one-story houses is shown in Figure 14-4 (flat roof and concrete floor slab) and Figure 14-5 (pitched roof and crawl space). Figure 14-6 shows barriers

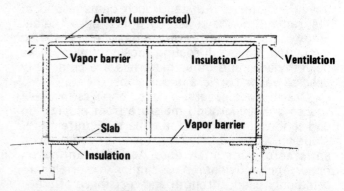

**Location of vapor barriers and insulation in
concrete slab and flatdeck roof
Figure 14-4**

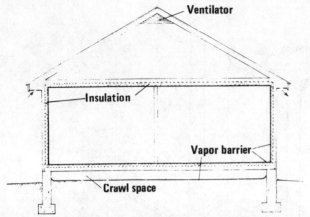

Location of vapor barriers and insulation in crawl space of another one-story house
Figure 14-5

and insulation in a 1½ story house with a full basement. Figure 14-7 depicts a two-story house with a full basement. Other combinations of slabs, crawl spaces, and basements in houses with 1, 1½, or 2 stories, should follow the same general recommendations. Detailed descriptions in the use of vapor barriers will be covered in the following sections.

Concrete Slabs

Every house or room addition constructed over a concrete slab must be protected from soil moisture which may enter the slab. Protection is normally provided by a vapor barrier, which completely isolates the concrete and perimeter insulation from the soil. Thermal insulation of some type is required around the house perimeter in the colder climates, not only to reduce heat loss

but also to minimize condensation on the colder concrete surfaces. Some types of rigid insulation impervious to moisture absorption should be used. Expanded plastic insulation such as polystyrene is commonly used.

One method of installing this insulation is shown in Figure 14-8. Another method consists of placing it vertically along the inside of the foundation wall. Both methods require insulation at the slab notch of the wall. If the insulation is placed vertically, it should extend a minimum of 12 inches below the outside finish grade. In the colder climates a minimum 24 inch width or depth should be used.

In late spring or early summer, periods of high humidity may cause surface condensation on exposed concrete slabs or on coverings such as resilient tile before the concrete has reached normal temperatures. A fully insulated slab or a wood floor installed over wood furring strips minimizes if not eliminates such problems.

Because the vapor barriers slow the curing process of the concrete, final steel troweling of the surface is somewhat delayed. Do not punch holes through the barrier to hasten the curing process as this will destroy its effectiveness!

Ventilation

Ventilation of attics and crawl spaces is essential in all houses located where the average January temperature is 35° F or lower. Vapor barriers help to control moisture problems, but there are always places, such as around utility pipes, where some moisture escapes. In the older house that does not have proper vapor barriers, ventilation is especially important.

Enclosed crawl spaces require some protection

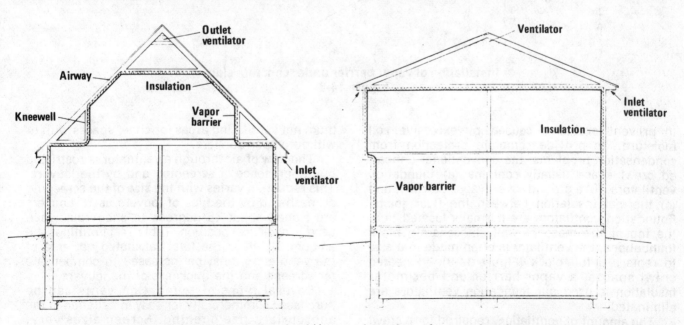

Location of vapor barriers and insulation in 1½-story house with basement
Figure 14-6

Location of vapor barriers and insulation in full two-story house with basement
Figure 14-7

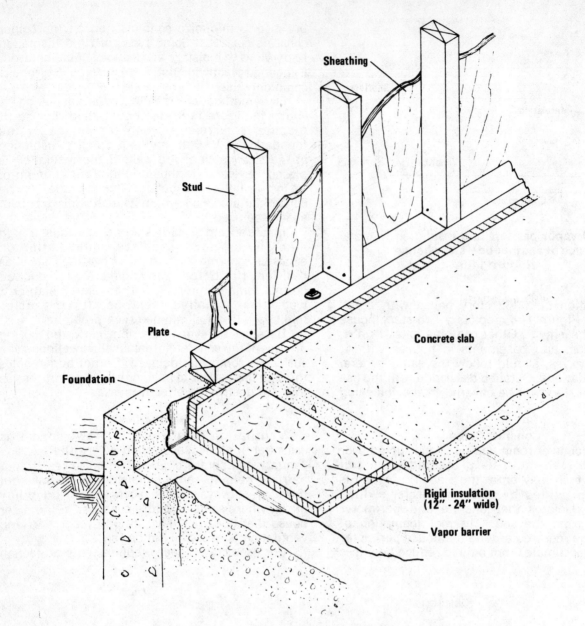

Stud

Sheathing

Plate

Concrete slab

Foundation

Rigid insulation
(12" - 24" wide)

Vapor barrier

**Installation of vapor barrier under concrete slab
Figure 14-8**

to prevent problems caused by excessive soil moisture. To provide complete protection from condensation problems, the conventional unheated crawl space usually contains (a) foundation ventilators, (b) a ground cover (vapor barrier), and (c) thermal insulation between the floor joists. Foundation ventilators are normally located near the top of the masonry wall. In concrete block foundations, the ventilator is often made in a size to replace a full block (Figure 14-9). In heated crawl spaces, a vapor barrier and perimeter insulation is used but foundation ventilators are eliminated.

The amount of ventilation required for a crawl space is based on the total area of the house in square feet and the presence of a vapor barrier soil cover. Table 14-10 lists the recommended mini-

mum net ventilating areas for crawl spaces with or without vapor barriers.

The flow of air through a ventilator is restricted by the presence of screening and by the louvers. This reduction varies with the size of the screening or mesh and by the type of louvers used. Louvers are sloped about 45 degrees to shed rain when used in a vertical position. Table 14-11 outlines the amount by which the total calculated net area of the ventilators must be increased to compensate for screens and the thickness of the louvers.

Several types of foundation vents can be purchased commercially for easy installation in the appropriate size opening. Screen sizes vary, depending on whether they are insect-proof or rodent-proof. Manufactured vents usually have a statement of free area. Often it may be sufficient

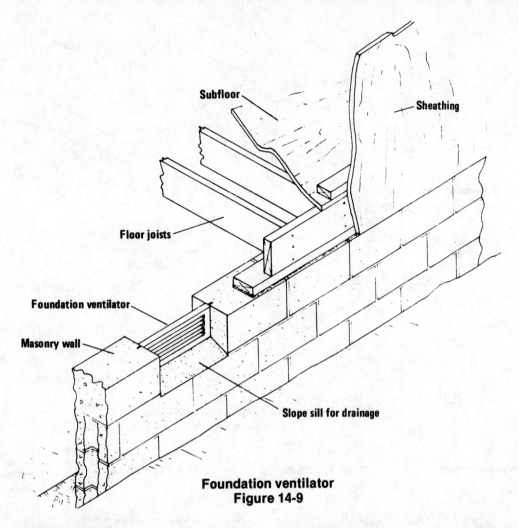

Foundation ventilator
Figure 14-9

to attach a screen over an opening rather than to purchase a manufactured vent.

In placing the vapor barrier over the crawl space soil, adjoining edges should be lapped slightly and ends turned up on the foundation wall (Figure 14-12). To prevent movement of the barrier, it is good practice to weight down laps and edges with bricks or other small masonry sections.

Attic and Roof

Moisture escaping from the house into the attic tends to collect in the coldest part of the attic. Relatively impermeable roofing, such as asphalt shingles or a built-up roof, complicates the problem by preventing the moisture from escaping to the outside. The only way to get the moisture out is to ventilate the attic. Attic ventilation also helps keep a house cool during hot weather.

Crawl space	Ratio of total net ventilating area to floor area[1]	Minimum number of ventilators[2]
Without vapor barrier	1/150	4
With vapor barrier	1/1500	2

[1] The actual area of the ventilators depends on the type of louvers and size of screen used—see Table 14-11.

[2] Foundation ventilators should be distributed around foundation to provide best air movement. When two are used, place one toward the side of prevailing wind and the other on opposite side.

Crawl-space ventilation
Table 14-10

Obstructions in ventilators—louvers and screens[1]	To determine total area of ventilators, multiply required net area in square feet by[2]
1/4 inch mesh hardware cloth	1
1/8 inch mesh screen	1¼
No. 16 mesh insect screen (with or without plain metal louvers)	2
Wood louvers and 1/4 inch mesh hardware cloth[3]	2
Wood louvers and 1/8 inch mesh screen[3]	2¼
Wood louvers and No. 16 mesh insect screen[3]	3

[1] In crawl-space ventilators, screen openings should not be larger than 1/4 inch; in attic spaces no larger than 1/8 inch.

[2] Net area for attics determined by ratios in Figures 14-13, 14-14, and 14-15.

[3] If metal louvers have drip edges that reduce the opening, use same ratio as shown for wood louvers.

Ventilating area increase required if louvers and screening are used in crawl spaces and attics
Table 14-11

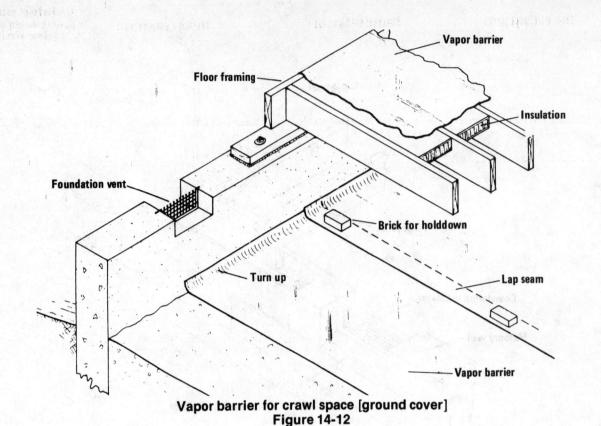

Vapor barrier for crawl space [ground cover]
Figure 14-12

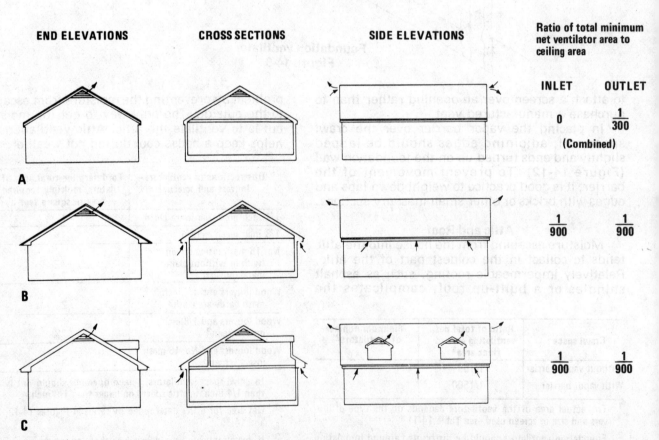

Ventilating area of gable roofs: A, louvers in end walls; B, louvers in end walls with additional openings at eaves; C, louvers at end walls with additional openings at eaves and dormers. Cross section of C shows free opening for air movement between roof boards and ceiling insulation of attic room
Figure 14-13

222

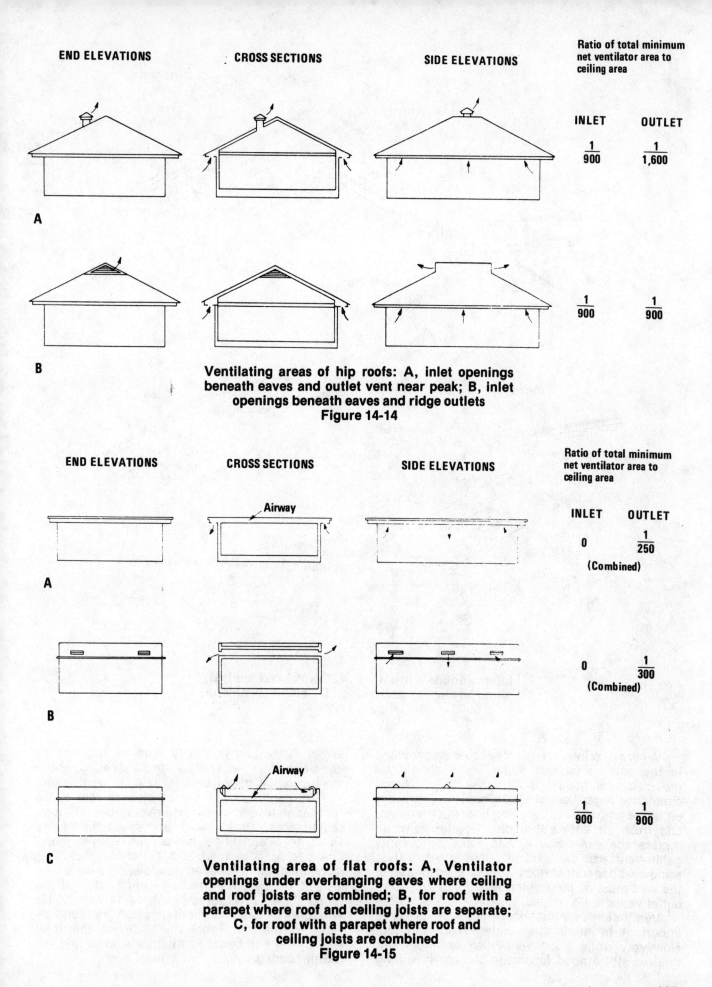

ratio of total minimum net ventilator area to ceiling area

END ELEVATIONS   CROSS SECTIONS   SIDE ELEVATIONS

A

INLET	OUTLET
$\frac{1}{900}$	$\frac{1}{1,600}$

B

INLET	OUTLET
$\frac{1}{900}$	$\frac{1}{900}$

Ventilating areas of hip roofs: A, inlet openings beneath eaves and outlet vent near peak; B, inlet openings beneath eaves and ridge outlets
Figure 14-14

END ELEVATIONS   CROSS SECTIONS   SIDE ELEVATIONS

Ratio of total minimum net ventilator area to ceiling area

Airway

A

INLET	OUTLET
0	$\frac{1}{250}$
	(Combined)

B

INLET	OUTLET
0	$\frac{1}{300}$
	(Combined)

Airway

C

INLET	OUTLET
$\frac{1}{900}$	$\frac{1}{900}$

Ventilating area of flat roofs: A, Ventilator openings under overhanging eaves where ceiling and roof joists are combined; B, for roof with a parapet where roof and ceiling joists are separate; C, for roof with a parapet where roof and ceiling joists are combined
Figure 14-15

223

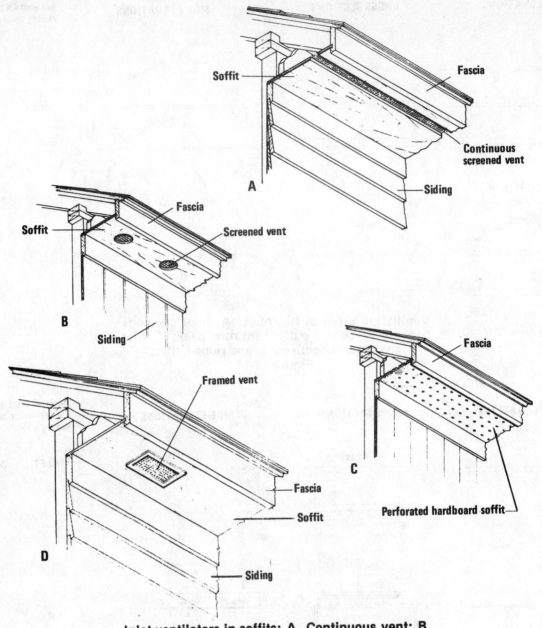

Inlet ventilators in soffits: A, Continuous vent; B, round vents; C, perforated; D, single ventilator
Figure 14-16

Where possible, inlet vents should be provided in the soffit area and outlet vents should be provided near the ridge. This results in natural circulation regardless of wind direction. The warm air in the attic rises to the peak, goes out the vents, and fresh air enters through the inlet vents to replace the exhausted air. In some attics only gable vents can be used. Air movement is then somewhat dependent upon wind. The open area of the vent must be larger than where both inlet and outlet vents are provided.

Ventilation of attic spaces and roof areas is important in minimizing water vapor buildup. However, while good ventilation is important, there is still a need for vapor barriers in ceiling areas. This is especially true of the flat or low-slope roof where only a 1 to 3 inch space above the insulation might be available for ventilation.

The minimum amount of attic or roof space ventilation required is determined by the total ceiling area. These ratios are shown in Figure 14-13, 14-14 and 14-15, for various types of roofs. The use of both inlet and outlet ventilators is recommended whenever possible. The total net area of ventilators is found by application of the ratios shown in Figures 14-13, 14-14 and 14-15. The total area of the ventilators can be found by using the data in Table 14-11. Divide this total area by the number of ventilators used to find the recommended square foot area of each.

For example, a gable roof similar to Figure 14-13, B with inlet and outlet ventilators has a minimum required total inlet and outlet ratio of 1/900 of the ceiling area. If the ceiling area of the house is 1,350 square feet, each net inlet and outlet ventilating area should be 1,350 divided by 900 or 1½ square feet.

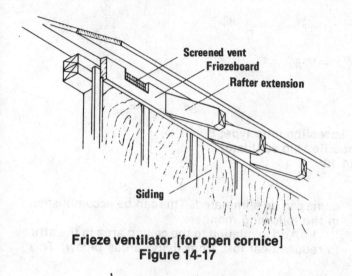

Frieze ventilator [for open cornice]
Figure 14-17

If ventilators are protected with Number 16 mesh insect screen and plain metal louvers (Table 14-11) the minimum gross area must be 2 x 1½ or 3 square feet. When one outlet ventilator is used at each gable end, each should have a gross area of 1½ square feet (3 divided by 2). When distributing the soffit inlet ventilators to three on each side, for a small house (total of 6), each ventilator should have a gross area of 0.5 square feet. For long houses, use 6 or more on each side.

Hip roofs cannot have gable vents near the peak, so some other type of outlet ventilator must be provided (Figure 14-14). This can be either a

ventilator near the ridge, or a special flue provided in the chimney with openings into the attic space. Both types require inlet vents in the soffit area. The hip roof can also be modified to provide a small gable for a conventional louvered vent.

Flat roofs with no attic require some type of ventilation above the ceiling insulation. If this space is divided by joists, each joist space must be ventilated. This is often accomplished by a continuous vent strip in the soffit. Drill through all headers that impede passage of air to the opposite eave. Other methods are illustrated in Figure 14-15.

Cathedral ceilings require the same type of ventilation as flat roofs. A continuous ridge vent is also desirable because even with holes in the ridge rafter, air movement through the rafter space is very sluggish without a ridge vent.

Inlet ventilators in the soffit may consist of several designs. It is good practice to distribute them as much as possible to prevent "dead" air pockets in the attic where moisture might collect. A continuous screened slot (Figure 14-16, A) satisfies this requirement. Small screened openings might also be used (Figure 14-16, B). Continuous slots or individual ventilators between roof members should be used for flat roof houses where roof members serve as both rafters and ceiling joists. Locate the openings away from the wall line to minimize the possible entry of wind driven snow. A soffit consisting of perforated hardboard (Figure 14-16, C) can also be used to advantage but holes should be no larger than 1/8 inch in diameter. Small metal frames with screened openings are also available and may be used in soffit areas (Figure 14-16, D). For open cornice design, the use of a frieze board with screen ventilating slots would be satisfactory (Figure 14-17). Perforated hardboard might also be used for this purpose. The recommended minimum inlet ventilating ratios shown in Figure

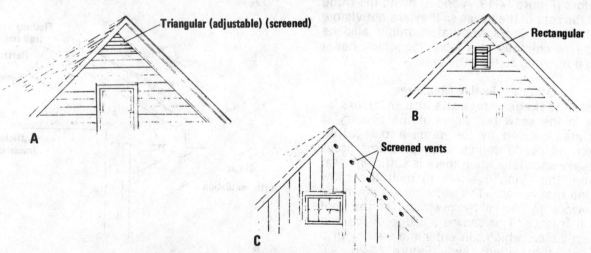

Gable outlet ventilators: A, Triangular gable end ventilator; B, rectangular gable end ventilator; C, soffit ventilators
Figure 14-18

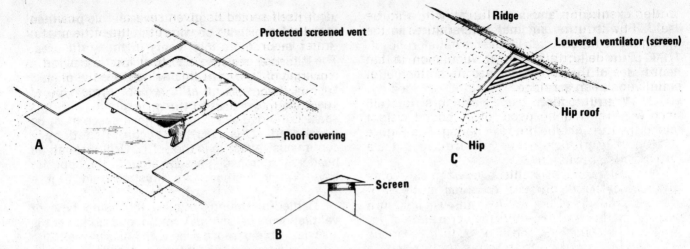

Ridge outlet ventilators: A, Low silhouette type; B pipe ventilator type; C, modified hip ventilator
Figure 14-19

14-13, 14-14 and 14-15 should be followed in determining total net ventilating areas for both inlet and outlet ventilators.

Outlet ventilators to be most effective should be located as close to the highest portion of the ridge as possible. They may be placed in the upper wall section of a gable roofed house in various forms as shown in Figure 14-18, A and B. In wide gable-end overhangs with ladder framing, a number of screened openings can be located in the soffit area of the lookouts. (Figure 14-18, C). Ventilating openings to the attic space should not be restricted by blocking. Outlet ventilators on gable or hip roofs might also consist of some type of roof ventilator (Figure 14-9, A and B). Hip roofs can utilize a ventilating gable (modified hip) (Figure 14-19, C). Protection from blowing snow must be considered, which often restricts the use of a continuous ridge vent. Locate the single roof ventilators (Figure 14-19, A and B) along the ridge toward the rear of the house so they are not visible from the front. Outlet ventilators might also be located in a chimney as a false flue which has a screened opening to the attic area.

Other Protective Measures

Water leakage into walls and interiors of houses in the snow belt areas of the country is sometimes caused by ice dams and is often mistaken for condensation. Such problems occur after heavy snowfalls when there is sufficient heat loss from the living quarters to melt the snow along the roof surface. The water moves down the roof surface to the colder overhang of the roof where it freezes. This causes a ledge of ice and backs up water, which can enter the wall or drip down onto the ceiling finish (Figure 14-20, A).

Several methods can be used to minimize this problem caused by melting snow. By reducing the attic temperatures in the winter so that they are only slightly above outdoor temperatures, most ice dams can be eliminated. This can be accomplished in the following manner:

1. Add insulation to the ceiling area in the attic to reduce heat loss from living areas below. This

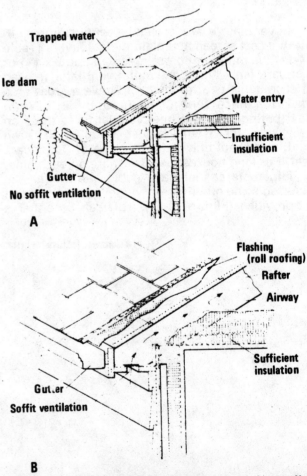

Ice dams: A, insufficient insulation and ventilation can cause ice dams and water damage; B, good ventilation, insulation, and roof flashing
Figure 14-20

added insulation and ventilation will also be helpful by reducing summer temperatures in the living areas below.

2. Provide additional inlet ventilation in the soffit area of the cornice as well as better outlet ventilation near the ridge.

3. When reroofing, use a flashing strip of 36 inch wide roll roofing paper of 45 pound weight along the eave line before reshingling. See Figure 14-20. While this does not prevent ice dams, it is a worthwhile precaution.

4. Under severe conditions, or when only some portions of a roof produce ice dams (such as at valleys), the use of electric-thermal wire laid in a zig-zag pattern and in gutters may prove effective. The wire is connected and heated during periods of snowfall and at other times as needed to maintain channels for drainage.

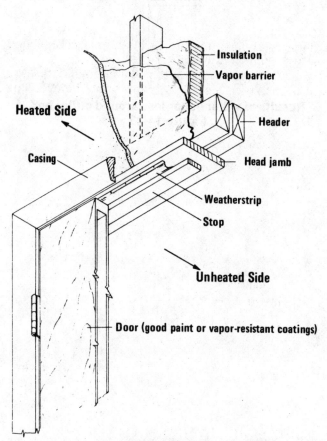

Insulating door to unheated attic space
Figure 14-21

Walls and doors to unheated areas such as attic spaces should be treated to resist water vapor movement as well as to minimize heat loss. This includes the use of insulation and vapor barriers on all wall areas adjacent to the cold attic (Figure 14-21). Vapor barriers should face the warm side of the room. In addition, some means should be used to prevent heat and vapor loss around the perimeter of the door. One method is through some type of weather strip (Figure 14-21). The

door itself should be given several finish coats of paint or varnish which will resist the movement of water vapor.

If further resistance to heat loss is desired, a covering of ½ inch or thicker rigid insulation such as insulation board or foamed plastic can be attached to the back of the door.

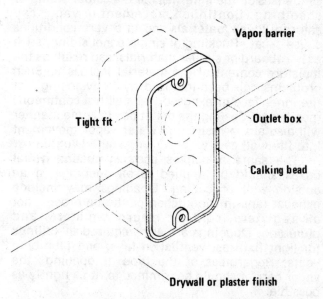

Protection around outlet boxes in exposed walls
Figure 14-22

Outlet or switch boxes or other openings in exposed (cold) walls often are difficult to treat to prevent water vapor escape. Initially, whether the vapor barrier is a separate sheet or part of the insulation, as tight a fit as possible should be made when trimming the barrier around the box (Figure 14-22). This is less difficult when the barrier is separate. As an additional precaution, a bead of caulking compound should be applied around the box after the drywall or the plaster base has been installed (Figure 14-22). The same caulking can be used around the cold air return ducts or other openings in exterior walls. This type of sealing may appear unnecessary, but laboratory tests have shown that there is enough moisture loss through the perimeter of an outlet box to form a large ball of frost on the back face during extended cold periods. Melting of this frost can adversely affect the exterior paint films. In the colder areas of the country and in rooms where there is excess water vapor, such as the bath and kitchen, this added protection is good insurance from future problems. Some switch and junction boxes are more difficult to seal than others because of their makeup. A simple polyethylene bag or other enclosure around such boxes will provide some protection.

Condensation problems caused by water vapor movement through unprotected outlet box areas in exposed walls are often due to poor workmanship during application of the insulation. Figure 14-23 shows a section of exterior wall with the vapor barrier loosely stapled to the face of the studs. Because of poor application, a small space is sometimes left at the top and bottom of the insulation in the stud space. Water vapor escaping through the unprotected outlet box travels by convection, on the warm side, to the top of the wall, where it moves to the cold side and condenses on the inner face of the colder siding or sheathing. Continued movement of vapor can saturate these materials and in severe conditions cause decay. Buckling of single panel siding, such as hardboard or similar materials, can result as the moisture content of the material increases. Such problems can be minimized by ''enveloping'' of the inner face of exposed walls with a continuous membrane. Sealing the outlet box in some manner will also aid in restricting water vapor movement into the wall cavity.

The same principles used in sealing outlet boxes should be applied to all openings in an outside wall or ceiling. Openings may include exhaust fans in the kitchen or the bathroom, hot air registers, cold air return registers, and plumbing. Openings are also required in ceilings for light fixtures, ventilation fans, and plumbing vents. Regardless of the type of opening, the vapor barrier should be trimmed to fit as tightly as possible.

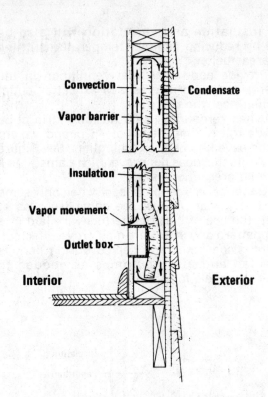

Results of water vapor loss around outlet box
Figure 14-23

Chapter 15
Fireplaces and Chimneys

The fireplace has been an important feature in homes for hundreds of years. Very few fireplaces serve as primary heat sources today, but they continue to have an appeal that goes far beyond their utility. This chapter will suggest ways traditional masonry fireplaces and chimneys can be repaired and how new fireplaces, both masonry and prefabricated, can be added. Including a fireplace in plans for a new family room or den can be a big sales point, especially in homes that do not already have a wood-burning fireplace.

Making Repairs

Masonry fireplaces seldom need repairs but chimneys can develop cracks that allow rain, smoke, and sparks to escape into the house interior. Flues for gas or oil heaters also develop leaks that may become serious. Defective chimneys are difficult to repair, so it may be best to replace a poor chimney. When gas or oil heaters are used, only small metal vents may be required. Check with local code authorities for type and allowable length of run. These can sometimes be placed in the stud space of an interior partition for low-capacity heaters. Then the defective chimney can be eliminated. If fuels requiring chimneys are used, fabricated metal chimneys can usually be installed at a much lower cost than the conventional masonry chimney. Check the local code for requirements for prefabricated chimneys.

Well built chimneys have flue linings which keep the flue tight even though there are cracks in the masonry. Where flues are not lined, stainless steel flue lining can be installed. This lining can be purchased in 2 or 2½ foot lengths and in cross sections to fit most standard sized chimneys. Sections are connected together and inserted from the top, with additional sections being added until the required length is achieved. Note that this type of lining can only be used with a reasonably straight flue. The flue lining will assure safe operation, but cracks in mortar should also be repaired by regrouting.

Fireplaces have several requirements for good draft and smoke-free operation. If they were not built with proper proportions, a metal extension across the top of the fireplace opening will sometimes improve draft. This improvement can be tested by holding a board against the fireplace just above the opening and observing the change in draft. Draft can sometimes be improved by extending the height of the flue or adding a chimney cap for venturi action. Another possible solution where draft is inadequate is to install a fan in the chimney for forced exhaust.

In cold climates, a damper that can be closed when the fireplace is not in use is quite important. Where there is no damper, an asbestos board can be cut to fit the flue opening at the top of the fireplace and supported on some type of brackets. It would have to be completely removed when the fireplace is in use, which is slightly inconvenient; but it would eliminate heat loss or cold drafts from the flue.

Chimney cleaning usually is not necessary. But should it become necessary, vacuuming by a commercial cleaning firm is the best and cleanest method.

If there is not too great an offset in the chimney, you can dislodge soot and loose material by pulling a weighted sack of straw up and down in the flue. Seal the front of a fireplace when cleaning the flue to keep soot out of the room.

Chemical soot removers are not particularly recommended. They are not very effective in removing soot from chimneys and they cause soot to burn, which creates a fire hazard. Some, if applied to soot at high temperatures and in sufficient quantity, may produce uncontrollable combustion and even an explosion. Common rock salt is not the most effective remover, but it is widely used, because it is cheap, readily available, and easy to handle. Use 2 or 3 teacupfulls per application.

Creosote may form in chimneys, especially when wood is burned and in cold weather. It is very hard to remove. The only safe method is to chip it from the masonry with a blade, and you must be careful not to knock out mortar joints or damage the flue lining.

Building Chimneys

All fireplaces and fuel-burning equipment such as stoves and furnaces require some type of chimney (Figure 15-1). The chimney must be designed and built so that it produces sufficient draft to supply an adequate quantity of fresh air to the fire and to expel smoke and gases emitted by the fire or equipment.

A chimney located entirely inside a building

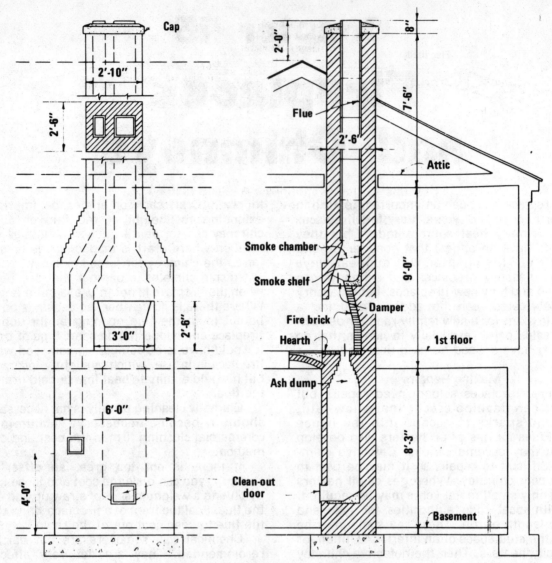

**Diagram of an entire chimney such as is commonly
built to serve the house-heating
unit and one fireplace
Figure 15-1**

has better draft than an exterior chimney, because the masonry retains heat longer when protected from cold outside air.

Proper construction of the flue is important. Its size (area), height, shape, tightness, and smoothness determine the effectiveness of the chimney in producing adequate draft and in expelling smoke and gases. Soundness of the flue walls may determine the safety of the building should a fire occur in the chimney. Overheated or defective flues are one of the chief causes of house fires.

Manufacturers of fuel-burning equipment usually specify chimney requirements, including flue dimensions, for their equipment. Follow their recommendations.

Local codes may call for slightly different construction of fireplaces and chimneys than is given in this book. In such cases, local code requirements should be followed.

Height

A chimney should extend at least 3 feet above flat roofs and at least 2 feet above a roof ridge or raised part of a roof within 10 feet of the chimney. A hood (Figure 15-2), should be provided if a chimney cannot be built high enough above a ridge to prevent trouble from eddies caused by wind being deflected from the roof. The open ends of the hood should be parallel to the ridge.

Low cost metal pipe extensions are sometimes used to increase flue height, but they are not as durable or as attractive as terra cotta chimney pots or extensions. Metal extensions must be securely anchored against the wind and must have the same cross sectional area as the flue. They are available with a metal cowl or top that turns with the wind to prevent air from blowing down the flue.

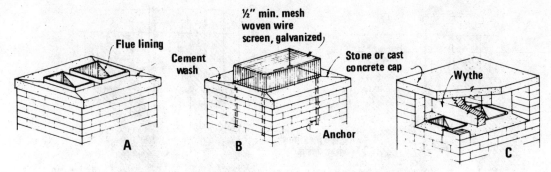

Top construction of chimneys: A, Good method of
finishing top of chimney, flue lining extends 4
inches above cap. B, spark arrester or bird screen;
C, hood to keep out rain
Figure 15-2

Support

The chimney is usually the heaviest part of a building and it must rest on a solid foundation to prevent differential settlement in the building. Concrete footings are recommended. They must be designed to distribute the load over an area wide enough to avoid exceeding the safe load-bearing capacity of the soil. They should extend at least 6 inches beyond the chimney on all sides and should be 8 inches thick for one-story houses and 12 inches thick for two-story houses having basements.

If there is no basement, pour the footings for an exterior chimney on solid ground below the frostline.

If the house wall is of solid masonry at least 12 inches thick, the chimney can be built integrally with the wall and, instead of being carried down to the ground, it can be offset from the wall enough to provide flue space by corbelling. The offset should not extend more than 6 inches from the face of the wall—each course projecting not more than 1 inch—and should be not less than 12 inches high.

Chimneys in frame buildings should be built from the ground up, or they can rest on the building foundation or basement walls if the walls are of solid masonry 12 inches thick and have adequate footings.

Flue Lining

Chimneys are sometimes built without flue lining to reduce cost, but those with lined flues are safer and more efficient.

Lined flues are usually required for brick chimneys. If the flue is not lined, mortar and bricks directly exposed to the action of flue gases disintegrate. This disintegration plus that caused by temperature changes can open cracks in the masonry, which will reduce the draft and increase the fire hazard.

Flue lining must withstand rapid fluctuations in temperature and the action of flue gases. Therefore, it should be made of vitrified fire clay at least five-eighths of an inch thick. Both rectangular and round shaped linings are available. Rectangular lining is better adapted to brick construction, but round lining is more efficient.

Each length of lining should be placed in position—set in cement mortar with the joint struck smooth on the inside—and then the brick laid around it. If the lining is slipped down after several courses of brick have been laid, the joints cannot be filled and leakage will occur. In masonry chimneys with walls less than 8 inches thick, there should be space between the lining and the chimney walls. This space should not be filled with mortar. Use only enough mortar to make good joints and to hold the lining in position.

Unless it rests on solid masonry at the bottom of the flue, the lower section of lining must be supported on at least three sides by brick courses projecting to the inside surface of the lining. This lining should extend to a point at least 8 inches under a smoke pipe thimble or flue ring.

Flues should be as nearly vertical as possible. If a change in direction is necessary, the angle should never exceed 45 degrees (Figure 15-3). An angle of 30 degrees or less is better, because sharp turns set up eddies which affect the motion of smoke and gases. For structural safety the amount of offset must be limited so that the center line XY, (Figure 15-3, A), of the upper flue will not fall beyond the center of the wall of the lower flue. Start the offset of the left wall (Figure 15-3A) of an unlined flue two brick courses higher than the right wall so that the area of the sloping section will not be reduced after plastering. Figure 15-3B shows the method of cutting the lining to make a tight joint. Cut the lining before it is built into the chimney; if cut after, it may break and fall out of place. To cut the lining, stuff a sack of damp sand into it and then tap a sharp chisel with a light hammer along the desired line of cut.

When laying lining and brick, draw a tight-fitting bag of straw up the flue as the work progresses to catch material that might fall and block the flue.

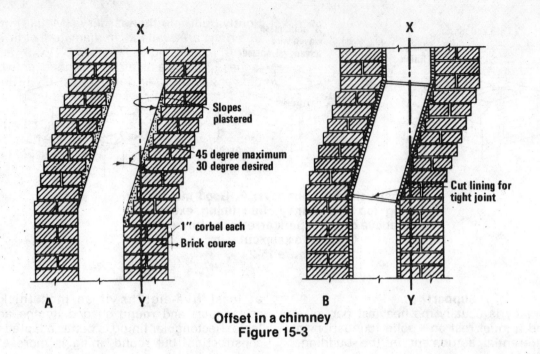

Offset in a chimney
Figure 15-3

Labels in figure: Slopes plastered; 45 degree maximum 30 degree desired; 1" corbel each; Brick course; Cut lining for tight joint; A; B; X; Y

Walls

Walls of chimney not more than 30 feet high with lined flues should be at least 4 inches thick if made of brick or reinforced concrete and at least 12 inches thick if made of stone.

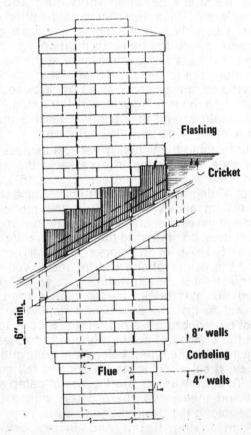

Labels in figure: Flashing; Cricket; 6" min.; 8" walls; Corbeling; Flue; 4" walls

Corbelling of chimney to provide 8-inch walls for the section exposed to the weather
Figure 15-4

Flue lining is recommended, especially for brick chimneys, but it often can be omitted if the chimney walls are made of reinforced concrete at least 6 inches thick or of unreinforced concrete or brick at least 8 inches thick. A minimum thickness of 8 inches is recommended for the outside wall of a chimney exposed to the weather.

Brick chimneys that extend up through the roof may sway enough in heavy winds to open up mortar joints at the roof line. Openings to the flue at that point are dangerous, because sparks from the flue may start fires in the woodwork or roofing. A good practice is to make the upper walls 8 inches thick by starting to offset the bricks at least 6 inches below the underside of roof joists or rafters (Figure 15-4).

Chimneys may contain more than one flue. Building codes generally require a separate flue for each fireplace, furnace or boiler. If a chimney contains three or more lined flues, each group of two flues must be separated from the outer single flue or group of two flues by brick divisions or wythes at least 3¾ inches thick (Figure 15-5). Two flues grouped together without a dividing wall should have the lining joints staggered at least 7 inches and the joints must be completely filled with mortar.

If a chimney contains two or more unlined flues, the flues must be separated by a well-bonded wythe at least 8 inches thick.

Brickwork around chimney flues and fireplaces should be laid with cement mortar; it is more resistant to the action of heat and flue gases than lime mortar. A good mortar to use in setting flue linings and all chimney masonry, except firebrick, consists of 1 part portland cement, 1 part hydrated lime (or slaked-lime putty), and 6 parts clean sand, measured by volume. Firebrick should be laid with fire clay.

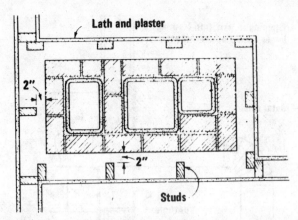

Plan of chimney showing proper arrangement of three flues. Bond division wall with sidewalls by staggering the joints of successive courses. Wood framing should be at least 2 inches from brickwork
Figure 15-5

Soot Pocket and Cleanout

A soot pocket and cleanout are recommended for each flue (Figure 15-6). Deep soot pockets permit the accumulation of an excessive amount of soot, which may take fire. Therefore, the pocket should be only deep enough to permit installation of a cleanout door below the smoke pipe connection. Fill the lower part of the chimney—from the bottom of the soot pocket to the base of the chimney—with solid masonry.

The cleanout door should be made of cast iron and should fit snugly and be kept tightly closed to keep air out. A cleanout should serve only one flue. If two or more flues are connected to the same cleanout, air drawn from one to another will affect the draft in all.

Smoke Pipe Connection

No range, stove, fireplace, or other equipment should be connected to the flue for the central heating unit. In fact, as previously indicated, each unit should be connected to a separate flue, because if there are two or more connections to the same flue, fires may occur from sparks passing into one flue opening and out through another.

Smoke pipes from furnaces, stoves, or other equipment must be correctly installed and connected to the chimney for safe operation.

Gas-fired house heaters and built-in unit heaters can be connected to metal flues instead of to a masonry chimney. The flues should be made of corrosion-resistant metal not lighter than 20 gauge and should be properly insulated with asbestos or other fireproof material that complies with the recommendations of Underwriters' Laboratories, Incorporated. The flues must extend through the roof.

A smoke pipe should enter the chimney horizontally and should not extend into the flue (Figure 15-6). The hole in the chimney wall should be lined with fire clay, or metal thimbles should be

tightly built into the masonry. (Metal thimbles or flue rings are available in diameters of 6, 7, 8, 10, and 12 inches and in lengths of 4½, 6, 9, and 12 inches.) To make an airtight connection where the pipe enters the wall, install a closely fitting collar and apply boiler putty, good cement mortar or stiff clay.

A smoke pipe should never be closer than 9 inches to woodwork or other combustible material. If it is less than 18 inches from woodwork or other combustible material, cover at least the half of the pipe nearest the woodwork with fire-resistant material. Commercial fireproof pipe covering is available.

Soot pocket and cleanout for a chimney flue
Figure 15-6

If a smoke pipe must pass through a wood partition, the woodwork must be protected. Either cut an opening in the partition and insert a galvanized iron, double wall ventilating shield at least 12 inches larger than the pipe (Figure 15-7) or install at least 4 inches of brickwork or other noncombustible material around the pipe. Smoke pipes should never pass through floors, closets, or concealed spaces or enter the chimney in the attic.

Every flue should be tested as follows before being used and preferably before the chimney has been furred, plastered, or otherwise enclosed. Build a paper, straw, wood, or tar-paper fire at the base of the flue. When the smoke rises in a dense column, tightly block the outlet at the top of the chimney with a wet blanket. Smoke that escapes through the masonry indicates the location of leaks. This test will show bad leaks into adjoining flues, through the walls, or between the lining and the wall. Correct defects before the chimney is used. Since such defects may be hard to correct, you should check the initial construction carefully.

Insulation

No wood should be in contact with the chimney. Leave a 2 inch space (Figure 15-5)

233

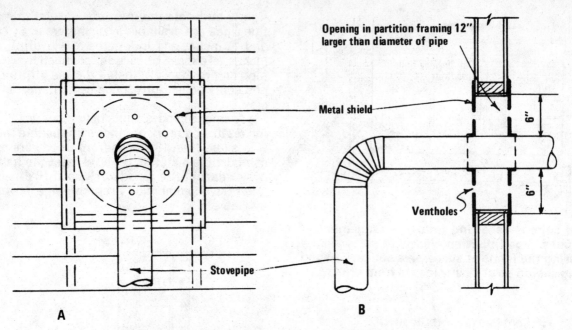

One method of protecting a wood partition when a smoke pipe passes through it. A, elevation of protection around the pipe, B, sectional view
Figure 15-7

between the chimney walls and all wooden beams or joists (unless the walls are of solid masonry 8 inches thick, in which case the framing can be within one-half inch of the chimney masonry).

Fill the space between wall and floor framing with porous, nonmetallic, noncombustible material, such as loose cinders (Figure 15-8). Do not use brickwork, mortar, or concrete. The filling forms a firestop and also prevents the accumulation of shavings or other combustible material. Flooring and subflooring can be within three-

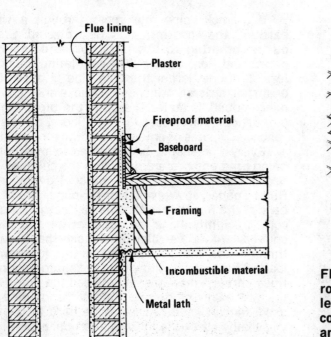

Method of insulating wood floor joists and baseboard at a chimney
Figure 15-8

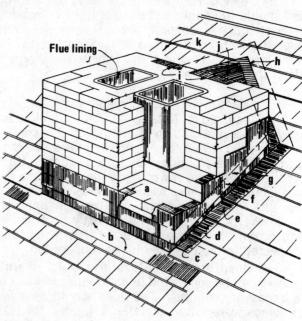

Flashing at a chimney located on the slope of a roof. Sheet metal [h], over cricket [j], extends at least 4 inches under the shingles [k], and is counterflashed at l in joint. Base flashings [b, c, d, and e] and cap flashings [a, f, and g] lap over the base flashings to provide watertight construction. A full bed of mortar should be provided where cap flashing is inserted in joints
Figure 15-9

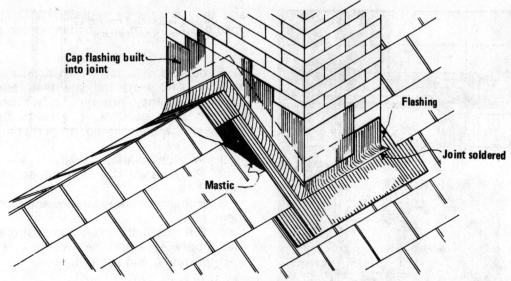

Flashing at a chimney located on a roof ridge
Figure 15-10

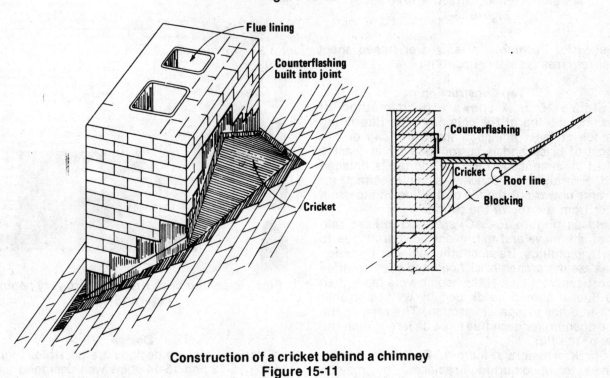

Construction of a cricket behind a chimney
Figure 15-11

fourths inch of the masonry. Wood studding, furring, or lathing should be at least 2 inches from chimney walls. (Plaster can be applied directly to the masonry or to metal lath laid over the masonry, but this is not recommended because settlement of the chimney may crack the plaster.) A coat of cement plaster should be applied to chimney walls that will be encased by wood partition or other combustible construction.

If baseboards are fastened to plaster that is in direct contact with the chimney wall, install a layer of fireproof material, such as asbestos, at least one-eighth inch thick between the baseboard and the plaster (Figure 15-8).

Connection With Roof

Where the chimney passes through the roof, provide a 2 inch clearance between the wood framing and the masonry for fire protection and to permit expansion due to temperature changes, settlement, and slight movement during heavy winds.

Chimneys must be flashed and counterflashed to make the junction with the roof watertight (Figures 15-9 and 15-10). When the chimney is located on the slope of a roof, a cricket (j, Figure 15-9) is built as shown in Figure 15-11 high enough to shed water around the chimney. Corrosion resistant metal, such as copper, zinc, or lead, should

A well-designed, attractive fireplace
Figure 15-12

be used for flashing. Galvanized or tinned sheet steel requires occasional painting.

Top Construction

Figure 15-2, A shows a good method of finishing the top of the chimney. The flue lining extends at least 4 inches above the cap or top course of brick and is surrounded by at least 2 inches of cement mortar. The mortar is finished with a straight or concave slope to direct air currents upward at the top of the flue and to drain water from the top of the chimney.

Hoods (Figure 15-2, C) are used to keep rain out of chimneys and to prevent downdraft due to nearby buildings, trees, or other objects. Common types are the arched brick hood and the flat-stone or cast-concrete cap. If the hood covers more than one flue, it should be divided by wythes so that each flue has a separate section. The area of the hood opening for each flue must be larger than the area of the flue.

Spark arresters (Figure 15-2, B) are recommended for wood burning fireplaces. They may be required when chimneys are on or near combustible roofs, woodland, lumber, or other combustile material. They are not recommended when burning soft coal, because they may become plugged with soot.

Spark arresters do not entirely eliminate the discharge of sparks, but if properly built and installed, they greatly reduce the hazard. They should be made of rust resistant material and should have screen openings not larger than five-eighths inch nor smaller than five-sixteenths inch. (Commercially made screens that generally last for several years are available.) They should completely enclose the flue discharge area and must be securely fastened to the top of the chimney. They must be kept adjusted in position

and they should be replaced when the screen openings are worn larger than normal size.

Fireplaces

Fireplaces are not an economical means of heating. And tests indicate that, as ordinarily constructed, they are only about one-third as efficient as a good stove or circulator heater. However, a well designed, properly built fireplace can:

1. Provide additional heat.
2. Provide all the heat necessary in mild climates.
3. Enhance the appearance and comfort of the room.
4. Burn as fuel certain combustible materials that otherwise might be wasted—for example, coke, briquets, and scrap lumber.

Floor-to-ceiling brick fireplaces are very popular
Figure 15-13

Design

Varied fireplace designs are possible. Figures 15-12, 15-13 and 15-14 show well designed units. A fireplace should harmonize in detail and proportion with the room in which it is located, but safety and utility should not be sacrificed for appearance.

Location of the fireplace within a room should be made on the basis of safety and convenience. Make sure the chimney will not present a fire hazard in the location you select. A fireplace should not be located near doors.

Fireplace openings are usually made from 2 to 6 feet wide. The kind of fuel to be burned can suggest a practical width. For example, where cordwood (4 feet long) is cut in half, an opening 30 inches wide is desirable; but where coal is burned, a narrower opening can be used.

Height of the opening can range from 24 inches

This attractive fireplace well illustrates the variations in design possible
Figure 15-14

for an opening 2 feet wide to 40 inches for one that is 6 feet wide. The higher the opening, the more chance of a smoky fireplace.

In general, the wider the opening, the greater the depth. A shallow opening throws out relatively more heat than a deep one, but holds smaller pieces of wood. You have the choice, therefore, between a deeper opening that holds larger, longer-burning logs and a shallower one that takes smaller pieces of wood, but throws out more heat. In small fireplaces, a depth of 12 inches may permit good draft, but a minimum depth of 16 inches is recommended to lessen the danger of brands falling out on the floor. Suitable screens should be placed in front of all fireplaces to minimize the danger from brands and sparks.

Second-floor fireplaces are usually made smaller than first-floor ones, because of the reduced flue height.

Construction

Fireplace construction is basically the same regardless of design. Figure 15-15 shows construction of a typical fireplace. Table 15-16 gives recommended dimensions for essential parts or areas of fireplaces of various sizes.

Hearth

The fireplace hearth should be made of brick, stone, terra cotta, or reinforced concrete at least 4 inches thick. It should project at least 20 inches from the chimney breast and should be 24 inches wider than the fireplace opening (12 inches on each side).

The hearth can be flush with the floor so that sweepings can be brushed into the fireplace or it can be raised. Raising the hearth to various levels and extending in length as desired is presently common practice, especially in contemporary design.

In buildings with wooden floors, the hearth in front of the fireplace should be supported by masonry trimmer arches or other fire-resistant construction (Figure 15-15). Wood centering under the arches used during construction of the hearth and hearth extension should be removed when construction is completed.

Figure 15-17 shows the recommended method of installing floor framing around the hearth.

Where a header is more than 4 feet long, it should be doubled as shown. If it supports more than four tail beams, its ends should be supported in metal joist hangers. The framing may be placed one-half inch from masonry chimney walls 8 inches thick.

Walls

Building codes generally require that the back and sides of fireplaces be constructed of solid masonry or reinforced concrete at least 8 inches thick and be lined with firebrick or other approved noncombustible material not less than 2 inches thick or steel lining not less than one-fourth inch thick. Such lining may be omitted when the walls are of solid masonry or reinforced concrete at least 12 inches thick.

Jambs

The jambs of the fireplace should be wide enough to provide stability and to present a pleasing appearance. For a fireplace opening 3 feet wide or less, the jambs can be 12 inches wide if a wood mantel will be used or 16 inches wide if they will be of exposed masonry. For wider fireplace openings, or if the fireplace is in a large room, the jambs should be proportionately wider. Fireplace jambs are frequently faced with ornamental brick or tile.

No woodwork should be placed within 6 inches of the fireplace opening. Woodwork above and projecting more than 1½ inches from the fireplace opening should be placed not less than 12 inches from the top of the fireplace opening.

Lintel

A lintel must be installed across the top of the fireplace opening to support the masonry. For fireplace openings 4 feet wide or less ½ by 3 inch flat steel bars, 3½ by 3½ by ¼ inch gangle irons, or specially designed damper frames may be used. Wider openings will require heavier lintels.

If a masonry arch is used over the opening, the fireplace jambs must be heavy enough to resist the thrust of the arch.

Throat

Proper construction of the throat area (ff, Figure 15-15) is essential for a satisfactory fireplace. The sides of the fireplace must be

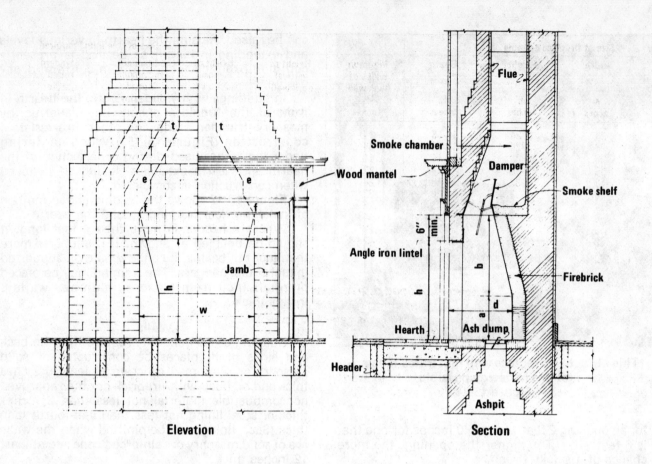

Elevation

Section

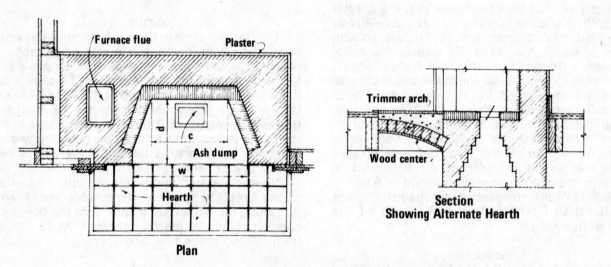

Furnace flue

Plaster

Ash dump

Hearth

Plan

Trimmer arch

Wood center

**Section
Showing Alternate Hearth**

Construction details of a typical fireplace. [The
letters indicate dimensions in Table 15-16 and
specific features discussed in the text]. The lower
right-hand drawing shows an alternate method of
supporting the hearth
Figure 15-15

vertical up to the throat, which should be 6 to 8
inches or more above the bottom of the lintel. Area
of the throat must be not less than that of the
flue—length must be equal to the width of the
fireplace opening and width will depend on the
width of the damper frame (if a damper is

installed). Five inches above the throat (at ee,
Figure 15-15), the sidewalls should start sloping
inward to meet the flue (at tt, Figure 15-15).

Damper
A damper consists of a cast iron frame with a

Size of fireplace opening						Size of flue lining required	
Width w Inches	Height h Inches	Depth d Inches	Minimum width of back wall c Inches	Height of vertical back wall a Inches	Height of inclined back wall b Inches	Standard rectangular (outside dimensions) Inches	Standard round (inside diameter) Inches
24	24	16-18	14	14	16	8½ x 13	10
28	24	16-18	14	14	16	8½ x 13	10
30	28-30	16-18	16	14	18	8½ x 13	10
36	28-30	16-18	22	14	18	8½ x 13	12
42	28-32	16-18	28	14	18	13 x 13	12
48	32	18-20	32	14	24	13 x 13	15
54	36	18-20	36	14	28	13 x 18	15
60	36	18-20	44	14	28	13 x 18	15
54	40	20-22	36	17	29	13 x 18	15
60	40	20-22	42	17	30	18 x 18	18
66	40	20-22	44	17	30	18 x 18	18
72	40	22-28	51	17	30	18 x 18	18

**Recommended dimensions for fireplaces and size
of flue lining required. [Letters at heads
of columns refer to Figure 15-15]
Table 15-16**

hinged lid that opens or closes to vary the throat opening. Dampers are not always installed, but they are definitely recommended, especially in cold climates. Dampers of various designs are on the market. Some are designed to support the masonry over fireplace openings, thus replacing ordinary lintels.

Responsible manufacturers of fireplace equipment usually offer assistance in selecting a suitable damper for a given size fireplace. It is important that the full damper opening equal the area of the flue.

Smoke Shelf and Chamber

A smoke shelf (Figure 15-15) prevents downdraft. It is made by setting the brickwork at the top of the throat back to the line of the flue wall for the full length of the throat. Depth of the shelf may be 6 to 12 inches or more, depending on the depth of the fireplace.

The smoke chamber is the area from the top of the throat (ee, Figure 15-15) to the bottom of the flue (tt, Figure 15-15). The sidewalls should slope inward to meet the flue.

The smoke shelf and the smoke-chamber walls should be plastered with cement mortar at least one-half inch thick.

Flue

Proper proportion between the size (area) of the fireplace opening, size (area) of the flue, and height of the flue is essential for satisfactory operation of the fireplace.

The area of a lined flue 22 feet high should be at least one-twelfth of the area of the fireplace opening. The area of an unlined flue or a flue less than 22 feet high should be one-tenth of the area of the fireplace opening.

Table 15-16 lists dimensions of fireplace

openings and, in the last two columns, indicates the size of flue lining required. From this table, you can determine the size of lining required for a given size fireplace opening and also the size of opening to use with an existing flue.

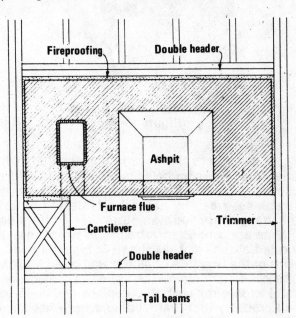

**Installation of floor framing around a chimney
and hearth
Figure 15-17**

Fireplace Forms

Fireplace forms are manufacturered fireplace units, made of heavy metal and designed to be set in place and concealed by the usual brickwork or other construction. They contain all the essential fireplace parts—firebox, damper, throat, and smoke shelf and chamber. In the completed installation, only grilles show. Figure 15-18 shows one of the several designs of fireplaces available.

Fireplace forms offer two advantages:

1. The correctly designed and proportioned firebox provides a ready made form for the masonry, which reduces the chance of faulty construction and assures a smokeless fireplace.

2. When properly installed, the better designed units heat more efficiently than ordinary fireplaces.

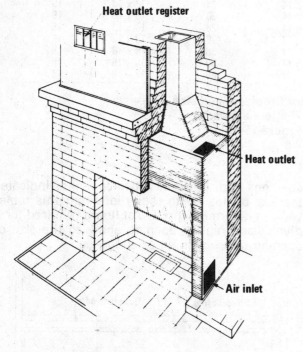

A fireplace form installed
Figure 15-18

In a fireplace form, air is drawn through the inlet from the room being heated. It is heated by contact with the metal sides and back of the fireplace, rises by natural circulation, and is discharged through an outlet. The inlets and outlets are connected to registers which may be located at the front (as shown in Figure 15-18) or ends of the fireplace or on the wall of an adjacent or second-story room. The use of a fireplace form can increase the cost of a fireplace (although you will probably find that labor and materials saved offset any additional cost).

Even a well designed fireplace form will not operate properly if the chimney is inadequate. Proper chimney construction is as important with a fireplace form as with any fireplace you build.

Prefabricated Fireplaces and Chimneys

Factory built, all steel fireplaces are designed in both free standing and built-in units, each with assembled firebox and insulated flue sections which interconnect with a prefabricated metal ceiling housing. Most are available in firebox

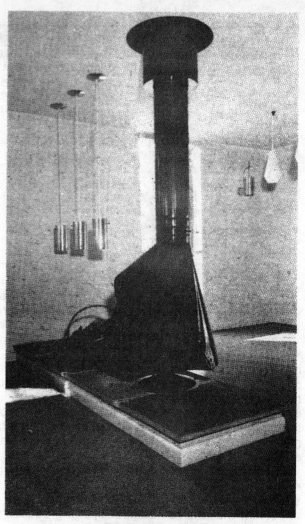

Prefabricated fireplace
Figure 15-19

opening sizes of 28, 36, 42 and 48 inches. Note Figure 5-19. Such units offer these features:

1. Wide selection of styles, shapes, and colors.

2. Pretested design that is highly efficient in operation.

3. Easy and versatile installation.

4. Light in weight.

5. Lower cost than comparable masonry units.

They can be flush mounted in a wall, placed against a wall, used as a room divider, placed in a corner or used as an island. This total flexibility of placement is possible because better fireplace units are designated "zero clearance fireplaces" meaning that they can be safely installed butted against a combustible wall, partition or wood floor without the need of insulating materials. These Underwriters Laboratories listed fireplaces make use of air spaces and factory-installed insulation around the firebox and throat to prevent excessive heat build-up.

Installation is relatively simple. Most of the required materials are supplied in the base fireplace assembly package. The only additional supplies needed would be 2 x 4 studs for framing in the fireplace, metal chimney sections and material for enclosing and trimming out the unit. Chimney sections are available in two and three foot lengths. They consist of an outer metal shell, an inner layer of insulation, and a stainless steel inner liner—constructed into one easy to handle unit. Chimney sections are installed simply by stacking one upon another and giving them a twist. A locking ring device at the end of each section tightly joins them together to provide a smoke free joint.

The "built-in" units contain all of the elements of a masonry fireplace including fire chamber, throat, damper, smoke shelf and smoke chamber. When the all-steel housing has been boxed in and decorative facing applied in a noncombustible material such as brick, stone or tile, the fireplace appears to be custom designed and built.

Free standing fireplaces are much simpler to install since they require only a fireproof base, an opening for the flue, and placement three feet from a wall. Floor protection under the unit can be anything such as gravel, stone or brick. Since they are light in weight, they do not require the heavy footing required for masonry fireplaces.

Prefabricated chimneys can be used for furnaces, heaters, and incinerators as well as for prefabricated fireplaces. The chimneys are tested and approved by Underwriters Laboratories, Incorporated, and other nationally recognized testing laboratories, and are rapidly being accepted for use by building codes in many communities.

Chapter 16
Porches and Decks

Outdoor living is a way of life for the American family. Most home owners would like to have an improved and covered outdoor area adjacent to the home for family enjoyment in pleasant weather. In the past, porches located at entrances served as the outdoor living area. In older homes the porch is usually covered with a roof which is an extension of the main structure roof. Modern structures more frequently use decks with more open covers which do not share a common roof line with the main structure. This outdoor living area adds spaciousness to a home at modest cost. A deck or remodeled porch can expand or frame a view to increase a homeowner's enjoyment. It can serve as an adult entertainment center by night and a children's play area by day — being easily adapted to the activity or degree of formality desired. Decks, which may offer the only means of providing outdoor living areas for steep hillside homes, have gained popularity for homes on level ground—as a way of adding charm, style, and livability. You should consider the possibility of remodeling a porch or adding a deck on most remodeling jobs, especially where the homeowner is interested in giving the structure a modern look.

Making Repairs

On some jobs repair of the porch is all that is required. The porch is usually the first part of a structure to deteriorate and is usually the most seriously deteriorated because it is exposed to the elements and is close to the ground. If examination of the porch shows all parts to be in a generally deteriorated condition, complete removal and replacement is recommended. However, it may be possible to replace components, such as steps, floor, posts, or roof, where other components are in good condition.

Steps

When wood steps are used, the bottom step and carriage should not be in contact with soil. A concrete step can be cast on the ground to support the carriage or it can be supported on a treated wood post. Apply a water-repellent preservative to all wood used in the steps.

If only one step is required and treated posts are used for support, the step can be supported on the same post that supports the edge of the porch (Figure 16-1). Use treated posts of 5 to 7 inch diameter embedded in the soil at least 3 feet. Nail and bolt the crossmember to the post, and block the inner end to the floor framing with a short 2 by 4.

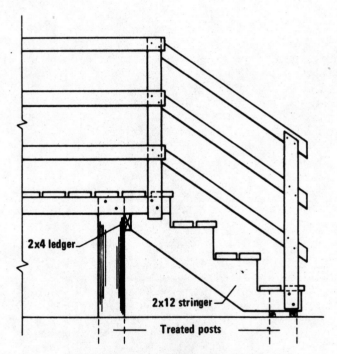

Step stringer supported by porch framing and posts
Figure 16-2

Where more steps are required, use a 2 by 12 stringer at each end of the steps with the lower end of each stringer bolted to a treated post (Figure 16-2). The upper end can be attached to the porch framing. Concrete or masonry piers can be used in place of treated posts. The important thing is that the stringer be kept from contact with the soil.

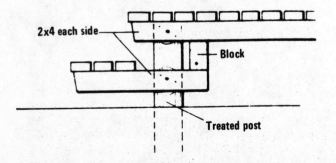

Single porch step supported on a treated post
Figure 16-1

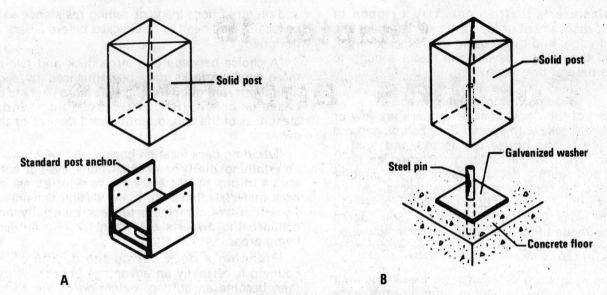

Base for post: A, standard post anchor for resistance to uplift; B, galvanized washer and pin where resistance to uplift is not critical
Figure 16-3

Floor

If floor framing is decayed, it should be completely replaced, and all replacement members should be treated with a water-repellent preservative. Framing members should be at least 18 inches above the ground, and good ventilation should be provided under the porch. Framing should be installed to give the finished floor an outward slope of at least 1/8 inch per foot.

Porch flooring is 1 by 4 inch lumber, dressed and matched. It is blind-nailed at each joist in the same manner as regular flooring. Apply a good stain or deck paint as soon as possible after installation.

Posts

Where a post rests directly on the porch floor, the base of the post may be decayed. The best solution may be to replace the post. When this is done, provide some way to support the post slightly above the porch floor. Standard post anchors can be purchased for this purpose (Figure 16-3, A). Another way to accomplish this is by using a small 3/8 or 1/2 inch diameter pin and large galvanized washer. Drill a hole for the pin in the end of the post and a matching hole in the floor (Figure 16-3, B). Apply a mastic caulk to the area and position with the pin inserted and the washer between the post and the floor. This will allow moisture to evaporate from the end of the post and prevent decay. This pinned method should be used only on small porches that will not have a major wind uplift load.

If the existing porch posts are ornamental, it may be desirable to cut off the decayed portion near the base and save the good portion. Perhaps a base slightly larger than the post can be added to

replace the decayed portion (Figure 16-4). Some method of ventilating the base of the post should be provided to avoid further decay. One solution is to use a pin and large washer, as described above for a new post.

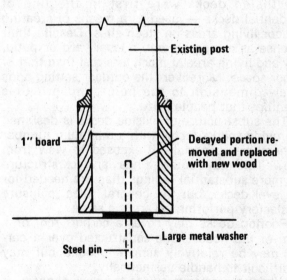

Replacement of decayed end of porch post
Figure 16-4

Roof

A porch roof can be repaired in much the same manner as repairing the house roof. The difference may be that porch roofs frequently have a very low slope and are often covered with roll roofing. Simply apply an underlayment of asphalt-saturated felt and follow with mineral surfaced half-lap roll roofing applied in accordance with manufacturer's instructions. Use a ribbon of asphalt roof cement or lap seal material under the lapped edge, and avoid using any exposed nails. Extend the roofing beyond the fascia enough to form a natural drip edge.

Adding Outdoor Living Area

Most of the decks considered here are low or high level decks with spaced floor boards and are attached to the house for access and partial support. There are, in addition, detached low level decks and rooftop decks. The latter are simpler than others in some respects since they rely on the roof for primary support but do introduce a need to prevent leakage to the space below. Solid decks may be made of caulked planking or of plywood with a waterproof coating such as an elastomeric wearing surface.

Low level wood decks may be chosen for their non-reflective and resilient qualities in preference to a paved patio. More frequently, the wood deck is chosen because of its design versatility and adaptability to varied use.

Low level decks can be simply supported on concrete piers or short posts closely spaced, thereby simplifying the main horizontal structure. However, drainage can be a problem on low or level ground and provision to insure good drainage should be made before the deck is built. Good drainage not only keeps the ground firm to adequately support the deck but avoids dampness that could encourage decay in posts or sills.

Hillside decks were first in the line of residential decks — used as a means of creating outdoor living areas on steep sites. Despite their expense as compared with a level yard or patio, they add living area at much less cost than that of indoor space. Moreover, the outdoor setting adds a new dimension to the home and provides amenities that people prize.

The substructure of hillside decks is designed to provide solid support with a minimum number of members, especially if exposed to view from below. This may require a heavier deck structure and more substantial railings than are needed for low level decks, but the general rules to insure satisfactory performance are the same.

Rooftop decks may cover a carport roof or a room of the house. One constructed over a carport may be relatively simple to build but may be difficult to handle aesthetically.

Where a rooftop provides the deck support, it may also serve as the floor, particularly if the deck is included in the plan for a room addition. The roof must then be designed as a floor to support the deck loads. If the deck is added to a completed house, it is more common to construct a separate deck floor over the roof.

Planning a Deck

A first step in planning a deck is to determine the requirements and limitations of the local building code. Limitations on height and width and required floor loads or railing resistance vary by locality and need to be checked before a deck is designed.

A choice between one large deck and two or more smaller decks may be influenced by code limitations, although the choice is more likely to be based on orientation, view, prevailing winds, steepness of the site, or anticipated desires of the owner.

Deciding deck location goes a long way toward determining the type of deck but leaves a wide choice in design. Where wood is selected as the deck material, there are many design considerations that can contribute to deck durability and enhance the owner's enjoyment of this outdoor living area.

Planning a deck during the design of an addition is certainly an advantage because it can then become an outdoor extension of the added living area. A deck can also be designed as an outdoor portion of an existing dining room or kitchen with access through sliding doors or other openings. It is also desirable to take advantage of prevailing breezes with space for both sun and shade areas during the day. Sun shades can be used as a substitute for the natural shade provided by trees.

Providing a deck for an existing house is sometimes more difficult because the rooms may not be located to provide easy access to a deck. However, introduction of a new doorway in the house and a pleasant walkway to the deck area may provide a satisfactory solution to even the most difficult problem.

Materials

Although wood and wood products are the primary materials used in the construction of exposed decks, other materials such as fastenings and finishes are also important. Footings used to anchor the posts which support the deck proper are usually concrete. The proper combinations of all materials with good construction details will insure a deck which will provide years of pleasure.

Many lumber species will provide good service in a wood deck. However, some are more adequate for the purpose than others. To select lumber wisely, one must first single out the key requirements of the job. Then it is relatively easy to check the properties of the different wood to see which ones meet these requirements. For example, beams or joists require wood species that are high in bending strength or stiffness; wide boards in railing or fences should be species that warp little; posts and similar members that are exposed to long wet periods should be heartwood of species with high decay resistance. The classification of woods commonly used in the United States according to their characteristics is given in Table 16-5. Follow the recommendations in this Table in selecting wood for a specific use in the outdoor structure.

Kind Of Wood	Working and behavior characteristics							Strength properties			Freedom from pitch
	Hardness	Freedom from warping	Ease of working	Paint holding	Nail holding	Decay resistance of heartwood	Proportion of heartwood	Bending strength	Stiffness	Strength as a post	
Ash	A	B	C	C	A	C	C	A	A	A	A
Western red cedar	C	A	A	A	C	A	A	C	C	B	A
Cypress	B	B	B	A	B	A	B	B	B	B	A
Douglas-fir larch	B	B	B-C	C	A	B	A	A	A	A	B
Gum	B	C	B	C	A	B	B	B	A	B	A
Hemlock, white fir[1]	B-C	B	B	C	C	C	C	B	A	B	A
Soft pine[2]	C	A	A	A	C	C	B	C	C	C	B
Southern pine	B	B	B	C	A	B	C	A	A	A	C
Poplar	C	A	B	A	B	C	B	B	B	B	A
Redwood	B	A	B	A	B	A	A	B	B	A	A
Spruce	C	A-B	B	B	B	C	C	B	B	B	A

A - among the woods relatively high in the particular respect listed; B - among woods intermediate in that respect; C - among woods relatively low in that respect. Letters do not refer to lumber grades.

1. Includes west coast and eastern hemlocks.
2. Includes the western and northeastern pines.

**Broad classification of woods according to
characteristics and properties
Table 16-5**

Plywood is adaptable for use in wood decks and is often recommended for solid deck coverings. Plywood is made in two types — Exterior and Interior. Only exterior type is recommended where any surface or edge is permanently exposed to the weather. Interior type plywood, even when made with exterior glue and protected on the top surface, is not recommended for such exposures.

The size of lumber is normally based on green sawn sizes. When the lumber has been dried and surfaced, the finish size (thickness and width) is somewhat less than the sawn size. The lumber sizes used in this chapter are those established by the American Lumber Standards Committee.

Nominal (inches)	Dry (inches)	Green (inches)
1	3/4	25/32
2	1-1/2	1-9/16
4	3-1/2	3-9/16
6	5-1/2	5-5/8
8	7-1/4	7-1/2
10	9-1/4	9-1/2
12	11-1/4	11-1/2

For example, a nominal 2 by 4 would have a surfaced dry size of 1½ by 3½ inches at a maximum moisture content of 19 percent. Moisture content of wood during fabrication and assembly of a wood frame structure is important. Ideally, it should be about the same moisture content it reaches in service. If green or partially dried wood is used, wood members usually shrink, resulting in poorly fitting joints and loose fastenings after drying has occurred.

Although not as important for exterior use as for interior use, the moisture content of lumber used and exposed to exterior conditions should be considered. The average moisture content of wood exposed to the weather varies with the season, but kiln dried or air dried lumber best fits the mid-range of moisture contents that wood reaches in use.

Plywood Specifications

For solid deck applications with direct exposure to the weather, plywood marked C-C Plugged Exterior, or Underlayment Exterior (C-C Plugged) may be specified. Higher grades, such as A-C or B-C Exterior, may also be used. These grades coated with a high performance wearing surface are commonly used for residential deck areas.

High Density Overlay (HDO) plywood having a hard, phenolic-resin impregnated fiber surface is often used for boat decks with a screened, skid-resistant finish specified. HDO may be painted with standard deck-type paints, if desired, but is usually used without further finish.

Medium Density Overlay (MDO) plywood having a softer resin-fiber overlay requires either a high performance deck paint, or an elastromeric deck coating system, depending on the intended use.

For premium deck construction, Plyron (plywood with a tempered hardboard face) may be used in conjunction with an elastromeric deck coating.

Decay Resistance of Wood

Every material normally used in construction has its own distinctive way of deteriorating under adverse conditions. With wood it is decay. Wood will never decay if kept continuously dry (at less than 20 percent moisture content). Because open decks and other outdoor components are exposed to wetting and drying conditions, good drainage, flashings, and similar protective measures are more important in decks than in structures fully protected by a roof.

To provide good performance of wood under exposed conditions, one or more of the following measures should be taken:

1. Use the heartwood of a decay-resistant species.

2. Use wood that has been given any good perservative treatment.

3. Use details which do not trap moisture and which allow easy drainage.

A combination of (1) and (3) above is considered adequate. (2) is satisfactory alone, but usually at increased cost if pressure treatment is used, or at the expense of increased maintenance if dip or soak treatments are used. Detailing that allows quick drying is always desirable. Frequently, it is cheaper and easier to use a good connection design than to use an inferior detail with a decay-resistant wood.

Deck Finishing and Treating

The best treatment of wood to assure long life under severe conditions is a pressure preservative treatment. However, most of the wood parts of a deck are exposed only to moderate conditions except at joints and connections. There are two general methods of treating wood with preservative, (a) pressure processes applied commercially, which provide lasting protection; and (b) non-pressure processes which normally penetrate the ends and a thin layer of the outer surfaces and require frequent maintenance. The non-pressure processes refer to treatments with water-repellent wood preservatives.

Two general types of preservatives are recommended for severe exterior conditions: (a) oils, such as creosote, or pentachlorophenol in oil or liquified gas carriers, and (b) nonleachable salts, such as the chromated copper arsenates or ammoniacal copper arsenite applied as water solutions.

Treatment of poles and posts which are in contact with the soil should comply with the latest Federal specification TT-W-571i. Insist that the wood material you use for these purposes has been treated according to these recommended practices.

Wood used under severe conditions such as with ground or water contact may be pressure treated as recommended for poles and posts. However, if cleanliness or paintability is a factor, creosote or pentachlorophenol in heavy oil should not be used. Instead, use wood treated with non-leachable water-borne preservatives or pentachlorophenol in light or volatile petroleum solvents.

Where preservative treatment is desirable because the more decay resistant woods are not available, the use of a more easily applied, less expensive but less effective non-pressure treatment may be considered for joints, connections, and other critical areas. A pentachlorophenol solution with a water repellent is one of the more effective materials of this type. It is available at most lumber or paint dealers as a clear, water-repellent preservative. Sears, Roebuck & Company's "Wood Tox" is one of many products available.

These materials should be applied by soaking, dipping, or flooding so that end grain, machine cuts, and any existing checks in the wood are well penetrated. Dipping each end of all exterior framing material in water-repellent preservative is recommended, and this should be done after all cutting and drilling is completed. Drilled holes can be easily treated by squirting preservative from an oil can with a long spout. Dry wood absorbs more of these materials than partially dry wood and, consequently is better protected.

Preservative treatments for plywood decks are covered by the same standards as those for lumber and may be applied by pressure or superficial soaking or dipping. Several types of treatments perform well on plywood; but some deck applications require specific treatments, and the compatibility of the treatment with finish materials should always be checked.

Light, oil-borne preservatives and water-borne preservatives, such as those recommended for lumber, should be used when a clean, odorless, and paintable treatment is required. For maximum service from plywood decks, these preservatives should be pressure applied.

Exterior Finishes For Wood

Exterior finishes which might be considered for wood components exposed to the weather include natural finishes, penetrating stains, and paints. In general, natural finishes containing a water repellent and a preservative are preferred over paint for exposed flat surfaces. The natural finishes penetrate the wood and are easily renewed, but paint forms a surface film that may rupture under repeated wetting and drying. Exposed flat surfaces of decks, railings, and stairways are more vulnerable to paint film rupture than are vertical sidewall surfaces. Therefore, a completely satisfactory siding paint may be suitable for a deck.

Natural finishes (lightly pigmented) are often used for exposed wood decks, railings, and stairways, not only because they can be easily renewed but because they enchance the natural color and grain of the wood. Such finishes can be obtained in many colors from a local paint dealer. Light colors are better for deck surfaces subject to traffic, as they show the least contrast in grain color as wear occurs and appearance is maintained longer.

One type of natural finish contains paraffin wax, zinc stearate, penta concentrate, linseed oil, mineral spirits, and tinted colors. Such finishes are manufactured by many leading producers of wood stains and are generally available from paint or lumber dealers.

Penetrating stains (heavily pigmented) for rough and weathered wood may be used on the large sawn members such as beams and posts. These are similar to the natural finishes just described but contain less oil and more pigment. They are also produced by many companies.

Paint is one of the most widely used finishes for wood. When applied properly over a paintable surface with an initial water-repellent preservative treatment, followed by prime and finish coats, paint is a highly desirable finish for outdoor structures or as an accent color when used with natural finishes. Exposed flat surfaces with end or side joints are difficult to protect with a paint coating unless there is no shrinking or swelling of the wood to rupture the paint film. A crack in the paint film allows water to get beneath the film where it is hard to remove by drying. Retention of such moisture can result in eventual decay.

Good painting practices include an initial application of water-repellent preservative. After allowing two sunny days for drying of the preservative, a prime coat is applied. This can consist of a linseed oil base paint with pigments that do not contain zinc oxide.

The finish coats can contain zinc oxide pigment and can be of the linseed oil, alkyd, or latex type. Two coats should be used for best results. A three coat paint job with good quality paint may last as long as 10 years when the film is not ruptured by excessive shrinking or swelling of the wood.

Coverings and Coatings for Plywood Decks

Tough, skid resistant, elastomeric coatings are available for plywood deck wearing surfaces. These coatings include liquid neoprene, neoprene/Hypalon, and silicone or rubber-based materials. Plywood joints for these systems are usually sealed with a high performance caulk such as a silicone or Thiokol (silicone caulks require a primer). Joints may also be covered with a synthetic reinforcing tape, prior to application of the final surface coat, when an elastomeric coating system is used.

Silicone or Thiokol caulks are applied to ¼ inch gaps between plywood panels over some type of filler or "backer" material — such as a foam rod. The caulk "bead" is normally about one-fourth inch in diameter. An alternate method is to bevel the panel edges first, then fill the joint with the caulk before the finish coating is applied.

For a premium quality joint, reinforcing tape is sometimes applied as a flashing over sealed joints. Reinforcing tape can also be used over cant strips at wall-to-deck corner areas and over unsealed plywood joints. For these applications, the tape flashing is embedded in a base coat of the elastomeric deck coating.

Where plywood must be installed under wet conditions, the primer or first coat may be applied under shelter at the site prior to installation of the panels. In general, the first coat of coating systems for plywood should be applied to a dry, fresh wood surface. Where preservatives are used, the surface should be scraped or sanded to removed any residue produced by the preservative before the prime coat is applied. Finish coats of most systems require a dry clean surface, for best results.

If outdoor carpeting is to be used on plywood exposed to the weather, it is advisable to use pressure preservative treated plywood, with the underside well ventilated, for both low and elevated decks. Since carpeting is relatively new as an exterior surface material, specific information on its long term performance when used on plywood under severe exposure is not available. Carpet may be readily applied to untreated plywood deck areas that are not subject to repeated wetting.

Canvas is sometimes used as a wearing surface on plywood. It should be installed with a waterproof adhesive, under dry conditions. A canvas surface well fused to the plywood may be painted with regular deck paints.

Framing Spans and Sizes

The allowable spans for decking, joists, and beams and the size of posts depend not only on the size, grade, and spacing of the members but also on the species. Species such as Douglas fir, southern pine, and western larch allow greater spans than some of the less dense pines, cedars, and redwood, for example. Normally, deck members are designed for about the same load as the floors in a dwelling.

The arrangement of the structural members can vary somewhat because of orientation of the deck, position of the house, slope of the lot, etc. However, basically, the beams are supported by the posts (anchored to footings) which in turn support the floor joists (Figure 16-6). The deck boards are then fastened to the joists. When beams are spaced more closely together, the joists

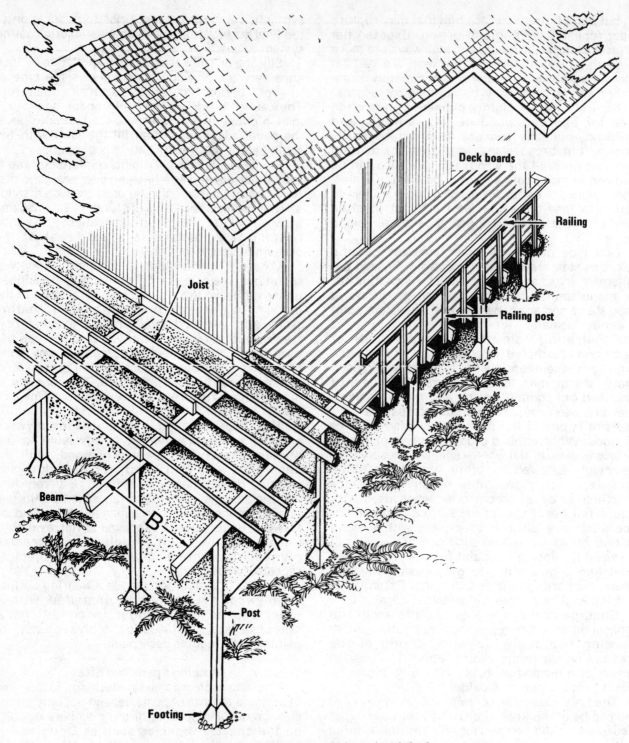

Member arrangement in a wood deck
Figure 16-6

can be eliminated if the deck boards are thick enough to span between the beams. Railings are located around the perimeter of the deck if required for safety (low level decks are often constructed without edge railings). When the deck is fastened to the house in some manner, the deck is normally rigid enough to eliminate the need for post bracing. In high free-standing decks, the use of post bracing is good practice.

Common sizes for wood posts used in supporting beams and floor framing for wood decks are 4 by 4, 4 by 6, and 6 by 6 inches. The size of the post required is based on the span and spacing of the beams, the load, and the height of the post. Most decks are designed for a live load of 40 pounds per square foot with an additional allowance of 10 pounds per square foot for the weight of the material. The suggested sizes of

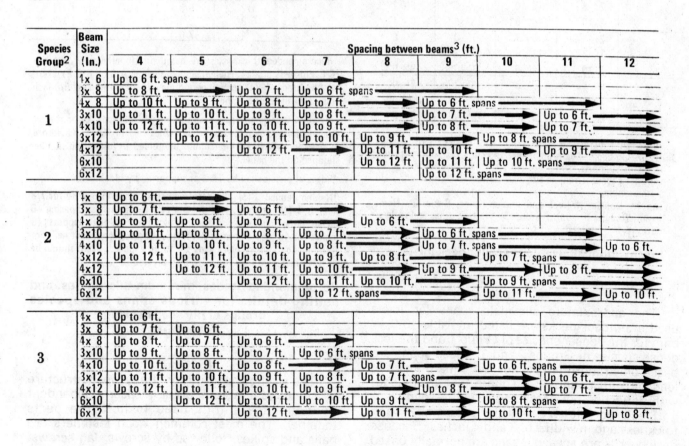

Species group[2]	Post Size (Inch)	Load area[3] beam spacing x post spacing (sq. ft.)									
		36	48	60	72	84	96	108	120	132	144
1	4x4	Up to 12 ft. heights →				Up to 10 ft. heights →			Up to 8 ft. heights →		
	4x6					Up to 12 ft. heights →			Up to 10 ft. →		
	6x6									Up to 12 ft. →	
2	4x4	Up to 12 ft. →		Up to 10 ft. hts. →		Up to 8 ft. heights →					
	4x6			Up to 12 ft. hts. →			Up to 10 ft. heights →				
	6x6						Up to 12 ft. heights →				
3	4x4	Up to 12	Up to 10 ft. →		Up to 8 ft. hts. →		Up to 6 ft. heights →				
	4x6		Up to 12 ft. →		Up to 10 ft. hts. →		Up to 8 ft. heights →				
	6x6				Up to 12 ft. hts. →						

[1] Based on 40 p.s.f. deck live load plus 10 p.s.f. dead load. Grade is Standard and Better for 4x4 inch posts and No. 1 and Better for larger sizes.

[2] Group 1 - Douglas fir-larch and southern pine; Group 2 - Hem-fir and Douglas-fir south; Group 3 - Western pines and cedars, redwood, and spruces.

[3] Example: If the beam supports are spaced 8 feet, 6 inches, on center and the posts are 11 feet, 6 inches, on center, then the load area is 98. Use next larger area 108.

Minimum post sizes [wood beam supports][1]
Table 16-7

Species Group[2]	Beam Size (In.)	Spacing between beams[3] (ft.)								
		4	5	6	7	8	9	10	11	12
1	4x 6	Up to 6 ft. spans →								
	3x 8	Up to 8 ft. →		Up to 7 ft.	Up to 6 ft. spans →					
	4x 8	Up to 10 ft.	Up to 9 ft.	Up to 8 ft.	Up to 7 ft. →		Up to 6 ft. spans →			
	3x10	Up to 11 ft.	Up to 10 ft.	Up to 9 ft.	Up to 8 ft. →		Up to 7 ft. →		Up to 6 ft. →	
	4x10	Up to 12 ft.	Up to 11 ft.	Up to 10 ft.	Up to 9 ft. →		Up to 8 ft. →		Up to 7 ft. →	
	3x12		Up to 12 ft.	Up to 11 ft	Up to 10 ft.	Up to 9 ft. →		Up to 8 ft. spans →		
	4x12			Up to 12 ft. →		Up to 11 ft.	Up to 10 ft. →		Up to 9 ft. →	
	6x10					Up to 12 ft.	Up to 11 ft.	Up to 10 ft. spans →		
	6x12						Up to 12 ft. spans →			
2	4x 6	Up to 6 ft. →								
	3x 8	Up to 7 ft. →		Up to 6 ft. →						
	4x 8	Up to 9 ft.	Up to 8 ft.	Up to 7 ft.		Up to 6 ft. →				
	3x10	Up to 10 ft.	Up to 9 ft.	Up to 8 ft.	Up to 7 ft. →		Up to 6 ft. spans →			
	4x10	Up to 11 ft.	Up to 10 ft.	Up to 9 ft.	Up to 8 ft. →		Up to 7 ft. spans →			Up to 6 ft.
	3x12	Up to 12 ft.	Up to 11 ft.	Up to 10 ft.	Up to 9 ft.	Up to 8 ft. →		Up to 7 ft. spans →		
	4x12		Up to 12 ft.	Up to 11 ft.	Up to 10 ft. →		Up to 9 ft. →		Up to 8 ft. →	
	6x10			Up to 12 ft.	Up to 11 ft.	Up to 10 ft. →		Up to 9 ft. spans →		
	6x12				Up to 12 ft. spans →		Up to 11 ft. →			Up to 10 ft.
3	4x 6	Up to 6 ft.								
	3x 8	Up to 7 ft.	Up to 6 ft.							
	4x 8	Up to 8 ft.	Up to 7 ft.	Up to 6 ft. →						
	3x10	Up to 9 ft.	Up to 8 ft.	Up to 7 ft.	Up to 6 ft. spans →					
	4x10	Up to 10 ft.	Up to 9 ft.	Up to 8 ft.		Up to 7 ft. →		Up to 6 ft. spans →		
	3x12	Up to 11 ft.	Up to 10 ft.	Up to 9 ft.	Up to 8 ft.	Up to 7 ft. spans →			Up to 6 ft. →	
	4x12	Up to 12 ft.	Up to 11 ft.	Up to 10 ft.	Up to 9 ft. →		Up to 8 ft. →		Up to 7 ft. →	
	6x10		Up to 12 ft.	Up to 11 ft.	Up to 10 ft.	Up to 9 ft. →		Up to 8 ft. spans →		
	6x12			Up to 12 ft. →		Up to 11 ft.	Up to 10 ft. →		Up to 8 ft.

[1] Beams are on edge. Spans are center to center distances between posts or supports. (Based on 40 p.s.f. deck live load plus 10 p.s.f. dead load. Grade is No. 2 or Better; No. 2, medium grain southern pine.)

[2] Group 1 - Douglas fir-larch and southern pine; Group 2 - Hem-fir and Douglas-fir south; Group 3 - Western pines and cedars, redwood, and spruces.

[3] Example: If the beams are 9 feet, 8 inches apart and the species is Group 2, use the I0 ft. column; 3x10 up to 6 ft. spans, 4x10 or 3x12 up to 7 ft. spans, 4x12 or 6x10 up to 9 ft. spans, 6x12 up to 11 ft. spans.

Minimum beam sizes and spans[1]
Table 16-8

posts required for various heights under several beam spans and spacings are listed in Table 16-7. Under normal conditions, the minimum dimension of the post should be the same as the beam width to simplify the method of fastening the two together. Thus a 4 by 8 inch (on edge) beam might use a 4 by 4 inch or a 4 by 6 inch post depending on the height, etc.

The nominal sizes of beams for various spacings and spans are listed in Table 16-8. These sizes are based on such species as Douglas fir, southern pine, and western larch for one group, western hemlock and white fir for a second group, and the soft pines, cedars, spruces, and redwood for a third group. Lumber grade is Number 2 or Better.

Species group[2]	Maximum allowable span (inches)[3]					
	Laid flat				Laid on edge	
	2x4	2x2	2x3	2x4	2x3	2x4
1	16	60	60	60	90	144
2	14	48	48	48	78	120
3	12	42	42	42	66	108

[1] These spans are based on the assumption that more than one floor board carries normal loads. If concentrated loads are a rule, spans should be reduced accordingly.

[2] Group 1 - Douglas-fir-larch and southern pine; Group 2 - Hem-fir and Douglas-fir south; Group 3 - Western pines and cedars, redwood, and spruces.

[3] Based on Construction grade or Better (Select Structural, Appearance, No. 1 or No. 2).

Maximum allowable spans for spaced deck boards[1] Table 16-10

Plywood species group[2]	Panel thicknesses in inches[3] [4]			
	For maximum spacings between supports (inches)			
	16	20	24	32 or 48
1	1/2	5/8	3/4	1-1/8
2 & 3	5/8	3/4	7/8	1-1/8
4	3/4	7/8	1	5

[1] Recommended thicknesses are based on Underlayment Exterior (C-C Plugged) grade. Higher grades, such as A-C or B-C Exterior, may be used. 19/32 inch plywood may be substituted for 5/8 inch and 23/32 inch for 3/4 inch.

[2] Plywood species groups are approximately the same but not identical to those shown for lumber in Tables 16-7 to 16-10. Therefore, in selecting plywood, one should be guided by the group number stamped on the panel.

[3] Edges of panels shall be T&G or supported by blocking.

[4] Nailing details: Size - 6d deformed shank nails, except 8d for 7/8 inch or 1-1/8 inch plywood on spans 24 to 48 inches. Spacing - 6 inches along panel edges, 10 inches along intermediate supports (6 inches for 48 inch on center supports). Corrosion resistant nails are recommended where nail heads are to be exposed. Nails should be set 1/16 inch (1/8 inch for 1-1/8 inch plywood).

Recommended grades, minimum thicknesses, and nailing details for various spans and species groups of plywood decking[1] Table 16-11

Species group[2]	Joist Size (Inches)	Joist spacing (inches)		
		16	24	32
1	2x 6	9'- 9"	7'-11"	6'- 2"
	2x 8	12'-10"	10'- 6"	8'- 1"
	2x10	16'- 5"	13'- 4"	10'- 4"
2	2x 6	8'- 7"	7'- 0"	5'- 8"
	2x 8	11'- 4"	9'- 3"	7'- 6"
	2x10	14'- 6"	11'-10"	9'- 6"
3	2x 6	7'- 9"	6'- 2"	5'- 0"
	2x 8	10'- 2"	8'- 1"	6'- 8"
	2x10	13'- 0"	10'- 4"	8'- 6"

[1] Joists are on edge. Spans are center to center distances between beams or supports. Based on 40 p.s.f. deck live loads plus 10 p.s.f. dead load. Grade is No. 2 or Better; No. 2 medium grain southern pine.

[2] Group 1 - Douglas fir-larch and southern pine; Group 2 - Hem-fir and Douglas-fir south; Group 3 - Western pines and cedars, redwood, and spruces.

Maximum allowable spans for deck joists[1] Table 16-9

The approximate allowable spans for joists used in outdoor decks are listed in Table 16-9 — both for the denser species of Group 1 and the less dense species of Groups 2 and 3. These spans are based on strength (40 pounds per square foot live load plus 10 pounds per square foot dead load) with deflection not exceeding 1/360 of span.

Deck boards are mainly used in 2 inch thickness and in widths of 3 and 4 inches. Because deck boards are spaced, spans are normally based on the width of each board as well as its thickness. (Roof decking, with tongue and groove edges and laid up tight, has greater allowable spans than spaced boards.) Decking can also be made of 2 by 3 inch or 2 by 4 inch members placed on edge, or of 1 by 4 inch boards. Deck board spans are listed in Table 16-10.

Spans for plywood decks are shown in Table 16-11.

Fasteners

The strength and utility of any wood structure or component are in great measure dependent upon the fastenings used to hold the parts together. The most common wood fasteners are nails and spikes, followed by screws, lag screws, bolts, and metal connectors and straps of various shapes. An important factor for outdoor use of fasteners is the finish selected. Metal fasteners should be rust-proofed in some manner or made of rust-resistant metals. Galvanized and cadmium plated finishes are the most common. Aluminum, stainless steel, copper, brass, and other rust-proof fasteners are also satisfactory. The most successful for such species as redwood are hot-dip

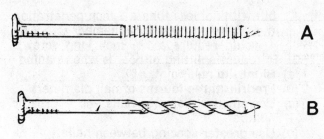

Deformed shank nails. A, annular grooved [ring shank]; B, spirally grooved
Figure 16-12

galvanized, aluminum, or stainless steel fasteners. These prevent staining of the wood under exposed conditions. A rusted nail, washer, or bolt head is not only unsightly but difficult to remove and replace. They are often a factor in the loss of strength of the connection.

Among the nails, smooth shank nails often lose their holding power when exposed to wetting and drying cycles. The best assurance of a high retained withdrawal resistance is the use of a deformed shank nail or spike. The two general types most satisfactory are (a) the annular grooved (ring shank) and (b) the spirally grooved nail (Figure 16-12). The value of such a nail or spike is its capacity to retain withdrawal resistance even after repeated wetting and drying cycles. Such nails should be used for the construction of exposed units if screws, lag screws, or bolts are not used.

The following tabulation lists the sizes of common nails ordinarily used in construction of outdoor wood structures. (Note: Sinker and cooler nails are one-eighth to one-fourth inch shorter.)

	Nail size (penny)	Nail length (inches)
	4	1½
	6	2
	7	2¼
	8	2½
	10	3
	12	3¼
	16	3½
	20	4
Usually classed as spikes	30	4½
	40	5
	50	5½
	60	6

Wood screws may be used if cost is not a factor in areas where nails are normally specified. Wood screws retain their withdrawal resistance to a great extent under adverse conditions. They are also superior to nails when end-grain fastening must be used. Because of their larger diameter, screw length need not be as great as a deformed shank nail. The flathead screw is best for exposed surfaces because it does not extend beyond the

surface (Figure 16-13), and the oval head protrudes less than the round head screw. This is an important factor in the construction of tables and benches. The use of a lead hole about three-fourths the diameter of the screw is good practice especially in the denser woods to prevent splitting. Screws should always be turned in their full length and not driven part way. The new variable speed drills (with a screwdriver bit) are excellent for applying screws.

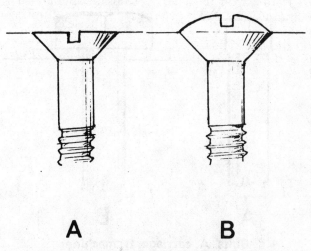

A   **B**

Wood screws. A, flat head; B, oval head
Figure 16-13

Lag screws are commonly used to fasten a relatively thick member, such as a 2 by 6 to a thicker member (3 or more inches) where a through bolt cannot be used. Lead holes must be used, and the lag screw turned in its entire length. Use a large washer under the head. Lead holes for the threaded portion should be about two-thirds the diameter of the lag screw for the softer woods such as redwood or cedar, and three-fourths the diameter for the dense hardwoods and for such species as Douglas fir. The lead hole for the unthreaded shank of the lag screw should be the same diameter as that of the lag screw.

Bolts are one of the most rigid fasteners in a simple form. They may be used for small connections such as railings-to-posts and for large members when combined with timber connectors. The two types of bolts most commonly used in light frame construction are the carriage bolt and the machine bolt (Figure 16-14). When obtainable, the step bolt is preferred over the carriage bolt because of its larger head diameter.

The carriage bolt is normally used without a washer under the head. A squared section at the bolt head resists turning as it is tightened. Washers should always be used under the head of the machine bolt and under the nut of both types. Bolt holes should be the exact diameter of the bolt. When a bolt-fastened member is loaded, such as a beam to a post, the bearing strength of the wood under the bolt is important, as well as the strength of the bolt. A larger diameter bolt or several smaller diameter bolts may be used when the softer woods are involved. Crushing of wood under the head of a carriage bolt or under the washer of any bolt should always be avoided. The use of larger washers and a washer under the carriage bolt head is advisable when the less dense wood species are used.

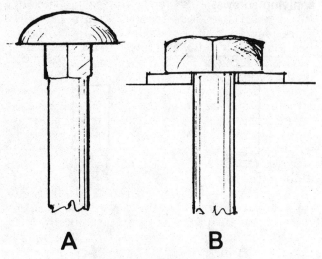

Bolts. A, carriage; B, machine
Figure 16-14

Miscellaneous fastening methods in addition to the nail, screw, and bolt are also used for fastening wood members together, or to other materials. Although split ring connectors and similar fasteners are normally used for large beams or trusses, other connectors may be used to advantage in the construction of a wood deck. These include metal anchors for connecting posts to concrete footings; angle irons and special connectors for fastening posts to beams; joist hangers and metal strapping for fastening joists to beams; and others. While research has not advanced far enough as yet, the new mastic adhesives are showing promise for field assembly of certain wood members. Such materials used alone or with metal fasteners will likely result in longer lived connections.

General Guide for Fastening
1. Use non-staining fasteners.
2. Always fasten a thinner member to a thicker member (unless clinched nails are used).
   - (a) A nail should be long enough to penetrate the receiving member a distance twice the thickness of the thinner member but not less than 1½ inches (i.e., for a ¾ inch board, the nail should penetrate the receiving member 1½ inches. Use at least a 7-penny nail).
   - (b) A screw should be long enough to penetrate the receiving member at least the thickness of the thinner (outside) member

but with not less than a 1 inch penetration (i.e., fastening a ¾ inch member to a 2 by 4 would require a 1¾ inch long screw).
3. To reduce splitting of boards when nailing-
   - (a) Blunt the nail point.
   - (b) Predrill (three-fourths of nail diameter).
   - (c) Use smaller diameter nails and a greater number.
   - (d) Use greater spacing between nails.
   - (e) Stagger nails in each row.
   - (f) Place nails no closer to edge than one-half of the board thickness and no closer to end than the board thickness.
   - (g) In wide boards (8 inches or more), do not place nails close to the edge.
4. Use a minimum of two nails per board — i.e., two nails for 4 and 6 inch widths and three nails for 8 and 10 inch widths.
5. Avoid end grain nailing. When unavoidable, use screws or a side grain wood cleat adjacent to the end grain member (as a post).
6. Lag screw use—
   - (a) Use a plain, flat washer under the head.
   - (b) Use a lead hole and turn in the full distance; do not overturn.
   - (c) Do not countersink (reduces wood section).
7. Bolt use—
   - (a) Use flat washers under the nut and head of machine bolts and under the nut of a carriage bolt. In softer woods, use a larger washer under carriage bolt heads.
   - (b) Holes must be the exact size of the bolt diameter.

Preparing The Site
Site preparation for construction of a wood deck is often less costly than that for a concrete terrace. When the site is steep, it is difficult to grade and to treat the backslopes in preparing a base for the concrete slab. In grading the site for a wood deck, you must normally consider only proper drainage, disturbing the natural terrain as little as possible. Grading should be enough to insure water runoff, usually just a minor leveling of the ground.

Often, absorption of the soil under an open deck with spaced boards will account for a good part of a moderate rainfall. If the deck also serves as a roof for a garage, carport, or living area below, drainage should be treated as a part of the house drainage, whether by gutters, downspouts, or drip and drain pockets at the ground level. In such cases, some form of drainage may be required to carry water away from the site and prevent erosion. This can usually be accomplished with drain tile laid in a shallow drainage ditch (Figure 16-15). Tile should be spaced and joints covered with a strip of asphalt felt before the trench is filled. The tile can lead to a dry well or to a drainage field beyond the site. Perforated cement or plastic tile is also available for this use.

There may also be a need for control of weed growth beneath the deck. Without some control or

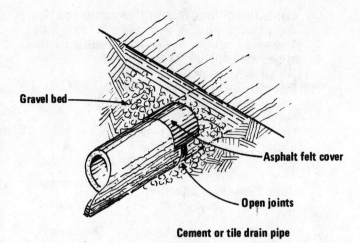

Gravel bed

Asphalt felt cover

Open joints

Cement or tile drain pipe

**Drain tile
Figure 16-15**

deterrent, such growth can lead to high moisture content of wood members and subsequent decay hazards where decks are near the grade. Common methods for such control consist of (a) the application of a weed killer to the plants or (b) the use of a membrane such as 4 or 6 mil polyethylene or 30 pound asphalt saturated felt. Such coverings should be placed just before the deck boards are laid. Stones, bricks, or other permanent means of anchoring the membranes in place should be used around the perimeter and in any interior surface variations which may be present. A few holes should be punched in the covering so that a good share of the rain will not run off and cause erosion.

Footings

Some type of footing is required to support the posts or poles which transfer the deck loads to the ground. In simplest form, the bottom of a treated pole and the friction of the earth around the pole provide this support. More commonly, however, some type of masonry, usually concrete, is used as a footing upon which the poles or posts rest. Several footing systems are normally used, some more preferred than others.

Footings required for support of vertical members such as wood poles or posts must be designed to carry the load of the deck super-structure. In a simple form, the design includes the use of pressure-treated posts or poles embedded to a depth which provides sufficient bearing and rigidity (Figure 16-16). This may require a depth of 3 to 5 feet or more, depending on the exposed pole height and applied loads. This type is perhaps more commonly used for pole structures such as storage sheds or barns. In areas where frost is a problem, such as in the Northern Sates, an embedment depth of 4 feet is commonly a minimum. But a lesser depth may be adequate in warmer climates. Soil should be well tamped around the pole.

Concrete footings below the surface are normally used for treated posts or poles. Two such types may be used. The first consists of a pre-poured footing upon which the wood members rest (Figure 16-17). Embedment depth should be only enough to provide lateral resistance, usually 2 to 3 feet. The exception is in cold climates where frost may penetrate to a depth of 4 feet or more. Minimum size for concrete footings in normal soils should be 12 by 12 by 8 inches. Where spacing of the poles is over about 6 feet, 20 by 20 by 10 inches or larger sizes are preferred. However, soil capacities should be determined before design.

Another type of below grade footing is the poured-in-place type shown in Figure 16-18. In such construction, the poles are pre-aligned, plumbed, and supported above the bottom of the

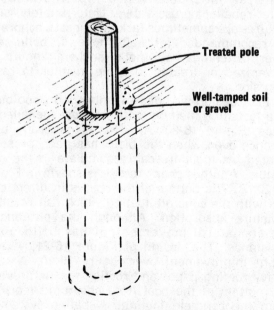

Treated pole

Well-tamped soil or gravel

**Pole without footing
Figure 16-16**

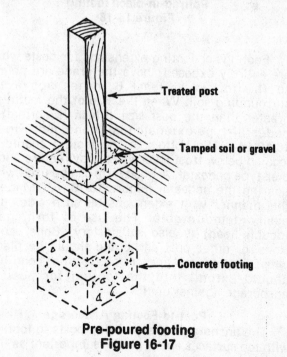

Treated post

Tamped soil or gravel

Concrete footing

**Pre-poured footing
Figure 16-17**

253

excavated hole. Concrete is then poured below and around the butt end of the pole. A minimum thickness of 8 inches of concrete below the bottom of the pole is advisable. Soil may be added above the concrete when necessary for protection in cold weather. Such footings do not require tamped soil around the pole to provide lateral resistance. All poles or posts embedded in the soil should always be pressure treated for long life.

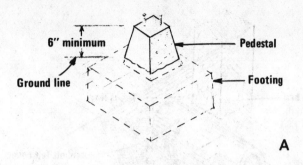

A

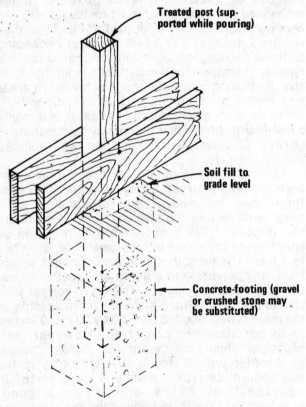

Poured-in-place footing
Figure 16-18

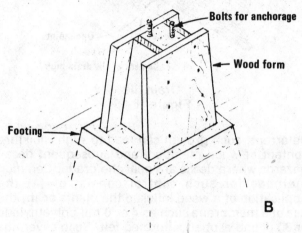

B

Pedestal footing extension. A, pedestal for post;
B, form used for pouring
Figure 16-19

Footings or footing extensions for posts which are entirely exposed above the grade are poured so the top is at least 6 inches above the surrounding soil. When the size of the footing is greater than the post size (which is normal), a pedestal-type extension is often used (Figure 16-19, A). The bottom of the footing should be located below frost level which may require a long pier-type pedestal. A wood form can be used when pouring the pedestal (Figure 16-19, B). Made in this manner with extensions on each side, it is easily demountable. The use of form nails (double-head) is also satisfactory. Bolts, angle irons, or other post anchorage should be placed when pouring, and anchor bolts or other bond bars should extend into the footing for positive anchorage against uplift.

Post-to-Footing Anchorage
The anchorage of supporting posts to footings with top surfaces above grade is important as they

should not only resist lateral movement but also uplift stresses which can occur during periods of high winds. These anchorages should be designed for good drainage and freedom from contact of the end-grain of the wood with wet concrete. This is advisable to prevent decay or damage to the bottom of the wood post. It is also important that the post ends be given a dip treatment of water-repellent preservative. Unfortunately, such features are sometimes lacking in post anchorage. As recommended for nails, screws, bolts, and other fastenings, all metal anchors should be galvanized or treated in some manner to resist corrosion.

Poor design includes an embedded wood block as a fastening member with the post toenailed in place (Figure 16-20, A). This is generally poor practice even when the block has been pressure treated, as moisture can accumulate in the post bottom. Another poor practice is shown in Figure 16-20, B. The bottom of the post is in direct contact with the concrete footing which can result in moisture absorption. Although the pin anchor resists lateral movement, it has little uplift resistance. The design of Figure 16-21, A is a slight improvement over Figure 16-20, A as a heavy roofing paper and roofing mastic prevents the bottom of the post from absorbing moisture from the concrete footing. A better system of anchoring small 4 by 4 inch posts is shown in Figure 16-21, B. In such anchorage, a galvanized

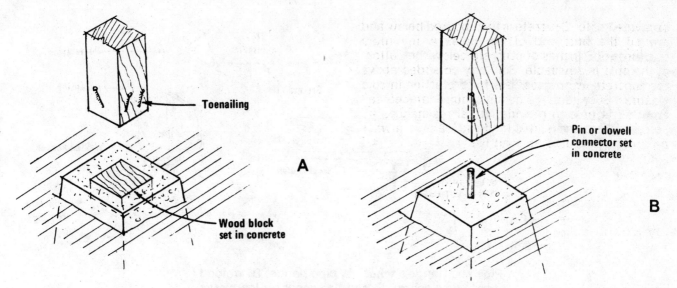

Post-to-footing anchorage. A and B, poor practice
Figure 16-20

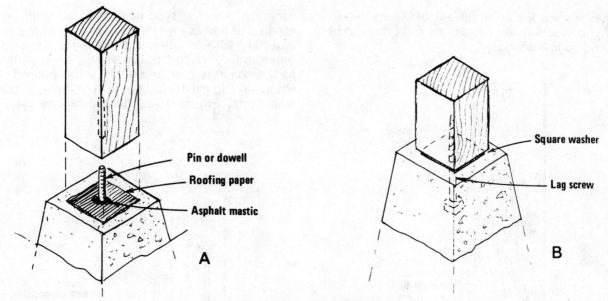

Post-to-footing anchorage. A and B,
improved practice
Figure 16-21

lag screw is turned into the bottom of the post with a large square washer (about 3 by 3 by ¼ inch thick for a 4 by 4 inch post) placed for a bearing area. The post is then anchored into a grouted pre-drilled hole or supported in place while concrete is poured. The washer prevents direct contact with the concrete and prevents moisture wicking into the bottom of the post, and the lag screw head provides some uplift resistance.

Good design is an anchorage system for supporting small posts, beams, stair treads, and similar members, utilizing a small steel pipe (galvanized or painted) with a pipe flange at each end (Figure 16-22, A). A welded plate or angle iron can be substituted for the pipe flange (Figure 16-22, B). The pipe flange or plate-to-post

connection should be made with large screws or lag screws. The flange can be fastened to the post bottom and turned in place after the concrete is poured (Figure 16-22, A). When an angle iron is used, the entire assembly is poured in place. A good anchor for beams used in low decks is shown in Figure 16-22, C.

Other post anchors can be obtained (or made up) for anchoring wood posts to a masonry base. Such anchors are normally used for solid 4 by 4 inch or larger posts. All are designed to provide lateral as well as uplift resistance. Some means such as a plate or supporting angle is provided to prevent contact of the post with the concrete, thus reducing the chances for decay. All holes drilled into posts for the purpose of anchorage should be

255

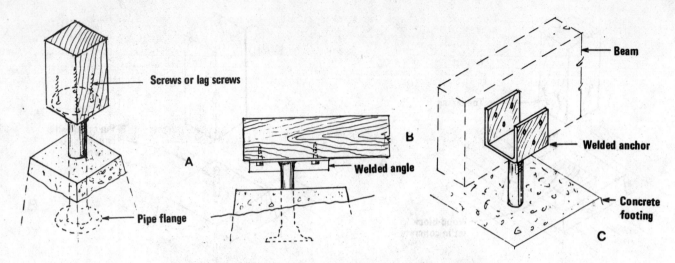

Pipe and flange anchor. A, pipe flange; B, welded
angle [low decks]; C, saddle anchor for low decks
Figure 16-22

flushed with a water-repellent preservative to
provide protection. An oil can is a good method of
applying such materials.

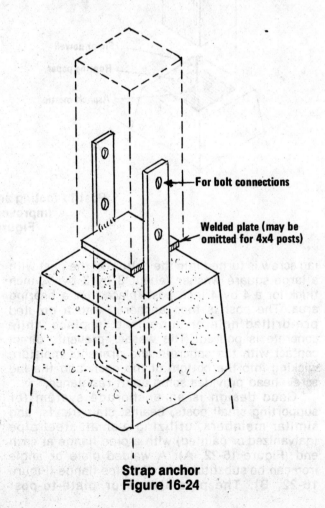

Step-flange anchor
Figure 16-23

One type of anchor is shown in Figure 16-23.
Post support is supplied by the anchor itself. This
step-flange anchor is positioned while the concrete
is being poured and should be located so that the
bottom of the post is about 2 inches above the

concrete. Another type of anchor for solid posts
consists of a heavy metal strap shaped in the form
of a "U" with or without a bearing plate welded
between (Figure 16-24). These anchors are placed
as the concrete pier or slab is being poured. As
shown in Figure 16-23, the post is held in place
with bolts. Figure 16-25 illustrates one type of

Strap anchor
Figure 16-24

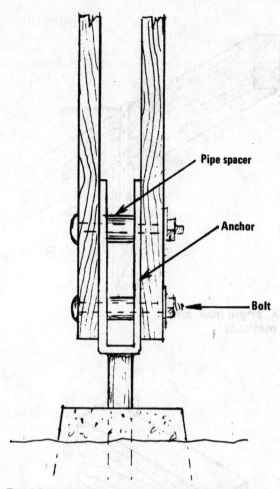

Double post anchor [without bearing plate]
Figure 16-25

anchor that may be used with double posts. In this and similar cases, the anchor in the concrete is positioned during the pouring operation.

Beam-to-Post Connection

Beams are members to which the floor boards are directly fastened or which support a system of joists. Such beams must be fastened to the supporting posts. Beams may be single large or small members or consist of two smaller members fastened to each side of the posts. When a solid deck is to be constructed, the beams should be sloped at least 1 inch in 8 to 10 feet away from the house.

Single beams when 4 inches or wider usually bear on a post. When this system is used, the posts must be trimmed evenly so that the beam bears on all posts. Use a line level or other method to establish this alignment.

A simple but poor method of fastening a 4 by 4 inch post to a 4 by 8 inch beam, is by toenailing (Figure 16-26). This is poor practice and should be avoided. Splitting can occur which reduces the strength of the joint. It is also inadequate in resisting twisting of the beam. A better system is by the use of a 1 by 4 inch lumber or plywood (Exterior grade) cleat located on two sides of the

post (Figure 16-26, B). Cleats are nailed to the beam and post with 7-penny of 8-penny deformed shank nails.

A good method of post-to-beam connection is by the use of a metal angle at each side (Figure 16-27, A). A 3 by 3 inch angle or larger should be used so that fasteners can be turned in easily. Use lag screws to fasten them in place. A metal strap fastened to the beam and the post might also be used for single beams (Figure 16-27, B). A 1/8 by 3 inch or larger strap, pre-formed to insure a good fit, will provide an adequate connection. Use 10-penny deformed shank nails for the smaller members and ¼ inch lag screws for larger members.

A good method of connection for smaller posts and beams consists of a sheet metal flange which

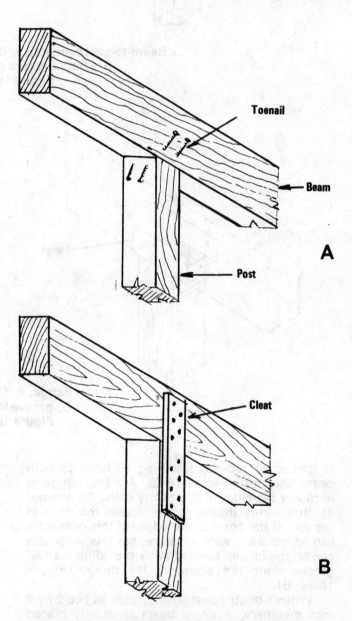

Beam-to-post connection. A, toenailing, a poor practice; B, better practice is to use cleat
Figure 16-26

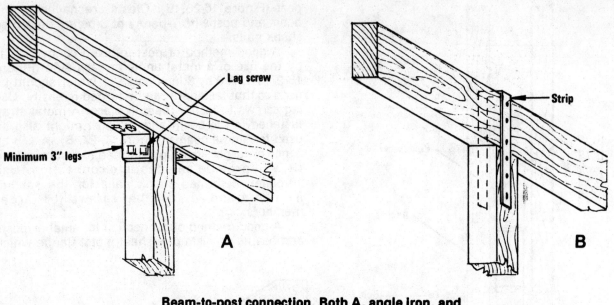

**Beam-to-post connection. Both A, angle iron, and
B, strapping, are good methods
Figure 16-27**

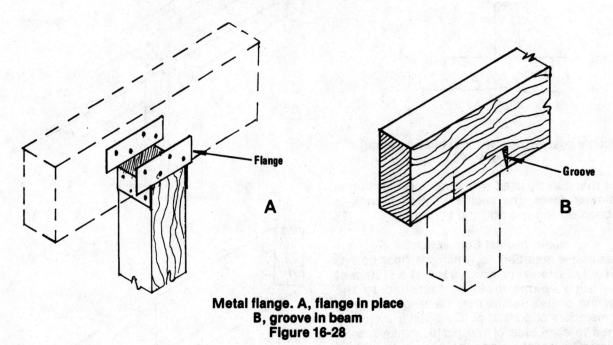

**Metal flange. A, flange in place
B, groove in beam
Figure 16-28**

is formed to provide fastening surfaces to both beam and post (Figure 16-28, A). The flange is normally fastened with 8-penny nails. To prevent splitting, nails should not be located too close to the end of the post. Upper edges of this connector can collect and retain moisture, but this weakness can be minimized somewhat by providing a small groove along the beam for the flange (Figure 16-28, B).

When a double post is used, such as two 2 by 6 inch members, a single beam is usually placed between them. One method of terminating the post ends is shown in Figure 16-29, A. This is not fully satisfactory as the end grain of the posts is

exposed. Some protection can be provided by placing asphalt felt or metal flashing over the joint. Fastening is done with bolts or lag screws. Another method of protection is by the use of cleats over the ends of the posts (Figure 16-29, B).

Double or split beams are normally bolted to the top of the posts, one on each side (Figure 16-30, A). As brought out previously, the load capacity of such a bolted joint depends on the bolt diameter, the number of bolts used, and the resistance of the wood under the bolts. Thus larger diameter bolts should be used to provide greater resistance for the less dense woods (i.e., ½ inch rather than 3/8 inch diameter). Notching the top of

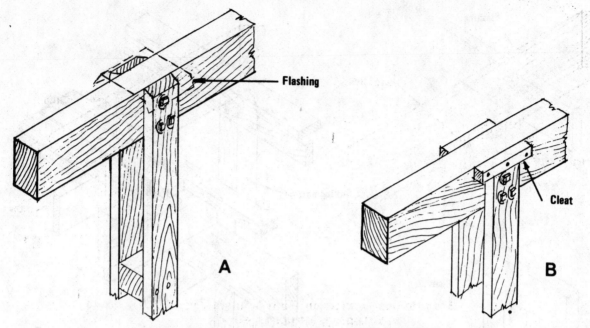

**Double post to beam. A, post connection with
flashing; B, post connection with cleat
Figure 16-29**

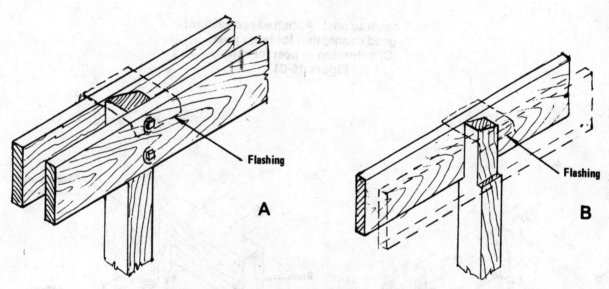

**Double beam to post. A, bolted joint with flashing;
B, notched and bolted joint with flashing
Figure 16-30**

the beam as shown in Figure 16-30, B provides greater load capacity. A piece of asphalt felt or a metal flashing over the joint will provide some protection for the post end.

Small single beams are occasionally used with larger dimension posts (i.e., 4 by 8 inch beam and 6 by 6 inch post). In such cases, one method of connection consists of bolting the beam directly to the supporting posts (Figure 16-31, A). Some type of flashing should be used over the end of the post.

Another method of connecting smaller beams to larger posts is shown in Figure 16-31, B. A short

section of angle iron is used on each side of the post for anchorage and a wood cleat is than placed to protect the exposed end grain of the larger post.

It is sometimes advantageous to use the post which supports the beam as a railing post. In such a design, the beam is bolted to the post which extends above the deck to support the railing members (Figure 16-31, C).

Beam-and-Joist-to-House Connections
When the deck is adjacent to the house, some method of connecting beams or joists to the house

259

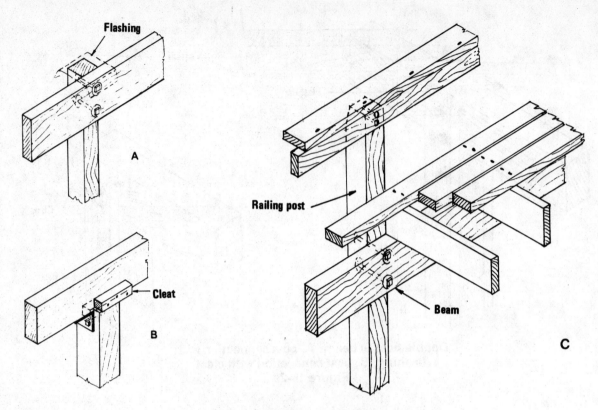

Small beam to post. A, bolted connection;
B, good connection for large post;
C, extension of post for rail
Figure 16-31

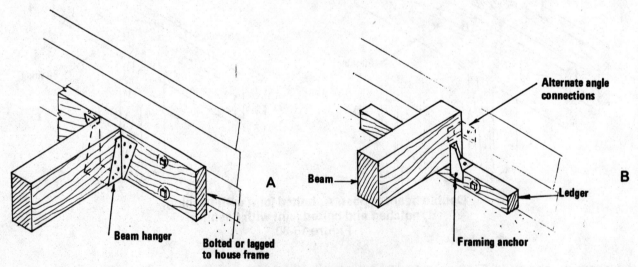

Beam to house. A, beam hanger;
B, ledger support
Figure 16-32

is normally required. This may consist of supporting such members through (a) metal hangers, (b) wood ledgers or angle irons, or (c) utilizing the top of the masonry foundation or basement wall. It is usually good practice to design the deck so that the top of the deck boards are just under the sill of the door leading to the

deck. This will provide protection from rains as well as easy access to the deck.

One method of connecting the beam to the house consists of the use of metal beam hangers (Figure 16-32, A). These may be fastened directly to a floor framing member such as a joist header or to a 2 by 8 inch or 2 by 10 inch member which has

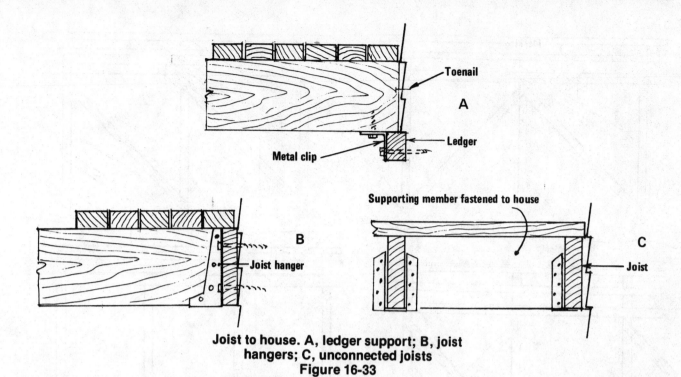

Joist to house. A, ledger support; B, joist hangers; C, unconnected joists
Figure 16-33

been bolted or lag-screwed to the house framing. Use 6-penny or longer nails or the short, large diameter nails often furnished with commercial hangers for fastening. Hangers are available for all beams up to 6 by 14 inches in size. In new construction, beam pockets or spaces between floor framing headers can be provided for the deck beam support. Beams can also be secured to the house proper by bearing on ledgers which have been anchored to the floor framing or to the masonry wall with expansion shields and lag screws. The beam should be fastened to the ledger or to the house with a framing anchor or a small metal angle (Figure 16-32, B).

When joists of the deck are perpendicular to the side or end of the house, they are connected in much the same manner as beams except that fasteners are smaller. The use of a ledger lag-screwed to the house is shown in Figure 16-33A. Joists are toenailed to the ledger and the house (header or stringer joists) or fastened with small metal clips.

Joists can also be fastened by a 2 by 8 inch or 2 by 10 inch member (lag-screwed to the house) by means of joist hangers (Figure 16-33, B). Six-penny nails or 1¼ inch galvanized roofing nails are used to fasten the hangers to the joist and to the header. When joists or beams are parallel to the house, no ledger or other fastening member is normally required (Figure 16-33, C). If they are supported by beams, the beams, of course, are then connected to the house, as previously illustrated.

Bracing

On uneven sites or sloping lots, posts are often 5 or more feet in height. When the deck is free (not attached to the house), it is good practice to use bracing between posts to provide lateral resistance. Treated poles or posts embedded in the soil or in concrete footings usually have sufficient resistance to lateral forces, and such construction normally requires no additional bracing. However, when posts rest directly on concrete footings or pedestals, and unsupported heights are more than about 5 feet, some system of bracing should be used. Braces between adjacent posts serve the same purpose as bracing in the walls of a house.

Special bracing in the horizontal plane is normally not needed for residential decks of moderate area and height. Decks can be braced efficiently in the horizontal plane by installing galvanized steel strap diagonals just under the deck surface. These should be in pairs in the direction of both diagonals and securely fastened at both ends. An alternative is to use flat 2 by 4 inch or 2 by 6 inch members across one diagonal, securely nailed to the underside of the deck members. In the case of a very large, high deck, it is advisable to consult a design engineer for an adequate bracing procedure.

Bracing should be used on each side of a "free" deck to provide racking resistance in each direction. Single bracing (one member per bay) should consist of 2 inch dimension material. When brace length is no greater than 8 feet, 2 by 4 inch members can be used; 2 by 6 inch braces should be used when lengths are over 8 feet. Fastenings should normally consist of lag screws or bolts (with washers) to fasten 2 inch braces to the posts.

One simple system of single bracing is known as the "W" brace which can be arranged as shown in Figure 16-34, A. Braces are lag-screwed to the post and joined along the centerline. When

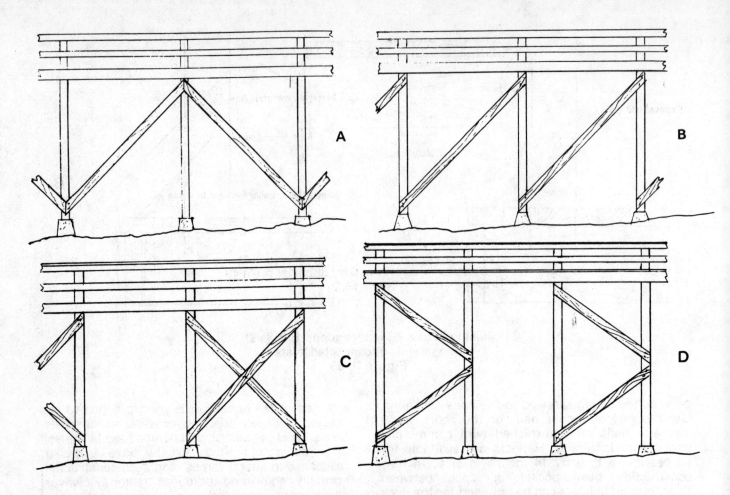

Bracing. A, "W" brace; B, single direction brace
C, cross brace; D, bracing for high posts
Figure 16-34

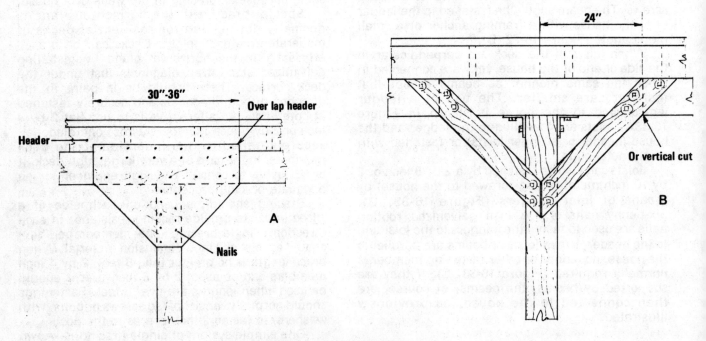

Partial braces. A, plywood gusset;
B, lumber brace
Figure 16-35

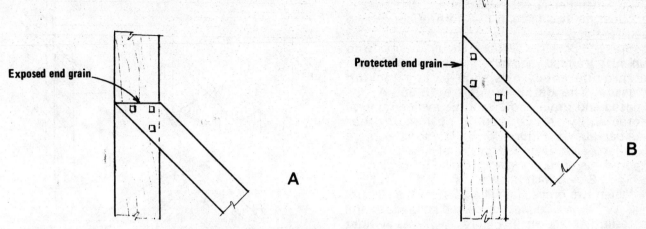

Brace cuts. A, poor practice; B, better practice
Figure 16-36

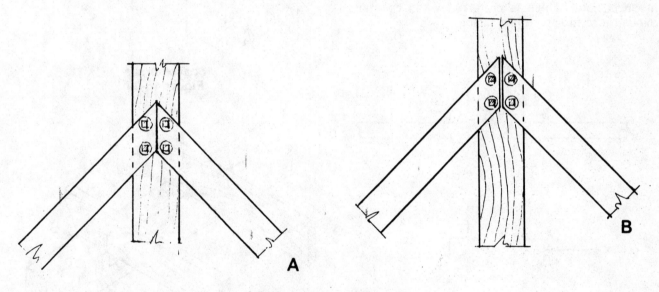

Joint at post. A, tight joint;
B, open joint [preferred]
Figure 16-37

desired and when space is available, braces can be placed on the inside of the posts.

Another single bracing method between posts is shown in Figure 16-34, B. Braces are located from the base of one post to the top of the adjacent posts. Braces on the adjacent side of the deck should be placed in the opposite direction.

Another system of bracing used between posts is the "X" or cross brace (Figure 16-34, C). When spans and heights of posts are quite great, a cross brace can be used at each bay. However, bracing at alternate bays is normally sufficient. A bolt may be used where the 2 inch braces cross to further stablize the bay. One inch thick lumber bracing is not recommended as it is subject to mechanical damage such as splitting at the nails.

When posts are about 14 feet or more in height, which could occur on very steep slopes, two braces might be required to avoid the use of too long a brace. Such bracing can be arranged as shown in Figure 16-34, D.

A plywood gusset brace, or one made of short lengths of nominal 2 inch lumber, can sometimes be used as a partial brace for moderate post heights of 5 to 7 feet. A plywood gusset on each side of a post can also serve as a means of connection between a post and beam (Figure 16-35, A). Use ¾ inch exterior type plywood and fasten to the post and beam with two rows of 10-penny nails. The top edge of the gusset should be protected by an edge or header member which extends over the plywood.

A partial brace made of 2 by 4 inch lumber can be secured to the beam and posts with lag screws or bolts as shown in Figure 16-35, B. Some member of the deck, such as the deck boards or a parallel edge member, can overlap the upper ends to protect the end grain from moisture. When an overlap member is not available and the area is sufficient for two fasteners, a vertical cut can be used for the brace.

Brace-to-post connections should be made to minimize trapped moisture or exposed end grain yet provide good resistance to any racking stresses. The detail in Figure 16-36, A has exposed end grain and should be avoided unless protected by an overlapping header or other member above. Figure 16-36, B shows a more acceptable cut. No end grain of the brace is exposed. use two lag screws (or bolts) for 2 by 4 inch and 2 by 6 inch braces.

When two braces join at a post, such as occurs in a ''W'' brace, connection should be made on the centerline as shown in Figure 16-37, A. A tight joint provides the resistance of all fasteners when one brace is in compression, but there is some hazard in trapped moisture. Figure 16-37, B shows a spaced joint which is preferred when constant exposure to moisture is a factor.

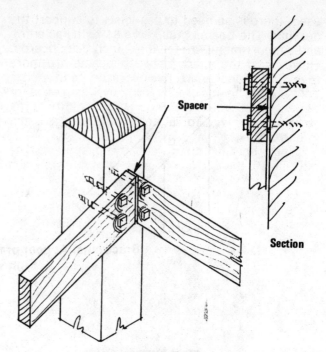

Spaced brace
Figure 16-39

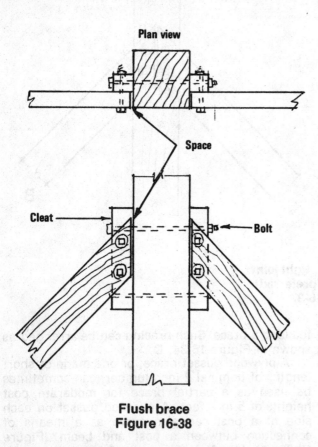

Flush brace
Figure 16-38

A flush brace may be used if desired from the standpoint of appearance (Figure 16-38). This type connection requires that a backing cleat be lagged or bolted to each side of the post. The braces are then fastened to the cleats as shown.

The use of large, galvanized washers or other means of isolating the brace from the post will provide a smaller area for trapping moisture behind the brace (Figure 16-39). Such a spacer at each bolt or lag screw might be used when the less decay-resistant wood species are involved.

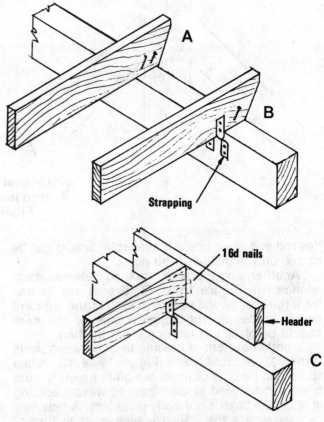

Joist-to-beam connection. A, toenail; B, strapping; C, connection with header
Figure 16-40

Joist-to-Beam Connections
When beams are spaced 2 to 5 feet apart and 2 by 4 inch Douglas fir or similar deck boards are

used, there is no need to use joists to support the decking. The beams thus serve as both fastening and support members for the 2 inch deck boards. However, if the spans between beams are more than 3½ to 5 feet apart, it is necessary to use joists between the beams or 2 by 3 or 2 by 4's on edge for decking (see Table 16-9). To provide rigidity to the structure, the joists must be fastened to the beam in one of several ways.

Joists bearing directly on the beams may be toenailed to the beam with one or two nails on each side (Figure 16-40, A). Use 10-penny nails and avoid splitting. When uplift stresses are inclined to be great in high wind areas, supplementary metal strapping might be used in addition to the toenailing (Figure 16-40, B). Use 24 to 26 gauge galvanized strapping and nail with 1 inch galvanized roofing nails. When a header is used at the joist ends, nail the header into the end of each joist (Figure 16-40, C). Have the header overhang the beam by one-half inch to provide a good drip edge.

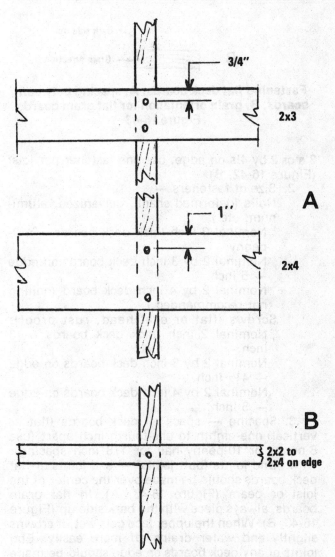

Joists between beams. A, ledger support; B, joist hanger support; C, joist hangers
Figure 16-41

Joists located between beams and flush with their tops may be connected in two manners. One utilizes a 2 by 3 inch or 2 by 4 inch ledger which is spiked to the beam. Joists are cut between beams and toenailed to the beams at each end (Figure 16-41, A). The joint can be improved by the use of small metal clips.

Another method utilizes a metal joist hanger (Figure 16-41, B). The hanger is first nailed to the end of the joist with 1 to 1¼ inch galvanized roofing nails and then nailed to the beam. Several

types of joist hangers are available (Figure 16-41, C).

Fastening Deck Boards

Deck boards are fastened to floor joists or to beams through their face with nails or screws. Screws are more costly to use than nails from the standpoint of material and labor but have greater resistance to loosening or withdrawal than the nail. A good compromise between the common smooth shank nail and the screw is the deformed shank nail. These nails retain their withdrawal resistance even under repeated wetting and drying cycles. Both nails and screws should be set flush or just below the surface of the deck board.

Fastening deck boards. A, flat deck boards;
B, deck boards on edge
Figure 16-42

Some good rules in fastening deck boards to the joist or beams are as follows:

1. Number of fasteners per deck board — use two fasteners for nominal 2 by 3 inch and 2 by 4 inch decking laid flat (Figure 16-42, A). For 2 by

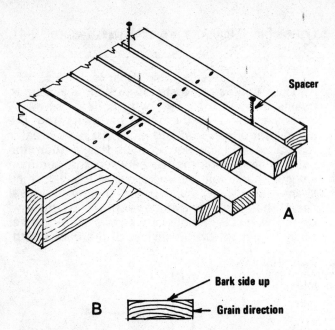

Spacer

Bark side up

B

Grain direction

Fastening flat deck boards. A, spacing between boards; B, grain orientation for flat grain boards
Figure 16-43

3's or 2 by 4's on edge, use one fastener per joist (Figure 16-42, B).

2. Size of fasteners —
   Nails (deformed shank, galvanized, aluminum, etc.):
      Nominal 2 inch thick deck boards — 12-penny.
      Nominal 2 by 3 inch deck boards on edge — 5 inch.
      Nominal 2 by 4 inch deck board (nailing not recommended.)
   Screws (flat or oval head, rust proof):
      Nominal 2 inch thick deck boards — 3 inch.
      Nominal 2 by 3 inch deck boards on edge — 4½ inch.
      Nominal 2 by 4 inch deck boards on edge — 5 inch.

3. Spacing — space all deck boards (flat or vertical) one-eighth to one-fourth inch apart (use 8-penny or 10-penny nail for 1/8 inch spacing).

4. End joints (butt joints) — end joints of flat deck boards should be made over the center of the joist or beam (Figure 16-43, A). In flat grain boards, always place with the bark side up (Figure 16-43, B). When the upper face gets wet, it crowns slightly and water drains off more easily. End joints of any deck boards on edge should be made over a spaced double joist (Figure 16-44, A), a 4 inch or wider single beam, or a nominal 2 inch joist with nailing cleats on each side (Figure 16-44, B).

When deck boards are used on edge, spacers between runs will aid in maintaining uniform spacing and can be made to effect lateral support between runs by using lateral nailing at the spacers. Spacers as shown in Figure 16-44, C are recommended between supports when spans

exceed 4 feet and should be placed so that no distance between supports or spacers exceeds 4 feet. An elastomeric construction adhesive or penta-grease on both faces of each spacer prevents water retention in the joints.

Always dip ends of deck boards in water-repellent preservative before installing.

Always pre-drill ends of 2 by 3 inch or 2 by 4 inch (flat) deck boards of the denser species, or when there is a tendency to split. Pre-drill when screws are used for fastening. Pre-drill all fastening points of 2 by 3 inch or 2 by 4 inch deck boards placed on edge.

To provide longer useful life for decks made of low to moderate decay-resistant species, use one or more of the following precautions:

1. Use spaced double joists or beams and place end joints between the joists (Figure 16-45).

2. Lay a strip of building felt saturated with a wood preservative over the beam or joist before installing deck boards.

3. Apply an elastomeric glue to the beam or joist edge before installing the deck boards.

4. Treat end joints of deck boards made over a support with yearly applications of a water-repellent preservative. (A plunger-type oil can will work well.)

Fastening Plywood

Plywood panels should generally be installed with a minimum 1/16 inch space between edge and end joints, using the support spacing and nailing schedule indicated in Table 16-11. When caulking is used, a joint space of at least one-fourth inch is usually required.

To avoid unnecessary moisture absorption by the plywood, seal all panel edges with an exterior primer or an aluminum paint formulated for wood. The panel edge sealant can be most conveniently applied prior to installation, while the plywood is still in stacks. Build some slope into the deck area to provide for adequate drainage. A minimum slope of 1 inch in 8 to 10 feet should be provided when installing the joists or beams.

Provide ventilation for the underside of the deck areas in all cases. For low level decks, this can be done by leaving the space between the joists open at the ends and by excavating material away from the support joists and beams. For high level decks over enclosed areas, holes can be drilled in the blocking between joists.

Railing Posts

Low level decks located just above the grade normally require no railings. However, if the site is sloped, some type of protective railing or system of balusters might be needed, because of the height of the deck.

The key members of a railing are the posts. Posts must be large enough and well fastened to provide strength to the railing. Some type of vertical member such as the post can also serve as a part of a bench or similar edge structure of the

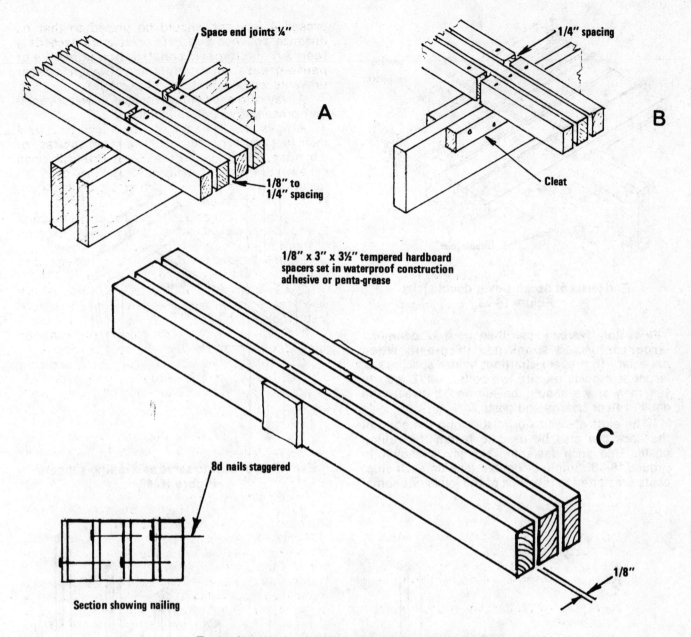

Space end joints ¼"

1/4" spacing

A

B

1/8" to
1/4" spacing

Cleat

1/8" x 3" x 3½" tempered hardboard
spacers set in waterproof construction
adhesive or penta-grease

C

8d nails staggered

1/8"

Section showing nailing

Fastening "on edge" deck boards. A, installing
over double joist or beam; B, installing over single
joist or beam; C, spacers for 2x4 decking on edge
Figure 16-44

deck. Railings should be designed for a lateral
load of at least 20 pounds per lineal foot. Thus,
posts must be rigid and spaced properly to resist
such loads.

One method of providing posts for the deck
railing is by the extension of the posts which
support the beams (Figure 16-46). When single or
double beams are fastened in this manner, the
posts can extend above the deck floor and serve for
fastening the railing and other horizontal members. Railing heights may vary between 30 and 40
inches, or higher when a bench or wind screen is
involved. Posts should be spaced no more than 6
feet apart for a 2 by 4 horizontal top rail and 8 feet
apart when a 2 by 6 or larger rail is used.

When supporting posts cannot be extended
above the deck, a joist or beam may be available to
which the posts can be secured. Posts can then be
arranged as shown in Figure 16-47, A. Such posts
can be made from 2 by 6's for spans less than 4
feet, from 4 by 4's or 2 by 8's for 4 to 6 foot spans,
and from 4 by 6's or 3 by 8's for 6 to 8 foot spans.
Each post should be bolted to the edge beam with
two 3/8 inch or larger bolts determined by the size
of the post. This system can also be used when the
railing consists of a number of small baluster type
posts (Figure 16-47, B). When such posts are
made of 2 by 2 or 2 by 3 inch members and spaced
12 to 16 inches apart, the top fastener into the
beam should be a ¼ or 3/8 inch bolt or lag screw.

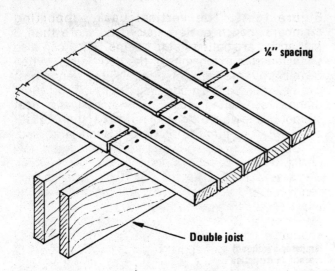

End joists of decking over double joist
Figure 16-45

The bottom fastener can then be a 12-penny or larger deformed shank nail. Pre-drill when necessary to prevent splitting. Wider spacings or larger size posts require two bolts. A 1/8 inch to 1/4 inch space should be allowed between the ends of floor boards and posts.

The ends of beams or joists along the edge of the deck can also be used to fasten the railing posts. One such fastening system is shown in Figure 16-48. Single or double (one on each side) posts are bolted to the ends of the joists or beams.

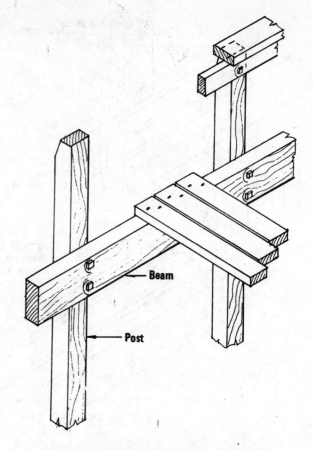

Extension of post to serve as a railing support
Figure 16-46

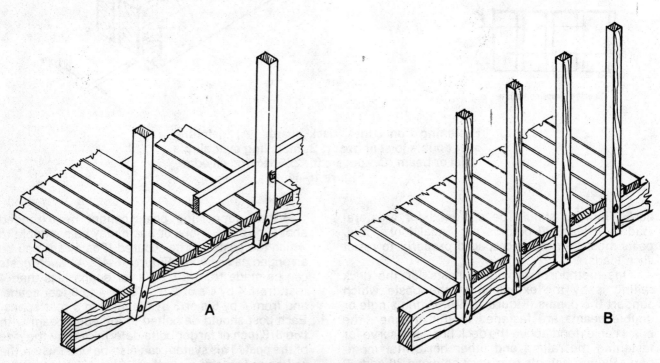

Railing posts fastened to edge of deck member.
A, spaced posts [4 feet and over];
B, baluster type posts
Figure 16-47

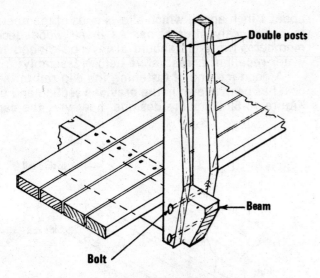

**Double railing posts at beam or joist ends
Figure 16-48**

Space the bolts as far apart as practical for better lateral resistance.

The practice of mounting posts on a deck board should be avoided. Not only is the railing structurally weak, but the bottom of the post has end grain contact with a flat surface. This could induce high moisture content and possible decay.

Deck Benches

At times there is an advantage in using a bench along the edge of a high deck, combining utility with protection. One such design is shown in

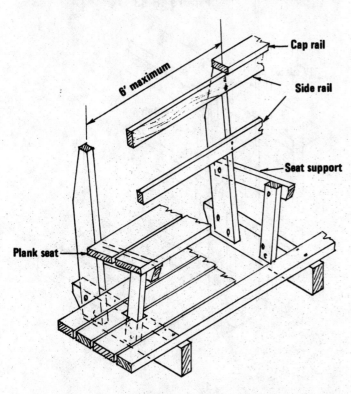

**Deck bench
Figure 16-49**

Figure 16-49. The vertical back supporting members (bench posts), spaced no more than 6 feet apart, are bolted to the beams. They can also be fastened to extensions of the floor joists. When beams are more than 6 feet apart, the bench post can be fastened to an edge joist in much the same manner as railing posts. The backs and seat supports should be spaced no more than 6 feet apart when nominal 2 inch plank seats are used.

Benches can also be used along the edge of low decks. These can be simple plank seats which serve as a back drop for the deck. Such bench seats require vertical members fastened to the joists or beams with cross cleats (Figure 16-50). For nominal 2 inch plank seats, vertical supports should be bolted to a joist or beam and be spaced no more than 6 feet apart. A single wide support (2 by 10) (Figure 16-50, A) or double (two 2 by 4's) supports (Figure 16-50, B) can be used. Cleats should be at least 2 by 3 inches in size.

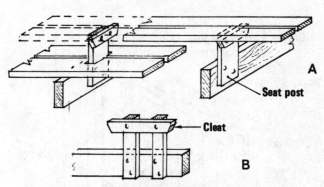

**Bench seats. A - single support and cross cleat;
B - double support
Figure 16-50**

Such member arrangements can also be used as a step between two decks with elevation differences of 12 to 16 inches. Many other bench arrangements are possible; but spans, fastenings, and elimination of end grain exposure should always be considered.

Railings

The top horizontal members of a railing should be arranged to protect the end grain of vertical members such as posts or balusters. A poorly designed railing detail is shown in Figure 16-51. Such details should be avoided, as the end grain of the baluster-type posts is exposed. Figure 16-52 is an improvement, as the end grain of the balusters is protected by the cap rail.

The upper side rail, which is usually a 2 by 4 inch or wider member, should be fastened to the posts with a lag screw or bolt at each crossing. The cap rail then can be nailed to the edge of the top rail with 12-penny deformed shank nails spaced 12 to 16 inches apart.

When railing posts are spaced more than about 2 feet apart, additional horizontal members may

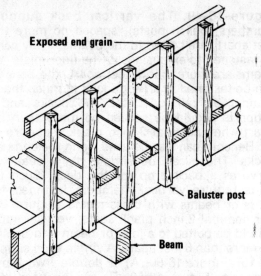

Poorly designed railing detail
Figure 16-51

Exposed end grain

Baluster post

Beam

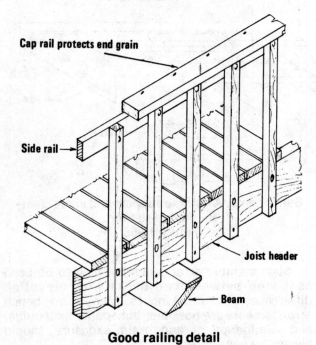

Cap rail protects end grain

Side rail

Joist header

Beam

Good railing detail
Figure 16-52

about 1 inch apart, which allows ends of members to dry quickly after rains. As in all wood deck members, the ends should always be dipped in water-repellent preservative before assembly.

A good method of fastening the cap rail to the post has been shown in the previous section and in Figure 16-52. In some designs, however, the cap

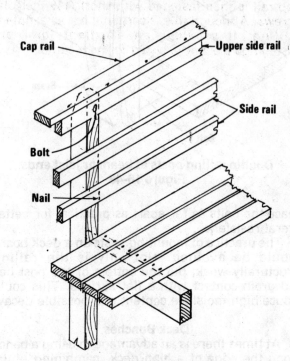

Cap rail

Upper side rail

Side rail

Bolt

Nail

Side rails for deck railing
Figure 16-53

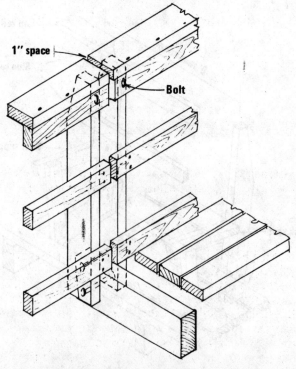

1" space

Bolt

Spaced rail joints - good practice
Figure 16-54

be required as a protective barrier (Figure 16-53). These side rails should be nominal 2 by 4 inch members when posts are spaced no more than 4 feet apart. Use 2 by 6's when posts are spaced over 4 feet apart.

When the upper side rail is bolted to the post (Figure 16-53), the remaining rails can be nailed to the posts. Use two 12-penny deformed shank nails at each post and splice side rails and all horizontal members at the center-line of a post. Posts must be more than 2 inches in thickness to provide an adequate fastening area at each side of the center splice.

A superior rail termination uses a double post (Figure 16-54). Horizontal members are spaced rail without additional members may be specified. An unsatisfactory method of connecting a cap rail to the post is by nailing (Figure 16-55, A). End grain nailing is not recommended in such connections. A better method is shown in Figure 16-55, B. Short lengths of galvanized angle irons are fastened to the post with lag screws or bolts. The cap rail is then fastened with short (1½ inch) lag screws. Although this is certainly not as simple as nailing, it provides an excellent joint and fastenings are not exposed to the weather.

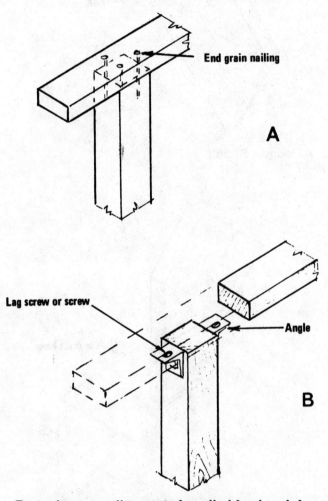

Fastening cap rail to post. A, nailed [end grain], a poor practice; B, angle iron connection, a good practice
Figure 16-55

There may be occasions in the construction of a railing of a deck to use members between the posts rather than lapping the posts. This might be in construction of an adjoining wind screen or mid-height railings between posts. Such connections might also be adaptable to fences where horizontal members are located between posts. The connection to the post is the important one, as it must be rigid as well as minimize areas where moisture could be trapped. Dado cuts for a 2 inch rail are shown in Figure 16-56, A and 16-56, B.

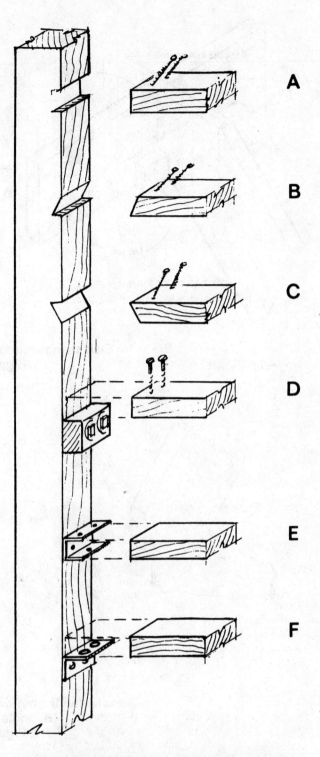

Rail-to-post connections. A-C, dado cuts [not recommended]; D, wood block support; E, connector; F, angle iron
Figure 16-56

Although these are reasonably good structurally, moisture could be retained at the end grain of the bottom cut. Figure 16-56, C shows the notch reversed. This will not retain moisture as much as the previous cuts, but the member must be cut precisely to provide a rigid joint. A wood block lag screwed to the side of the post serves as a good

271

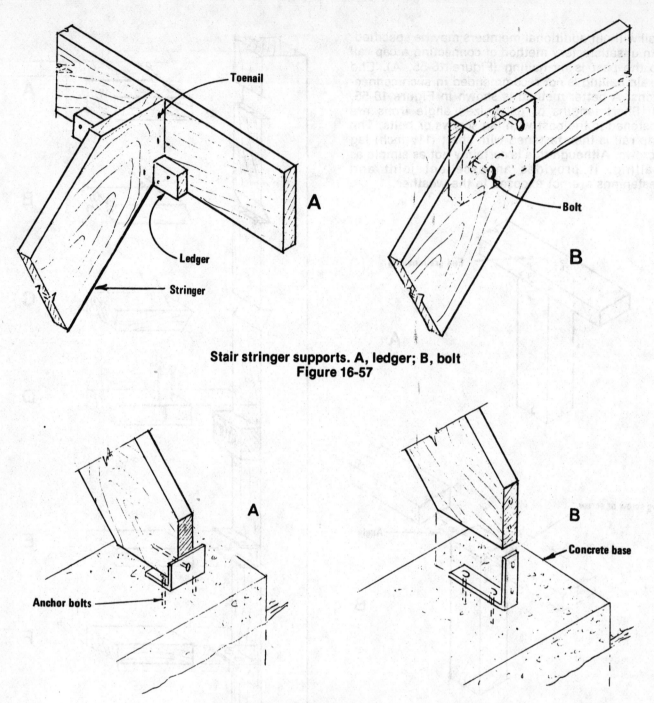

Stair stringer supports. A, ledger; B, bolt
Figure 16-57

Fastening the bottom of stair stringer. A and
B, angle iron anchors
Figure 16-58

fastening area for the rail (Figure 16-56, D). This is a good connection when the rail is spaced slightly away from the post. The rail should be fastened to the block with screws.

A commercial-type bracket is shown in Figure 16-56, E. This connector can also be used to advantage for 1 inch members used in a fence or a wind screen. Another good method utilizes a small angle iron lagged to the post (Figure 16-56, F). The rail is then fastened to the angle with lag screws from below.

Stairways

There is often a need for a stairway as an access to a deck or for use between decks with different levels. Exterior stairs are much the same as stairs within a house, except that details which avoid trapped moisture or exposed end grain of the members should be used.

Research has indicated that for woods with moderate to low decay resistance, a three minute dip in a water repellent preservative for all members at least tripled the average service life of

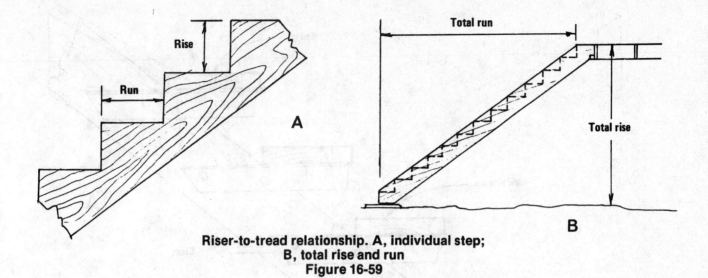

**Riser-to-tread relationship. A, individual step;
B, total rise and run
Figure 16-59**

exterior stairways and their parts. Use of all heartwood of decay resistant species or of pressure treated wood will insure even longer life.

A basic stair consists of stair stringers (sometimes called the stair carriage) and treads. Additional parts include balusters and side cap rails and, on occasion, risers. The supporting members of a stair are the stringers. Stringers are used in pairs spaced no more than 3 feet apart. They are usually made of 2 by 10 inch or 2 by 12 inch members. Stringers must be well secured to the framing of the deck. They are normally supported by a ledger or by the extension of a joist or beam. A 2 by 3 inch or 2 by 4 inch ledger nailed to the bottom of an edge framing member with 12-penny nails supports the notched stringer (Figure 16-57, A). Toenailing or small metal clips are used to secure the carriage in place. Stair stringers can also be bolted to the ends of joists or beams when they are spaced no more than about 3 feet apart (Figure 16-57, B). Use at least two ½ inch galvanized bolts to fasten the stringer to the beam or joist.

The bottom of the stair stringers should be anchored to a solid base and be isolated from any moisture source. Two systems frequently used consist of metal angles anchored to a concrete base (Figures 16-58, A and B). The angles should be thick enough to raise the stringer off the concrete, which should also be sloped for drainage. They might also be fastened to a treated wood member anchored in the concrete or in the ground.

The relation of the tread width ("run") to the riser height is important in determining the number of steps required. For ease of ascent, the rise of each step in inches times the width of the tread in inches should equal 72 to 75 (Figure 16-59, A). Thus, if the riser is 8 inches (considered maximum for stairs), the tread would be 9 inches. Or if the riser is 7½ inches, the tread should be about 10 inches. Thus, the number of risers and treads can be found when the total height of the

stair is known. Divide the total rise in inches by 7½ (each riser) and select the nearest whole number. Thus, if the total rise is 100 inches, the number of risers would be 13 and the total run, about 120 inches (Figure 16-59, B).

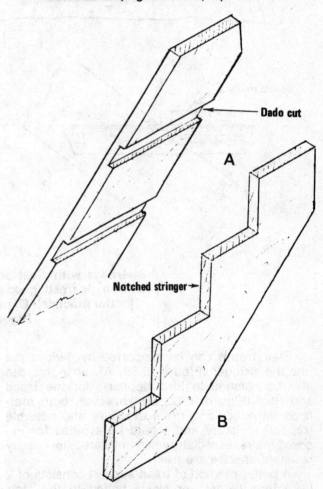

**Treated supports [not recommended]. A, dadoed
stringer [poor practice]; B, notched
[better practice]
Figure 16-60**

273

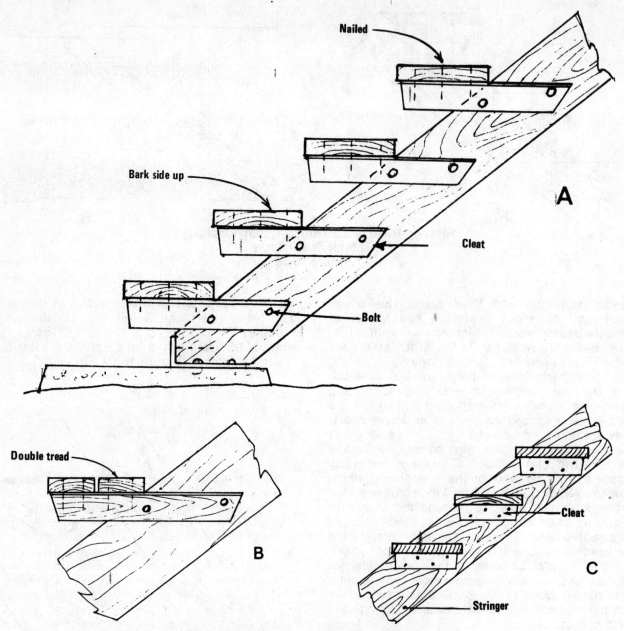

Stairways with cleat support. A, extended cleat with single tread [good practice]; B, double tread [better practice]; C, nailed cleat [poor practice]
Figure 16-61

Stair treads can be supported by dadoes cut into the stringer (Figure 16-60, A). Stringers can also be notched to form supports for the tread and riser (Figure 16-60, B). However, both methods introduce end grain exposure and possible trapped moisture and should be avoided for exposed stairs, especially when untreated, low decay resistant species are used.

A better method of tread support consists of 2 by 4 inch ledgers or cleats bolted to the stair stringers and extended to form supports for the plank treads (Figure 16-61, A). The ledgers can be sloped back slightly so that rain will drain off the treads. Ledgers might also be beveled slightly to

minimize tread contact. Nail 2 by 10 inch or 2 by 12 inch treads to the ledgers with three 12-penny deformed shank nails at each stringer. Rust proof wood screws 3 inches in length can also be used. Always place plank treads with bark side up to prevent cupping and retention of rain water. Treads can also be made of two 2 by 6 inch planks, but the span must be limited to 42 inches for the less dense woods (Figure 16-61, B).

Another method of fastening the stair cleats is by nailing them directly to the stair stringers (Figure 16-61, C). Use 2 by 3 inch or 2 by 4 inch cleats and fasten with three or four 12-penny deformed shank nails. Treads are then nailed to

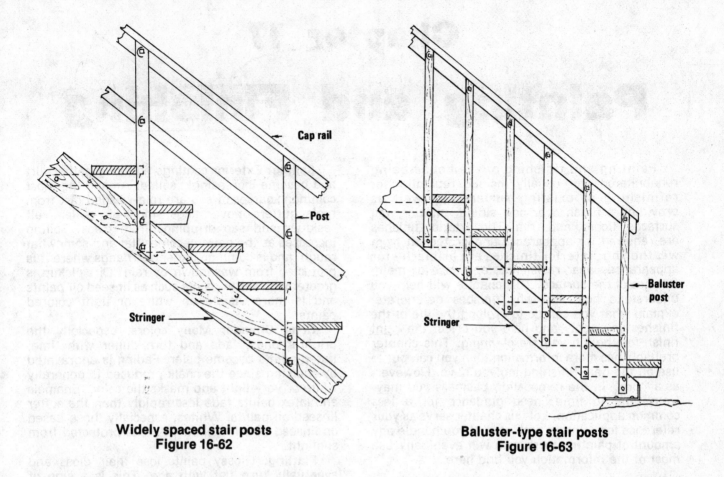

Widely spaced stair posts
Figure 16-62

Baluster-type stair posts
Figure 16-63

the cleats in a normal manner. This method is not as resistant to exposure as the extended cleat shown in Figure 16-61, A, because there are more areas for trapped moisture. However, with the use of a decay resistant species and water repellent preservative treatment, good service should result.

On moderate to full height stairs with one or both sides unprotected, some type of railing is advisable. Railings for stairs are constructed much the same as railings for the deck. In fact, from the standpoint of appearance, they should have the same design. Railings normally consist of posts fastened to stair stringers and supplementary members such as top and intermediate rails.

One method similar to a deck rail uses widely spaced posts and protective railings, (Figure 16-62). Posts are 2 by 4 inch members when spacing is no more than 3 feet and 3 by 4 inch or 2 by 6 inch members for spacings from 3 to 6 feet.

Longitudinal cap rails, top and intermediate, are normally 2 by 4 inch or wider members. Assembly should be with bolts or lag screws. The cap rail can be nailed to the top rail with 12-penny deformed shank nails spaced 12 to 16 inches apart.

The design shown in Figure 16-63 has closely spaced posts which serve as balusters. Each should be bolted to the stringer and to a top rail. The cap rail which also protects the baluster ends can then be nailed to the adjoining rail.

A single cap railing can also be used for such stairs, but it is advisable to fasten it to the posts with metal clips or angles to eliminate unreliable end grain fastening.

Many other variations of post and rail combination can be used. All designs should consider safety and utility as well as a pleasing appearance. A well designed deck, railing, and stairway combination with care in details will provide years of pleasure with little maintenance.

Chapter 17

Painting and Finishing

Painting and finishing of the house being rehabilitated will usually include repainting or refinishing of existing surfaces as well as providing a finish over new siding, interior wall surfaces, floors, and trim. The exterior finishes are required for appearance and protection from weathering; interior finishes are primarily for appearance, wear resistance, and ease of maintenance of the surface. This chapter will help you understand how and why finishes deteriorate, explain what you can do to prolong the life of the finishes you apply and help you select the right finish for the job you are planning. This chapter probably has more information than you can put to use on the next finishing job you have. However, as a "pro" in the remodeling business you may, from time to time, need guidance for a less common application. Let this chapter serve as your reference for painting problems. If you handle any amount of painting work, you will eventually use most of the information you find here.

Normal Deterioration of Exterior Coatings

Paints are not indestructible. Even properly selected protective coatings properly applied on well prepared surfaces will gradually deteriorate and eventually fail. The rate of deterioration under such conditions, however, is slower than when improper painting procedures are followed. Repainting at the right time avoids the problems resulting from painting either too soon or too late. Painting before it is necessary is uneconomical and eventually results in a heavy film buildup leading to abnormal deterioration of the paint system. Painting too late results in costly surface preparation and may be responsible for damage to the structure, which then may require expensive repairs. You should know when to recommend painting and when to recommend against it.

Paints which are exposed outdoors normally proceed through two stages of deterioration: (1) a general change in appearance followed by gradual degradation, and (2) if repainting is not done in time, disintegration of the paint followed ultimately by deterioration of the material under the paint (the "substrate").

The first stage of deteriortion shows up as a change in appearance of the coating with no significant effect on its protective qualities. This change in appearance may result from any one or a combination of the following, depending on the type and color of the paint used and the conditions of exposure:

Soiling: Exterior coatings normally gather dirt and become increasingly soiled. Among the most common sources of soil are rain-washed dirt from roofs, gutters or overhangs, smoky air, pollen, salt residues and sap drippings from trees. Soiling increases as the paint becomes flat and somewhat rough, and is common under overhangs where it is protected from washdown by rain. Dirt pickup is greater with softer paints such as linseed oil paints and is more visible on white or light colored paints.

Color Change: Many colors, especially the brighter ones, fade and turn duller with time; tinted paints become paler. Fading is aggravated by chalking since the chalk produced is generally white or very light and masks the color. Enamels and latex paints fade less rapidly than the softer linseed oil paints. Whites, especially those based on linseed oil will yellow in areas protected from sunlight.

Flatting: Glossy paints lose their gloss and eventually turn flat with age. This is a sign of initial breakdown of the paint composition at the surface. Loss of gloss is soon followed by chalking. Enamels flatten (and chalk) less rapidly than the softer linseed oil paints.

The second stage of normal deterioration occurs after continued exposure. The coating begins to break down, first at the surface, then, unless repainted, gradually through the coating and down to the substrate. There are two types of degradation which may take place—chalking, and checking and cracking; the degree of either depends on the type of paint and the severity of exposure.

Chalking is the result of weathering of the paint at the surface of the coating. The vehicle is broken down by sunlight and other destructive influences, leaving behind loose powdery pigment which can easily be rubbed off with your finger. Chalking takes place more rapidly with softer paints such as those based on linseed oil. Chalking is most rapid in areas exposed to large amounts of sunshine. For example, chalking will be most apparent on the south side of a building. On the other hand, little chalking will take place in areas protected from sunshine and rain, such as under eaves or overhangs. Controlled chalking can be an asset, especially in white paints, since it is a self cleaning process and helps to keep the surface clean and white. Furthermore, by gradually wearing away, it reduces the thickness of the coating, thus allowing repeated painting without making the coating too thick for satisfactory service. Chalked paints are also generally easier to repaint since the underlying paint is in good condition, and, generally, little surface preparation is required. This is not the case when water-thinned paints are to be applied. Their adhesion to chalky surfaces is poor.

Severe cracking, curling and flaking
Figure 17-1

Checking and cracking are breaks in the paint film which are formed as the paint becomes hard and brittle. See Figure 17-1. Temperature changes cause the substrate and overlying paint to expand and contract. As the paint becomes hard, it gradually loses its ability to expand without breaking to some extent. Checking is described as tiny breaks which take place only in the upper coat or coats of the paint film without penetrating to the substrate. The pattern is usually similar to a crowsfoot. Cracking describes larger and longer breaks which extend through to the substrate. Both are a result of stresses in the paint film which exceed the strength of the coating. Whereas checking arises from stresses within the paint film, cracking is caused by stresses between the film and the substrate. Cracking will generally take place to a greater extent on wood than on other substrates because of its grain. When wood expands, it expands much more across the grain than along the grain. Therefore, the stress in the coating is greatest across the grain causing cracks to form parallel to the grain of the wood. Checking and cracking are aggravated by excessively thick coatings because of their reduced elasticity.

As the coating degrades, it finally reaches the point of disintegration. As chalking continues, the entire coating wears away or erodes and becomes thinner. Eventually, it becomes too thin to hide the substrate. Patches of substrate are laid bare. For example, the grain of wood substrates begins to show through. See Figure 17-2. If the cracks are

Moderate erosion
Figure 17-2

Severe checking and crumbling
Figure 17-3

relatively small, the moisture penetrating through the coating will cause small pieces of the coating to lose adhesion and fall off the substrate. See Figure 17-3. If the cracks are large, the eventual result is the most rapid method of deterioration—flaking and peeling. The penetrating moisture loosens relatively large areas of the coating. The paint then curls slightly, exposing more of the substrate and finally flakes off. Peeling is an aggravated form of flaking in which large strips of paint can be easily removed. See Figure 17-1.

When large areas of substrate become exposed, the coating has reached the point of complete deterioration. Such surfaces require extensive and difficult preparation before repainting. All of the old coating may have to be removed to be sure that it does not create problems by continuing to lose adhesion, taking the new coating with it. Furthermore, complete priming of the exposed substrate will also be required, thus adding to cost and time. Continued neglect may also lead to deterioration of the structure resulting in expensive repairs in addition to painting costs.

The important thing to remember here is that all painting surfaces age. Repainting should be done before disintegration (erosion, cracking, crumbling) begins. Advise your clients about the importance of halting the normal aging process before it makes expensive surface preparation necessary.

Normal Deterioration of Internal Coatings

Interior finishes change in apearance while aging, though not as rapidly as exterior finishes. The changes are somewhat similar to exterior surfaces but for different reasons.

Soiling: All painted areas will become soiled to some extent from dust, smoke, fingerprints, fumes and residues.

Flatting: Glossy finishes will gradually lose some of their gloss over a long period of time, especially when they are cleaned often. However, they do not actually become flat or lose their washability.

Color Change: Interior finishes will change color slowly. This is generally not noticeable except when areas which were covered are compared with the surrounding area, e.g., behind pictures or furniture.

Deterioration is a relatively minor problem with interior coatings. It generally is confined to relatively small areas. Enamels on woodwork may become brittle with age and crack, especially when the total coating thickness is excessively high. Cracking may also show up in wall paints when the building settles slightly. The cracks usually are quite fine and may be easily repaired and touched up. Areas around switches or door handles may be cleaned often during the life of the paint in order to remove fingerprints. Eventually, the paint will be removed by abrasion of the cleaner.

When to Paint over Normal Deterioration

It is time to repaint when the coating is becoming thin but has not yet reached the point of disintegration. This way, little surface preparation is needed and only one or two coats are required. In general, you should repaint the south side of any structure when it has begun to chalk heavily or when about 50 percent of the area is covered with cracking or checking.

For wood surfaces, remove the disintegrated paint by scraping, wire brushing and sanding. Sand exposed wood smooth. Wipe off all dust and loose chalked paint. Wash off dirty areas and lightly sand glossy areas (under overhangs).

Prime the exposed substrate. When dry, apply one coat of topcoat if the paint is generally in good condition or two coats if the paint shows signs of considerable chalking or any erosion.

On iron railings and exposed metals, watch for signs of local rusting or corrosion. Spot paint as soon as possible before general surface preparation and painting are required. Remove disintegrated paint and clean the area well using the best method for the conditions and type of paint used. Proper surface preparation is extremely important to prevent rusting or corrosion under the new coating. Prime cleaned areas immediately.

Concrete and masonry substrates usually present less of a problem than wood or steel under normal conditions since they neither expand excessively nor corrode. The necessity for repainting is usually determined by the condition of the paint itself. Remove disintegrated paint with a wire brush and wipe of all dust and loose chalk. Then apply one coat of masonry paint on the cleaned area followed by a complete coat over the entire surface to be painted. No special primers are required under normal conditions.

Interior coatings generally do not require repainting as a result of normal deterioration. The most common reason for painting is to improve the appearance. Cleaning, rather than frequent repainting will often be quite effective, saving cost and time. It also will prevent excessive paint buildup though it may not satisfy the desire of your client for a "new look."

Avoiding Abnormal Deterioration

You will seldom be called on to repaint because normal deterioration is causing a surface to fail. You should, however, know when normal deterioration should be stopped. Most often you will repaint over a relatively new paint job that has failed after a short period. Most important, you should know the causes of abnormal deterioration so you can avoid mistakes that can cause premature failure.

When coatings deteriorate sooner than they should or in an abnormal manner, the cause of the premature failure must be found and corrected before repainting. The cause may be due to the type of material, moisture in the structure, the environment, the paint, or the technique you use.

Painting Wood

Many painted materials have individual characteristics which can present problems if not corrected or eliminated before or during painting. Many types of wood are used in construction, some of which vary considerably in their characteristics. Redwood and cedar are brown and rather uniform in grain, while pine and fir are light colored and vary considerably in grain structure. Both redwood and cedar contain soluble dyes which can dissolve in moisture absorbed by the wood. The dye solution will rise to the surface of the paint, then appear as pink or brown colored streaks or spots. Staining can be eliminated by preventing moisture from getting to the wood. Prime new lumber with a good sealing paint such as an oil primer rather than a relatively porous latex paint. Once the moisture is removed, no further staining should occur. The stain on the surface should eventually be washed off by rainfall.

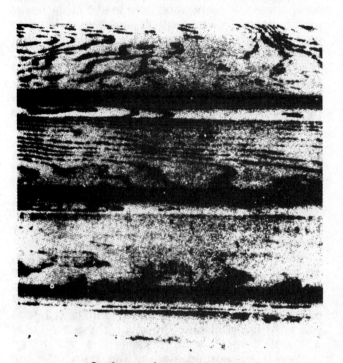

Spring and summer wood
Dark areas show failure over summer-wood in flat grain southern yellow pine
Figure 17-4

Trees grow more rapidly in the spring than during the summer. Consequently, the spring-wood tends to be relatively soft with wide bands, whereas summerwood is harder and has narrower bands. Each type absorbs water and expands to a different degree causing stresses at the junction of the two bands. When cracking does take place, it generally starts along these junction lines. Adhesion will be poorer on the more dense summerwood so that peeling will start in this area actually showing the grain pattern. See Figure 17-4.

The method of sawing the lumber will determine the pattern of the wood produced. If the saw cuts radially, facing the center of the log, it will cut directly across the growth bands forming an edge grain, which shows up as parallel lines or bands. If the saw cuts at right angles to the radial lines, band widths will vary considerably throughout each piece of lumber. This flat grain pattern is more interesting for furniture but is less useful for painting. The larger the grain pattern, the greater will be the problem with differential absorption and ultimate cracking along the grain junction

lines with subsequent flaking and peeling. Southern yellow pine is a marked example of this problem, which is exaggerated even further because of its high resin content.

All trees have branches which start well within the trunk. Therefore when boards are cut, especially flat grained, they will contain cross sections of these branches or knots. This is more of a problem with pine which is cut from smaller trees with many branches as compared with redwood which comes from very large trees with few branches. These knots contain resinous material, which, under the heat of the sun, will melt and bleed through the paint. The discolored area also becomes brittle from the resin, and cracks long before the rest of the coating. To overcome this, remove all paint from the knots and surrounding area down to the wood. Seal with knot sealer and repaint with at least two coats of the same paint as used in the surrounding area.

Some pine, especially of lower grades, contains pockets of pitch or resin similar to that found around knots. This resin will rise to the surface and discolor and eventually degrade the paint in that area. Such areas should be cleaned, sealed with knot sealer and repainted. If the pitch pocket is below the surface, a hole should be drilled to allow drainage, and then puttied and sealed before painting. Small isolated spots of pitch, which appear on the surface and have not harmed the paint, can be removed by scraping and washing with mineral spirits.

Fresh lumber contains a considerable amount of water. Most of this must be removed before use, not only to prevent shrinkage after installation but to prevent blistering, cracking and loss of adhesion of the applied paint. Be sure that all lumber that is to be painted has been properly dried and kept dry before painting.

Permitting wood to weather naturally without protection of any kind is, of course, very simple and economical. Wood fully exposed to all elements of the weather, rain and sun being the most important, will wear away at the approximate rate of only ¼ inch in a century. The time required for wood to weather to the final gray color will depend on the severity of exposure. Wood in protected areas will be much slower to gray than wood fully exposed to the sun on the south side of a building. Early in the graying process, the wood may take on a blotchy appearance because of the growth of mico-organisms on the surface. Migration of wood extractives to the surface also will produce an uneven and unsightly discoloration, particularly in areas that are not washed by rain.

Unfinished lumber will warp more than lumber protected by paint. Warping varies with the wood density, width and thickness of the board, basic wood structure, and species. Warp increases with density and the width of the board. The width of boards should not exceed eight times the thickness. Flat grain boards warp more than vertical grain lumber. Bald cypress, the cedars, and redwood are species which have only a slight tendency to warp.

Water Repellent Preservative Finishes

A simple treatment of an exterior wood surface with a water repellent preservative markedly alters the natural weathering process. Most pronounced is the retention of a uniform natural tan color in the early stages of weathering and a retardation of the uneven graying process which is produced by the growth of mildew on the surface.

Water repellent finishes generally contain a preservative (usually pentachlorophenol), a small amount of resin, and a very small amount of a water repellent that is frequently wax or waxlike in nature. The water repellency imparted by the treatment greatly reduces the tendency toward warping, and excessive shrinking and swelling which lead to splitting. It also retards the leaching of extractives from the wood and staining from water at the ends of boards.

This type of finish is quite inexpensive, easily applied, and very easily refinished. Water repellent preservatives can be applied by brushing, dipping, and spraying. Rough surfaces will absorp more solution than smoothly planed surfaces and the treatment will be more durable on them. It is important to thoroughly treat all lap and butt joints and ends of boards. Many brands of effective water repellent preservatives are on the market.

The initial applications may be short lived (1 year), especially in humid climates and on species that are susceptible to mildew, such as sapwood and certain hardwoods. Under more favorable conditions, such as on rough cedar surfaces which will absorb large quantities of the solution, the finish will last more than 2 years.

When blotchy discolorations of mildew start to appear on the wood, remove blotches and rain spatters, and lighten dark areas by steel brushing with the grain. Retreat the surface with water repellent preservative solution. If extractives have accumulated on the surface in protected areas, clean these areas by mild scrubbing with a detergent of trisodium-phosphate solution.

The continued use of these water repellent preservative solutions will effectively prevent serious decay in wood in above ground installation. This finishing method is recommended for all wood species and surfaces exposed to the weather.

Effect of Impregnated Preservatives on Painting

Wood treated with the water soluable preservatives in common use can be painted satisfactorily after it has redried. The coating may not last quite as long as it would on untreated wood, but there is no vast difference. Certainly, a slight loss in durability is not enough to offer any practical objection to using treated wood where preservation against decay is necessary, protection against weathering desired, and appearance of painted wood important. Coal-tar creosote or other dark oily preservatives tend to stain through paint unless the treated wood has been exposed to the weather for many months before it is painted.

Penetrating Pigmented Stain Finishes

The penetrating stains also are effective and economical finishes for all kinds of lumber and plywood surfaces. They are especially well suited for rough sawn, weathered, and textured wood and plywood. Knotty wood boards and other lower quality grades of wood which would be difficult to paint can be finished successfully with penetrating stains.

These stains penetrate into the wood without forming a continuous film on the surface. Because there is no film or coating, there can be no failure by cracking, peeling, and blistering. Stain finishes are easily prepared for refinishing and easily maintained.

The penetrating pigmented stains form a flat and semi-transparent finish. They permit only part of the wood grain to show through. A variety of colors can be achieved including shades of brown, green, red, and gray. The only color which is not available is white. This color can be provided only through the use of white paint.

Stains are quite inexpensive and easy to apply. To avoid the formation of lap marks, the entire length of a course of siding should be finished without stopping. Only one coat is recommended on smoothly planed surfaces. It will last 2 to 3 years. After refinishing, however, the second coat will last 6 to 7 years because the weathered surface has absorbed more of the stain than the smoothly planed surface.

Two coat staining is possible on rough sawn or weathered surfaces, but both coats should be applied within a few hours of each other. When using a two coat system, the first coat should never be allowed to dry before the second is applied, because this will seal the surface and prevent the second coat from penetrating. A finish life of up to 10 years can be achieved when two coats are applied to a rough or weathered surface.

A stained surface should be refinished only when the colors fade and bare wood is beginning to show. A light steel-wooling or steel brushing with the grain and hosing with water to remove surface dirt and mildew is all that is needed to prepare the surface. Re-stain after the surfaces have thoroughly dried.

Clear Film Finishes

Clear finishes based on varnish, which form a coating or film on the surface, should not be used on wood exposed fully to the weather. These finishes are quite expensive and often begin to deteriorate within 1 year. Refinishing is a frequent, difficult, and time consuming process.

Exterior Paints

Of all the finishes, paints provide the widest selection of color. When properly selected and applied, paint will provide the most protection to wood against weathering.

Best paint durability will be achieved on the select high grades of vertical grain western red cedar, redwood, and low density pines. Exterior grade plywood which has been overlaid with medium density resin treated paper is another wood base material on which paint will perform very well.

Follow these three simple steps when painting wood:

1. Apply water repellent preservative to all joints by brushing or spraying. Treat all lap and butt joints, ends and edges of lumber, and window sash and trim. Allow 2 warm days of drying before painting.

2. Prime the treated wood surface with an oil base paint free of zinc-oxide pigment. Do not use a porous low luster oil paint as primer on wood surfaces. Apply sufficient primer so the grain of the wood cannot be seen. Open joints should be caulked after priming.

3. Apply two topcoats of high quality oil, alkyd, or latex paint over the primer. The south side has the most severe exposure, so two topcoats are particularly important on that side of the house.

Painting Metals

All metals are much more uniform than wood. They expand uniformly in all directions so that adhesion loss because of uneven stresses is much less of a problem than with wood. Some types of metals do present certain problems which can cause abnormal deterioration.

Iron and steel rust when exposed unprotected. If moisture penetrates through thinly coated sharp corners or breaks in the film, rust is formed. This rust will increase in area, lifting the edge of the film around the break, then creep underneath the film and continue the process. Thus, the paint deteriorates quite rapidly around each area of exposed metal. Rusting is accelerated in humid atmospheres and even more so in marine atmospheres. Rusting will also spread under the paint film in areas which have been insufficiently cleaned. Poorly cleaned surfaces leave rusted areas in which moisture and air can be trapped when painted. The area should be adequately cleaned depending on the coating to be applied.

Galvanized steel such as chain link fence is sheet coated with zinc and then treated with chemicals to prevent white rust (a white deposit which forms when zinc is exposed in humid areas). The combination of the zinc metal and chemical treatment often creates problems of adhesion of applied coatings after exposure. If the incorrect paint system is used, extreme flaking and peeling may take place after a year or so of exposure, especially when wide temperature changes take place. Allow galvanized steel to weather, if at all possible, and use an appropriate primer. If earlier painting is necessary, first wash the surface with a vinegar solution and rinse it thoroughly. This will remove any manufacturing residue and stain inhibitors. Apply a special primer before painting. Rust and loose paint can usually be removed from old metal surfaces with sandpaper or with a stiff wire brush. Chipping may be necessary in severe cases. Chemical rust removers are available.

Oil and grease may be removed with a solvent such as mineral spirits. Rinse the surface thoroughly.

The most common non-ferrous metals which are painted are aluminum and copper. Although both of these metals do corrode, their corrosion products do not tend to expand as rapidly as in the case of iron and steel. They should be cleaned thoroughly to obtain optimum adhesion. Since non-ferrous metals are relatively soft and thin, this must be done with care to avoid damaging the material.

Painting Concrete and Masonry

Concrete, stucco, masonry, and plaster have three things in common. They are hard; they all contain lime and other soluble salts and they may be relatively porous. The surface of new concrete or plaster may be somewhat rough and porous or very smooth and slick, depending on the type and degree of troweling used to finish the surface. Very smooth concrete can create a problem with loss of adhesion thus causing rapid flaking and peeling. The surface should be etched before painting to prevent this problem.

Fresh concrete, mortar, stucco and plaster are highly alkaline. Alkalinity can cause premature failure of applied coatings unless they are alkali resistant. Oil paints, for example, should not be used on alkaline surfaces. The alkali will saponify the oil to form a soap which has no binding qualities. Latex and rubber based paints are not harmed by alkali in the substrate.

Concrete, stucco, masonry, and plaster contain water soluable salts which dissolve in moisture carried through the substrate and then crystallize on the exposed surface. If the paint is water permeable, such as latex paint, the solution will pass through the coating and discolor the surface in a non-uniform spotty manner. If the coating is not permeable, the salts may be deposited under the paint film and cause it to lose adhesion in spots. See Figure 17-5. All efforescence must be removed before repainting, and the cause eliminated.

New concrete should weather for several months before being painted. If earlier painting is necessary, first wash the surface with a solvent such as mineral spirits to remove oil or grease.

Patch any cracks or other defects in masonry surfaces. Pay particular attention to mortar joints. Clean both new and old surfaces thoroughly before painting. Remove dirt, loose particles, and efflorescence with a wire brush. Oil and grease may be removed by washing the surface with a commercial cleaner or with a detergent and water. Loose, peeling, or heavily chalked paint may be

Severe efflorescence on brick wall
Figure 17-5

removed by sandblasting.

If the old paint is just moderately chalked but is otherwise "tight" and nonflaking, coat it with a recommended sealer or conditioner before you repaint with a water base paint. Some latex paints are modified to allow painting over slightly chalked surfaces. Follow the manufacturer's directions. After cleaning the surface, wash or hose it—unless efflorescence was pesent.

Improper proportioning, mixing, placing and curing of concrete, stucco and plaster create areas which may be of different porosities. This will result in uneven absorption of the applied coatings, which shows up as uneven gloss of the paint. Deterioration will also be more rapid over these areas.

Abnormal Environments

Unusual conditions of exposure are a major cause of abnormal deterioration of coatings. Moisture may cause abnormal deterioration in two ways: it may cause flatting or formation of mildew. If moisture, in the form of fog, rain or dew lies on the surface of newly applied paint before it is thoroughly dry, it may cause a spotty or complete loss of gloss of the paint. This is primarily an appearance problem which makes a new paint job look inferior. Paint coatings exposed in humid climates or in warm damp rooms may be attacked by mildew, a fungus which feeds on the coating. Mildew will grow and become quite unsightly, and eventually will accelerate degradation of the coating. See Figure 17-6. In its early stages it looks like dirt, but it cannot be washed off as easily. The presence of mildew can be determined by using

household bleach; this will bleach mildew, whereas it has no effect on dirt. Hard drying paints such as enamels, or paints containing zinc oxide, are more resistant to mildew. Use specially formulated moisture resistant and mildew resistant paints for these exposures.

One type of mildew [magnified]
Figure 17-6

Smoke and fumes can adversely affect paint coatings causing discoloration and rapid failure. Sulfur-containing gases, such as sulfur dioxide and hydrogen sulfide, will discolor coatings, especially those containing lead or iron. They will also accelerate chalking and erosion. Wind driven dust will accelerate dirt collection especially on softer drying paints such as those based on linseed oil. Salt-laden atmosphere in coastal areas will accelerate deterioration of coatings which are not resistant to salt.

Sudden changes in temperature can create unexpected problems. A rapid drop overnight, just after painting, may cause a heavy dew or even frost to deposit on the paint film with consequent flatting. It may also retard drying so that dirt and insects can become embedded in the coating. Wrinkling can occur if the coating is excessively thick. A rapid increase in temperature may cause air entrapped in a porous substrate to increase in pressure and form dry blisters in the paint film. See Figure 17-7.

Excessive wind velocity during painting makes application extremely difficult. It may also cause

**Temperature blistering
Figure 17-7**

the paint to dry too rapidly on the surface thus forming a skin which prevents through drying. This can lead to recoating problems and to solvent entrapment. In any case, durability is impaired. Do not paint when the wind velocity is above 15 miles per hour. Winds also carry dirt, tending to impinge the dirt particles on the painted surface, especially when it is fresh or soft. Grit carried by high velocity winds can also abrade cured painted surfaces.

Incompatible Preservatives and Paints

The entire coating must be compatible through each layer, from the substrate to the surface, to achieve durability. Any incompatibility between the material painted and the paint, and between coats, will reduce adhesion and accelerate deterioration, such as lifting and peeling.

Some wood preservatives affect paints applied over them. They may either retard drying, affect adhesion or bleed through and discolor the paint. Preservatives containing creosote or copper nephthenate, for example, may bleed. Zinc naphthenate or pentachlorphenol can be used with no adverse effects.

It is always safest to recoat surfaces with the same kind of paint previously used, unless experience shows that the new paint is compatible with the old paint. Incompatibility may result in the following defects, all of which affect adhesion and ultimate service life of the paint system:

Lifting: This is an affect produced by the solvent in the applied paint acting as a paint remover on the coating underneath. The result is a softening, swelling and lifting of the coating. It can happen when paints containing strong solvents such as xylene are applied over relatively soft paints such as oil paints. Lifting is more likely to occur when a second or third coat is applied over an undercoat which has not dried hard enough. Always be sure that the coating is not only dry but fairly hard before applying the next coat. Test a small area if you are not sure.

Alligatoring: Alligatoring describes a pattern in a coating which looks like the hide of an alligator. It is caused by uneven expansion and contraction of a relatively hard topcoat over a relatively soft or slippery undercoat. See Figure

17-8. Alligatoring can be caused by:
1. Applying an enamel over an oil primer.
2. Painting over bituminous paint, asphalt, pitch or shellac.
3. Painting over grease or wax.

Crawling: Crawling occurs when the new coating fails to wet and form a continuous film over the preceding coat. Examples are applying latex paints over high gloss enamel or applying paints on concrete or masonry treated with a silicone water repellent. See Figure 17-9.

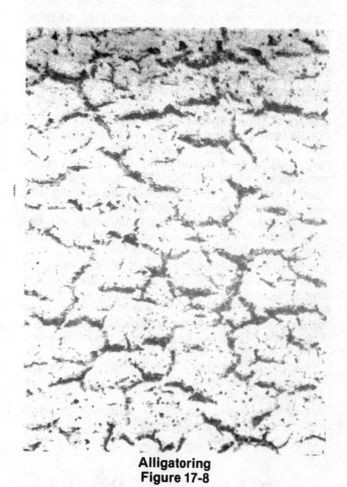

**Alligatoring
Figure 17-8**

**Crawling
Figure 17-9**

Intercoat Peeling: The loss of adhesion caused by the use of incompatible paints may not be obvious until after a period of time has elapsed. Then, the stresses in the hardening film will cause the two coatings to separate and the topcoat will then flake and peel.

Poor Painting Practice

You can prevent or cause most types of deterioration by the way you paint and the materials you use. Always use the recommended primer if a primer is necessary. Be sure that the surface is properly prepared and that painting conditions are within the limits recommended by the manufacturer as to temperature and humidity. Follow application directions exactly. Short cuts or disregard of instructions are bound to accelerate deterioration of the applied coatings.

The adhesion of all the paint depends on direct contact of the first coat with a clean substrate. If the surface contains wax, grease or oil, the paint may dry very slowly, crawl or alligator. In any case, flaking and peeling will take place. Holes and cracks which are not filled and sealed will allow moisture to get in behind the coating and cause blistering and film degradation. If paints are thinned or applied in too thin a coat, they will not last as long. If too little primer is used, especially on porous substrates, then gloss and color will be uneven, and adherence of topcoats may be affected. In any case, any chalking and erosion which takes place will wear through a thin film faster and result in the necessity for repainting earlier than normal. Too much paint is just as bad as too little paint. Too heavy a coat may cause any of the following problems:

Sagging: The paint may curtain on vertical surfaces thus affecting its appearance and dry film thickness.

Drying: Drying, especially "through drying" may be retarded considerably. This may cause lifting when recoated.

Severe wrinkling
Figure 17-10

Cross grain cracking
Figure 17-11

Wrinkling: This may occur either in cold weather when the thickened paint is improperly applied or in hot weather when the topcoat dries quickly but the paint underneath is still wet. The resulting stresses cause the paint to wrinkle. See Figure 17-10.

Cracking: The film may not show any defects initially, but the extreme stresses present in a thick hardening film may cause cracking after exposure. This is especially true in multicoat work. See Figure 17-11.

Blistering: In hot weather the uneven drying of the thick film may cause solvent entrapment with subsequent blistering.

Rushing a job may also speed up its failure as a result of loss of adhesion or improper cure. If a coat is not thoroughly dry, the next coat may cause trapping of the solvent or lifting. Trapped solvent must come out eventually and will cause either pinholing, blistering or a reduction in adhesion.

Moisture

The major cause of abnormal deterioration of coatings, especially those exposed outdoors, is moisture. This moisture may either come from external sources or be developed within the structure. This moisture can produce abnormal deterioration of applied coatings such as wood stain, mildew, blistering and loss of adhesion, resulting in a poor appearance and eventual deterioration by flaking and peeling. See Figures 17-12 and 17-13. A prime reason for this problem is that the major construction materials used, i.e.,

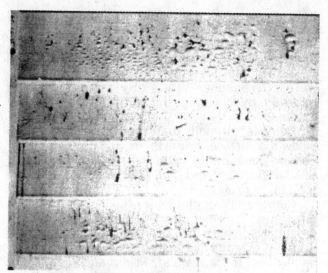

**Blistering from moisture
Figure 17-12**

**Loss of adhesion from moisture
Figure 17-13**

wood, concrete, stucco, masonry and plaster, are essentially porous and will allow moisture to pass through. If the walls are wet and the surface is warmed, as by sunlight, the moisture will tend to move to the outside atmosphere. If nonpermeable coatings are used (most paints other than latex paints or cement paints) this moisture will be trapped. Increased pressure will eventually cause the coating to either blister or lose adhesion. The problem is much less serious with metals, but increased contact with moisture does reduce service life of coatings applied on metals.

Poor quality construction which allows moisture to enter behind painted woodwork, masonry or plaster is a major reason for discoloration and abnormal deterioration of both interior and exterior coatings. It will also eventually cause wood decay if not corrected.

All of the following defects either trap moisture in the walls, allow moisture to enter the walls, or trap moisture vapor which will condense on cold walls:

1. Use of green lumber or building during rainy weather so that the structure was wet when originally painted.

2. Poorly fitted windows, door trim and joints allowing water to enter.

3. Omission of drips or gutters at eaves, or omission of eaves and overhangs thus increasing the flow of water down the walls of the structure.

4. Lack of flashing or improper installation of flashing around chimneys, roof, corners, doors and windows allowing rain to penetrate walls.

5. Lack of a vapor barrier behind trim, around basement walls and in crawl spaces.

6. Lack of ventilation in attics, basements and crawl spaces allowing moisture to condense and collect on walls.

7. Direct contact of wood walls with ground or shrubbery.

8. Use of non-galvanized ferrous nails which

will eventually rust and loosen, allowing water to enter.

9. Painting plaster when still wet.

10. Inadequate use of caulking compound allowing rain to enter openings around windows and doors.

The humidity within a structure should be kept fairly low especially during the cold weather when outside walls are cold. Otherwise, moisture will collect and eventually work its way into and through the walls unless the interior paint on the walls is impermeable. This usually is not a problem unless humidifiers are used with heating equipment.

There are two relatively low cost solutions to blistering and peeling problems if the source of moisture is by normal use and no structural defects are involved:

1. Seal the inside surface of exterior walls with aluminum paint or enamel and apply breathing-type paints such as latex paints to the outside surface.

2. Vent the outside walls by the use of vents or breathing spaces in the siding.

Structures not kept in good repair will eventually allow moisture to enter the walls and cause paint failures. Some examples are as follows:

1. Leaking roofs caused by loosened, curled or missing shingles.

2. Plumbing leaks.

3. Corroded flashing.

4. Broken, leaky or clogged gutters and downspouts.

5. Cracked or missing caulking and glazing compound.

6. Allowing water to collect in basements.

7. Loose siding.

All of these conditions must be corrected before painting is started.

Preparing the Surface

Proper preparation of the surface prior to painting is essential to achieve maximum life of the coating. The best quality paint will not perform effectively if applied on a poorly prepared surface. The initial cost of adequate surface preparation is more than compensated for by increased durability, minimum repairs and repainting. It is your obligation as a reliable professional to be sure that surfaces to be painted are prepared adequately.

Many surface contaminants reduce adhesion and cause blistering, peeling, flaking, and underfilm rusting. Among these contaminants are: dirt, grease, rust, rust scale, chemicals, moisture and efflorescence. In addition, the following surface defects will affect adhesion adversely: irregular areas, burrs, crevices, sharp edges, knots, splinters, nail holes, loose aggregates and old paints in various stages of failure.

In general, a surface that is to be painted should be firm, smooth, and clean. With oil base paint, it must also be dry. Latex or water base paint can be applied to a damp surface (but not to a wet one). The paint can label may contain additional or special instructions for preparing the surface.

Old paint may be removed by sanding, scraping, burning, or with chemical paint remover. Scraping is the simplest but hardest method. Sanding is most effective on smooth surfaces. Chemical paint remover can be expensive for large areas.

Cleaning by Hand

Hand cleaning will remove loose surface contaminants. These include rust scale and loosely adhering paint. Hand cleaning cannot be expected to do more than remove major surface contamination. As such, it is primarily recommended for spot cleaning in areas where corrosion is not a serious factor.

When multiple coats of paint have built up on interior walls or woodwork, they may present an unsightly appearance due to the presence of numerous cracks, peeling or uneven areas where loose paint has often been removed and then repainted. In this case, it is best to remove all of the old paint down to the base material. The best procedure is to apply a water-rinsable paint and varnish remover (non-sag type) and allow to stand for fifteen minutes. Then apply steam by means of a wallpaper steamer. Move the steamer slowly and scrape the steamed surface with a wide bladed scraper. Repeat the process, if necessary. It is very important to avoid allowing the steamer pan to remain too long in one spot because this may cause water logging of the base material. It is much better to keep the steamer pan moving slowly but continuously, and repeat the process, if necessary, rather than to concentrate the steam in any area for a prolonged period of time.

Smooth any rough spots in wood with sandpaper or other abrasive. Dust the surface just before you paint it. Old surfaces in good condition—just slightly faded, dirty, or chalky—may only need dusting before being repainted. Very dirty surfaces should be washed with a mild synthetic detergent and rinsed thoroughly with water. Grease or other oily matter may be removed by cleaning the surface with mineral spirits.

Remove all nail rust marks. Set nailheads below the surface, prime them, and putty the hole. Fasten loose siding with non-rusting type nails. Fill all cracks. Sand the area smooth after the compound dries.

Remove all rough, loose, flaking, and blistering paint. Spot prime the bare spots before repainting. Where the cracking or blistering of the old paint extends over a large area, remove all old paint down to bare wood. Prime and repaint the old surface as you would a new wood surface. Sand or "feather" the edges of the sound paint before you repaint.

Cleaning with Power Tools

Power tool cleaning methods provide faster and more adequate surface preparation than hand tool methods. Wire brushes (cup or radial) are used for removing loose scale, old paint, and dirt deposits. Grinders and sanders are used for complete removal of old paint, rust or scale on small surfaces and for smoothing rough surfaces. As with hand tools, care must be exercised with grinding tools not to cut too deeply into the surface, since this may result in burrs that are difficult to protect satisfactorily. Care must also be taken when using wire brushes to avoid polishing metal surfaces and thus prevent adequate adhesion of the subsequent coatings. Power tool cleaning should be preceded by solvent or chemical treatment and painting must be started and completed as soon after power cleaning as possible.

Blast cleaning abrades and cleans through the high velocity impact of sand, metal or synthetic grit or other abrasive particles on the surface. Blast cleaning is most often used on metal surfaces but may also be used, with caution, on masonry and stucco. It is, by far, the most thorough of all mechanical treatments.

Cleaning with Chemicals

Solvent cleaning is a procedure for removing oil, grease, dirt, chemical paint stripper residues and other foreign matter from the surfaces and other foreign matter from the surfaces prior to painting or mechanical treatment. Solvents clean by dissolving and diluting to permit contaminants to be wiped or washed off the surface. The simplest procedure is to first remove soil, cement spatter and other dry materials with a wire brush. The surface is then scrubbed with brushes or rags saturated with solvent. Clean rags are used for rinsing and wiping dry. More effective methods include immersing the work in the solvent or spraying solvent over the surface. In either case, the solvent quickly becomes contaminated, so it is essential that several clean solvent rinses be applied to the surface. Mineral spirits is an effective solvent for cleaning under normal conditions. Toxic solvents and solvents with low flash points represent hazards to health and safety.

Alkali cleaning is more efficient, less costly and less hazardous than solvent cleaning, but is more difficult to carry out. Alkaline cleaners are disolved in water and used at relatively high temperature (150° F - 200° F) since cleaning efficiency increases with temperature. Alkalis attack oil and grease, converting them into soapy residues that wash away with water. Other active ingredients contained in the alkali cleaners aid in removing surface dirt and other contaminants such as mildew. These cleaners are also effective in removing old paint by saponifying the dried vehicle. The most commonly used alkaline cleaners are trisodium phosphate, caustic soda and silicated alkalis. They can be applied by brushing, scrubbing, spraying or by immersion. Thorough water rinses are absolutely necessary to remove the soapy residue as well as all traces of alkali, to avoid reactivity with the applied paint. Otherwise, cleaning may do more harm than good. The water should be hot and, preferably, applied under pressure.

Steam or hot water under pressure is used for some cleaning. A detergent should be included for adding effectiveness. The steam or hot water removes oil and grease by liquefying them because of the high temperature, then emulsifying them and diluting them with water. When used on old paint, the vehicle is swelled so that it loses its adhesiveness and is easily removed. Steam or hot water alone is commonly used to remove heavy dirt deposits, soot and grime. Wire brushing and brush-off blast cleaning may be necessary to augment the steam cleaning process by removing remaining residues.

Acid cleaning is used for cleaning iron, steel, concrete and masonry by treating them with an acid solution. Iron and steel are treated with solutions of phoshoric acid containing small amounts of solvent, detergent and wetting agent. (Do not use on aluminum or stainless steel). Such cleaning effectively removes oil, grease, dirt and other foreign contaminants. In addition (and unlike alkali cleaners), it also removes light rust and faintly etches the surface to assure better adhesion of applied coatings. There are many types of phosphoric acid metal cleaners and rust removers, each formulated to perform a specific cleaning job. Concrete and masonry can be washed with 5 to 10 percent muriatic (hydrochloric) acid to remove efflorescence and laitance, to clean the surface, to remove any glaze, and to etch the surface. Laitance is fine cement powder which floats to the surface after concrete is poured. Coatings applied over loose deposits of efflores-

cence or laitance will loosen prematurely and result in early coating failure. Remove as much of the loose efflorescence or laitance as possible using a clean, dry, wire or stiff fiber brush. Putty knives or scrapers may also be used. All oil and grease must also be removed prior to acid cleaning, either by solvent wiping or by steam or alkali cleaning. To acid cleaning these surfaces, thoroughly wet the surface with clean water, then scrub with a 5 percent solution (by weight) of muriatic acid, using a stiff fiber brush. In extreme cases, up to a 10 percent muriatic acid solution may be used, and may be allowed to remain on the surface up to five minutes before scrubbing. Work should be done on small areas, not greater than four square feet in size. Immediately after the surface is scrubbed, wash the acid solution completely from the surface by thoroughly sponging or rinsing with clean water. Rinsing must be done immediately to avoid formation of salts on the surface which are difficult to remove. The above procedure is also used when you want to etch the surface, (e.g., to remove the glaze). Often, when concrete surfaces are steel troweled they may become so dense, smooth and even glazed that paint will not adhere to the surface. A simple method to determine whether etching is required is to pour a few drops of water on the surface. If water is quickly absorbed, etching is unnecessary. In addition to this acid washing, glaze may be removed by rubbing with an abrasive stone, lightly sandblasting, or allowing the surface to weather for six to twelve months. It may also be removed by treatment with a solution of 3 percent zinc chloride plus 2 percent phosphoric acid. This is not flushed off but is allowed to dry to produce a paintable surface. It may be necessary, in certain instances, to use acid cleaning methods to neutralize concrete and masonry surfaces before applying a coating which is sensitive to alkali.

Paint and varnish removers generally are used for small areas. Solvent type removers or solvent mixtures are selected according to the type and condition of the old finish as well as the nature of the substrate. Removers are available as flammable or non-flammable types, also liquid or semi-plate in consistency. While most paint removers require scraping or steel wool to physically remove the softened paint, types are available that allow the loosened finish to be flushed off with steam or hot water. Many of the flammable and non-flammable removers contain paraffin wax to retard evaporation. It is absolutely essential that this residue be removed from the surface prior to painting to prevent loss of adhesion of the applied coating. In such instances, follow the manufacturer's label directions or use mineral spirits to remove any wax residue.

Special Repainting Problems

Some surfaces cannot be repared without taking special precautions. Note especially the following:

1. Complete removal of old paint is required in the following cases, and a new covering material may be more practical:

    (a) Kalsomine over sand-textured plaster.

    (b) Kalsomine applied alternately with oil base paint.

    (c) Noncompatible paints that have resulted in intercoat peeling.

    (d) Paints applied in kitchens or other areas over a grease film.

2. All plaster cracks should be repaired with fiberglass tape and plaster patching.

3. Old varnish should be cleaned with strong trisodium and then painted soon after drying. Heavily alligatored varnish must be removed before cleaning.

4. Old radiators painted gold or aluminum should be coated with bronzing liquid before painting.

5. Plastered chimneys may have creosote soaking through the plaster resulting from an accumulation in the chimney. This will continue to soak through, so the wall should be framed out around the chimney before adding a new covering material.

Conditioners, Sealers and Fillers

Latex (water thinned) paints do not adhere well to chalky masonry surfaces. To overcome this problem, conditioner is applied to the chalky masonry before the latex paint is applied. The entire surface should be vigorously wire brushed by hand or power tools, then dusted to remove all loose particles and chalk residue. The conditioner is then brushed on freely to assure effective penetration and allowed to dry. This surface conditioner is not intended for use as a finish coat.

Sealers are used on bare wood to prevent resin exudation (bleeding) through applied paint coatings. Freshly exuded resin, while still soft, may be scraped off with a putty knife and the affected area solvent cleaned with alcohol. Hardened resin may be removed by scraping or sanding. Since the sealer is not intended for use as a priming coat, it should be used only when necessary, and applied only over the affected area. When previous paint on pine lumber has become discolored over knots, the sealer should be applied over the old paint before the new paint is applied.

Fillers are used on porous wood, concrete, and masonry to fill the pores to provide a smoother finish coat. Wood fillers are used on open-grained hardwoods. In general those hardwoods with pores larger than in birch should be filled. When filling is necessary, it is done after any staining operations. Stain should be allowed to dry for 24 hours before filler is applied. If staining is not warranted, natural (uncolored) filler is applied directly to the bare wood. The filler may be colored with some of the stain in order to accentuate the grain pattern of the wood. To apply, first thin the filler with mineral spirits to a creamy consistency, then liberally brush it across the grain, followed by light brushing along the grain. Allow this to stand five to ten minutes until most of the thinner has evaporated. By then the finish will have lost its glossy appearance. Before it has a chance to set and harden, wipe the filler of across the grain using burlap or other coarse cloth, rubbing the filler into the pores of the wood while removing the excess. Finish by stroking along the grain with clean rags. It is essential that all excess filler be removed. Knowing when to start wiping is important; wiping too soon will pull the filler out of the pores, while allowing the filler to set too long will make it very difficult to wipe off. A simple test for dryness consists of rubbing a finger across the surface. If a ball is formed, it is time to wipe. If the filler slips under the pressure of the finger, it is still too wet for wiping. Allow the filler to dry for 24 hours before applying finish coats.

Masonry fillers are applied by brush to bare and previously prepared (all loose, powdery, flaking material removed) rough concrete, concrete block, stucco or other masonry surfaces, both new and old. The purpose is to fill the open pores by brushing the filler into the surface to produce a fairly smooth finish. If the voids on the surface are large, it is preferable to apply two coats of filler, rather than one heavy coat, to avoid mud-cracking. Allow 1 to 2 hours drying between coats. Allow the final coat to dry for 24 hours before painting.

Repair of Surfaces

All surfaces must be in good condition before painting. Repair or replace degraded wood, concrete, masonry, stucco, metal, plaster and wallboard. Remove and replace all loose mortar in brickwork. Replace broken windows and loose putty or glazing compound. Securely fasten or replace loose gutter hangers and downspout bands. Fill all cracks, crevices and joints with caulking compound, or sealants. Drive all exposed nail heads below the surface. Patch all cracks or holes in wood, masonry and plaster. The final surface should be smooth, with no openings or defects of any kind. These preparatory procedures eliminate the major areas for the entrance of moisture which can lead to blistering and peeling of the paint film.

Caulking Compounds and Sealants. Caulking compounds are oil or resin based. They are used in fixed joints of wood, metal, or masonry, also in joints with very limited movement. Sealants on the other hand, are elastomeric, rubber-like compounds. They are intended for use in expansion or other movable joints. Sealants are available as one or two component compounds. When using caulking and sealants, all surfaces must be clean and dry to obtain good adhesion. Remove all oil, grease, soot, dirt or loose paint, or old materials. Be sure the crevice openings are large enough to allow an adequate amount of material to be inserted. Prime the substrate when recommended by the manufacturer. If the opening is deep, first insert back-up materials such as oakum, foamed plastic or rubber, fiberglass, or fiberboard.

Caulking compounds are used to fill joints and crevices around doors and windows in wood, brick, concrete and other masonry surfaces. They are supplied in two grades, a gun grade and a knife grade. The gun grade is most popular because it can be applied with a caulking gun. The knife grade must be applied by hand using a putty knife. The gun grade is supplied in two forms, in bulk and in factory prefilled cartridges. The cartridge type fits directly into a caulking gun. Triggering the gun extrudes the caulking compound directly into the crevice. A variety of different shaped tips aid in speeding up the work. Caulking compounds tend to dry on the surface but remain soft and tacky within the crevice. The applied caulking should be painted each time the surrounding area is painted to help extend its life.

When applying gun grade caulking, move the gun along the crevice while triggering so that the compound is extruded directly into the crevice. Move the gun slowly and steadily, so as to push the bead into the crevice rather than pull it away. Allow the compound to overlap the opening slightly for a better seal, and to allow sufficient surface area for adhesion. The best position to hold the gun in is at a slight angle with the bevel parallel to the work. The compound should finally be tooled to insure close contact with the joint surfaces.

When applying knife grade caulking or putty, use a putty knife and press firmly into cracks or holes until full. Then smooth with the flat side of the knife by sliding it across the surface. The exposed area should be sightly convex to allow for shrinkage. Putty is used to fill nail holes, cracks and imperfections in wood surfaces. Putty is not flexible and should not be used for joints or crevices. It dries to a harder surface than caulking compound.

Sealants are an advanced and much more durable type of caulking compound. Compared with caulking compounds, they have better adhesion to the walls of the crevices, have better extensibility so that they do not pull away when the walls contract in cold weather, and they remain flexible for much longer periods of time. Although they are considerably more expensive than caulking compounds, their longer life is often well worth the difference in cost. There are many types of sealants, and they may be divided into three groups:

One Component: These are similar to caulking compounds in general handling. All are available in bulk and some are also available in cartridges. They are applied in a manner similar to caulking compounds.

Two Component: These are supplied only in bulk since they must be pre-mixed before use. Their useful, or pot-life when mixed is normally not less than three hours. High temperatures may drastically shorten this time. The two components react to form a tough rubbery seal with excellent adhesion, extensibility and durability.

Preformed Tapes: These sealants are supplied in an extruded bead so that they can be applied simply by pressing the tape into crevices by hand without the use of tools.

Glazing Compounds

Glazing compounds are used on both interior and exterior wood and metal window sash either as bedding or face glazing. They are used to cushion glass in metal or wood frames and are not intended to keep or hold the glass in position. Glazing compounds set firmly, but not hard, and have some limited flexibility. They are more flexible than putty. They tend to harden upon exposure, with life expectancy estimated to be approximately ten years if they are properly applied. Painting over glazing compounds will extend their useful life. Glazing compounds are relatively inexpensive though more costly than putty.

For face glazing, apply a generous quantity of glazing compound into the glazing rabbet, and gently press the glass into the rabbet leaving a bed of back glazing material of approximately 1/16 inch. Apply glazing points to hold the glass in place. Strip surplus glazing compound at an angle to allow for run-off of condensation. Apply additional glazing compound to the face and tool into place with the aid of a putty knife, applying sufficient pressure to completely fill the void. Tool face glazing approximately 1/16 inch short of the sight line to allow paint to overlap onto the glass.

For bead or channel glazing, apply a generous amount of the compound to the fixed (stationary) side and the bottom of the channel. Place nonporous resilient spacer shims (such as vinyl floor tile) at points around the perimeter of the channel to position glass and prevent squeezing out of the compound (keep spacer shims below the edge of the channel). Press the glass into place until intimate contact with spacer shims is made. Spread compound on the removable bead and gently press it into place. Insert spacer shims between the glass and the removable bead (opposite spacer shims on the fixed side of the channel) and apply pressure to the removable bead until intimate contact with the spacer shims is made. Fasten the bead in place and strip the excess compound. When the glazing compound has attained a surface skin, apply paint, slightly overlapping the sight line.

Patching Materials

Cracks, holes and crevices in masonry, plaster, wallboard and wood are filled with patching material. It is supplied either ready for use or as a dry powder to which water is added before use. There are a variety of types depending on the surface and its conditions.

Patching plaster is used for repairing large areas in plaster. It is similar to ordinary plaster except that it hardens quickly. It is supplied as a powder. Spackling compound is used to fill cracks and small holes in plaster and wallboard. It is very easy to work with and sands very well after it hardens. It is supplied both as a paste and as a powder.

Joint cement is used primarily to seal the joints between wallboards. It can also be used to repair large cracks. It is supplied as a powder and is used in conjunction with perforated tape which gives it added strength.

Portland cement grout is used to repair cracks in concrete and masonry. Hydrated lime is often added to slow up its cure time and lengthen its working life.

Plastic wood is a filler suitable for such repair work as filling gouges and nail holes. It is also used for building up and filling in wood patterns and joiner work. It is applied in a manner similar to putty. Sand plastic wood smooth after it has completely dried before applying paint.

When using any of the above patching materials (except plastic wood) on masonry, plaster, or wallboard, the crack should first be opened with a putty knife or wall scraper so that weak material is removed and the patching compound can be forced in completely. Dampen these areas with clear water and apply the compound with a putty knife or trowel depending on the size of the hole. Level and smooth off the surface allowing it to be slightly convex to allow for shrinkage. Follow the manufacturer's instructions explicitly if they are available. None of these materials requires attention during drying, except for the Portland cement grout which should be kept damp at least one, and preferably two or three days for optimum cure. When the surface is dry and hard, sand it (except Portland cement) until it is smooth and level with the surrounding area.

Paint Materials

A knowledge of the types of materials used for painting is useful in determining their capabilities and limitations. There are sound reasons for each type of coating. This information is presented to aid you in determining which product should be used for your job and to explain why the product is best suited for the particular combination of conditions present.

Most paints are based on a film former or binder which is either dissolved in a solvent or emulsified in water. Upon application of the product in a relatively thin film, it will dry or cure to form a dry, tough coating. Solutions of such binders in solvent may be called by various names — clear finishes, varnishes if they dry by oxidation, or lacquers if they dry by evaporation. If opaque pigments or colors are dispersed in the binder, the product, which will produce an opaque white or colored film, is called a paint. Pigment concentration can also be varied to produce a high gloss, a semigloss or lustreless (flat) finish. Special pigments such as red lead and zinc

chromate can be used to provide corrosion resistance in primers. Metallic pigments can be added to varnishes to produce metallic coatings such as aluminum paints. The major performance characteristics of the coating depend generally on the type of binder used. The principal binders used in the paint are described in this chapter. These descriptions are concerned only with the reasons for their use, their superior characteristics and their deficiencies. No attempt has been made to discuss their analytical composition in any detail. The paint binders discussed are listed in alphabetical order for ease of reference.

Alkyd

Alkyd binders, of the type covered in this handbook, are oil modified phthalate resins which dry by reacting with oxygen from the surrounding air. Alkyd finishes are usually of the general purpose type, are economical and are available as clear or pigmented coatings. The latter are available in flat, semigloss and high gloss finishes, and in a wide range of colors. They are easy to apply, and, with the exception of fresh (alkaline) concrete, masonry and plaster, may be used on most surfaces which have been moderately cleaned. Alkyd finishes have good color and gloss, and retain these characteristics well in normal interior and exterior environments. Their durability is excellent in rural environments, but only fair in mildly corrosive environments. Alkyd finishes are also available in odorless formulations for use in hospitals, kitchens, sleeping quarters and other areas where odor during painting might be objectionable.

Cement

Portland cement mixed with several ingredients acts as a paint binder when reacted with water. The paint is supplied as a powder to which the water is added before use. Cement paints are used on rough surfaces such as concrete, masonry and stucco. They dry to form hard, flat, porous films which permit water vapor to pass through readily. Since cement paints are powders, they can also be mixed with masonry sand and less water to form filler coats to smooth rough masonry before applying other paints. Cement paints can be used on fresh masonry and are economical. The surface must be damp when they are applied, and must be kept damp for a few days to obtain proper curing. They should not be used in arid areas. When properly cured, cement paints of good quality are quite durable; when improperly cured, they chalk excessively on exposure, and then may present problems in repainting.

Epoxy

The epoxy binders covered in this handbook are made up of two components, an epoxy resin and a polyamide hardener, which are pre-mixed before use. When mixed, the two ingredients react to form the final coating. These paints have a limited working or pot life, usually a working day. Anything left at the end of the day must be discarded. Epoxy paints can be used on any surface and can be applied at high solids, thus producing high film build per coat applied. The cured film has outstanding hardness, adhesion, flexibility and resistance to abrasion, alkali and solvents, as well as being highly corrosion resistant. Their major uses are as tile-like glaze coatings for concrete and masonry, and for the protection of structural steel in corrosive environments. Their cost per gallon is high, but this is offset by the reduced number of coats required to get adequate film thickness. Epoxy paints tend to chalk on exterior exposure so that low gloss levels and fading can be anticipated; otherwise, their durability is excellent.

Coal tar is often added as an ingredient of epoxy paints, resulting in a significant decrease in cost with relatively minor effect on corrosion resistance. Color choice is limited because of the black color of the coal tar. It is used primarily for interior and submerged surfaces.

Inorganic

The major inorganic binders used in paints are sodium, potassium, lithium and ethyl silicates. These binders are used in zinc dust pigmented primers, in which they react with the fine zinc metal to form very hard films. These films are extremely resistant to corrosion in humid or marine environments. Many of these primers also contain substantial concentrations of lead oxides which react with the silicates in conjunction with the zinc to form an even more corrosion resistant coating.

Latex

Latex paints are based on aqueous emulsions of three basic types of polymers: polyvinyl acetate, polyacrylic, and polystyrene-butadiene. They dry by evaporation of the water, followed by coalescence of the polymer particles to form tough, insoluble films. They have little odor, are easy to apply, and dry very rapidly. Interior latex paints are generally used either as a primer or finish coat on interior walls and ceilings whether made of plaster or wallboard. Exterior latex paints are used directly on exterior (including alkaline) masonry or on primed wood. They are non-flammable, economical, and have excellent color and color retention. Latex paint films are somewhat porous so that blistering due to moisture vapor is less of a problem than with solvent thinned paints. They do not adhere readily to chalked or dirty surfaces, nor to glossy surfaces as under eaves. Therefore, careful surface preparation is required for their use. Latex paints are very durable in normal environments, at least as durable as oil paints.

Oil

Linseed oil is the major binder in oil house paints. These paints are the oldest type of coatings in use and have the longest history of performance. They are used primarily on exterior wood and metal since they dry too slowly for interior use and are sensitive to alkaline masonry. Oil paints are easy to use and give high film build per coat. They also wet the surface very well so that surface preparation is less critical than with other types of paints for metal. They are recommended for hand cleaned iron and steel. Oil paints are not particularly hard or resistant to abrasion, chemicals or strong solvents, but they are durable in normal environments.

Oil-Alkyd

Linseed oil binders are often modified with alkyd resins in order to reduce drying time; to improve leveling, hardness, gloss and gloss retention; and to reduce fading; and yet maintain the brushability, adhesion and flexibility of the oil. One use is in trim paints which are applied to exterior windows and doors. Since these areas are relatively small and painted in solid colors rather than tints, they require better leveling, gloss retention and fade resistance than the rest of the exterior walls. Also, these areas are subject to some handling, and therefore require faster drying and harder finishes. Oil-alkyd paints are also used on structural steel when faster drying finishes are desired. However, somewhat better surface preparation is required than with oil paints.

Oleoresinous

These binders are made by processing drying oils with hard resins. They generally are used either as spar varnishes or as mixing vehicles to be added to aluminum paste to produce aluminum paints. Alkyd finishes are often called oleoresinous because a drying oil is combined with the alkyd (phthalate) resin. Alkyd finishes usually are preferred where better color retention is desired.

Phenolic

Phenolic binders are made by processing a drying oil with a phenolic resin and are thus a class of oleoresinous binders. They may be used as clear finishes or pigmented in a range of colors in flat (lustreless) and high gloss finishes. The clear finishes may be used on exterior wood and as mixing vehicles for producing aluminum paints. The durability of the clears is very good for this class of finishes (1 to 2 years); the durability of the aluminum paints is excellent. Phenolic paints are used as topcoats on metal for extremely humid environments and as primers for fresh water immersion. These paints require the same degree of surface preparation as alkyds but are slightly higher in cost than alkyds. Phenolic coatings have excellent resistance to abrasion, water, and mild chemical environments. They are not available in white or light tints because of the relatively dark color of the binder. Furthermore, phenolics tend to darken during exposure.

Phenolic-Alkyd

Phenolic and alkyd binders are often blended to combine the hardness and resistance properties of the phenolics with the color and color retention of the alkyds. This may be done either by blending phenolic varnish with the alkyd vehicle or by addition of phenolic resin during processing of the alkyd resin.

Rubber-Base

So-called rubber-base binders are solvent thinned and should not be confused with latex binders which are often called rubber-base emulsions. Four types are available: chlorinated rubber, styrene-butadiene, vinyl toluene-butadiene and styrene-acrylate. They are lacquer type products and dry rapidly to form finishes which are highly resistant to water and mild chemicals. Recoating must be done with care to avoid lifting by the strong solvents used. Rubber-base paints are available in a wide range of colors and levels of gloss. They are used for exterior masonry, also for areas which are wet, humid or subject to frequent washing such as swimming pools, kitchens and laundry rooms. Styrene-butadiene, when combined with chlorinated plasticizers and silicone resins, is used to produce high heat-resisting ready-mixed aluminum paints.

Silicone

Silicone binders are used in two ways, for water repellents and for heat resistant finishes. Dilute solutions (5 percent solids) of silicone resin are of temporary help in reducing water absorption when applied to unpainted concrete or masonry such as brick or stone. They usually do not affect the color or appearance of the treated surface. Cracks and open joints must be repaired before water repellents are applied. Heat-resistant organic finishes containing a high concentration of silicone resins and when pigmented with aluminum, have the ability to withstand temperatures up to 1200° F.

Silicone Alkyd

The combination of silicone and alkyd resins results in an expensive but extremely durable coating for use on smooth metal.

Urethane

Two types of urethane finishes are covered in this manual, oil-modified urethanes and oil-free moisture-curing urethanes. Both are used as clears but the oil free type is also available pigmented.

Oil-modified urethanes are similar to phenolic varnishes, although more expensive, but have better initial color and color retention, dry more rapidly, are harder, and have better abrasion resistance. They can be used as exterior spar varnishes or as tough floor finishes. Oil modified urethanes can be used on all surfaces. In common with all clear finishes, they have limited exterior durability.

Moisture curing urethanes are the only organic products presently available which cure by reacting with moisture from the air. They also are unique in having the performance and resistance properties of two-component finishes yet are packaged in single containers. Moisture-curing urethanes are used in a manner similar to other one package coatings except that all containers must be kept full to exclude moisture during storage. If moisture is present in the container, they will gel.

Vinyl

Lacquers based on modified polyvinyl chloride resins are used on steel where the ultimate in durability under abnormal environments is desired. They are moderate in cost but have low solids and require the most extensive degree of surface preparation to secure a firm bond. Because of their low solids, vinyl finishes require numerous coats to achieve adequate dry film thickness so that the total cost of painting is higher than with most other paints. Since vinyl coatings are lacquers, they are best applied by spray. They dry quickly, even at low temperatures. Recoating must be done with care to avoid lifting by the strong solvents which are present. In addition, these solvents, present an odor problem. Vinyls can be used on metal or masonry but are not recommended for use on wood. They have exceptional resistance to water, chemicals and corrosive environments but are not resistant to strong solvents.

Vinyl-Alkyd

The combination of vinyl and alkyd resins offers a compromise between the excellent durability and resistance of the vinyls with the lower cost, higher film build, ease of handling, and adhesion of the alkyds. They can be applied by brush or spray and are widely used on structural steel in marine and moderately severe corrosive environments.

Table 17-14 summarizes the outstanding properties of the most important paint bases. Properties of the following binders are not included but can be estimated to be similar to those listed as follows:

Oil-Alkyd	= Oil + Alkyd
Oleoresinous	= Similar to alkyds but with less color retention
Phenolic-Alkyd	= Phenolic + Alkyd
Oil-Modified Urethane	= Phenolic + Alkyd
Vinyl-Alkyd	= Vinyl + Alkyd

Types of Paint

Paints can be described by type as well as by the base or vehicle. The various types of paint are listed and described below.

	Alkyd	Cement	Epoxy	Latex	Oil	Phenolic	Rubber	Moisture Curing Urethane	Vinyl
Ready for use	Yes	No	No[3]	Yes	Yes	Yes	Yes	Yes	Yes
Brushability	A	A	A	+	+	A	A	A	−
Odor	+[1]	+	−	+	A	A	A	−	−
Cure normal temp.	A	A	A	+	−	A	+	+	+
Cure low temp.	A	A	−	−	−	A	+	+	+
Film build/coat	A	+	+	A	+	A	A	+	−
Safety	A	+	−	+	A	A	A	−	−
Use on wood	A	−	A	A	A	A	−	A	−
Use on fresh conc.	−	+	+	+	−	−	+	A	+
Use on metal	+	−	+	−	+	+	A	A	+
Corrosive service	A	−	+	−	−	A	A	A	+
Gloss - choice	+	−	+	−	A	+	+	A	A
Gloss - retention	+	X	−	X	−	+	A	A	+
Color - initial	+	A	A	+	A	−	+	+	+
Color - retention	+	−	A	+	A	−	A	−	+
Hardness	A	+	+	A	−	+	+	+	A
Adhesion	A	−	+	A	+	A	A	+	−
Flexibility	A	−	+	+	+	A	A	+	+
Resistance to:									
Abrasion	A	A	+	A	−	+	A	+	+
Water	A	A	A	A	A	+	+	+	+
Acid	A	−	A	A	−	+	+	+	+
Alkali	A	+	+	A	−	A	+	+	+
Strong solvent	−	+	+	A	−	A	−	+	A
Heat	A	A	A	A	A	A	+[2]	A	−
Moisture permeability	Mod.	V. High	Low	High	Mod	Low	Low	Low	Low

+ = Among the best for this property   A = Average
− = Among the poorest for this property   X = Not applicable
[1] Odorless type
[2] Special types
[3] Two component type

Comparison of paint binders principal properties
Table 17-14

Lacquer

Strictly speaking, all coatings which dry solely by evaporation of the solvents may be described as lacquers. Thus, the rubber-base coatings and vinyl solution coatings covered in this handbook can be considered to be lacquers. Lacquers dry rapidly even at low temperatures, and brushing may be difficult. Their solids content usually is relatively low so that numerous coats are often necessary in order to obtain adequate film thickness. Recoating, especially by brush, must be done with care to avoid lifting by the strong solvents used.

Varnish

Varnishes are solutions of oil-modified alkyds or oil-modified resins (oleoresinous) in solvents, with driers added so that they will dry by oxidation. The film produced is clear so that the term "clear finishes" is often used to not only classify varnishes but other non-pigmented finishes as well, such as lacquers and moisture-curing urethanes. Varnishes are available in a variety of types as follows:

Spar Varnish. The term "spar" refers to the spars on a ship, hence spar varnishes are intended for exterior use in normal or marine environments. However, this material has limited exterior durability.

Aluminum Mixing Varnish. Aluminum paint is often made by mixing aluminum paste with a varnish before use. The varnish used is specially formulated to produce optimum leafing of the aluminum, (metallic brilliance) and to retain this leafing for at least a few days after mixing.

Sealer. Clear sealers are varnishes to which additional solvent has been added to make them quite thin in viscosity. Thus, they penetrate and seal the substrate rather than form a relatively thick film on its surface. Sealers are used to prevent grain raising, and to fill the pores of porous substrates such as plywood to avoid excessive loss of binder when topcoats are applied. Another use is to seal wood floors without leaving any significant glossy film on the surface which could be marred by subsequent abrasion under use.

Flat Varnish. Varnishes normally dry with a high gloss. Often, a lower gloss or even flat finish is desired for reasons of appearance or to reduce glare. The varnish finish, if hard enough, can be dulled by hand rubbing with a very mild abrasive such as rottenstone. A much simpler method is to use a varnish whose gloss has been reduced by the addition of transparent but highly efficient flatting pigments such as certain synthetic silicas. These pigments are dispersed in the varnish to produce a clear finish, which will dry to a low gloss, but will still be transparent so that the surface underneath (for example, the grain of the wood) will not be obscured.

Paint and Enamel

White and colored pigments are dispersed in paint binders to produce paints and enamels. If the pigmented product is relatively easy to brush and is used on large areas such as walls, it is called a paint. If it is relatively fast drying, levels out to a smooth, hard finish and is used on relatively small areas or smooth materials such as woodwork, it is called an enamel. Paints can be rolled whereas enamels rarely are. The term "paint" is also used in the broad sense to cover all types of pigmented opaque finishes and also their application.

The extent of pigmentation of the paint or enamel determines its gloss. Generally, gloss is reduced by adding non-opaque, lower cost pigments called extenders. Typical extenders are calcium carbonate (whiting), magnesium silicate (talc), aluminum silicate (clay), and silica. The level of gloss is achieved by varying the ratio of pigment to binder.

The maximum gloss in paints and enamels is obtained by omitting all extenders. This characteristics produces maximum washability and durability and is referred to as "high gloss." Low gloss paint may be called flat, dull or lustreless. Pigmentation is highest to reduce gloss to a minimum. Low gloss minimizes surface imperfections on interior walls. In using low gloss finishes, washability and durability are sacrificed to some degree. Semigloss is a compromise between the appearance of the flat finish and the performance of the high gloss finish.

Primer for Paint and Enamel

Generally two types of paints are required. The coat of paint next to the material painted, called the primer, is formulated to perform other than as a finish coat. The topcoat is used to produce the desired finished appearance or to protect the primer and consequently the substrate beneath it. Various types of primers are used, depending on the substrate and, in some cases, the topcoat as well.

Primer-sealer is used to seal a porous or alkaline substrate so that the topcoat is unaffected by loss of or damage to the binder. Some paints, such as interior latex wall paints, are self-priming and usually do not require a special primer.

Enamel undercoater is a coating which dries to a smooth hard finish that can be sanded. When sanded, the perfectly smooth surface is ideal for obtaining the best leveling and appearance of the subsequently applied enamel topcoat.

Primers based on anti-corrosive pigments, such as red lead, zinc chromate, lead silico chromate and zinc dust, are applied to iron and steel wherever corrosion is a problem. They retard corrosion of the metal but must usually be protected by other types of coatings.

"House Paint"

"House paint" is the commercial term for exterior paints mixed with many different formulations. It is the most widely used type of paint. Formulations are available for use on all surfaces and for all special requirements such as chalk or mildew resistance. White is the most popular color. House paint comes in both oil base and latex (water base) paint.

Aluminum Paint

Metallic paints such as aluminum paints are available in two forms: ready mixed and ready-to-mix. Ready mixed aluminum paints are supplied in one package and are ready for use after normal mixing. They are made with vehicles which will retain leafing qualities or metallic brilliance after moderate periods of storage. They are more convenient to use and allow for less error in mixing than the ready-to-mix form.

Ready-to-mix aluminum paints are supplied in two packages, one containing the clear varnish and the other the required amount of aluminum paste (usually ⅔ aluminum flake and ⅓ solvent). They are mixed just before use by slowly adding the varnish to the aluminum paste while stirring. The mixed paint will usually retain its leafing for a few days. Since leaf retention during storage is no problem, ready-to-mix aluminum paints allow a wider choice of vehicles and present less of a problem with storage stability. Leafing also is generally better when this paint is freshly mixed.

A potential problem with aluminum paints is moisture. If moisture is present in the container, it may react with the aluminum flake to form hydrogen gas and create pressure buildup in the closed container. This can cause bulging and even popping of the cover of the container. Check all ready mixed paints for bulging. If present, puncture the can cover carefully before opening. Be sure to use dry containers when mixing two package paints. Always brush out a test area before use to be sure that leafing is satisfactory.

Oil Stain

Oil stains are based on a drying oil such as linseed oil in which color pigments are dispersed and which is then thinned to a very low consistency to obtain maximum penetration when applied to wood. Interior stains are applied generously to the sanded, dust free surface,

	Aluminum	Cement base paint	Exterior clear finish	House paint	Metal roof paint	Porch-and-deck paint	Primer or undercoater	Rubber base paint	Spar varnish	Transparent sealer	Trim-and-trellis paint	Wood stain	Metal primer
Wood													
Natural finish	–	–	P	–	–	–	–	–	P	–	–	P	
Porch floor	–	–	–	–	–	P	–	–	–	–	–	–	
Shingle roof	–	–	–	–	–	–	–	–	–	–	–	–	
Shutters and trim	–	–	–	P+	–	–	P	–	–	–	P+	–	
Siding	P	–	–	P+	–	–	P	–	–	–	–	–	
Windows	P	–	–	P+	–	–	P	–	–	–	P+	–	
Masonry													
Asbestos cement	–	–	–	P+	–	–	P	P	–	–	–	–	
Brick	P	P	–	P+	–	–	P	P	–	P	–	–	
Cement & Cinder block	P	P		P+	–	–	P	P	–	P	–	–	
Cement porch floor	–	–	–	–	–	P	–	P	–	P	–	–	
Stucco	P	P	–	P+	–	–	P	P	–	P	–	–	
Metal													
Copper	–	–	–	–	–	–	–	–	P	–	–	–	
Galvanized	P+	–	–	P+	–	–	P	–	P	–	P+	–	P
Iron	P+	–	–	P+	–	–	–	–	–	–	P+	–	P
Roofing	–	–	–	–	P+	–	–	–	–	–	–	–	P
Siding	P+	–	–	P+	–	–	–	–	–	–	P+	–	P
Windows, aluminum	P	–	–	P+	–	–	–	–	–	–	P+	–	P
Windows, steel	P+	–	–	P+	–	–	–	–	–	–	P+	–	P

"P" indicates preferred coating for this surface.

"P+" indicates that a primer or sealer may be necessary before the finishing coat or coats (unless the surface has been previously finished).

**Exterior paint selection chart
Table 17-15**

allowed to dry for a short period of time, and the excess then wiped off so that only the stain which has penetrated the wood remains.

Selecting Exterior Paints

Use Table 17-15 to select the preferred coating for exterior coatings. Some specific suggestions ae as follows:

For wood siding, oil base house paint is still first choice. However, latex paint is preferred by many users.

For wood trim, windows, shutters, and doors, exterior trim (or trim-and-trellis) paint is the favorite finish. It is an oil-alkyd-resin base enamel. Its properties include rapid drying, high gloss, good color and gloss retention, and good durability. House paint may be used, but it does not retain its gloss as long. Also, any chalking may discolor adjacent surfaces.

Exterior latex masonry paint is a standard paint for masonry. Cement-base paint may be used on nonglazed brick, stucco, cement, and cinder block. Rubber-base paint and aluminum paint with the proper vehicle may also be used.

Ordinary house or trim paints may be used for the finish coats on gutters, downspouts, and hardware or grilles. A specially recommended primer must be used on copper or galvanized steel. Use house paint, aluminum paint, or exterior enamel on steel or aluminum windows. Paint window screens with a special screen enamel.

For concrete or wood porches and steps, use porch-and-deck paint. On wood, a primer coat is applied first. On concrete, an alkali-resistant primer is recommended. Rubber-base paints are excellent for use on concrete floors. Hard and glossy concrete surfaces must be etched or roughened first.

Selecting Interior Paints

Interior finishes for wood and drywall or plaster surfaces are usually intended to serve one or more of the following purposes:
1. Make the surface easy to clean.
2. Enhance the natural beauty of wood.
3. Achieve a desired color decor.
4. Impart wear resistance.

	Aluminum paint	Casein	Cement base paint	Emulsion paint (including latex)	Enamel	Flat paint	Floor paint or enamel	Floor varnish	Interior varnish	Metal primer	Rubber base paint (not latex)	Sealer or undercoater	Semigloss paint	Shellac	Stain	Wax (emulsion)	Wax (liquid or paste)	Wood sealer
Floors:																		
Asphalt tile																	P+	
Concrete																P+	P+	
Linoleum							P							P		P	P	
Vinyl and rubber							P	P								P	P	
Wood							P+	P+									P	
Masonry:																		
Old	P	P	P	P	P+	P+					P	P	P+					
New			P	P	P+	P+					P	P	P+					
Metal:																		
Heating ducts	P				P+	P+				P	P		P+					
Radiators	P				P+	P+				P	P		P+					
Stairs:																		
Treads							P	P						P	P			
Risers					P+	P+			P			P	P+	P	P			
Walls and ceilings:																		
Kitchen and bathroom				P	P+						P	P	P+					
Plaster		P		P		P+					P	P	P+					
Wallboard		P		P		P+					P	P	P+					
Wood paneling				P+		P+			P									
Wood trim				P+	P+	P+			P		P	P	P+	P	P		P	P
Windows:																		
Aluminum	P				P+	P+				P	P		P+					
Steel	P				P+	P+				P	P		P+					
Wood sill					P+				P			P			P			

"P" indicates preferred coating for this surface.

"P+" indicates that a primer or sealer may be necessary before the finishing coat or coats (unless the surface has been previously finished).

Interior paint selection chart
Table 17-16

The type of finish depends largely upon type of area and the use to which the area will be put. The various interior areas and finish systems employed in each are summarized in Table 17-16. Wood surfaces can be finished either with a clear finish or a paint. Plaster-base materials are painted.

Wood Floors

Hardwood floors of oak, birch, beech, and maple are usually finished by applying two coats of wood seal, also called floor seal, with light sanding or steel wooling between coats. A final coat of paste wax is then applied and buffed. This finish is easily maintained by rewaxing. The final coat can also be a varnish instead of a sealer. The varnish finishes are used when a high gloss is desired.

When floors are to be painted, an undercoater is used, and then at least one topcoat of floor and deck enamel is applied.

Wood Paneling and Trim

Wood trim and paneling are most commonly finished with a clear wood sealer or a stain-sealer combination and the topcoated, after sanding, with at least one additional coat of sealer or varnish. The final coat of sealer or varnish can also be covered with a heavy coat of paste wax to produce a surface which is easily maintained by rewaxing. Good depth in a clear finish can be achieved by finishing first with one coat of high-gloss varnish followed with a final coat of semigloss varnish.

Wood trim of nonporous species such as pine can also be painted by first applying a coat of primer or undercoater, followed with a coat of latex, flat, or semigloss oil base paint. Semigloss and gloss paints are more resistant to soiling and are more easily cleaned by washing than the flat oil and latex paints. Trim of wood species such as oak and mahogany, which are porous, requires filling before painting.

Drywall and Plaster

Plaster and drywall surfaces which account for the major portion of the interior area are finished with two coats of either flat oil or latex paint. An initial treatment with size or sealer will improve holdout (reduce penetration of succeeding coats) and thus reduce the quantity of paint required for good coverage.

Allow new plaster to age at least thirty days before painting if any oil based paint is to be applied. Latex paint can be applied after 48 hours, though a 30 day wait is generally recommended. Fill narrow cracks or small holes with spackle; larger openings or deep holes should be filled in thin layers with patching plaster. Allow spackle to dry four hours and plaster or joint cement 24 hours, then sand smooth before painting. On old work, remove all loose or scaling paint, then sand lightly, especially edges of surrounding painted areas. Wash off all dirt, oil and stains, then allow to dry thoroughly if solvent-thinned paints are to be applied.

Kitchen and Bathroom Walls

Kitchen and bathroom walls which normally are plaster or drywall construction are finished best with a coat of undercoater and two coats of semigloss enamel. This type of finish wears well, is easy to clean, and is quite resistant to moisture.

Chapter 18
Managing Your Home Improvement Business

Business management is usually one of the last things remodeling contractors have time to study. Meeting today's challenges and cleaning up yesterday's problems may be about all many remodelers have time for. Business management is what goes on at I.B.M. and General Motors, not in your operation. Right? Well, not exactly. The plain truth is that every year many remodeling contractors go into business, take on a few jobs, and come to realize that they can't make enough money to support themselves. Other remodelers seem to get all the breaks. They develop growing, prosperous businesses that pay a substantial salary to the owner and the businesses become more prosperous year after year. What is the difference between the growing operation and the business going nowhere? Sometimes it really is just good luck or being in the right area, getting started at the right time or having a sales program that the market can't resist. But more often, the business that grows and prospers year after year is managed better than the competition. It's been said that businesses don't compete, managements compete. In your business you will be in competition with the managers (owners) of other businesses. The more you know about the management of your business, the better your chance for survival.

O.K., you say, I have to understand business management, whatever that is. But what is there to managing my business that I don't know already? Unfortunately, there may be plenty about doing business as a remodeler that you have not seen or heard about. If you once worked for a large, well run home improvement firm or construction company and saw first hand and at close range how they handled estimating, sales, finances and job control, you probably know as much about managing a construction company as you will learn by reading this book. Most home improvement contractors don't begin business with a background like that though. If you are like most remodeling contractors, you have a relative who is or was in building and you worked on construction or home improvement projects before you decided to go into business for yourself. You probably became very good at carpentry, selling, estimating or installing flooring, equipment, tile, roofing or some other part of the whole job. But it isn't very likely that you handled the financial records, employee reports and government requirements that you have as a manager of your own business. Most important, you never had responsibility for putting the whole operation together and planning for profits. That's really what management is; handling sales, production, costs and finances so that you make a profit on the work you do. Understand business management as increasing sales, controlling the work, planning costs and handling finances so that you make money. You are managing your business every time you make a decision or fail to make a decision about sales, production, costs and finances. The better you are at managing your business, the more money you will make.

The last four chapters in this book cover the four major phases of managing a home improvement business. Several earlier chapters have touched on sales and production as they related to bathroom, kitchen and roofing jobs. The four chapters here will apply to almost any type of home improvement work, though some specialties may need more precise management in some areas than in others.

Making It In Home Improvement

If you have been in the home improvement business for any length of time, you know that one contractor will get more work and make more money while another in about the same business, in the same area, and with about the same skills won't do nearly as well. The remainder of this chapter compares two home improvement operations; one fails and the other continues to grow. Both firms are in the same business, roofing, and did their work in the same metropolitan area of a growing southern state. The success and failure really happened. The names and some circumstances have been changed, but the facts are true. These contractors were in roofing, but they might have been in bathroom or kitchen remodeling, room additions or any other part of the home improvement industry. The outcome would have been about the same.

As you read over these short histories you will be reminded of mistakes you made or have seen made and, hopefully, some outstanding successes you have participated in. Ask yourself what you would have done to avoid failure in the first case and how you can use some of the principles the second contractor used to succeed.

Morgan Roofing Company

Craft experience of the owners is considered an important asset in the home improvement industry. However, it is not the only or the most important key to success. In many cases failure is not the result of the lack of craft experience but of incompetence in management. The case of Morgan Roofing Company proves the point. It shows a typical example of failure due to the lack of planning on the part of management.

The owners had a number of years of on-the-job experience prior to starting out on their own. As roofing mechanics, they were superior; as managers they lacked even average ability. The initial success of the company was due to an external factor, a favorable economic climate in the area, rather than to good management. As soon as competition got tough and the company desperately needed expert guidance, the management faltered and the company failed.

The Morgan Roofing Company and its subsidiary the ABC Tile Company were both started in 1971. From the beginning of operations until bankruptcy in 1976, the Morgan Company specialized in residential roofing, mostly within about 20 miles of their office in Capitol City. The ABC Company made roofing tile, with over 90 percent of the production sold to the parent roofing company. Tile roofing was very popular on homes in the area. Four brothers, aged between 31 and 47, were the officers and stockholders of both companies. All four had at least 12 years of practical work experience as roofing mechanics prior to forming the two companies. Their formal education, however, had suffered because all had started working as roofers at an early age. The brothers first worked for their father, who had also been a roofer for most of his adult life. In 1963 they formed a partnership, which subsequently was incorporated in 1968. This corporation was later dissolved and reorganized under the name of Morgan Roofing Company in 1974 because the father was getting old and wanted to retire from the business.

The father of the four brothers furnished most of the capital for the new partnership in 1963. The beginning was very modest. Assets of the company were almost nonexistent and most contracts were performed by the family's own manpower. Big time was not hit until the brothers met Mr. Craig, president of the City Supply Company at the time the Morgans started to deal with that firm in 1966. Mr. Craig introduced the Morgan brothers to the owner of a large home rehabilitation contractor (Rehab, Inc.) which began to subcontract most of its reroofing jobs to Morgan Roofing. Rehab Inc. gave work to Morgan Roofing at negotiated but competitive prices. Rehab Inc. paid Morgan Roofing with checks made out jointly to Morgan Roofing and City Supply Company. City Supply advanced roofing supplies to Morgan Roofing knowing that payment was secured by the requirement for joint endorsement on the checks.

From the time Morgan Roofing acquired Rehab, Inc. as a prime customer, the brothers' business advanced rapidly in volume, reaching a high of over $525,000 in 1972. At the same time, the tile company sold over $140,000 worth of roofing tile. The combined annual sales of the two companies amounted to over $650,000 in 1972. Net profit before taxes totaled $39,120, or 5.8 percent of net sales. Things looked rosy indeed and the brothers were enthusiastic. Expansion outside of the county was planned. These plans were carried out in the next year and land was purchased in Centerville in neighboring Jefferson County. A total of $40,000 was invested in land and a building. Centerville was experiencing a minor building boom then and the brothers thought they could capitalize on it. However, the boom was soon followed by a bust and not enough business was generated to warrant the investment. Meanwhile, cash reserves were drained from the parent company, and working capital was seriously short, a condition from which the company never recovered. A debtor's petition for corporate reorganization was filed and approved in January 1976. However, it was too late to save anything by then. Both corporations were adjudicated bankrupt in April 1976.

Capital and Resources

When Morgan Roofing Company was formed (1974), it took over the assets of the predecessor corporation, consisting mostly of rolling stock and equipment such as ladder loaders, but very little cash. This, however, did not stop the brothers from further investment. During the past years, roofing tile had occasionally been in short supply. The brothers decided to operate their own plant so that work would never have to be stopped because of lack of tile. The Culver Tile Company was purchased, and a separate corporation was set up to handle this operation. The total purchase price was $52,000. This included land and a building ($44,000), and tile-making equipment ($8,000), which included various forms, kilns, drying racks, etc. The plant was not fully mechanized and production of tile was essentially a hand operation. A standing inventory of ready-mix tile (about $7,000 worth) was taken over on a consignment basis.

Financing of the purchase was accomplished in the following manner: Mr. Culver was given a $40,000 first mortgage on the property. The loan carried an interest rate of nearly 10 percent. The remaining $12,000 was paid in cash in several installments within a six months' period, the initial down payment being $6,000. The money to make the down payment and the installment payments was obtained by "using the supply houses" as was explained by one of the brothers.

He meant that the Morgans would not pay for supplies upon receipt of payment for jobs where the supplies had been used. By deferring payment they were able to accumulate money for the down payment on the Culver property. The suppliers were paid eventually with money earned from subsequent jobs. Similar procedures were repeated later and were largely responsible for the eventual bankruptcy of the business.

When the decision was made to engage in roofing business in Centerville, Jefferson County, only a small warehouse for tools and a limited amount of supplies was planned. That was in February 1973. When the building was finished about August 1973, the warehouse had increased in size considerably and three apartments had been built on a second floor at a total cost of about $40,000. This was an unwarranted investment not appropriate for the size and character of the business. The brothers "spread themselves too thin" without any cash reserves. Apparently they believed that business would continue to expand and take care of their increasing fixed expesnes and mortgage payments. However, this was not the case and a slump in business during 1974 coupled with heavy tax liens by the Federal Government for back taxes led to trouble.

Mode of Operation

The two firms were located at an outlying commercial-residential section in Capitol City. The tile manufacturing plant occupied approximately 4,000 square feet in a one-story concrete block structure with open yard storage. Offices for both companies were maintained in a separate one-story concrete block building measuring approximately 800 square feet.

During the entire period of operations activities were concentrated on residential reroofing work, and very little commercial roofing was undertaken. The largest volume was with Rehab, Inc. on subcontracts. There was very little reroofing done on direct contracts with individual home owners. It was only in 1974 and 1975, when the volume of subcontract work declined, that work on individual houses became more prominent in the company's operations.

Employment of the roofing company fluctuated sharply depending on the amount of work on hand. At times there were up to 50 roofers, including helpers, on the payroll. The company used union labor until the middle of 1974, then switched over to non-union roofers, mostly because management preferred to have its men work overtime in order to finish jobs quickly. At union rates the overtime had become too expensive. Constant and immediate need of cash to pay suppliers and creditors was the chief reason for this and most other major decisions the brothers made during 1974.

The tile company, when in full production, employed about 15 tile makers (non-union) steadily. The active operations of this venture,

however, did not last long. In the latter part of 1973 three larger tile companies in this area became fully mechanized. Production at these mechanized plants was sufficient to satisfy demand for the entire county. Machine-made tile undersold hand-made tile by at least twenty dollars per square (100 square feet). It became more profitable for the Morgans to buy tile from outside sources than to produce their own. Consequently, the tile plant was shut down. Mechanization at a cost of about $75,000 was considered. However, by this time it was already clear, even to the Morgan brothers, that such an investment would be beyond the financial resources of the company, especially since the branch facilities for the roofing company in Centerville were nearing completion about this time.

During the short period of active operations the Tile Company produced highly satisfactory profits. For the fiscal year ending October 31, 1972, net profits after taxes amounted to $18,295, or 12.7 percent of net sales. The tile was sold to Morgan Roofing at prevailing market rates. In 1973 the company stopped production, but stayed active as a holding company, owning the land and buildings in Capitol City.

All four brothers participated in the management of both companies. Duties of each were not clearly defined. Management decisions were made more or less on the spur of the moment by any of the officers who happened to be present at the time when a problem came up. In addition to negotiating for new business, making purchases, and supervising field crews, the brothers themselves worked as roofers whenever time permitted them to do so. They had a reputation in the trade as extremely hardworking men, capable of outproducing any average journeyman roofer. They also worked long hours, often 10 to 12 hours a day, seven days a week, as long as there was work to be done. The quality of their roofs was satisfactory by any measure.

A full-time bookkeeper was employed to take care of the paperwork in the office. An outside accountant prepared financial statements whenever they were needed for the purpose of negotiating a bank loan, and also filed the income tax returns for both companies. Books were kept separately for each company. However, intercompany relations were quite complicated. Loans, advances, and personnel were shuffled between the two companies to such an extent that for all intents and purposes the two enterprises were actually operated as one. For example, the roofing company had a very high accident rate among its employee who did the "hot work", that is, mopped asphalt on the roofs. The insurance rate for the roofing company was extremely high. The rate for ABC Company's tile makers, on the other hand, was low. Because of this, it was decided that the truckers and tile-layers should be kept on ABC's payroll so that the business could take advantage of that company's low insurance rate. Consequently, the tile company showed a paper loss every year after 1972 although no business was transacted. The profits of the roofing company looked high because of the losses A.B.C. Company was piling up. After a while the books became so confusing that proper tax returns could not be filed. The Internal Revenue Service audited the books of both companies in June 1974 and discovered that the roofing company had at times "subcontracted" tile-laying work to individual tile layers and their helpers. In these instances employee payroll deductions had not been made. The I.R.S. auditors determined that these men were not independent subcontractors at all, but in fact were employees of the Morgan Roofing Company. The roofing company was then considered responsible for the tax deductions from their wages. The company was assessed for back employee taxes (withholding) for three years, with a 20 percent penalty and 6 percent interest added to the total amount. At the time when bankruptcy proceedings started late in 1975, the total tax indebtedness of both corporations amounted to $93,496.25, enough by itself, regardless of other liabilities, to make recovery hopeless.

Bidding and New Business

In the beginning, when business was small, the Morgans used F. W. Dodge reports to uncover leads for new starts in residential construction. Later, when business expanded and the company became well known to builders in Capitol City, personal contacts and telephone calls were deemed sufficient to bring in new business. Very little advertising was done and no real effort was made to sell directly to homeowners, though the brothers could reroof homes as quickly and as well as anyone in Capitol City.

Bidding on jobs was done on the basis of squares (100 square feet) to be covered. Profits were satisfactory in 1972 and 1973. Total volume in 1973 fell only slightly below the all-time high in 1972. In 1974, however, annual contract income declined to about the $400,000 level and the company either lost money or barely broke even that year.

The reason for this was that the firm desperately needed money to meet outstanding financial obligations. The company was deeply in installment debt as a result of overinvestment in fixed assets. In pressing hard for more volume, the Morgans had to lower prices substantially. The brothers soon established a reputation in the trades for underbidding everybody in sight. The going price per square of gravel roof in 1974 was $35.00 or more. Morgan Roofing often quoted prices as low as $30.00 per square. At a cost of $27.00 for themselves (material and labor), the gross profit per square was only $3.00, or 10 percent of net sales. Such a profit margin was not even enough to cover overhead expenses, which had increased because of the burden of debt

service. The brothers were handling a large volume of business though. Plenty of cash was coming in and the brothers assumed that they were making money.

Purchasing

Until 1973 the company was able to, and generally did, pay bills promptly to take advantage of purchase discounts. Trouble started, however, in the latter half of 1973 because of the building in Centerville and a general decline in building. The drain on the company's assets proved to be severe. No longer was it possible to make large purchases at advantageous prices for cash or pay bills within the 10-day period to get a 2 percent discount. The discount loss on Morgan's volume of purchase amounted to $4,000 to $6,000 a year.

Furthermore, trouble developed with supplier creditors because of late payments. Some cut off credit to Morgan Roofing completely and sold only on a cash and carry basis. Some started billing general contractors directly for supplies sold to the Morgans, but charged a little more for the extra work involved.

Late in 1973 the brothers decided to consolidate the $25,000 in outstanding balances owed to suppliers. City Supply Company was willing to take their account because of the working arrangement with Rehab, Inc. However, a few months after City Supply became the Morgan brothers' largest creditor, Rehab, Inc. began demanding greater price concessions and grew slower and slower in paying its bills. Money available for construction loans was drying up and less home improvement work was available. This further compounded the squeeze on Morgan Roofing. The Morgans needed work from Rehab, Inc. but they needed work with at least a reasonable profit margin and most important, they needed to be paid on time. Morgan Roofing refused some work from Rehab because the price offered was too low. At one point Morgan refused all work from Rehab and threatened legal action. Rehab then brought their account up to a 60 day basis but relations between the two companies were now less than friendly.

By early 1975 the Morgan brothers were doing almost no business with Rehab, Inc. and their overall volume had declined to less than half what it had been in 1972. With the regular checks from Rehab, Inc. cut off, Mr. Craig at City Supply was no longer willing to carry the Morgan brothers. Mr. Craig refused requests for additional roofing supplies on credit and eventually forced Morgan Roofing into bankruptcy by suing to collect on the $25,000 loan.

Morgan Roofing had a good reputation for professional work in late 1975 and the brothers discovered that there were plenty of reroofing jobs available at prices of $50.00 per square or more when they could sell directly to homeowners. Selling roofing services was new to the brothers, however, because they had been reluctant to take on "little pick-up jobs" when large volume (at low markup) was their objective. The brothers worked hard at selling small roofing jobs and by late 1975 Morgan Roofing was doing a reasonably good business in reroofing individual houses. However, it was clear to nearly everyone that the brothers would never be able to pay the $200,000 they owed City Supply, the I.R.S. and the mortgage holders.

Reasons for Failure

The main reason why the Morgans failed in business was the nearly complete lack of clear headed planning by management. The affairs of the company were run in a haphazard manner.

This is shown with particular clarity in the case of the Centerville investment. The idea of expansion was definitely premature and the company was certainly not ready for it financially. Despite the good profits of 1972, danger signals were already evident indicating that the company was becoming top heavy with liabilities.

Another unfavorable factor was excessive overhead. The schedule of depreciable assets as of April 30, 1973 reveals that the roofing company owned 14 trucks, mostly one or two years old, and 5 late model, expensive automobiles. Most of this equipment was being paid for on the installment basis. In addition, mortgage payments at high interest rates had to be met. Consequently, both companies always labored under working-capital shortages. In late 1974 salaries to officers were about the only obligations that the firm met on schedule.

In addition to being hard workers, the brothers also enjoyed a reputation as being fast spenders. Initial success of both companies apparently had an adverse effect on their business judgment. Before they realized what was happening and stopped spending heavily on land, buildings, trucks, cars and high salaries, it was almost too late to save anything, even if more experienced management could have taken over the firm.

At no time had fraud been suggested in this case. It seems rather that this is a typical failure where the owner-managers had sufficient technical know-how to run the production end of the business but were woefully lacking in business or managerial ability.

Dade Roofing Company

Dade Roofing Company presents the example of an operator in the home improvement field who survived throughout a turbulent era that saw businesses like Morgan Roofing fail. By foregoing some of the quick profits which can be made in boom periods, and by wisely cashing in on the opportunities which exist in the area's large home repair market, the company was able to grow into a well established concern of medium size, in good financial standing, with a fine reputation for quality work.

The business was originally established in 1938 as an individual proprietorship and is thus one of the oldest roofing companies in the state. It was incorporated in 1963.

Mr. Dade, the founder of the firm, is in his sixties, and has spent his entire adult life in the roofing business. Prior to entry into business for himself, he worked for several years as a roofing mechanic in the maintenance division of a large manufacturing company in Detroit. The firm is a family enterprise; two sons of Mr. Dade, in their twenties, also work in the business. The sons are already thoroughly familiar with roofing work. The boys worked as apprentices during summers and, after graduating from high school, as roofing mechanics and estimators. Tom, the older son, is now in charge of the office and shares the management load with his father. Bob, five years younger, still works as supervisor, salesman and estimator, but spends more and more of his time in the office, learning the routine functions. By the time the elder Dade is ready to retire, the management turnover should create no problems.

Business Development

Dade Roofing Company is located on a principal thoroughfare in a commercial district of the northwest section of Capitol City. The company rents the one-story concrete block building of about 70 x 100 feet from Mr. Dade, Sr.

The building houses the offices (three rooms), the warehouse, and a small sheet metal shop run by the company. Additional warehouse space is rented across the street. There is sufficient room behind the building to park company cars, trucks, and kettles.

Mr. Dade started the business in 1938 with about $6,000 in capital, his personal savings. His gross sales were between $40,000 - $50,000 per annum, on which he used to net around 12-15 percent, or $6,000 per year as contrasted to the margins of today, which rarely exceed 6 to 8 percent.

The firm managed to survive boom and bust in the 1950's, although it lost money in several years. Real prosperity did not come until the mid 1960's. The company engaged profitably in defense contracts when these were available, doing roofing and sheet metal work for air bases and army camps in several parts of the state. During the 1950's business remained prosperous overall. The company participated in the housing boom and gross annual volume rose as high as $700,000. For various reasons, Mr. Dade decided to curtail operations in 1968 and to specialize in the reroofing phase of the business. Since then, annual volume has been smaller, around $500,000.

In 1975 the roofing division reported sales at $503,000 and had $16,000 (after depreciation) worth of equipment and rolling stock. The latter figure includes 3 cars and 6 trucks (some over 8 years old, but all in good condition), carried on the books at a value of $7,900 after depreciation. Sheet metal shop machinery and kettles are carried at a depreciated value of $4,400, again a conservative valuation, and furniture and fixtures at only $482. The assets of the company are worth considerably more than the reported book value.

Capitalization and Resources

The company has an authorized capital of 50 shares of no par value common stock, all of which have been issued and are held by members of the Dade family. The common stock has been valued at $25,000. The earned surplus in December 1976 amounted to $63,797, adding up to $88,797 as total net worth, or $1,776 book value per share. Actual value per share should be considerably higher since the total assets of the company seem to be undervalued. For instance, the company's balance sheet indicates an inventory of $37,000 but the inventory is insured for $60,000. Secondly, as explained before, the fixed assets are also estimated at a very conservative figure.

The working capital position of the company is very strong. The available financial statements show the following picture:

Year Dec. 31	Current Assets	Current Liabilities	Ratio of Current Assets to Current Liabilities
1974	$ 92,869	$18,268	5.1 to 1
1975	91,738	26,911	3.4 to 1
1976	101,641	25,707	4.0 to 1

With long-term liabilities not exceeding $1,100 in those years, it is clear that the business is in very sound financial condition. For 1976, cash in a savings account alone covered the total debt.

Mode of Operation and Position in the Industry

The company is divided into three separate units as follows:

	Percent of Annual Sales
Roofing Division	50
Sheet Metal Division	15
Supply Division	35

Books are kept separately for each division for control purposes only. Federal income tax is paid as one corporation.

The sheet metal division manufactures gutters, ventilators, skylights, ducts, stainless steel table tops, etc. The supply company has the wholesale dealership for a well known line of roofing products. In the future, the roofing division is expected to produce at least 70 percent of the annual sales volume. The other two divisions will be deemphasized. Insufficient profits is the reason given by management. It is felt that too much supervisory effort and office paperwork goes into the operation of these other divisions and that the returns do not justify the time spent on their operation.

A routine working day begins with the crews reporting to the office for daily assignments. Each work crew consists of one mechanic, two apprentices, one kettleman, and two or three laborers. If the job is small (14 to 16 squares), a smaller crew may be used. The mechanic or journeyman roofer acts as foreman. He gets the instructions on location of work and job requirements and draws the materials needed from the warehouse.

The typical job is a 4-ply built-up roof over a plywood deck. In reroofing, the first job is to rip off the old cover. Then the deck is inspected for firmness and any deteriorated parts are replaced. Most mechanics employed by Dade Roofing Company also qualify as carpenters. Small repairs on wood decks are therefore handled by the company and are not subcontracted.

After the deck has been restored to a sound condition, 30-pound asphalt felt is applied over the deck. Next, two 15-pound felts are mopped in place. On top of these the required amount of asphalt is poured. Then a layer of gravel is spread over the asphalt to form a protective cover against the sun.

Dade Roofing Company is one of the leading companies in the county specializing in reroofing and roofing repairs. Its policy of maintaining quality workmanship and guaranteeing all work performed has assured the company an excellent reputation and consequently a steady volume of business.

Bidding and New Business

The bulk of the business is obtained by finding and bidding on jobs on the open market. Homeowners shop around, usually contact at least two or three companies and compare prices before entering into a contract for a new roof. Commercial jobs have been obtained mainly because the company is well known in the field and is asked to participate in bidding for contracts.

Because the company concentrates on reroofing and individual homeowners make up most of this market, it is important to keep the company's name constantly before the eyes of the general public. Mr. Dade advertises extensively. He places ads in the two local newspapers twice a week and regularly spends money on such miscellaneous gift items as calenders, notebooks, pens and pencils.

Between 1.0 and 1.5 percent of the total sales dollar is spent on advertising. Over 2 percent of roofing income is spent on advertising. The company employs two field estimator-salesmen to follow leads and sell reroofing jobs. For purposes of easier work distribution, the Capitol City area has been divided into two sections. Each estimator has a specific territory.

The field estimator-salesman performs a very important part in the company's business. He signs the contract for the company and his judgment as to the total cost involved on any given job will determine the profitability of the contract. Experience in estimating reroofing work is most important here. Just knowing the cost of labor and

303

materials is insufficient. A good estimator must be familiar with the capabilities of the men working for the firm and must be able to judge by visual inspection of a roof just what difficulties may be encountered and how much time the job will take. After taking all these factors into consideration, he computes the cost per square and then quotes a price for the job. Because of variables such as the pitch of the roof, number of valleys, and quality of materials to be used, the prices quoted per square vary, even if the total area to be covered is the same. The prices per square also fluctuate from week to week because of the effect of competition. Dade roofing has lowered its target price to meet competition on some jobs but passes up many jobs that might be unprofitable.

Generally, management sells roofing at a mark-up of 33 percent over cost to cover overhead and profit. This means 25 percent of the sales dollar goes for overhead and profit. If overhead runs 15 percent, profit before taxes is 10 percent. The company has not been able to maintain a 10 percent margin in recent years for their entire operation. Lower profits in the supply division and the sheet metal division have reduced overall profitability to about 5 percent in most years. Because the profit margin is thin, Mr. Dade insists that his estimator-salesmen take into account all administrative overhead and indirect costs when bidding jobs. Every job has to carry it's own share of all company expenses including all salaries and overhead. Whenever prices are reduced to meet competition, an effort is made to reduce the overhead or other expense on that job or increase the profit margin on another job that estimator figures. Consequently, Dade Roofing has a reputation for being fairly selective about the jobs they take.

Management and Labor Relations

Mr. Dade has remained in control of the company since the beginning of operations. He takes care of the customer relations for the firm, negotiates the larger contracts, decides on the amount of money to be spent on advertising and the mediums to be used, negotiates bank loans, and sets the salaries. He also determines the amount and type of materials to be purchased for the supply division and checks job estimates.

Son Tom expedites the office work, checks contracts, and generally shares in the management of the company with his father in a way that has given him a chance to familiarize himself with every aspect of the business. Besides two salaried field estimators, a field superintendent is employed on an annual salary. This man has been with the company for 27 years. He supervises all field jobs by daily trips to the job sites. Two girls are engaged full time in bookkeeping, billing, typing, and secretarial work. The clerical force has been with the company for many years and very little supervisory effort is needed. This allows Tom to devote most of his time to the more important aspects of the company management.

Two workers are employed on a full time basis in the sheet metal shop. Another employee operates the supply division, receiving and issuing materials. As many as 30 workers have been employed in the roofing end of the business, but the number fluctuates depending on the amount of work at hand. Three are mechanics or journeyman roofers, functioning as foremen on the jobs. The rest are apprentices, kettlemen, and helpers.

Scheduling of production represents no serious problem. New jobs are set up two or three days in advance. Tom says that his experience allows him to estimate labor time within one to two hours on most jobs. He checks the estimates and job records regularly and compares notes with the estimator-salesmen. Orders to issue materials for the job are checked and signed by Tom every morning before crews go out. Tom also checks materials returned to the warehouse at the end of the day.

Quality is controlled by municipal inspectors. There are two inspections; the first after felt is applied, the second upon completion. In addition, the company itself insists on high professional standards. The foreman on the job is directly responsible for the quality of work. He is checked by the superintendent, who makes daily trips, and by Tom, who makes unannounced spot checks. Thus the company is able to issue a guarantee for every job it undertakes and make good on any roof that fails. So far, losses from such guarantees have been very small. However, Tom has found that considerable time has been spent on service calls. After examination into the cause of leaks, it is often determined that Dade Roofing Company was not responsible. Often the walls or windows leak or an air-conditioning unit on top of the roof has been installed improperly after the roofing was finished. Tom now records the roof attachments on the roof after each job is finished and notes the name of the journeyman roofer assigned to each job. If a homeowner calls about a leak, Tom sends out the journeyman who did the job originally to investigate. Most complaints are not the result of faulty work. Many complaints in fact result in an additional fee for Dade Roofing.

Subcontractors are used for sandblasting on commercial jobs. Occasionally Tom subcontracts out some sheet metal work when Dade's shop is too busy or when he gets a good bid on a difficult job. In addition, solar heating systems and decks, if large enough, have been subcontracted.

Mr. Dade has used union labor from the beginning of operations in 1938 until March 1974. On that date he dropped out of the union. He claims that union wage rates became too high and productivity did not rise correspondingly. The company started to lose money on some jobs and was regularly underbid by certain contractors. For comparison, a small crew for a reroofing job receiving union wages (including all fringe benefits) as of July 1, 1976 was as follows:

Journeyman roofer	$10.90 an hour
Apprentice (1st six months)	6.20 an hour
Apprentice (2nd six months or more, average)	8.40 an hour
Kettle tender (average)	7.80 an hour
Roofer's helpers (2 @$7.50 an hour)	15.00 an hour
Total	$48.30 an hour

Average cost of non-union labor to Dade for the same size crew:

Mechanic	$9.80 an hour
Apprentices (2@$6.50 an hour)	13.00 an hour
Kettleman	7.20 an hour
Helpers (2@$5.75 an hour)	11.50 an hour
Total	$41.50 an hour

The above is $6.80 an hour or $54.40 a day less than union wages. However, being a non-union shop, the company has on occasion paid more than union wages to a particularly good man.

There are no bonuses or incentive payments to labor other than wage increases, if a man does a good job. As part of fringe benefits, the company contributes 50 percent of the premium of Employee Group Life, Health & Hospitalization Insurance. This insurance also covers the workers' families. Because the company has been selective in its choice of workers, the labor morale has been good and the company has never experienced any labor difficulties. Dade Roofing is usually able to keep most of its roofing crews busy for most of the year.

The salaried personnel, including estimators and office girls, participate in a year-end bonus. The amount divided as bonuses fluctuates from year to year, depending on the earnings of the company. The bonus paid is not uniform, but is scaled according to the position held and length of service with the company.

Purchasing and Credit Risks

As a dealer for nationally known roofing products, the company gets a dealer discount of 7 percent plus 5 percent per carload. No stock is carried on consignment. Dade Roofing Company absorbs the freight cost. The credit terms to Dade are 2/10 - net 30. Dade Roofing almost always pays within 10 days to take the two percent discount.

Mr. Dade is his own best customer for roofing products. About one-third of all the purchases are used by Dade Roofing Company. Mr. Dade sells wholesale to his own company and to any other roofer, charging the same prices, without discrimination.

The supply division applies a mark-up over cost of 20 percent if roofing is sold in small lots. Only 7 percent is added to roofing sold in whole car lots. The only profit in whole car lot sales is the additional 5 percent dealer discount which doesn't quite pay the cost of handling and storing the roofing. Customers receive the same credit terms that the Dade Company gets, namely 2/10 - net 30.

Bad debts have been considerable in the past years in the roofing supplies division. Losses have caused the management to review all accounts and weed out the poor risks. At present the company has only about 25 good accounts left. As a result, sales volume will go down in 1978. To sell more, Mr. Dade would have to compete more vigorously. Due to the competitive market situation (many other brands of roofing materials are available), he will have to sacrifice even some of the 5 percent dealer discount he is now earning. In the management's opinion, this does not warrant the effort involved. The credit risks would increase and lead to more collection troubles.

In the reroofing end of the business, collections have been very good and the losses from bad debts have been minimal. The reason for this is the method Mr. Dade uses to finance his accounts. About 70 percent pay cash upon completion. (Half of those use FHA financing. The bank makes the loan, pays Dade, and takes care of its own collections.) About 30 percent are extended credit for 60 days as follows: (1) Down payment not less than one-third of the total upon delivery of material to the job. (2) After 30 days, a payment of not less than one-third of the total. (3) After 60 days the balance is due.

Mr. Dade uses the Consumer Credit Bureau occasionally, but gets most of his credit reports from banks. These reports are confidential, given over the phone only.

Accounting and Cost Control

Each month a profit and loss statement is made up and a balance sheet prepared. Mr. Dade has a good idea of how profitable each month was about 10 days after the end of the month. A computer service bureau prepares the balance sheet and profit and loss for about $40 a month from check stubs and deposit records that are kept in the office. One of the girls prepares the records for the service bureau in about 2 hours on the last day of each month. Mr. Dade has a file of monthly reports on his firm that goes back several years. By comparing the sales and expenses for the current month with the same month of previous years, he can evaluate the performance of the company very precisely. At the end of each year a Certified Public Accountant firm prepares the year end profit and loss, balance sheet and tax returns. Employee payroll and withholding taxes are handled by the bank Dade Roofing deals with at a monthly cost of about $15.

At the monthly staff meeting with the two estimator-salesmen, sons and the field superintendent, Mr. Dade reports on the previous month, discusses estimating and sales problems and reviews upcoming job requirements. At the most recent meeting Tom commented on a small job that had been done the previous month. A 1500 square foot reroofing job had been done for $920 including the tear-off. The costs were as follows:

Labor	$218.82	
Workers' Compensation Insurance	26.27	
	$245.09	26.6%
Materials and equipment	$321.93	
Sales Tax	9.66	
	$331.59	36.0%
Total	$576.68	
Overhead at 20% of bid	$184.00	
Total Cost	$760.68	82.7%
Profit	$159.32	17.3%

Tom could tell from the figures on his desk that the job had been a good one for the company and he wanted to know why it had gone so quickly. The labor cost was less than $16.50 per square and Dade Roofing usually figured a little more than $20 per square on jobs like that.

General Trends in the Industry
While the total volume of the roofing industry has grown over the years due to increased building activity, Dade Roofing Company's operations have become smaller. This, however, is not necessarily a sign of weakness of the company. The decision to specialize in a segment of the industry, namely in reroofing, was made because experience had shown that this portion of roofing work proved to be most profitable.

The future outlook for this company continues to be good. The company is specialized and has an excellent reputation as to quality of workmanship. The so-called "reroofing cycle" has turned in its favor. Most of the homes built during the 1955 to 1965 period need new roofs now. Furthermore, the quality of new roofs installed by some roofers in the last 10 years apparently has become progressively worse, according to Mr. Dade. His company has been installing new roofs on some tract homes only three or four years old. As far as the Dade Company is concerned, this means more business.

Reasons for Success
A main reason why this company has been successful is that a strong working capital position has been maintained at all times. This has enabled the firm to take advantage of purchase discounts and has lessened the need for outside borrowing, thus saving the finance charges on such loans. Furthermore, because of this strong financial condition, the company is under no pressure to seek volume regardless of profit. In fact, it is able to charge, on the average, higher prices than most competitors because it can afford to turn down less profitable jobs. Mr. Dade is content to earn a good living year after year and is not interested in becoming the largest roofer in Capitol City. Profits have not been spectacular, but Mr. Dade and his sons have paid themselves over $500,000 in bonuses, dividends and salary in the last 10 years.

This is as good and perhaps considerably better than they could have done working for some other roofing company. In addition, Mr. Dade and his sons continue to build the asset value of the company, even while drawing a good salary.

The management has been, perhaps, too conservative in the use of funds at the company's disposal. It is a good practice to salt away net profits in earned surplus to take care of potential future losses. However, overcapitalization is not a virtue in itself. Better uses for such capital might be found, and if the management does not see fit to expand present operations, surplus money might be invested in other ventures. The return on capital invested in this business in 1974 was 13.3 percent; and in 1975, 12.9 percent. Any future increase in net worth, provided that net income remains the same, will drive these percentages down. One may suspect that as long as the elder Mr. Dade retains the main voice in the conduct of this business, the company will not expand to the size that its resources might justify.

Failures and Successes
Morgan Roofing and Dade Roofing are not unusual among home improvement companies. Many failures are more spectacular than that of Morgan and some successes are more dramatic than Dade's. You probably can name several mistakes the Morgan brothers made in managing sales, estimating, finances and production. Contrast this with the techniques used by Mr. Dade and his sons. True, Dade Roofing had been in business for 35 years and had learned to avoid most serious management mistakes. But Dade Roofing had the same opportunity that the Morgan brothers had. Mr. Dade just handled his business better because he recognized the importance of careful management.

Mr. Dade may have succeeded because he managed to avoid failure for many years and through shear longevity accumulated enough capital, knowledge, reputation and talented employees to become a leading reroofer in the community. Any contractor may eventually develop the resources it takes to succeed, but the contractor who promptly fails may never have that opportunity. Statistics indicate that failure rates among contractors decline as the business grows older. One recent survey by Dun & Bradstreet revealed that nearly half of all contracting businesses that failed had been in operation five years or less. By contrast, less than one-quarter of the failures had been in business more than ten years. In short, it seems that the longer the business has been going, the better the chance it will keep going.

What then can be done to avoid a premature failure? In a sense, the problem is easier than asking what it takes to make a success. Success is usually a happy combination of many factors while failure quite often can be pinned on a single factor. Dun and Bradstreet regularly analyzes business failures and has concluded that most failures in construction contracting can be traced to one or more of ten causes: These causes of failure are explored in the remainder of this chapter.

Overextension is probably the most common cause of failure. No home improvement contractor wants to bypass opportunity or turn away profitable business, but accepting more business than you can handle or committing more resources than you can marshal can be disastrous. Capacity is a function of capital, equipment, personnel, past experience and business know-how. A heavy work load requires business skill, financial strength, careful cost control and coordinating skill. These are capabilities that come from business experience, not technical experience. The skilled craftsman first venturing into a contracting business can easily become overloaded. When delays occur, costs exceed estimates, sales slip or collections are slow because too much work was attempted, the craftsman-owner can sometimes use his technical skill to complete the job and keep the money coming in. The Morgan brothers tried to make up for management failures by working long hours on their jobs. But even physical endurance has limits beyond which it cannot be extended.

Another danger is overextension beyond financial capabilities. Without enough cash to carry the workload, an otherwise minor payment problem can be disastrous. As a rule of thumb, incomplete work should not be valued at more than 10 times working capital. Because prompt payment is critical to continued availability of credit, it is vital that cash balances and transactions be carefully planned so that payments can be made when due. Contractors and subcontractors must be sure that sufficient capital is available to meet payrolls and costs that come due before payment is received for work completed. Critical scheduling binds can occur if supplies cannot be paid for from internal funds and outside credit is not available. The Morgan Roofing Company case is a classic example of overextension of financial resources.

Unsophisticated record keeping rates high on everyone's list of reasons for failure. Actually this heading can be broken into two parts, estimating records and financial records. It is surprising how few contractors bother to keep a record of their costs on completed jobs. Contracting is one of the few industries where the selling price is established before the product is produced. Considering the number of factors that the builder can't control (weather, state of the economy, strikes, price fluctuations, etc.) it is extremely important that the home improvement contractor know what his costs are and prepare accurate bids. Upon signing a contract, you are legally bound to perform and cannot terminate the project short of completion without risking serious damage to your reputation and possible legal entanglements.

Poor estimating practice and poor bookkeeping usually go hand in hand as the Morgan Company case illustrates. The most common shortcoming is confusing the accounts among the various components of one job, accounts for several jobs, and business and personal funds. Poor or nonexistent business records may create tax problems and almost certainly will make it impossible for the business to obtain a bank loan. Federal law requires certain minimum business records for tax purposes and every employer is required to maintain accurate payroll records and make the appropriate tax deposits for employees. To operate effectively and control costs, every business should have a regular statement of profit and loss and a projected budget for the coming months. A balance sheet should be prepared quarterly because banks demand them for loans and a balance sheet is the quickest way to examine the financial health of a business. Mr. Dade used his monthly balance sheet and profit and loss statement to help him make decisions and isolate problems. The Morgan brothers had no way of knowing what the profit or losses were from month to month. They had a lot of money coming in and assumed, erroneously it turns out, that they were making money.

Failure to investigate the resources of the client. The common practice of payment upon completion and "pay when paid" creates a line of mutual interdependence from the owner or lending institution through the general contractor and subcontractors to the material suppliers and individual craftsmen. Payments can be withheld from the top level if a lower level fails to meet the specifications. Since money filters from one level to another upon reaching a common goal, it is important that each operator analyze the degree to which the others can perform and fulfill their assigned task. Many contractors flatly refuse to deal with other contractors or to work with certain individuals because of the risks involved.

Even when a series of partial payments is made, the feature of "retainage" places risk at each level of the construction chain. On most large jobs, the contractor receives only about 90 percent of his estimated costs for each billing period. The rest is withheld in retainage, payable under most contracts about 35 days after the job is completed and accepted by the owner. Any mechanics' liens or unpaid labor or material bills should be filed within 30 days of acceptance. State lien laws vary but all states offer protection to individuals supplying labor or materials for construction work.

Since retainage often exceeds profit, contractors who have several jobs in progress usually have a considerable amount of capital tied up in retainage. This problem is especially acute for contractors who finish their work long before the job is completed and accepted. Of course, if there is a dispute over acceptance, no matter whose fault, all contractors may have their retainage held up. The effect on subcontractors is often more serious because subs usually bear the main material and labor costs. There is a strong movement to limit retainage to 10 percent of the first half of the job. After the first half is completed, the contractor is paid for the full value of all work as it is done.

Most home improvement contractors work on a private job basis where no retainage is required. It is reasonable for the small contractor to request payment of one-third as of delivery of the materials and the start of work, one-third at completion and one-third 30 days after completion. Mr. Dade found that a similar pay schedule worked out very well. Many contractors ask for and get 10 percent down upon agreeing to do the work. On small remodeling jobs, where work is to be completed in 30 days or less, payment is sometimes made only upon completion. This practice is probably used less today than it was 20 years ago.

No builder ever starts a project he doesn't expect to get paid for. But anyone who has been in building for more than a few years and has not had trouble collecting at one time or another is very fortunate indeed. Rehab, Inc. survived their problems but would have sunk the Morgan brothers in mid-1974 if the brothers hadn't wisely decided to give up their best account. The remodeling contractor who has many small accounts is going to survive a few slow pay periods. If you are dependent on one large customer, your future is no brighter than that of your best customer.

Slow pay is fairly common in the construction industry and is responsible for many bankruptcies each year. As a rule of thumb, the cost of collection will be about one-third to one-half the total amount owed if you cannot collect through your own ingenuity. Where the debtor is nearing insolvency and may turn belly-up, debts are often settled for 10 cents to 25 cents on the dollar.

What can a remodeler do to make sure he will be paid? The answer lies in investigating the resources of those he plans to deal with. Reputation is usually the best indication of reliability. A reputation for meeting obligations when due is the best recommendation you could have for a prospective client. If you don't know about a prospect's reputation, ask for the names of people he deals with on a credit basis and talk with them personally. Most firms that extend credit are willing to share credit information. Banks will usually give you approximate bank balances if you have your prospect's account number. A simple credit application is a good device for learning the essential information about your prospect. Don't be embarrassed about asking for credit information and beware of the prospect who says he "obviously" does not have to fill out an application.

Lack of managerial know how. There is a definite "ladder" of personal growth within the construction industry. While not everyone follows these steps, many have and probably many more

will. The "ladder" goes something like this:
1. Craftsman
2. Subcontractor-Craftsman
3. Subcontractor
4. Contractor-Subcontractor
5. Contractor
6. Contractor-Manager
7. Manager

Many skip one or more steps and most never reach steps 6 or 7. While entrance into the industry can be made and is made from architecture, law, finance, insurance or retailing, the series of steps above are the steps of growth within the industry followed by most contractors. Notice that none of the traditional areas from which contractors are developed has a heavy emphasis on management. Yet, as a builder progresses up the ladder of succession he will need more and more managerial know-how and less and less technical skill. The craftsman turned subcontractor will continue to perform as a craftsman as he builds his business as a subcontractor. The smaller contractor will perform more as his own subcontractor than will the larger contractor. For each function, a volume level is reached where it makes better business sense to assign lower level functions to other individuals. Notice also that with each step the importance of managerial know-how increases, the builder's technical skill becomes less important and the degree of responsibility for project completion and chance for gain or loss increases.

The skilled craftsman who becomes a contractor usually establishes himself as a small subcontractor in his construction specialty. As such, his technical knowledge is of value. As his busness grows, he should be able to pick up the more sophisticated business skills that are required. With a thorough knowledge of his technical specialty and a familiarity with the construction "community" in the geographic area of operation, the new contractor is in a fair position to secure business. This is often a necessity since the new contractor has only his skill and reputation as a craftsman to recommend him for higher levels and types of work.

Home improvement contracting requires more business skill than technical skill (although technical skill is needed to supervise tradesmen and subcontractors). Estimating is less critical to the degree that the contractor uses some estimates prepared by subcontractors. Scheduling, however, is a major responsibility for the contractor. Keeping to a schedule is crucial to completing a job at the predetermined contractual cost, and a delay by any one of the many subcontractors, suppliers, and others whom the contractor is coordinating can force a major reorganization of the job schedule with a loss in time and money. As you branch out into general contracting you should be aware of the additional managerial requirements which such a venture entails. Just as the successful skilled craftsman does not necessarily make a successful subcontractor because of the difference in required skills, the successful subcontractor is not necessarily ready to become a contractor.

Inadequate profit margins. Competition in the construction industry tends to keep profit margins low as compared to other industries. Dun and Bradstreet reported that for the ten years between 1960 and 1970, general contractors showed a profit after taxes of only 1.2 percent of sales. This was less than half the profit for the average of all manufacturing business. Only 6 of the 71 manufacturing and construction industries showed a lower profit on sales.

The factors which make entry into the construction industry so attractive make earning an adequate profit so difficult. The small capital requirements and the absence of a ''General Motors'' of construction means that most of the firms in building are small. It is estimated that there are over 50,000 ''visible'' home improvement businesses with one or more employees. In addition, there are about 25,000 independent operators in home improvement.

Since competition in the home improvement business is largely price competition, the level of profit for everyone tends to be low. For a company that is thinly capitalized, a low profit margin can result in a disastrous loss if the job does not go as planned and estimated. In spite of the low overall profit margins, the growth potential is great because along with high risk goes the potential of high gain.

Though contractors may find the profit on sales to be low, the profit per dollar of net worth has traditionally been high. In fact, during the ten years between 1960 and 1970 contractors earned an after tax profit equivalent to over 9 percent of their net worth. This placed contractors 7th from the top among 71 manufacturing and construction industries. Profit per dollar invested in the company is so high because fairly few dollars have to be invested in order to do business.

What then can be done to improve profits? Better management is one answer. Good estimating helps. Being at the right place with the ability to perform as required is also important. But some contractors would report that the most consistently profitable companies have the smallest proportion of salaried employees and lowest overhead. Efficiency and maintaining a small staff seems to pay off in higher profits. Note also the effect of the business cycle on overhead. Most firms are slow to hire new employees as the market expands and they operate very efficiently while expanding. Profits rise. As the market contracts, unprofitable projects are abandoned more slowly and management will be slow to lay off excess employees. As a result losses begin to develop. In the long run the construction industry will expand at about 10 percent a year, but because of the cyclic nature of business, expansions will be in bursts followed by busts. Management has to be nimble to keep pace with trends and foresee developments.

Chapter 19
Estimating and Controlling Costs

Controlling costs is one of your four key functions as the manager of your remodeling business; sales, production and financial control being the other three. Accurate cost estimating is the most important part of controlling costs, and keeping good cost records is the best way to ensure accurate estimates. This chapter will help you set up a system for making quantity surveys and converting quantities of labor and material into dollars and cent costs. If you already have an estimating system that is well suited to the type of work you do, and if that system produces reliable estimates, you probably can skip over most of this chapter. If you do only certain types of work and seldom remodel bathrooms or kitchens and seldom take on room additions, the check lists and forms won't be very useful to you. Most home improvement contractors, however, can benefit from adopting some system for compiling estimates and nearly every estimator should use a checklist to suggest items that might be overlooked.

Cost Data File

Most successful remodelers would advise any less established home improvement contractor to use an orderly, systematic method when approaching estimating problems. List costs in the order in which they will be encountered. Use a notebook or some other orderly means of collecting estimating data. Probably the most convenient and practical book for this purpose is the loose-leaf kind, with pages 8½ by 11 inches. A number of sheets will be used on each estimate, but all sheets of the same estimate should bear the same estimate number. With such a loose-leaf form you can take out several sheets at any time and take them with you to the owner's home or office without being burdened with an entire book.

When an estimate is rejected or the job completed, you can remove that estimate, tie it together and file it away in numerical order. An index in your file or in the book itself simplifies finding any estimate at will. An added advantage of the loose-leaf form is that separate sections of the estimate may be changed at will to conform to modified plans or price changes without interfering with the remaining sheets. With items kept separate, a small change requires a simple

adjustment and the other figures can just be copied.

You will refer to the cost data in your estimating file to find your costs on work which you have completed. You should build up your own cost data file because the figures you develop and test yourself are the most reliable figures you will ever find for the type of work you do. Time required to shingle certain areas of roof, time for laying a floor per unit, time for setting doors, time for brick, etc., should be recorded and filed for ready reference on future estimating.

Carefully developed and maintained cost records will assist you in lowering future costs as well as in estimating. If your cost records, for example, show that a regular stair of a certain type takes two men 8 hours, and later a similar piece of work takes 12 hours with the same number of men, you can check on the conditions and the workmen to avoid the same set of circumstances in the future.

In figuring unit costs, bear in mind the season of the year when the work is to be carried out. A shingle roofer, for instance, will work faster and consequently cheaper on a warm sunny day than when the cold is stiffening his hands and interfering with his movements. A close record should be kept of the average output of labor for all seasons of the year, and account taken of these results in preparing estimates. Rainy seasons should also be recorded.

As suggested before, a most convenient form for your estimates is the loose-leaf book. You may wish to develop your own forms or use forms purchased from publishers. Each estimate should bear an estimate number on all sheets, and each sheet should be numbered consecutively. The essential data for the first page of the estimate sheet will probably include: Estimate Number; Sheet Number; Location; Owner; Date; Estimate Prepared By . . . and Checked By.

Home Improvement Estimating Costs

A construction estimate consists of two elements: the quantity survey and the cost estimate. The quantity survey is an organized list of the quantities of each type of material that goes into the structure. Quantity surveys are made from the construction plans and specifications. The cost

			Material		Labor		Actual Man Hours
Description	Quantity	Unit					

Owner

Phone

Date

Estimated By

Type of Job

Checked By

Sheet No.

Estimate No.

Quantity take-off sheet
Table 19-1

311

Excavation
- Backfilling
- Clearing the site
- Compacting
- Dump fee
- Equipment rental
- Equipment transport
- Establishing new grades
- General excavation
- Hauling to dump
- Pit excavation
- Pumping
- Relocating utilities
- Removing obstructions
- Shoring
- Stripping top soil
- Trenching

Demolition
- Cabinet removal
- Ceiling finish removal
- Concrete cutting
- Debris box
- Door removal
- Dump fee
- Dust partition
- Electrical removal
- Equipment rental
- Fixtures removal
- Flooring removal
- Framing removal
- Hauling to dump
- Masonry removal
- Plumbing removal
- Roofing removal
- Salvage value allowance
- Siding removal
- Slab breaking
- Temporary weather protection
- Wall finish removal
- Window removal

Concrete
- Admixtures
- Anchors
- Apron
- Caps
- Cement
- Columns
- Crushed stone
- Curbs
- Curing
- Drainage
- Equipment rental
- Expansion joints
- Fill
- Finishing
- Floating
- Footings
- Foundations
- Grading
- Gutters
- Handling
- Mixing
- Piers
- Ready mix
- Sand
- Screeds
- Slabs
- Stairs
- Standby time
- Tamping
- Topping
- Vapor barrier
- Waterproofing

Forms
- Braces
- Caps
- Cleaning for reuse
- Columns
- Cornice
- Cripples
- Door frames
- Dormers
- Entrance hoods
- Fascia
- Equipment rental
- Footings
- Foundations
- Key joints
- Layout
- Nails
- Piers
- Salvage value
- Slab
- Stair
- Stakes
- Ties
- Walers
- Wall

Reinforcing
- Bars
- Handling
- Mesh
- Placing
- Tying

Masonry
- Arches
- Backing
- Barbecues
- Cement
- Ceramic tile
- Chimney
- Chimney cap
- Cleaning
- Clean-out doors
- Dampers
- Equipment rental
- Fireplace
- Fireplace form
- Flashing
- Flue
- Foundation
- Glass block
- Handling
- Hearths
- Laying
- Lime
- Lintels
- Mantels
- Marble
- Mixing
- Mortar
- Paving
- Piers
- Repair
- Reinforcing
- Repointing
- Sand
- Sandblasting
- Sills
- Steps
- Stonework
- Tile
- Veneer
- Vents
- Wall ties
- Walls
- Waterproofing

Rough Carpentry
- Area walls
- Backing
- Beams
- Blocking
- Bracing
- Bridging
- Building paper
- Columns
- Fences
- Flashing
- Framing clips
- Furring
- Girders
- Gravel stop
- Grounds
- Half timber work
- Hangers
- Headers
- Hip jacks
- Insulation
- Jack rafters
- Joists, ceiling
- Joists, floor
- Ledgers
- Nails
- Outriggers
- Pier pads
- Plates
- Porches
- Posts
- Rafters
- Ribbons
- Ridges
- Roof edging
- Roof trusses
- Rough frames
- Rough layout
- Scaffolding
- Sheathing, roof
- Sheathing, wall
- Sills
- Sleepers
- Soffit
- Stairs
- Straps
- Strong backs
- Studs
- Subfloor
- Timber connectors
- Trimmers
- Valley flashing
- Valley jacks
- Vents
- Window frames

Finish Carpentry
- Baseboard
- Bath accessories
- Belt course
- Built-ins
- Cabinets
- Casings
- Caulking
- Ceiling tile
- Closet doors
- Closets
- Corner board
- Cornice
- Counter tops
- Cupolas
- Door chimes
- Door hardware
- Door jambs
- Door stop
- Door trim
- Doors
- Drywall
- Entrances
- Fans
- Flooring
- Frames
- Garage doors
- Hardware
- Jambs
- Linen closets
- Locksets
- Louver vents
- Mail slot
- Mantels
- Medicine cabinets
- Mirrors
- Molding
- Nails
- Paneling
- Rake
- Range hood
- Risers
- Roofing
- Room dividers
- Sash
- Screen doors
- Screens
- Shelving
- Shutters
- Siding
- Sills
- Sliding doors
- Stairs
- Stops
- Storm doors
- Threshold
- Treads
- Trellis
- Trim
- Vents
- Wallboard
- Watertable
- Window trim
- Wardrobe closets
- Weatherstripping
- Windows

Flooring
- Adhesive
- Asphalt tile
- Carpet
- Cork tile
- Flagstone
- Hardwood
- Linoleum
- Marble
- Nails
- Paste
- Rubber tile
- Seamless vinyl
- Slate
- Tack strip
- Terrazzo
- Tile
- Vinyl tile
- Wood flooring

Plumbing
- Bathtubs
- Bar sink
- Couplings
- Dishwasher
- Drain lines
- Dryers
- Faucets
- Fittings
- Furnace hookup
- Garbage disposers
- Gas service lines
- Hanging brackets
- Hardware
- Laundry trays
- Lavatories
- Medicine cabinets
- Pipe
- Pumps
- Septic tank
- Service sinks
- Sewer lines
- Sinks
- Showers
- Stack extension
- Supply lines
- Tanks
- Valves
- Vanity cabinets
- Vent stacks
- Washers
- Waste lines
- Water closets
- Water heaters
- Water meter
- Water softeners
- Water tank
- Water tap

Heating
- Air conditioning
- Air return
- Baseboard
- Bathroom
- Blowers
- Collars
- Dampers
- Ducts
- Electric service
- Furnaces
- Grilles
- Hot water
- Infrared
- Radiant cable
- Radiators
- Registers
- Relocation of system
- Thermostat
- Vents
- Wall units

Roofing
- Adhesive
- Asbestos
- Asphalt shingles
- Built-up
- Canvas
- Caulking
- Concrete
- Copper
- Corrugated
- Downspouts
- Felt
- Fiberglass shingles
- Flashing
- Gravel
- Gutters
- Gypsum
- Hip units
- Insulation
- Nails
- Ridge units
- Roll roofing
- Scaffolding
- Shakes
- Sheet metal
- Slate
- Tile
- Tin
- Vents
- Wood shingles

Sheet Metal
- Access doors
- Caulking
- Downspouts
- Ducts
- Flashing
- Gutters
- Laundry chutes
- Roof flashing
- Valley flashing
- Vents

Electrical Work
- Air conditioning
- Appliance hook-up
- Bell wiring
- Cable
- Ceiling fixtures
- Circuit breakers
- Circuit load adequate
- Clock outlet
- Conduit
- Cover plates
- Dimmers
- Dishwashers
- Dryers
- Fans
- Fixtures
- Furnaces
- Garbage disposers
- High voltage line
- Hood hook-up
- Hook-up
- Lighting
- Meter boxes
- Ovens
- Panel boards
- Plug outlets
- Ranges
- Receptacles
- Relocation of existing lines
- Service entrance
- Switches
- Switching
- Telephone outlets
- Television wiring
- Thermostat wiring
- Transformers
- Vent fans
- Wall fixtures
- Water heaters
- Wire

Plastering
- Bases
- Beads
- Cement
- Coloring
- Cornerite
- Coves
- Gypsum
- Keene's cement
- Lath
- Lime
- Partitions
- Sand
- Soffits

Painting and Decorating
- Aluminum paint
- Cabinets
- Caulking
- Ceramic tile
- Concrete
- Doors
- Draperies
- Filler
- Finishing
- Floors
- Masonry
- Paperhanging
- Paste
- Roof
- Sandblasting
- Shingle stain
- Stucco
- Wallpaper removal
- Windows
- Wood

Glass and Glazing
- Breakage allowance
- Crystal
- Hackout
- Insulating glass
- Mirrors
- Obscure
- Ornamental
- Plate
- Putty
- Reglaze
- Window glass

Indirect Costs
- Barricades
- Bid bond
- Builder's risk insurance
- Building permit fee
- Business license
- Cleaning floor
- Cleaning glass
- Clean-up
- Completion bond
- Debris removal
- Design feel
- Equipment floater insurance
- Equipment rental
- Estimating fee
- Expendable tools
- Field supplies
- Job phone
- Job shanty
- Job signs
- Liability insurance
- Local business license
- Maintenance bond
- Patching after subcontractors
- Payment bond
- Plan checking fee
- Plan cost
- Protecting adjoining property
- Protection during construction
- Removing utilities
- Repairing damage
- Sales commission
- Sales taxes
- Sewer connection fee
- State contractor's license
- Street closing fee
- Street repair bond
- Supervision
- Survey
- Temporary electrical
- Temporary fencing
- Temporary heating
- Temporary lighting
- Temporary toilets
- Temporary water
- Transportation of equipment
- Travel expense
- Watchman
- Water meter fee
- Waxing floors

Administrative Overhead
- Accounting
- Advertising
- Automobiles
- Depreciation
- Donations
- Dues and subscriptions
- Entertaining
- Interest
- Legal
- Licenses and fees
- Office insurance
- Office rent
- Office phone
- Office salaries
- Office utilities
- Pensions
- Postage
- Profit sharing
- Repairs
- Small tools
- Taxes
- Uncollectable accounts

Figure 19-2

estimate converts quantities of material into costs for labor, material, and equipment and adds subcontracted items, overhead, contingency, and profit to arrive at the total cost. Most estimators use ruled estimating forms and checklists to avoid errors and omissions. If you don't have a supply of estimating forms, take Figure 19-1 to an "instant printer" and have a hundred or more printed up. On the back of Figure 19-1 is a checklist that will help you remember to include all costs in your estimate.

The quantity survey, often called the "take-off", requires an ability to visualize the construction process. When plans, specifications or perspective drawings are available, look over the plans carefully to get a mental picture of the proposed building. If there are no plans to work from, picture in your mind the work that will have to be done and the materials that will be required. Before you begin to write, remember that the quantity survey requires precision and care. Most successful estimators consciously or unconsciously follow certain practices which tend to reduce the chance of error or miscalculation. The following points should be observed while developing estimates:

1. In most cases the materials should be listed in the order in which they will be used. For example, figure and list framing before paneling. In a like manner, figure loosening of the soil before hauling of the excavation.

2. List materials in the same order on every job. This habit of estimating according to a definite routine reduces to a minimum the chances of overlooking items.

3. Every estimate should be checked before it is completed. The estimate must be written in detail and every item must be shown. When checking, you should be able to see at a glance how you or another estimator arrived at the various figures. Where practical, sketches should be drawn on the estimate sheet showing how the dimensions were obtained. For example, in figuring the area of a roof, a small sketch could be drawn giving the essential dimensions.

4. Check your figures. Do not list a figure that you can see must be wrong.

5. Use decimal fractions rather than feet and inches. Wherever necessary, refer to a table of decimal equivalents. This will make your work much easier.

6. Keep separate elements separate. Remember that you are going to have to figure the labor and equipment required to erect each material used. Stairs require far more labor per board foot of lumber than wall framing does. List the stair framing lumber and the wall framing lumber on separate lines. Each item should be taken off separately, but small items should be grouped in such a way that each group or class can be found easily. The work of pricing small items is simplified by fewer operations of multiplication.

7. Each section of the estimate should be kept separate as far as possible. Total each section at the end of the section. In this way, should any errors occur or if a change in price occurs, changes can be made in the proper section or heading quickly and easily, without affecting other items or making it necessary to go through the entire schedule in detail.

8. Avoid rounding off dimensions until calculations have been completed. A concrete slab 40 feet square and 4¼ inches thick has 20.98 cubic yards of concrete, not the 19.75 cubic yards you would get if you rounded the 4¼ inch thickness down to 4 inches.

List first the heading for each section on a form similar to Figure 19-1. Take for example the heading "Excavation". Then, under this heading list each operation: Foundation, grading, backfilling, etc. Record next the quantity of soil to be moved in each operation and the unit of measure such as cubic yard, square foot, etc. Later, when you have completed the quantity, fill in the material cost per unit and the labor cost per hour. The material cost should include equipment rental unless this is figured separately on a lump sum basis. The labor cost is the hourly wage plus all fringe benefits and mandatory contributions. You may want to include the cost of supervision here or at the end of the estimate. Usually for a small job the supervisor is not on the job continually and it is easiest to add a few hours for supervision at the completion of the estimate. The labor cost per hour can include the "contractor burden" of about 20 percent of payroll, if not included in the labor cost per hour, the "contractor burden" must be included when the costs are summarized on the summary sheet. The contractor burden is discussed later under "The Cost Estimate."

Accounting for all the labor hours expended will be the most difficult and potentially most rewarding part of estimating. You may know, for example, that each "square" of roofing can be applied in two man hours on a typical job. Yet a two man crew may only be able to complete a 600 square foot job in 8 hours. The same crew might complete a similar 1,000 square foot job in the same time the next day. There are many subtle human factors to consider when estimating the labor requirement. Each man hour yields perhaps 50 minutes of productive work and part of that time is spent waiting on crew members, locating tools, and many other less than fully productive activities. As an estimator you must make an educated evaluation of what your crews can do. Your estimates will be much more valid if you base your figures on the record of work you have done previously.

Excavation

Before starting an estimate for any kind of excavation, visit the site where the proposed building will be erected and make notes on any trees or landscaping that will have to be removed, access problems that may develop, and any sign that water will be a problem. Form an opinion on the soil conditions you will encounter and determine what water, gas, sewer, phone, cable, sprinkler and electrical lines will have to be relocated on the building exterior.

Your excavation take-off should be divided into the sequence of operations as they actually take place in the field.

First: Preparing the site.

Second: Stripping and storing top soil.

Third: General excavating.

Fourth: Trench and pit excavating.

Fifth: Backfilling.

Sixth: Excavating and filling for new site grades.

Preparing the site: This includes the removal of standing buildings, paved areas, fences, poles, free standing walls and any other obstacle above the existing grades. Some of this work rightly comes under wrecking. When any extensive amount is involved, most estimators will get a bid from a wrecking contractor.

Stripping and storing topsoil: The topsoil may have to be stripped to a depth equal to the total thickness of the topsoil. This depth can be determined by digging a test hole to determine the actual ground condition. Stripping of the topsoil is usually figured to 5 feet outside the actual building lines.

General excavating: In considering excavating of large areas, never skimp on the dimensions of the cut required. Conditions in the field almost always require generous measurements to ensure sufficient coverage of the yardage to be removed. In general, it requires 2 feet outside the wall footings for working space for the installation of concrete forms plus about a 45 degree slope to keep the banks of the holes from sliding if the soil is loose or sandy. The stiffer or more stable the excavated material is, the less slope is required.

Look at the basement plan, if there is one, and see how much larger the excavation must be than the size of the building on account of the projection of the footings. Generally, one foot or two feet is allowed outside the foundation for working room, depending on the soil and kind of construction outside of the wall. Next, get the area of the building from the foundation plan. Then look at the elevations for the natural grades and the depth of the basement. Take the depth to the bottom of the concrete floor or cinders, if any, for the general basement level and put it down on the estimate sheet.

In finding the area covered by the building, take off the larger or main portions first; then take off the smaller portions, additions, or wings. Do not attempt to use "over all" dimensions on odd shaped buildings without first making sure that these do not include some part not excavated. For example, if you were to excavate a hole 100 feet long, 50 feet wide, and 10 feet deep, first add 2 feet to each side, making the dimensions 104 feet by 54 feet. Then add for the 1 to 1 slope. Visualize

this slope as being in the shape of a right angle along the side of the building with the base at the top of the excavation, the altitude parallel to the basement wall, and the hypotenuse lying along the 45 degree slope. Accordingly, you will have two triangles, one on each side of the building, which will form a square 10 feet by 10 feet in size. Now add the extra cut for the slope to your 104 feet by 54 feet and you will obtain the dimensions of 114 feet by 54 feet as the size of the actual hole to be dug.

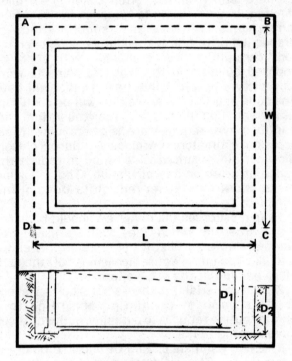

Excavation plan
Figure 19-3

Where the natural ground slopes, we have a slight change in the procedure. In this case the volume calculation must reflect the difference in the height D1 and D2 in Figure 19-3. The volume is equal to the length multiplied by the width by one half the sum of the heights. This may be written:

$$\text{Vol.} = \frac{LW \, (D1 + D2)}{2}$$

When the ground surface slopes in two directions, the points - A, B, C and D of Figure 19-3 are all unequal. In this case the individual heights of the four corners may be added and divided by four. Thus the volume in this case (if A, B, C and D are assumed to represent the height in feet at each corner) is given as:

$$\text{Vol.} = \frac{LW \, (A + B + C + D)}{4}$$

An earthwork contractor regards excavation for the basement or cellar of any ordinary house as a day's work, regardless of the size of the basement. It is seldom possible to use excavation equipment on two jobs in the same day. For this reason, the cost of excavating a cellar and many other small excavation jobs should be based on a day's use of the operator and machine rather than on a yardage basis. However, the quantities should be estimated for comparing the actual production time with future jobs.

Trench and pit excavating: Trench and pit excavations extend below or outside of the lines of the general excavation. These items may be estimated as their actual size plus 1 foot outside for working room. For walls that are to be waterproofed, figure 2 feet all around instead of 1 foot. Where shoring is required, it may be necessary to make the excavation 2 feet or more larger all around to provide room for placing and bracing the shoring. When figuring accurate costs of shoring and bracing, it is considered best to list the shoring for each class of excavation separately. The different depths should also be indicated under these groups.

On some jobs it may not be possible to get a backhoe to dig the footing trenches, so they must be dug by hand. Multipy the width of the trench by the depth, both in inches. Divide the result by 144 to get the cross-sectional area in square feet. Multiply this by the length of the footing in feet, and divide the result by 27. This is the volume in cubic yards to be dug.

Some drawings may not give the depth of excavation of piers and footings. In most cases the depth of footings can be assumed at below frost line, but in all cases this is dependent on soil conditions. Trenches more than 2 feet deep and less than 3 feet wide should be figured as 3 feet wide as it will require almost this width to give room for digging. Even if the trench is dug narrower, there are always chances for cave-ins.

To estimate excavation for piers and abutments, multiply the outside dimensions by the depth; no extra allowances are required unless working room is necessary. The bottoms of concrete footings, and sometimes the sides, may not require forms, depending upon conditions and the specifications. In case the piers are constructed of stone or brick, it is necessary to allow extra room on each side for placing materials and for performing the work properly.

Pits and pier holes more than 2 feet deep must be figured as having an area of at least 12 to 16 square feet, even though the piers do not require such an area. For example, if a footing 2 feet by 2 feet is to be carried down 5 feet below grade, it should be estimated as 12 square feet of area times the depth. It requires more than a space 2 feet by 2 feet to dig a hole 5 feet deep.

Backfilling is the filling of open spaces in the ground with earth that has been excavated. Backfill may be estimated by taking the total volume of the excavation, less the displacement or total volume of the structure below grade. The difference is the volume of earth to be handled for backfill. Backfill material may be something other than the excavated material. Usually the exterior backfill may be done with the original excavated material. The interior backfill may be sand, stone, loamy clay or other material suitable for compaction. Compaction is required in some types of work. Mechanical tampers are used to obtain the degree of compaction desired.

Where a space is allowed around buildings in excavating, the space between the outside of the wall and the bank of earth must be filled. This item also includes refilling trenches dug for laying pipes. Excavating and backfilling for sewers and drains is usually done by the plumber.

Figuring backfilling for ordinary basement walls (where either two feet or one foot is allowed around the building) is a very simple matter if edges of the excavation are reasonably straight. Multiply the distance around the excavation by the depth of the excavation and divide by 27. Normally, you can assume that the backfill will just about equal the volume of concrete or brick used in the foundation.

It is important that no backfilling be done until the owner or building inspector has inspected the outside of the foundation walls and given his permission to start backfilling. This is particularly true where the foundation is made of concrete blocks and is to be waterproofed below grade. If any waterproofing has been omitted, some opening may exist that will let water pass and result in a wet basement or house interior.

Establishing new site grades: New site grades are always required to prevent pooling of runoff and to provide slopes for paved and landscaped areas. Cuts and fills for new site grades vary from simple to complex. In the average case the amount of cut or fill required can be obtained by the simple grid method. This is done by superimposing a system of grids, similar to graph paper, upon the survey of the plot plan which shows the various elevations. Lay out a grid of squares on the plot plan. In the center of each square inscribe the average cut or fill as shown by the nearest elevation. Total the averaged cuts and fills shown in the grids to obtain the total cubic yards of cut and fill.

Excavating costs will vary with the kind of soil and the method used in loosening and removing. Except on very small jobs, all excavating today is done by power operated equipment. Small basements are dug with a bulldozer and a "high lift" tractor or power shovel, the latter being a tractor equipped with a shovel that may be elevated to load directly into a truck or over the sides of the excavation. Remember to include the cost of trucking all surplus to a legal dump site. Disposal costs can be a major portion of the excavation estimate.

315

Demolition

Investigate the actual conditions on the site carefully before beginning to figure demolition. Often you won't know how much demolition is necessary until the work actually begins. It can be very expensive to break out a few cubic feet of concrete, especially if you were not planning on removing any concrete. Where there is doubt about the actual quantity that must be removed, allow for removal of a certain amount of material and make it clear in your agreement that the removal of more is an "extra" for which a charge will be made. Your contract can even specify how much that charge will be per unit. In some cases it may be wise to charge for demolition by the hour rather than estimate the actual cost.

Many of the items you remove will have some salvage value. You should have a clear understanding as to who owns these salvaged materials if they are not going to be reused in the job. Also, allow more time for removing nearly any item if it is to be kept in a reusable condition. Don't forget the cost of hauling demolished materials to a legal dump and the dump fee.

Watch out for anything that must be removed during the course of construction and then reinstalled. Installing used materials is nearly always more time consuming than installing new materials. Damage to materials adjacent to items to be removed is unavoidable or nearly so in many cases. Try to visualize the demolition in process and make an intelligent guess about how much extra ceiling, wall or floor finish will have to be replaced because of unavoidable or negligent damage during demolition. Also, note that existing code deficiences often turn up during demolition. Naturally, you should stand ready to bring the structure up to code requirement, but not at your own expense. Make it clear in your contract and point out to your prospect that remedying existing deficiences discovered during work will result in extra charges.

Concrete

Most jobs can use ready-mix concrete. It eliminates the mixing, storage, material handling and most waste of material. In addition, it usually makes placing the concrete easier and faster. Ready-mix trucks often can be brought in close enough to the forms so that little or no handling is required. Some waste will occur no matter how the concrete is handled, especially where concrete has to be handled over some distance or pumped into place. Allow 5 percent for waste. In other words, order 1.05 cubic yards for every 1 cubic yard required by calculation.

On some jobs it may not be practical to use ready-mix concrete. Truck loads less than 6 cubic yards are usually considered less than a full load and a premium may be charged by the seller. Also, stand-by time and unloading time in excess of 4 minutes per cubic yard is usually charged to the buyer.

Foundations and Walls

First determine the width and thickness of footing required for main outside walls. If the footing has offsets or is in two or more layers, find the size of each. Put these thicknesses and widths in the proper columns marked on your estimate sheet. Now get the total length of this type of footing. Start at some convenient corner and add up all the dimensions. By this method, the corners are figured twice. The volume you arrive at will be sightly more than necessary.

Next, take all the cross walls and longitudinal walls that have the same sizes. Put these down as described above. Now look for wall footings of other sizes, putting each size down with its total length. Having listed all the footings for walls, the piers should come next. Starting with the outside piers, list the number and sizes; then do likewise with inside piers. Now take the foundation walls and area walls. Determine the size and length and compute the volume. Look for miscellaneous footings and walls and tabulate their size on the estimate sheet. After you have listed all the items carefully, go over the entire job again checking each item.

Openings are usually deducted unless very small. Deduction is not made for the space occupied by the reinforcement unless the amount is considerable. No deduction is made for pipes or openings having a sectional area of less than 1 square foot.

Slabs

Concrete flatwork such as sidewalks, floors, driveways, and patios involve four distinct operations: grading, setting the forms, handling the concrete into place, and finishing the surface. Where a separate topping is required or cinders or gravel are placed under the concrete, these should also be added.

On nearly every slab job it is necessary to do some grading to prepare the surface. Naturally, where extensive surface preparation is necessary, more time will be required. Where more than a few inches of fill are placed on the surface, the area will have to be compacted. This can be done by hand or with a rammer. For most soil types, compaction should be done while the soil is moist. If a fairly large amount of soil must be moved, a small bulldozer or tractor with a landscape bucket will be appropriate. Naturally, the time required will vary greatly with the type of equipment, soil type and how far the soil must be moved.

Where cinder, sand, gravel or slag fill is placed under concrete floors or walks, the cost will vary with the thickness of the fill, the job conditions and whether it is necessary to load the fill into wheelbarrows and wheel from the street or whether it can be spread from piles. It costs as much to grade and tamp a bed of slag or gravel 3 inches thick as one 9 inches thick because it is only the surface that is graded and tamped and the only additional labor on the thicker fill is for handling and spreading a larger quantity of material. When ordering material, remember that most types of fill shrink when compacted. Between 10 and 15 percent of the volume of sand, crushed rock or gravel is lost in compaction and up to 40 percent of the volume of cinders or slag is lost.

Find the concrete required for flatwork by multiplying the length, width and depth together (change all inches to decimals of a foot). This will give you the number of cubic feet in that portion. The results should be placed in the total column. Adding these totals will give you the total cubic feet in the job. Since concrete is usually measured by the cubic yard, divide the total by 27 (the number of cubic feet in a cubic yard).

Forms

Finding the amount of concrete required is a comparatively small part of estimating concrete work. Much of the cost in concrete construction is in the formwork. Contractors generally design and build their own forms. An understanding of form design principles is essential to estimating the cost of concrete construction.

Forms for concrete construction must support plastic concrete until it has hardened. Stiffness is an important feature in forms and failure to provide for this may cause unfortunate results. Forms must be designed for all the weight they are liable to be subjected to including the dead load of the forms, the plastic concrete in the forms, the weight of workmen, the weight of equipment and materials whose weight may be transferred to the forms, and the impact due to vibration. These factors vary with each project but none should be neglected. Ease of erection and removal are also important factors in the economical design of forms. If you have any doubt about the strength required of the forms you plan to use, use form design tables. A reference titled "Construction Manual: Concrete and Formwork" has good wood form design information and is offered on the final pages of this handbook.

When the forms are to be used for only one foundation, and the form material is to be reused in other parts of the building, the studs and walers are not cut to length but are left in stock lengths. The sheathing is nailed only enough to hold it in place. It can then be easily taken apart. After the walls have been stripped, the lumber should be cleaned of cement and piled in such a way as to prevent it from warping.

Usually 2 by 4 or 2 by 6 inch lumber is used for forms at each side of concrete walks, patios and driveways. The depth of the forms depends upon the thickness of the concrete. On ordinary sidewalk work, obtain the linear feet of walk and multiply by 2. The answer will be the linear feet of forms required. A sidewalk 4 inches thick will require 2 by 4 inch lumber; a 6 inch thickness needs 2 by 6 inch lumber. The cost of the forms may be obtained by calculating the cost of lumber required and dividing by the number of times it may be used on the job.

Where concrete floors are placed on the ground, it may be necessary to set wood screeds 6 feet to 8 feet on centers so the floor will be level or have a uniform pitch. These are usually of 2 by 4 lumber, with stakes placed 3 to 4 feet apart to hold the forms in place.

On many jobs a separate cement topping is applied to the concrete to provide a smoother, more easily finished surface. Usually a cement-rich mix is used with fine sand and no more than 5 gallons of water per sack of cement.

Concrete for sidewalks is usually of a very dry mix and stiff consistency so that the finish may be applied immediately after a section of the rough base has been laid. After allowing sufficient time for the topping to set, the finisher can return and trowel this section of the walk while the next section is being laid. Overtime may be materially reduced by using this method. Nearly all classes of finish require two to three men as finishers.

Reinforcing

Reinforcing steel is estimated by weight calculated by listing all bars of different sizes and lengths and reducing the total to pounds. Reinforcing bars may be purchased either in mill lengths with all cutting and bending done on the job, or it may be cut to length and bent, ready to place.

When estimating quantities and weights of reinforcing steel, all bars of each size should be listed separately to obtain the correct weight. Convert feet to pounds as follows:

Bar No.	Diameter	Wt./Ft.
3	3/8"	0.376
4	1/2"	0.668
5	5/8"	1.043
6	3/4"	1.502

Welded steel fabric is a popular and economical reinforcing for concrete work of all kinds, especially driveways and floors. It may also be used for temperature reinforcing, beam and column wrapping, pavement reinforcing, and cement gun work. It is usually furnished in square or rectangular mesh and sold at a certain price per square foot, depending upon the weight.

The quantity required will equal the total area to be reinforced with 5 to 10 percent more added for side and end laps. The fabric to be used is designated by specifying the style, such as 412-812. This style represents a rectangular fabric having longitudinal wires spaced 4 inches apart and transverse wires spaced 12 inches apart, using number 8 gauge longitudinal wires and number 12 gauge transverse wires. It has a weight of 25 pounds per 100 square feet. Many other styles are also on the market.

Masonry

Concrete block and brick should be estimated by the square foot of wall of any thickness and then multiplied by the number of units per 100 square feet.

When estimating quantities, always take exact measurements and do not count corners twice. Make deductions in full for all openings, regardless of size. The result will be the actual number of units required for the job.

The labor cost of laying the various types and sizes of brick, block, and tile will vary with the size and weight of the units, the class of work, and whether there are long straight walls or walls cut up with numerous openings. Masonry can be laid very economically if the walls are designed to make maximum use of full and half length units, thus minimizing cutting and fitting of units on the job. The length and height of wall, and the width and height of openings and wall areas between doors, windows, and corners should be examined to make sure they can use the full-size and half-size units which are usually available for block and tile. All horizontal dimensions should be in multiples of nominal full length masonry units, and both horizontal and vertical dimensions should be in multiples of 8 inches.

Concrete masonry has become increasingly important as a construction material because of significant developments in the manufacture and utilization of these units. Concrete masonry walls properly designed and constructed will satisfy varied building requirements including fire, safety, durability, economy, appearance, utility, comfort, and good acoustics.

A hollow load-bearing concrete block of 8 inch by 8 inch by 16 inch nominal size will weigh from 40 to 50 pounds when made with heavyweight aggregate such as sand, gravel, crushed stone or air cooled slag. The same block made with lightweight aggregate will weigh from 25 to 35 pounds and may be made with coal cinders, expanded shale, clay, slag, or natural lightweight materials such as volcanic cinders and pumice. Heavyweight and lightweight units are used for all types of masonry construction. The choice of units depends on availability and the requirements of the structure under consideration. Lightweight units usually cost more to buy but less to install.

Estimating the quantity of materials for the mortar for masonry may be divided into two parts: first, finding the quantity of mortar required; second, finding the quantity of materials for this amount of mortar. We must first know how much actual mortar is required regardless of what mixture is used. This will largely depend on the thickness of the mortar joint, the thickness of the wall and how many cores are grouted. In some cases it will depend on the method of bonding. It will also depend on the method of laying the masonry, whether all joints are filled or whether some of the vertical joints are left open.

Sometimes in common brickwork the outside 4 inch thickness of the wall will be thoroughly bedded with mortar and all joints filled. The inside of the wall will also be properly bedded and all joints filled, but the center of the wall is not always filled with mortar. Sometimes the space between each 4 inch thickness of brick is left open in order to form a sort of open space in the wall, a "hollow wall."

If strength is required, all joints and every second core must be properly filled. This will require more mortar. After the quantity of mortar is decided on, the materials required to make this quantity must be figured. This will depend upon the mixture specified. Tables have been worked out in different ways. Some give the quantity of mortar in cubic yards required per thousand brick or block. Others give the quantity of material required to make a cubic yard of mortar. In some tables the two operations have been combined, the quantity or volume of mortar required, and the materials for the mortar for laying 1000 brick. Thus, we have tables giving the amount of cement, lime and sand per thousand square feet of wall surface.

The proportions for lime mortar are sometimes determined on the job and depend on the quantity of sand. If there is too much lime in the mortar, the mortar will stick to the trowel. If there is too much sand, the mortar will be stiff and difficult to work. The man who mixes the mortar judges the workability of the mixture and adds sand or lime as required.

On practically all jobs of any size, the mortar is mixed in a mixer because it produces a more uniform mortar and one that is more easily spread. It not only enables a mason to lay more brick but also results in a saving of 10 to 20 percent of labor required in mixing mortar.

There are a number of special masonry cements on the market that are used instead of lime or Portland cement mortar. They reduce the number of materials you store and handle and make it easier to mix and maintain uniform mortar batches. The air entraining properties of mortar cement provide weathertight joints, highly resistant to freezing and thawing. They work easier under the trowel than pure Portland cement mortar. About 18 cubic feet of mortar mixed in the proportions of 1 part masonry cement to 2½ parts sand lays 1,000 bricks.

In figuring the cost of masonry work, considerable judgment must be used because labor varies greatly depending on the class of work and also the kind of building. It is quite apparent that on a straight thick wall, such as is found in large warehouses, many more bricks or blocks can be laid by a mason than on a high class residence remodeling job.

There are many factors that tend to influence the cost of labor on masonry — the kind of bonding, kind of joints, kind of mortar used, thickness and height of the walls, irregularity in walls, weather conditions, and labor conditions.

The estimator should carefully study the job to see what kind of work is required. If, for example, the job requires matching brick laid with shoved joints, the estimator should make allowance for this. It will take much more time to lay brick with shoved joints than it will take for ordinary work. The thickness of the joints and the kind of bond must also be considered. A Flemish or English bond, for example, will require much more labor than ordinary running bond. In fact, any pattern work will require more labor than ordinary work. Where pattern work is required in which the brick have to be cut, higher labor cost should be allowed. Extra thick joints also add materially to the cost of labor. It will be impossible to lay many courses in succession as the mortar does not set rapidly enough to allow a heavyweight to be placed on it. Joints as thick as ¾ inch will slow up the work very much. The flush cut joint is the most economical because the bricklayer cuts off the excess mortar with his trowel and the joint is finished. For concave, V-tooled, weathered and struck joints, it is necessary to cut off the excess mortar with the trowel and then another operation is required: a trowel for weathered or struck joints, a jointing tool for concave and V-tooled joints.

Extra high walls will require the use of scaffolding or hoisting machinery. The qualifications of the bricklayers must also be considered. Bricklayers, like other workmen, vary in skill and efficiency.

Rough Carpentry

One of the most important parts of estimating rough carpentry is arranging and listing the items in a comprehensive and systematic way. The items should be so marked and placed on the estimate sheet so that they can be found at any time without going through the whole plan or estimate. Often changes are requested which will require refiguring a portion of the work. If items are clearly marked, they can be revised without going through the entire estimate. The estimate should be so written that anyone can check each item with the actual amount of material and the cost for that particular part of the work. A well arranged estimate sheet is not only a money saver, it also tends to give the carpenter and owner confidence in the figures.

In all cases, the following information should be given on a material list for lumber:

1. The number of pieces of each size.
2. The thickness of the lumber.
3. The width of each piece.
4. The length of each piece.
5. The grade of lumber.
6. The kind of wood.
7. Information stating whether the lumber is to be rough, partly smooth, or surfaced on all four sides.

The most common errors made by estimators are crowding their estimates and not making each item clear. Ample room should be allowed for totaling up. Each article should be clearly marked with its size and amount required. This is a slow and tedious process, but it must be mastered if the estimator wants a clear picture of the job requirements.

The number of posts in the basement will be found on the foundation plan and can easily be counted. In putting these on your estimate sheet, list the length and size of each piece. Common sizes of wooden posts are 6 inches by 6 inches, 6 inches by 8 inches, and 8 inches by 8 inches. About 5 percent may be allowed for waste. In some cases cast iron, steel or reinforced concrete posts are used.

After listing the girders, tabulate the sills, beginning at one corner and working around the building. List the sizes, number and lengths on the estimate sheet. The result equals the linear feet of sill required. Sill plates, which may be 2 by 6, 2 by 8, 2 by 10 or 4 by 6 inches in size, are bored so the anchor bolts can pass through them.

Joists

After estimating the sills, examine the floor joists. Put down the largest joists first. These are usually under partitions or are trimmers and headers around stairs or chimney openings. Next take off the regular floor joists. Begin at one side or top of the plan according to the way the joists run and complete each bay before starting on the next.

Joists are bought in lengths of an even number of feet. The usual lengths are from 8 feet to 24 feet. Before deciding on the length to use, it is best to check the lengths carried by the lumber yard. Sometimes it is economical to use full length joists for rooms 22 to 24 feet in width. At other times it will be better to use two lengths for each span as the price of the long material may be more per thousand board feet.

Consider also the type of framing when figuring the length of joists. If a built-up girder is required, the joists are butted together in the middle and not overlapped. In other cases they will have to over-lap enough so they can be spiked together. Also, the details at the outside wall must be studied. In western framing, there is generally a header to which the joists frame. In balloon framing, the joists extend to the outside edge of the studding. For masonry walls, the joists must be set at least 4 inches into the masonry.

To find the number of joists required for any floor area, divide the spacing of the joists into the length of the wall that carries the floor joists and add one to allow for the extra joist required at the end of the span. Also, add one extra piece for every partition that runs parallel to the joists.

Ceiling joists are estimated using the same method as when estimating floor joists.

Subflooring

Rough flooring is laid over the joists to form a working floor and a base for the finish floor. It is usually, 1 by 4 or 1 by 6 S1S rough lumber, or 5/8 inch or thicker plywood. Since a square foot of lumber is 12 inches by 12 inches (or equivalent area) irrespective of its thickness, and rough flooring is 1 inch or less when milled, a square foot of lumber is the same as one board foot. Consequently, the square foot unit of measurement is used for convenience. Therefore, to estimate rough flooring, calculate the entire floor area and deduct the openings. To the net area must be added a certain amount of lumber due to end waste and shrinkage. Add 5 percent for waste on most jobs. About 120 board feet of 1 by 4 inch lumber covers 100 square feet and 114 board feet of 1 by 6 inch lumber covers 100 square feet.

There are several grades and kinds of building paper. The standard width of the paper is 36 inches and a roll usually contains 500 square feet. Special kinds of paper or felt vary in width, number of square feet and weight per roll. The type of building paper laid over rough flooring is known as 15 pound asphalt saturated felt. To find the number of rolls of standard paper, divide the area of the floor to be covered by 500. Part of a roll should be counted as a whole roll.

Wall Framing

Figure the number of linear feet of stud partition first. Measure the linear feet of inside stud partition running in one direction. Then measure the partition running at a right angle to those just figured. Outside walls should be figured in the same manner.

Rule for plain walls: Multiply the length of the wall or partition by ¾ or .75 and add one piece. The result equals the number of pieces of studding for studs 16 inches on center.

Rule for walls with openings: When spaced 16 inches on center, estimate one stud for each linear foot of wall or partition. This is usually sufficient to allow for doubling studs at corners and at window openings and trussing over the heads of openings. The length of these studs can be scaled on the elevation or wall section.

Rule for top and bottom plates: Multiply the linear feet of all walls and partitions by 2 if top and bottom plates are required. The result will be the number of linear feet of plate required. If a double plate consisting of 2 top members, and a single bottom plate are used, multiply the linear feet of walls and partitions by 3.

To find the linear feet of firestop required, subtract from the total linear feet of walls and partitions the total or combined width of all openings. The result is the linear feet of firestops required.

In balloon framing the studding extends from the sill to the top plates of the second story. There are slightly different lumber requirements in balloon framing for outside wall studding and the ribbon used to support second-floor joists. Estimate one full length stud for the total linear feet of outside wall. Order the next even foot length of lumber. Then add two studs for every opening. The length of these opening studs will be the same as the stud height for each story.

A balloon frame requires a piece of 1 inch stock notched into the inside of studs which are two stories high. This piece forms a support for the second floor joists and is called a ribbon. The width will vary though 1 by 4 inch or 1 by 6 inch lumber is commonly used. To figure the amount of ribbon, measure or scale the linear feet of outside supporting walls. The direction of the second floor or ceiling joists will determine which are supporting walls. Any wall parallel to the joists does not require a ribbon.

Sheathing

Wall sheathing is nailed on the outside of exterior framed walls. 1 by 6 inch lumber, insulating fireboard, pressed wood, plywood, or gypsum sheathing may be used. Insulating sheathing board is usually furnished ½ inch or 25/32 inch thick, 4 feet wide and 6 feet to 12 feet long, all having square edges.

For plain walls, multiply the height of the wall from the sill to the top plate by the wall length. If sheathing is laid straight, add 10 percent to allow for waste; if laid diagonally, add 17 percent to the actual area. The result is the board feet of sheathing required. For walls with openings, find the total wall area as above and deduct all openings of 20 square feet or more. The result equals the board feet of sheathing. For plywood or panel sheathing, add 5 to 10 percent for waste.

For gables, calculate the gable area by multiplying the roof rise by one half the span. If sheathing is laid straight or plywood is used, add 10 percent. If laid diagonally, add 20 percent. The result in each case is the board feet of sheathing needed.

Rafters

On a plain gable roof the number of common rafters is found in the same way as the joists. If spaced at 16 inch centers, multiply the length of the wall by ¾ and add one. If 20 inch centers, multiply by 3/5 and add one. If there are extra rafters required for the gable cornice, these should be added. This rule gives the number for one side of a gable roof. Double the result for the other side. Allow 5 percent for end waste on ordinary roof work.

The easiest way to find the rafter lengths on a pitched roof is to use an architect's scale and read the length of the common rafter on the plan. For more accurate estimating, Table 19-4 and 19-5 can be used. The length of a rafter for any given pitch can be found by merely multiplying the constant given by the amount of run for that particular rafter. The result is the rafter length. If the roof has a cornice, the overhang length must be added to this result. Then increase the length to the next standard length of lumber. This method is practical for the estimator. The carpenter usually

Roof slope			Roof slope		
Rise	Run	Ratio	Rise	Run	Ratio
3	12	1.4361	9	12	1.6008
4	12	1.4530	10	12	1.6415
4.5	22	1.4631	11	12	1.6853
5	12	1.4743	12	12	1.7321
6	12	1.5000	13	12	1.7815
7	12	1.5298	14	12	1.8333
8	12	1.5635	15	12	1.8875

Ratios of hip or valley length to run of common rafter for various slopes
Table 19-4

Rise	Run	Ratio	Rise	Run	Ratio
3	12	1.0308	9	12	1.2500
4	12	1.0541	10	12	1.3017
5.5	12	1.0680	11	12	1.3566
5	12	1.0833	12	12	1.4142
6	12	1.1180	13	12	1.4743
7	12	1.1577	14	12	1.5366
8	12	1.2019	15	12	1.6008

Ratios of common rafter length to run for various slopes
Table 19-5

lays out rafter lengths and cuts by means of his steel square or a rafter length book.

To obtain the length of jack rafters, divide the distance between the corner of the roof plate and the first common rafter into equal parts. The distance between jack rafters should be as near to the distance between the common rafters as possible. The length of the shortest jack rafter is found by dividing the length of the common rafter into the same number of spaces as there is between the jack rafters on centers. The second jack rafter will be twice as large as the first, the third three times as long and so on.

Saddle boards can be measured from the roof plan or from the elevations. Usually each run of saddle board is shown on two elevations.

The length of ridge board needed is found by scaling on the elevation or roof plan. For a gable roof, the length of a ridge equals either the length or width dimension of the building. For a hip roof, the ridge length equals the difference between the length and width measurements. Order the lumber to an even foot length.

Roof Areas

To obtain the area of a plain gable roof, multiply the length of the ridge by the length of the rafter. This will give you one-half of the roof. Multiply this by 2 to obtain the total square feet of roof surface.

To obtain the area of a hip roof, multiply the length of the eaves by ½ of the length of the rafter, then multiply this by 2. This will give the area of both ends. To get the sides, add the length of the eave to the length of the ridge and divide by 2. Multiply this by the length of the rafter. This gives the area of one side of the roof, and when multiplied by 2, gives the number of square feet on both sides of the roof. Add this to the area of the two ends to get the total area.

The area of a plain hip roof running to a point at the top is obtained by multiplying the length of eaves at one end by one-half the length of the rafter. This gives the area of one end of the roof. To obtain the area of all four sides, multiply by 4.

Where rafter lengths are not known, the roof area may be computed if the horizontal or plan area and the pitch are known. Multiply the appropriate factor in Table 19-6 by the horizontal or plan area (including overhang) to find the roof area.

Rise in 12"	Factor	Rise in 12"	Factor
3"	1.031	8"	1.202
3½"	1.042	8½"	1.225
4"	1.054	9"	1.250
4½"	1.068	9½"	1.275
5"	1.083	10"	1.302
5½"	1.100	10½"	1.329
6"	1.118	11"	1.357
6½"	1.137	11½"	1.385
7"	1.158	12"	1.414
7½"	1.179	--	--

Roof area from plan area
Table 19-6

From the roof area you can use Table 19-7 to check your calculations on lumber required for the top plate, rafters and the ridge board. If the figure you arrive at by using Table 19-7 does not very nearly equal the lumber quantity you computed, you should recheck your figures.

Roof Sheathing

Sheathing for wood shingles is sometimes 1 by 3 inch boards spaced 2 inches apart. For spaced sheathing, figure 83 board feet of 1 by 3 inch lumber for each 100 square feet of roof area.

If insulating rigid sheathing is used for roofs or walls, add about 6 percent to the net area for waste. The sheets come in widths of 4 feet and lengths of 6, 8, 10, and 12 feet. The most economical size sheets should be ordered. Plywood is also used for roof sheathing. It is applied in sheets 4 feet by 8 feet using 8-penny common nails 6 inches on center. It is estimated similar to rigid sheathing mentioned above.

Finish Carpentry

Almost every job has a large number of finish carpentry items to figure. Most interior and exterior wall finish and ceiling finish fall into this category. Shingle roofing also may be included in this category but is better covered under the roofing section. Figure 19-2 will help you recall all finish carpentry items that should be in your estimate.

Rafter size	Spacing center to center							
	12"		16"		20"		24"	
	Board feet per square foot of roof area	Nails lbs. per MBM	Board feet per square foot of roof area	Nails lbs. per MBM	Board feet per square foot of roof area	Nails lbs. per MBM	Board feet per square foot of roof area	Nails lbs. per MBM
2"x 4"	.89	17	.71	17	.59	17	.53	17
2"x 6"	1.29	12	1.02	12	.85	12	.75	12
2"x 8"	1.71	9	1.34	9	1.12	9	.98	9
2"x10"	2.12	7	1.66	7	1.38	7	1.21	7
2"x12"	2.52	6	1.97	6	1.64	6	1.43	6

Rafters including collar beams, hip and valley rafters, ridge boards

Note: End gable material must be added when required.

Pitched roof framing
Table 19-7

Most finish carpentry items are "millwork" or wood that has been "milled" into a finished condition. Usually millwork is easy to estimate because it is merely fitted into place on the job. However, accurate material and labor estimates for millwork require that each item be listed separately. The number of linear feet of wood base, chair rail, picture mold, the number of door jambs to set, the number of windows to fit and hang, the number of openings to be cased or trimmed, the number of wardrobes, linen cases, kitchen units to set, etc. Estimating labor costs by this method improves accuracy and allows you to check actual and estimated costs during construction.

Estimating the material and labor costs of millwork is merely a matter of listing the materials required in separate items and then computing the total cost of each item at a cost per unit which includes both the price of the material and the cost of erection. The customary order of listing is as follows:

1. Exterior moldings and details.
2. Window frames and door frames.
3. Windows, sash, and doors.
4. Interior doors.
5. Ordinary continuous trim.
6. Casings and standing trim.
7. Cabinets and cases.
8. Stairs.

Exterior moldings and details includes all ordinary continuous moldings on the exterior of the building such as cornices, watertable, belt course, corner boards and also the porches, entrances, columns and trellis work. On the estimate sheet all simple items can be named and the size and number of linear feet given. Complicated cornices, watertables, etc. will require sketches and descriptions. Exterior moldings are measured by the linear foot and sold by the same unit or 100 linear feet. The required amounts of each kind or size are found by scaling the drawings or adding the given dimensions.

Watertable can be measured very quickly from the floor plan. Set down on a piece of paper each length and add these for the total number of linear feet. The number of linear feet together with details such as inches of plain stock and moldings are recorded together on the estimate sheet. For every outside corner which is mitered, allow 12 inches extra.

On a box cornice, if the soffit is 12 inches, the frieze 8 inches and the fascia 4 inches, the combined width is 24 inches or 2 board feet to the linear foot of cornice. For each outside corner allow 12 inches extra.

Generally the millwork estimator treats window frames and door frames as complete units, constructed of finished material into which a window or exterior door is fitted and hung. Door and window frames should have the style, quality and dimensions stated. In some regions they may come knocked-down, and it is necessary for the carpenter to nail them together before setting. On some remodeling work the carpenter may be called upon to make the window and door frames because standard sizes will not fit the opening. It is then necessary to estimate the different materials needed such as jambs, stiles, balances, outside casings, bank moldings, blind stop, parting bead, and sills.

Door jambs are ordered by sets which include two side pieces and one head piece. The measurements for doors should always be written with the width first, followed by the height and the thickness.

Windows, doors, screens, storm windows, and shutters are usually furnished prefitted by the mill. The modern way to buy windows is to order them as units all ready to set in the wall. This saves the carpenter's time and avoids delays in case a frame is missing or a sash fails to fit the frame just right.

Interior doors are mill made, prefitted to the exact size, completely machined (bored for tubular locks, mortised and gained or beveled for hinges) and delivered ready to hang. They should be listed as to style, size, kind, quality and catalog number.

Ordinary continuous trim includes baseboards, base moldings, chair rails and picture moldings. It also includes, at times, all cornice, cornice moldings and ceiling beams as well as wainscoting and paneling. These items are all listed by the linear foot except the paneling which is priced by the square foot of surface measure.

In specifying windows, the width is always given first, followed by the height and then the thickness. Windows can be measured either by the glass size or the opening size. If glass measurement is used, the correct method is to give the size in inches only, with the width first, followed by the height in inches only, and then the number of lites and thickness, thus: 26 inches by 28 inches, 2-lite, 1-3/8 inch. If the opening size is given, the width is given first in feet and inches, followed by the height in feet and inches, and then by the thickness and the number of lites, thus: 2 feet 6 inches by 5 feet 2 inches, 1-3/8 inch, 2-lite.

Door and window trim are listed by writing down the number of ''sides'' required. A side consists of a complete set of material required to trim one side of an opening. Thus, an inside door requires two sides of trim while an outside door or a window requires only one side. In listing door and window trim, the width of the opening should always be given first, followed by the height, with the complete list of all the different pieces wanted.

Casings and standing trim are listed next if you do not wish to take off the casings in connection with the windows themselves.

Cabinet work and cases should include all bookcases, cabinets, china closets, linen closets, wardrobes, dressers, built-in seats, etc. Sometimes it includes paneling when it is a special part of the particular detail being listed. Cabinet work should be listed in detail with sketches if they are not shown in a plan. The sketch of a cabinet should show the width, height, and depth and indicate the number of drawers and doors required and location of the shelving.

Cases and cabinets can be made by the mill according to individual requirements if necessary. Standard ''stock'' cabinets are less expensive and should be used when they meet job requirements. They are usually delivered ready to install but some manufacturers deliver disassembled or ''knocked down'' units. Assembly can add an hour or more to the installation.

Stairways should be listed separately. Mills or cabinet shops that specialize in stair work will build all standard types of stairways. The stairway is delivered in cartons of heavy cardboard. Complete directions for fitting and the necessary fitting bolts are furnished with all easements. If desired, the shop will also install the complete stairway.

Wood Flooring

Finish wood flooring is purchased on the basis of the original size of the rough stock. Flooring 25/32 inch thick measures about ¾ inch less on its face than the rough measurement. The term ''face'', when applied to hardwood flooring, means the net surface width of the stock after it has been milled. A certain percentage must,

therefore, be added to the actual floor area. For 2¼ inch stock, add ⅓ to the actual floor area to be covered to get the number of feet ''board measure'' required in a room. For 1 by 4 inch fir or pine flooring, add ¼ to the floor area and add 1/6 to the floor area for 6 inch stock.

Roofing

Roofing materials are estimated by the **square**. A square is 10 feet by 10 feet or an area of 100 square feet, equal to 14,400 square inches. To figure the number of squares, divide the area of the roof in square feet by 100.

In estimating shingle roofing materials, the exposed roof area, or the area of the roofing material that is exposed to the weather, is the only part of the roof covering material that is considered. The lap, that part of the roofing material which is covered by the piece above it and is not exposed to the weather, should not be figured in the area that the roofing material is to cover. For instance, we will consider that the average wood shingle measures on the face 4 inches by 18 inches. When laid on a roof, the area of the same shingle exposed to the weather will be 4 inches by about 4 ½ inches or 18 square inches. Eight shingles will cover one square foot of roof area. The number of shingles in a square would be 14,400 divided by 18, or 800. Adding 10 percent for waste in cutting on hips, valleys and double courses, there are 880 shingles 4 inches by 18 inches with 4½ inches exposure in each square.

Wood shingles are sold by the square but are packed in bundles which contain about 250 shingles. Four bundles will cover a square when standard exposure is used.

Most asphalt shingles are packed two, three or four bundles to the square. One bundle of a two bundle square will cover 50 square feet. One bundle of a three bundle square will cover 33⅓ square feet. One bundle of a four bundle square will cover 25 square feet.

To estimate the number of 12 inch by 36 inch square butt strip shingles, proceed as follows:

1. Figure the roof area using Table 19-6. Increase the net area about 5 percent on gable roofs for waste.

2. Divide the area by 100 to find the number of squares.

3. Multiply the number of squares by the number of bundles per square mentioned above. The result equals the number of bundles needed.

4. Obtain the number of linear feet of hips, valleys and ridges to be covered with asphalt shingles and calculate the same as 1 foot wide. Asphalt starting strips may be laid along the eaves of a roof before laying the first row of shingles or the first row may be doubled.

Other Work

Most remodelers subcontract plumbing, heating sheet metal, electrical work, plastering and other specialties. When you get a bid for these items, refer to Figure 19-2 to make sure you are not forgetting any part of the job.

The Cost Estimate

When you have completed the quantity survey and checked to be sure it is accurate, you can begin calculating the labor cost. Many estimating manuals are available to help you estimate the labor required to install each material. Several man-hour manuals are offered on the final pages of this book. However, the best source of information on your man-hour requirements is your record of jobs you have completed. On the estimate sheet list your estimate of the labor hours required per unit of material (board foot, square foot, cubic yard, etc.). Multiply the number of units by the man-hours per unit to find the labor hours for each operation. Be sure to include the cost of supervision if it has not been figured into the labor hours. Multiply your cost for materials per unit by the units required to find the total cost of materials. Include sales tax and delivery expense where appropriate.

Even after the labor cost and material cost have been computed and added, you have not found your cost for the job. The "labor burden" will add between 15 percent and 25 percent to the labor cost. For every dollar of payroll you must pay an additional 15 cents to 25 cents for taxes and insurance to government agencies and insurance carriers. This 15 to 25 percent must be added to the gross hourly wage, including all fringe benefits and contributions. Many contractors add 25 percent to the estimated labor cost to cover taxes and insurance, and this figure will be adequate for most home improvement contractors.

Most states levy an unemployment insurance tax (S.U.I.) on employers based on the total payroll for each calender quarter. The actual tax percentage is usually based on the employer's history of unemployment claims and may vary from less than 1 percent of payroll to 4 percent or more. The Federal government levies an unemployment insurance tax based on payroll (F.U.T.A.). The tax has been about 0.6 percent of payroll. The Federal government also collects Social Security (F.I.C.A.) and Medicare taxes. Together these come to about 6 percent of payroll, depending on the earnings of each employee, and are collected from the employer each calender quarter.

States generally require employers to maintain Worker's Compensation Insurance to cover their employees in the event of job-related injury. Heavy penalties are imposed on employers who fail to provide the required coverage. The cost of the insurance is taken as a percentage of payroll and is based on the type of work each employee performs. Clerical and office workers have a very low rate classification and the employer's cost may be only a small fraction of one percent of payroll. Most light construction trades have a rate between 5 percent and 8 percent though roofing usually carries a classification of 10 percent or more. The actual cost of Worker's Compensation Insurance varies from one area to another and from one year to the next depending on the history of injuries for the previous period. Your insurance carrier will be able to give you the cost of coverage for the type of work your employees are doing.

Every contractor should maintain liability insurance to protect his business in the event of an accident. Liability insurance is also based on the total payroll and usually is about 1½ percent of payroll. Higher liability limits will cost more.

The total contractor burden can be itemized as shown below. The percentages listed are approximate maximums. Your accountant or bookkeeper will have more exact figures.

State Unemployment Insurance 4.0%
F.I.C.A. and Medicare 6.0%
F.U.T.A. 0.6%
Worker's Compensation Insurance .5.0% to 8.0%
Liability Insurance 1.5%
    Total contractor burden . . . . .17.1% to 20.1%

Every contractor who has a payroll is well advised to add this labor burden into the cost of labor on every estimate and make the insurance and tax deposits when due. No contractor can ignore these requirements and operate for long.

Overhead

There are many costs which the contractor must bear that are not associated with any particular trade or phase of construction but are incurred as a result of taking on a particular job. These costs are usually called indirect costs. The list below includes some of the items that are usually included as indirect job costs.

Fire Insurance, Surety Bonds, Pensions, Building Permit, Sidewalk Permit, Telephone, Water, Electricity, Sewer and water connection fee, Temporary Office, Repairs to adjoining property, and Job Toilets.

You can probably think of many more indirect cost items. Some contractors include in indirect costs or job overhead the cost of supervision and other nonproductive labor such as the cost of estimating the job. The time you spend on each job should be charged against each job. These are very real costs and must be included somewhere in the estimate. Since they are incurred as a result of taking each particular job, they should be included under indirect costs.

After everything is figured there are certain items of expense you must bear in conducting your business and which cannot be charged against any certain job. For example, office rent, telephone at the office, your salary as the business manager, office lighting, office staff, small tools, office insurance, printing, serve on your car, postage, and countless other items. These types of expenses are usually called overhead and are really general administrative costs.

Administrative costs will be about the same each month regardless of how much work you do because your overhead relates to administering your business rather than doing any particular job. You can change the overhead cost into a cost per business day. This cost has to be carried by the work you do each day. If you have two jobs going that day, figure that each carries an appropriate part of the overhead. You must add this cost per day into your estimate. Many home improvement contractors estimate their overhead for the year and divide it by the dollar volume of work they will do during the year. The result is a percentage that can be added to every job during the year. For example, if your business volume will be $200,000 for the year and your overhead will be $20,000, your overhead will be 10 percent of the total selling price of the work you do. Some contractors figure the overhead as a cost per productive man-hour. Other contractors have reduced overhead to a cost per square foot of floor, roof or wall area. Any one of these systems is good if it works for you. Keep a record of your indirect overhead and develop some method of dividing this cost among your jobs. Most of all, don't forget to include this important item in your bid. The total cost of direct and indirect overhead will be 10 percent or more of the total job cost for most light construction work. This 10 percent will make the difference between a profit and a loss on almost every job you have.

Notice that adding 10 percent to the job cost for overhead is not the same as taking 10 percent of your selling price for overhead. To get 10 percent of the selling price you have to add 11.1 percent to the total. To get 15 percent of gross for overhead you have to add 17.6 percent to the total.

Contingency and Escalation

Most contractors add a small amount to their bids to allow for any unanticipated conditions that may develop during the course of construction. Work is seldom done faster than planned and problems are frequently encountered which make the work more costly. The right amount to add for contingency depends on your experience and the job. Remodeling or repair work may require a larger allowance to meet difficulties which cannot be accurately forecast. Escalation is the increase in costs of labor, materials and equipment between the time the bid is submitted and the time work is actually done and paid for. Even though you are sure of the price of 1,000 board feet of framing lumber when you submit your bid, you may not know what you will pay for that lumber when it is actually purchased from the yard. Lumber, plywood and some metals change in value quite rapidly. The price of materials such as wallboard, concrete and glass change relatively little. If you can't secure firm quotes for materials to be delivered in the future, it is best to allow some amount for price increases or specifically exclude price increases from your bid.

Profit

The profit is the return on investment to the contractor. The contractor should be able to pay himself a wage for the work he performs and, in addition, pay himself a return on the money he has invested in his business. If a remodeler has $50,000 invested in his business he should receive a return on investment of $3,200 to $6,000 per year (8 percent to 12 percent of investment) in addition to a reasonable wage. This profit can be thought of as interest on the money invested in equipment, office, inventory, work in progress and everything else associated with running a construction business. How much then should you include in your estimate for profit? You will hear many conflicting figures. Some estimators say a 20 percent profit is a good target figure and they try to end up with a "profit" of 20 percent of the total contract price after all bills are paid. Some builders may operate efficiently enough to achieve a 20 percent profit. They certainly are the exception. The contractor who talks about a 20 percent profit may mean that after he has paid for his labor, material and equipment he has 20 percent left over for himself. This 20 percent is really his wage and, though it may be substantial, it is not a profit in the true sense. A profit is what remains after **all** costs are considered. The builder should include the cost of his own work under indirect costs (estimating, supervising and working on the job) and overhead (managing his business). What then is a realistic profit in the true sense? Dun and Bradstreet, the national credit reporting organization, has compiled figures on contractors for many years. They report the average net profit after taxes for all contractors sampled to be consistently between 1.2 percent and 1.5 percent of gross receipts. This includes many contractors who reported losses or became insolvent. A 1½ percent profit, even after taxes, is a fairly slim profit. Not many contractors, especially contractors on small home improvement projects, include so small a profit in their bid. On extremely large projects such as highways, power plants or dams, the contractor may allow only 1 percent or less for profit . . . especially if the amount to be received is based on the contractor's actual cost rather than a fixed bid. Remodeling and repair work traditionally carries a higher profit margin because the size of jobs is much smaller than other types of work and the risk of significant cost overruns is larger. Probably 8 percent to 10 percent profit is a reasonable expectation on most jobs, with very small remodeling jobs running to as much as 25 percent.

To get a 10 percent profit, the total cost of labor, materials, equipment and subcontracted work will usually be only about 60 percent of your selling price. In other words, your selling price must be about 1.7 times your out of pocket cost for the job. Many remodelers make the mistake of assuming that their selling price should be about 10 percent more than their cost. Look over the list of indirect costs and overhead in Table 19-2. These are all costs you have to carry if you want to stay in business. You can see why you must price the job at least 170 percent of your cost to get a 10 percent profit.

Of course, there is more to "profit" than just how much profit you would like to earn. Sharp competition will reduce the amount of profit you can figure into your estimate. If you include too much profit in your bids you will find yourself under-bid for the jobs you would like to have. If you have developed a specialty and you can do a particular type of work better than other contractors in the area and have enough work to keep you busy, then you may increase your profit by one or two percent or even more.

In practice there is no single profit figure which will fit all situations. For most home improvement work an 8 to 10 percent profit is a very nice expectation. A contractor who has all the work he needs and wants and is asked to bid on more work may figure a 15 percent profit is not excessive. At the end of each year the profit on your business should give you a reasonable return on the money you have invested in the business (after you have taken a reasonable wage for yourself). You should earn a profit equal to 8 percent to 12 percent of the "tangible net worth" of your business. The tangible net worth is the value of all the assets of your business less the liabilities (anything your business owes) and less any intangible items such as goodwill.

The small remodeler who has only a vehicle, some tools and a few hundred dollars working capital, may have a tangible net worth of less than $5,000. Any profit he shows, after he has taken a reasonable wage for himself, will very likely be used to buy additional equipment and increase his working capital. Still, he should include a profit in each job which will give him a return at the end of the year between 8 percent and 12 percent of his $5,000 tangible net worth.

Use Table 19-8 to total the estimate for each category of work and add overhead, escalation, contingency and profit. Using an estimate summary sheet such as Table 19-8 will save you time and give you a good picture of the job cost at a glance.

Cost Control

Estimating and controlling construction costs is only one part of your management responsibility for controlling costs. You also have to control the flow of materials and labor, office costs and selling costs if you are going to make a profit at the end of the year.

Many successful home improvement contractors draw up a "plan for profit" each year. They actually establish a cost target for each of the major expenses for the year. This is their budget for running the company just like the cost estimate is a budget for a particular job.

When the subject of budgeting comes up, some remodelers say, "That's for the big fellow. I know what my volume is and my bank account tells me how much money I have." These owners fail to realize that budgeting can help to eliminate errors of judgment made in haste or made on assumptions rather than facts.

The first thing you must know in budgeting is what your anticipated expenses are going to be for the period being budgeted. Then you have to make an estimate of how much in sales will be generated to pay these expenses? What will be left? You must try to determine the high and low points in your operations in order to provide the adequate amount of cash. A sales analysis of previous periods will indicate when the high and low points occur. This forecasting helps you to plan for financing the jobs you work on and for carrying your accounts receivable. Controlling your inventory of work completed and accounts receivable can help to take the strain off of your working capital.

The cash budget is the most effective tool for planning the cash requirements and resources of your business. With it you plan your financial operations and the cash you expect to take in and pay out. Your goal in budgeting is to maintain a satisfactory cash position for any contingency. When used to project the cash flow of the business, the cash budget will:

1. Provide efficient use of cash by timing cash disbursements to coincide with cash receipts. These actions may reduce the need for borrowing temporary additional working capital.

2. Point up cash deficiency periods so that predetermined borrowing requirements may be established and actual amounts determined to reduce excessive indebtedness.

3. Determine periods for repayment of borrowings.

4. Establish the practicability of taking trade discounts or not taking them.

5. Determine periods of surplus cash for investments or purchase of inventory and equipment.

6. Indicate the adequacy of or need for additional permanent working capital in the business.

The important thing to keep in mind in making a cash budget is the word "cash." Be as factual as you can. Try not to over-estimate sales or under-estimate expenses. Your sales forecast must be as accurate as possible because it is the basis for figuring your cash and expenses.

Your cash flow estimate will be based on the cash budget for all work you plan and the profit and loss projections for each month. For an existing business you will have profit and loss statements for a year or more to help you project your month to month cash needs for the coming year. For a new enterprise you should have a projected profit and loss statement for the first year or more. To develop this profit and loss statement, estimate what you feel are reasonable sales goals for the year. Then subtract from your

Home Improvement Estimate Summary

Owner _____   Estimate No. _____

Address _____

Estimate by _____   Tel. No. _____

Date _____   Type Of Job _____

Item	Material		Labor	Subcontract	Total		Actual Cost	
Excavation								
Demolition								
Concrete								
Forms								
Reinforcing								
Masonry								
Rough carpentry								
Finish carpentry								
Roofing								
Flooring								
Plumbing								
Heating								
Sheet metal								
Electrical work								
Plastering								
Paint & decorating								
Glass & glazing								
Cabinets								
Ceramic tile								
Counter tops								
Range & oven								
Kitchen equipment								
Bathroom fixtures								
Equipment rental								
Supervision								
Job cost								
Direct overhead								
Indirect overhead								
Allowance for callbacks								
Subtotal								
Escalation								
Subtotal								
Contingency								
Subtotal								
Profit								
Total Cost								

Table 19-8

Sales			
Subcontract revenue	$48,000		
Small shopping center	20,000		
Small jobs	24,000		
Repair work	8,000		
Total		$100,000	100%
Costs of production			
Materials	$30,000		
Labor (half to partners working as craftsmen)	40,000		
Total		$70,000	70%
Gross Profit		$30,000	30%
Overhead Expenses			
Wages (clerical help)	$2,000		
Rent	2,000		
Interest	500		
Taxes	2,000		
Bad debts	200		
Repairs	1,800		
Depreciation, amortization	2,500		
Other (licenses, vehicle costs, tool losses)	4,000		
Payments to partners and bonuses	3,000		
Total		$18,000	18%
Net operating profit before taxes		$12,000	12%

**Projected profit and loss statement
Table 19-9**

gross sales the cost of materials and labor used on each job. This leaves your "gross profit" which is used to pay all overhead and operating expenses.

For example, two carpenters are forming a partnership to operate a carpentry subcontracting business. Table 19-9 shows a "pro forma" or projected profit and loss statement for the first year. They project a total volume of $100,000 for their first year, almost half of which is for work done under contract with a home improvement contractor. Also, they will perform remodeling work on a small shopping center and carry out a series of smaller jobs plus some repair work.

Labor costs for the partnership come to $40,000 for the year, half of which goes to the partners themselves working as craftsmen in the execution of their contracts. They draw an additional $3,000 for the year from the firm and show a year end profit of $12,000. Therefore, in self-generated wages, draws and profits, the partners divide $35,000 for an income of $17,500 each before Federal Income Tax.

Notice the percentage figures in the right hand column of Table 19-10. Compare these figures with the percentages in the right hand column of Table 19-9. Table 19-10 will be explained in the next paragraph. Table 19-11 shows a projected profit and loss statement for a small contractor. This builder projects a business volume of $240,000 for the year. Work includes substantial remodeling of five homes and two stores and a series of smaller jobs (such as finishing unfinished basements and adding rooms to a home). On this volume the contractor anticipates a gross margin of one-quarter of $240,000 or $60,000 which, after

overhead is deducted, yields an income for the year of $17,600 before payment of Federal Income Taxes.

Notice that the carpentry subcontractor spent 70 percent of sales on labor and materials and the

	Contractors	Sub-contractors
Sales (contract revenues)	100%	100%
Materials and subcontract work	44	44
On site wages (excluding owner)	19	23
Gross profit	37	38
Controllable expenses		
Off site labor	2	1
Operating supplies	3	2
Repairs and maintenance	1	1
Advertising	1	1
Auto and truck	3	2
Bad debts	Less than .5	Less than .5
Administrative and legal	.5	.5
Miscellaneous expense	2	1
Total controllable expense	13	9
Fixed expenses		
Rent	1	1
Utilities	1	1.5
Insurance	1	1
Taxes and licenses	.5	1
Interest	Less than .5	Less than .5
Depreciation	2	2
Total fixed expenses	6	7
Total expenses	19	16
Net profit	18	17

**Typical expense ratios for small contractors
Table 19-10**

Sales		
New construction - 3 stores	$120,000	
Remodeling - 5 houses	60,000	
Remodeling - 2 stores	20,000	
Small jobs	40,000	
Total	**$240,000**	**100%**
Cost of production		
Materials	$50,000	
Labor (including payroll taxes)	40,000	
Subcontracts	90,000	
Total	**$180,000**	**75%**
Gross profit	**$60,000**	**25%**
Overhead expenses		
Salaries	$15,600	
Rent	1,800	
Interest	1,500	
Taxes	4,000	
Bad debts	500	
Repairs	4,000	
Depreciation, amortization	6,000	
Other	9,000	
Total	**$42,400**	**17-2/3%**
Net operating profit before taxes	**$17,600**	**7-1/3%**

Projected profit and loss statement
Table 19-11

	Percent of sales	Year total $ sales	January Projected*	January actual
Sales	100.00	$100,000	8,000	8,774
Cost of sales (labor and materials)	67.23	67,230	4,750	5,113
Gross profit	32.77	32,770	3,250	3,661
Controllable expense				
Office labor	1.15	1,150	90	78
Operating supplies	2.34	2,340	160	181
Repairs and maintenance	.59	590	40	23
Advertising	1.12	1,120	75	62
Auto and truck	2.04	2,040	60	118
Bad debts	.03	30	0	0
Administrative and legal	.48	480	23	18
Miscellaneous	1.03	1,030	120	107
Total controllable expense	8.78	8,780	568	587
Fixed expenses				
Rent	1.00	1,000	83	83
Utilities	1.42	1,420	120	113
Insurance	1.16	1,160	420	420
Taxes and licenses	.85	850	120	165
Interest	.10	100	0	0
Depreciation	1.65	1,650	137	135
Total fixted expenses	6.18	6,180	880	916
Total expenses	14.96	$14,960	1,448	1,503
Net profit (before income tax)	17.81	$17,810	1,802	2,158

*Project sales and expenses for 12 months in 12 columns.

Expense worksheet for a small subcontractor-remodeler
Table 19-12

home improvement contractor spent 75 percent on labor, material and subcontracted work. Overhead expense was about 18 percent for both contractors. These are key figures in estimating your operating budget for any year and will be a major point of interest to the lending officer if you go to your bank for a loan. Check your projections against Table 19-10. If you do business like most builders, your expense ratios should fall within one or two percent of the figures listed. As you accumulate more experience in your business you can use these expense ratios to evaluate your expenses and determine where money can be saved and where additional costs are justified.

Once you have a profit and loss statement for the year, you can develop an expense worksheet for each month. This becomes your working budget. See Table 19-12. Estimate your sales and expenses for each month. You probably have a pretty good idea of the months when your sales will be greatest and when sales will be the smallest. Your expenses won't be a constant percentage of all sales or sales each month. But try to anticipate as many expenses as possible and fill them in on your expense worksheet in the months where they fall. As each month goes by, fill in the actual dollars you spent in each expense category. Watch carefully any expenses that are exceeding expectations. You are now budgeting your expenses like most major corporations.

Material Cost Control

Now you have a budget for both your jobs and your company operations. Next, you have to work to stay within your estimates and budget to make the profit you plan on. Keeping material costs within budget will go a long way toward achieving your goal.

Your objective in controlling material costs is to keep the quantity of material used within the limits of your estimate. Several different control systems should be used for the different material classifications. The first classification might include material delivered to the site by a supplier or purchased in quantities no greater than the amount required for one job. The second classification of materials covers items purchased in quantity, stored, and allocated to each job. The third category includes materials purchased in quantity, stored, but not controlled by each job.

Material Delivered to the Site

The best system of control for material delivered to the site by the supplier is based on ordering from the take-off sheet. This type of controlling is done either in the field by your superintendent or in your office. The advantage of control in the field lies in its closeness to the point at which errors are made. The advantage of centering control in the office is that you may be better able to handle the information at less cost than the field superintendents.

Controlling the quantity of material implies controlling the cost. The price, however, need not be specified on the take-off sheet. When the materials for any given job are ordered, the construction superintendent, or whoever does the ordering, can merely check off the material on the estimate. Other information, such as date ordered or date to be delivered, may also be added.

If materials are ordered only once, you know the job is going according to plan and no further action is necessary. If a second order is necessary,

the quantity of materials needed for the specific job obviously exceeds your estimate. The reason for ordering additional materials should then be checked.

The second system of control for material delivered to the site by the supplier always centers in the office. There are two advantages: The system uses office time rather the field time and records are more easily handled in the office. This method can also be used as a check on the performance of the superintendent and can be used to provide cost information for future estimates. The disadvantages of the system are the time lag and the constant need for communication between the office and the job.

As the office orders materials for each job or as bills are paid, the information on quantity and cost is posted on the cost estimate. An entry is made for each purchase according to job number and material classification. The time of comparison of the estimated and actual costs can vary according to how much control you want. The minimum control is checking all the job costs after completion. If the costs agree with the estimate or are within a reasonable range, no action is necessary. A finer degree of control is obtained by using a cumulative total. Your signal for action is given whenever this running total exceeds the material cost estimate. Also, you should look at the cumulative cost total and, on the basis of your judgment and knowledge of each particular job, decide whether or not costs to date are in line with the estimate. An even finer degree of control would be obtained by making periodic checks during the various stages of completion. A percentage of total material cost (by material classification) is set up for a number of stages. Every time material to be used in a specific stage of the job is ordered, the total cumulative cost is compared with the estimate for that stage. This intermediate estimate is found by multiplying the estimated cost by the percentage of cost that should have been incurred. If the comparison shows a big difference, you should take action to eliminate the excess costs or revise your estimating methods.

Figure 19-13 is a material purchase order which should help you eliminate most disputes with suppliers and resolve the remaining disputes in your favor. It is an excellent means of controlling material costs and deliveries on your job. Two copies go to the supplier (he returns one to your office as an acknowledgement) and two copies are kept in your office. One office copy is filed with the job records. The other copy is filed under the supplier's name, compared with his invoice when received and then filed away with a copy of your check paying the invoice.

Inventory Materials Controlled by Each Job

Some materials are purchased in quantity and kept in stock to be used in specific quantities on certain jobs. One system of control for these

```
┌─────────────────────────┐          ┌────────────────────────────────┐
│     PURCHASE ORDER      │          │  This order number must appear on │
└─────────────────────────┘          │  all packages, B/L's, packages, etc.│
                                      │                                    │
                                      │            No.  0000               │
                                      └────────────────────────────────┘

HOME
IMPROVEMENT                            Job No. _____
COMPANY
                                      Cost Acct. No. _____

                                      Customer Extra No. _____


V _____    S _____
E _____    H
N                                      I _____
D _____    P
O                                        _____
R _____    T
                                       O
```

VENDOR CONTACT		ORDER DATE	BUYER	REQUISITION NO	REQUISITIONER

DATE REQUIRED	TERMS	SHIP VIA	FREIGHT INFO.	BACK CHARGE INFO.

ITEM	QUAN.	UNIT	DESCRIPTION	UNIT PRICE	TOTAL	A/C DISTRIB.
			TOTAL _____		$	

PLEASE RETURN THIS ACKNOWLEDGEMENT IMMEDIATELY

I have read and understand the terms and conditions on the
face and the reverse side of this order. I agree to be bound by
all the terms and conditions as stated hereon without exception.

SIGNED BY _____ TITLE _____

DATE _____

THIS ORDER IS PLACED SUBJECT
TO TERMS AND CONDITIONS NOTED
ON FACE AND REVERSE SIDE.

HOME IMPROVEMENT COMPANY

BY _____

Table 19-13

TERMS AND CONDITIONS OF PURCHASE

Quality of all materials and/or work furnished must be as specified on the face of this order. All material purchased will be subject to our inspection and approval after delivery unless otherwise agreed upon in writing. If materials are rejected, they will be held for disposition at the seller's risk and expense.

Billed price must not be greater than that shown on the face of this order or greater than that last quoted or paid without first notifying this office and obtaining written consent. The buyer reserves the right to return goods shipped at a higher price at the seller's expense.

Discounts will be calculated from the date an acceptable invoice (showing this P.O. number) with Bill of Lading attached is received by this office.

Packing, crating and drayage charges will not be paid by buyer unless otherwise previously agreed upon, but materials damaged because of inadequate or improer packing will be returned at seller's expense, including original shipping cost. Replacement of damaged materials shall be at the option of buyer.

Delivery must be made within the specified time. Otherwise the buyer reserves the right to cancel this order and purchase elsewhere.

Changes in conditions of this contract shall be agreed upon in writing with the buyer and such changes negotiated directly with the buyer's ultimate customer will not be valid without written approval from the buyer.

Mechanics liens: If this order calls for work to be performed on any premises owned or controlled by the buyer, the seller agrees to keep the premises free and clear of any mechanics liens and will furnish the buyer with a certificate and waiver as required by law.

Indemnity: The seller shall forever indemnify and save buyer from any loss or damage whatsoever and of every kind, nature or description due to or arising out of or claimed to arise out of any act or omission of seller or its subcontractors or any materialmen or suppliers, agents, servants or employees of seller, which indemnity shall include, but not limited to, attorneys fees, court costs, civil and criminal penalties.

Compliance and taxes: All materials and/or labor furnished under this order must have been produced and sold in compliance with all Federal, State and other laws and all labor taxes or levies, or other taxes or levies levied by reason of this order, including any claims brought under the Worker's Compensation Law of the State in which the work is performed, are the sole responsibility of the seller unless otherwise agreed upon in writing. The seller shall hold and save harmless the purchaser, the officers, agents, servants and users of its product from liability of any nature arising out of any patent litigation in connection with articles purchased.

Payment of invoice does not constitute acceptance of merchandise covered by this order (and is without prejudice to any and all claims of buyer against vendor).

To be valid this order must bear a written signature.

Figure 19-13 [continued]

materials is similar to that of ordering from the take-off sheet. The take-off sheet is the authority to draw from the inventory. The material is already in stock and need not be processed through the office. The control process is the same as for ordering from the take-off sheet for purchase of materials. This system has advantages and disadvantages similar to those mentioned for the purchase of materials. It requires the superintendent to authorize the use of each material, which may be a great disadvantage, especially if your job sites are scattered. This disadvantage, however, may be overcome by delegating control to various craftsmen, each of whom carries out the control process for his own part of the take-off sheet. This variation can be quite effective since the craftsman is close to the actual situation. The system may fail at least partly if the craftsmen are careless. A check may be made by a review of the costs of each job from the inventory records to see if the craftsmen are actually carrying out the control process.

Another method uses part of the take-off sheet but transfers control to a person who releases materials from the storage area. This has the advantage of putting in one place all the paper work related to control of materials drawn from inventory. The person who checks out materials has that part of the take-off sheet necessary for each job and checks off on the take-off sheet the materials drawn for the job. If excess materials are drawn, he notifies the craftsman, superintendent, or you so that the reasons for the use of excess materials can be investigated.

Control can also be centered in the office. When materials are drawn, the information is reported to the office; the quantity and cost are posted to the job number and classification. Performance is then compared with the estimate according to the degree of control desired, either when the job is completed or during the course of construction.

Inventory Materials Not Controlled by Each Job

It is not worth the trouble to control by each job some of the materials that are kept in inventory and that make up only a small part of the cost of construction. For example, nails, inexpensive hardware, and anything that is not estimated fairly precisely for each job would fall in this category. Some control over these materials is necessary, however. Your standard for control of these items should be the average cost of these materials per job, per unit installed of cabinet, per 1,000 board feet, or some other measure. If you feel pretty confident about your control of this type of material, you can reduce the control to a monthly comparison of units of work done and units of material used.

If the quantity used compares favorably with the estimate, no further investigation is necessary. If it does not, some remedial action, such as changing issuing methods, is required. This simple and inexpensive method of control is essential. Without it, you may not know why costs are increasing and why profit margins are being squeezed.

Inventory Records

Where materials are purchased and stored for later use, a great deal of detail is required to keep account of the stock on hand and of deliveries to the building locations. Your objective in inventory control is to be sure that every bit of material bought is eventually installed in one of your jobs. By keeping tab of the quantities of materials on hand and knowing where they are located, the planning of operations can be simplified. Also, it is extremely important for you to have a record of the cost of materials in stock so that actual costs of jobs can be more accurately kept. Further, an accurate accounting for materials may reveal that various items disappear from time to time or that the amount of wastage is excessive. Stock records will help you exercise more effective control over the use of materials.

Stock records should account for receipt and issuance of materials and supplies. The Stock Record, Form 19-14 has spaces provided for the receipt and issuance of materials. One of these cards should be used for each group of items which can be identified by size, grade, etc.

When material is obtained for stock it should be entered in the "received" and "on hand" sections of the record. The quantity should be stated in a measure which can be easily identified so that a price per unit may be determined. Any one of several methods may be used in assigning cost to materials moved out of stock. Perhaps the average cost procedure is the most convenient to use; however, the first-in, first-out (the first costs per unit are the first taken out) method may be employed. For tax purposes you should be consistent and follow the same procedure from year to year.

Upon issuance of materials for a job, a Materials Requisition, Form 19-15, should be prepared and used as the authority for the withdrawal and as the basis for entries in the records. The first entry to be made from the Materials Requisition is in the "issued" and "on hand" columns of the Stock Record. Here the date, requisition number, quantity, price, and total are entered in the "issued" column. The quantity and cost of materials issued are deducted from the preceding balance in the "on hand" column and the new balances of quantity and cost are extended on the same line as the "issued" column entry.

Under the procedure outlined above, the materials on hand, as shown by the Stock Records, must be checked periodically against an actual count of the materials stored (a physical inventory). Through the use of a physical inventory the items on hand may be verified and any errors in the records may be detected and corrected or

STOCK RECORD

RECEIVED					ISSUED					ON HAND			
DATE	PUR. ORDER	QT.	PRICE /UNIT	TOTAL	DATE	REQ. NO.	OT.	PRICE /UNIT	TOTAL	DATE	QT.	PRICE /UNIT	TOTAL

ITEM   DESCRIPTION   LOCATION

Form 19-14

MATERIALS REQUISITION

NO. _____

JOB NO. _____

DATE: _____   CLASSIFICATION: _____

QUANTITY	DESCRIPTION	PRICE PER UNIT	TOTAL

_____
FOREMAN

Form 19-15

material loss may be discovered. A complete physical inventory should be taken at least once each year and more often if time permits. Frequent counts of particular items of material that move fast will not prove to be burdensome and will provide a control over these items.

When materials are requisitioned for a job and are later returned to storage, certain adjustments must be made. A Materials Requisition, listing the quantities and prices of materials returned may be used as the basis for adjustments. It should be clearly marked to indicate that materials are being returned and not issued. Of course, a special form for this purpose, perhaps of a different color from the Materials Requisition, would be more appropriate if materials are being returned with any frequency. Information concerning the material returned should be entered in the "received" and "on hand" columns of the Stock Record and, if necessary, a new average cost should be computed.

Informal Controls

While there are informal methods for keeping material used within the planned quantity, these are not very comprehensive or satisfactory. Most informal controls start with a visual inspection. For example, you or your superintendent may notice broken sheet rock, wasted lumber, or excessive broken glass. Only through very close personal supervision, however, can costs be consistently kept to a minimum, and the time required for this might be put to other uses.

Other methods based upon inspection are fairly effective for certain kinds of material. For example, measuring the thickness of a slab will indicate the quantity of concrete actually poured. Checking spacing or lap of materials will indicate the quantity of materials being installed. Although this method will not work for all kinds of materials and does not provide as much information as may be desired, occasional checks are helpful. Where your estimate and a check of materials actually erected on the job agree with each other but not with the invoices or quantities released from stock, something should be done to find out what is happening. Naturally, there is some waste of nearly every material on every job. But excessive waste, breakage or disappearance should be discovered and investigated promptly.

How Much Inventory?

Many remodelers are tempted to go into the building materials business or carry large inventories because of the discounts often available for volume purchases. Certainly, many home improvement companies have cut material costs with volume buying. However, there are many costs of carrying inventory which tend to reduce the savings: storing, handling, taking inventory, added insurance, property taxes, interest on inventory loans, and normal shrinkage or waste. These costs may add up to a profit drain rather than a savings. Certainly, if you are short of working capital or have borrowed to meet current expenses, you should let someone else carry your inventory and buy only for the jobs at hand. If you consistently have idle funds and regularly take advantage of prompt payment discounts, consider increasing your inventory of key supplies, especially if it assures availability of materials and you have secure storage space that isn't being used. In general, your inventory should not be more than 10 percent of annual sales. If you have an inventory valued at more than 10 percent of annual sales, you are in the building materials business or are speculating in the price of building materials and should be aware of the possible consequences of those actions. Most home improvement contractors who are successful at home improvement leave handling big inventories to others and concentrate on the portion of their business which is most profitable.

Labor Costs Control

Your objective in controlling labor costs is to keep labor time within the estimate in order to protect your profit margin and sell every hour of time you pay for.

Your judgment in handling men is very valuable. But that in itself does not guarantee cost control when that control operates from observation alone. Building can be a complex manufacturing process. You cannot be at all places on the job observing all the men at all times to judge productivity, and note differences from estimated time and work.

In addition to keeping within the estimate, there is another possibility to bear in mind. There are actually **two profits** possible on any job: First, the profit actually estimated on the work; second, the profit to be made by reducing estimated costs through care in purchasing, efficient management, and supervision. This calls for accurate knowledge of every item of cost on the job, and its control through a practical, detailed system of cost finding records — or cost keeping. The home improvement contractor who has little or no payroll and personally watches every part of every job may feel that he has complete control of all his costs. In fact, he may be right. But as his business grows he must adopt some system for controlling labor costs. Successful remodelers recognize cost control as the essence of efficient operation and efficiency is essential to success in the home improvement business.

Labor cost keeping requires a system for recording the cost of each unit of production or work. Its chief objects are: (1) To enable you to analyze the lowest cost possible under existing conditions; (2) To provide data upon which to base future estimates. To have this information, you need daily reports covering the quantity of units installed, an outline of working conditions, hours of labor, overtime, and special conditions. With records like this you can work out definite costs for each unit of work done. An analysis of these unit costs and a comparison with similar jobs will identify inefficient labor or padded payrolls. Such comparisons of unit costs may also point out ways to reduce the number of men required to direct the workmen, emphasize the greater efficiency possible with special machinery as compared to handwork, and lead to a more thorough analysis of differences in costs and their causes.

The important thing to remember is that a cost-keeping system will not tell you what your costs should be; it simply tells you what your costs actually are on each job in question. You must analyze the situation to know whether you are getting the lowest possible labor costs consistent with past records and the conditions in hand.

While labor cost controls may make use of many of the figures or records of bookkeeping, they must not be confused. Cost controls deal with costs on a unit basis, while bookkeeping deals with accounts and with profits or losses directly. Labor cost controlling is a management function while bookkeeping is a clerical function.

Naturally, in the one man or small organization, the proprietor will act as both bookkeeper and cost accountant — and he will modify and simplify the cost keeping records. No matter how small your organization, divide each employee's or each crew's time into a charge against installing each quantity of material or doing each major task. When you have finished the job, determine the actual man hours per unit for each operation in the take-off and enter this figure in the last column on the take-off sheet. This will be your most valuable reference when compiling future estimates.

Labor cost recording can be done either in your office or in the field. Your supervisor can act as controller or you can compile the cost records in your office as a part of keeping time records for payroll purposes.

Field Control

Supervisory field control has the advantage of the closeness between the supervisor and the actual work. However, it means more paperwork on the job. The standard against which performance is measured is the time per operation according to your cost estimates. Only man-hours need be used in the comparison. It is unnecessary to break down the time check to very small operations or anything that takes less than ¼ hour. A time standard should be set for each major operation, such as framing partitions or hanging doors. The operation should be large enough to take in a fairly large portion of the day's work because your men's performance will vary during shorter periods. The time period should fit a particular operation so that reasons for failure to meet the estimate can be determined. The extent of subdivision of operations for this purpose will depend upon the precision of control you want. The standard is the time allocated in the cost estimate. Within a certain range, corrective action

isn't really necessary. The time standard is noted by the person who assigns the work. Performance is measured when the workman is ready for the next assignment by noting the time that has elapsed. A form such as Figure 19-16 will simplify the recording process. This and other control forms are available from the Frank R. Walker Company, 5030 N. Harlem Avenue, Chicago, Illinois 60656.

This type of control can be applied to all labor on the job. Reasonable tolerances should be set to avoid poor labor relations. Correction may be put off until several labor over-runs make action necessary.

Spot Control on the Job

There are several ways of spot checking the labor time on the job before unit times are available. One method is by time-and-motion study. Here, standards are set for very small segments of a job. This procedure can be used occasionally to identify poor performance and to discover better ways of doing things.

In another type of control employing spot supervision, measurement is made at varying intervals during occasional spot checks on each operation. This is a sampling technique. Unannounced spot checks on your jobs is about the best means you have of following progress, meeting problems before they stop production and being reasonably well informed on labor cost control.

If checks are made haphazardly, however, you may fail to discover opportunities for reduction of labor costs even though you have reduced supervisory control time. Spot supervision control may provide some information for future cost estimating, but it is not likely to be as precise or as complete as the information provided by your supervisor.

Direct Control in the Office

Another way to control labor costs places control responsibility in your office. Since the information received for control purposes is similar to that required for payroll, this method has the advantage of minimizing field records and of consolidating record keeping in your office. Of course, this method may also be used in addition to field controls.

Performance is measured against labor cost estimates showing time, hourly rate, and total cost. Although the standard is the cost estimate, some allowable range should be set up. Operations are grouped in the same way as under the field control system discussed earlier. Costs are first broken down by groups of operations. Broad or narrow classifications may be used, depending upon the degree of control you feel you need.

Measurement takes place after you get the payroll information. As each workman records his hours, he also indicates the number of the job worked and the operation performed. A separate entry is made for each operation and job. The

LABOR DISTRIBUTION

Job _____   Sheet No. _____

Class of Work _____   Week Ending _____   Job No. _____

OCCUPATION							HOURS	RATE	AMOUNT
1									
2									
3									
4									
5									
6									
7									
8									
Total									

	Quantity Work in Place	Pay Roll Costs	Labor Average Unit Cost	Average Quantity Per 8 Hour Day	Quantity Work in Place	Pay Roll Costs	Labor Average Unit Cost	Average Quantity Per 8 Hour Day
Previous								
This Week								
Total								

COST ANALYSIS RECORD

TOTAL LABOR HOURS   **UNIT**

	1	2	3	4	5	6	7	8
Previous								
This Week								
Total								

LABOR HOURS PER UNIT   **UNIT**

	1	2	3	4	5	6	7	8
Previous								
This Week								
Total								

REMARKS

Labor distribution form
Figure 19-16

information is posted to the quantity take-off or card file. If cards are used, the card shows job number and operation and is filed first by job number.

Controls For Indirect Labor

Indirect labor is work done that can not be charged against any one job, work on a number of jobs simultaneously or work on a specific job for very short periods. Those involved in this kind of work include supervisory personnel, office personnel, and some field personnel who go from job to job or perform a wide variety of work.

Just as it does not pay with some materials to

control the item by the job, it is usually not practical to allocate short periods of time to specific jobs. If your planned sales volume is met and actual indirect costs do not exceed estimates, employee performance is obviously up to standard. If operations exceed estimates and no more indirect labor than was planned has been used, performance has been above standard. If overall production is below estimates, then the cost of indirect labor per job has increased and either personnel should have been reduced or existing personnel should have been used to greater advantage.

The spot-check method may be used in controlling indirect labor. For example, time spent in delivery of materials may be checked by occasionally noting duplicate trips or partially loaded trucks.

Controls for Subcontract Costs

The object in controlling subcontracting costs is to keep the payments to subcontractors within the amounts called for by the cost estimate. There are three points at which you should be careful to control costs: (1) when bids are accepted, (2) when the work is ordered, and (3) when the work is billed.

Bids are usually accepted by you personally. You are in a position to know if the amount of the bid is in line. Beware of any unreasonably low bid and any bid from a sub that does not have your confidence. Note especially that the work you expect the sub to do is the same as he submits a price for. Spotting an error in your sub's bid can save both of you a lot of trouble.

When the work is ordered, the sub should be checked as to timely and accurate completion. Most important, you should compare the amount invoiced for the work with the amount entered on the estimate summary.

Some subcontract work can be done without a bid when you have fairly routine work for key subs to handle and you have a good working relation with those subs. Ask your regular subcontractors to give you target unit prices for their most common work. Have them agree to work at that target rate (perhaps plus or minus 5 percent on any given job) unless they inform you that they cannot do the work at the target rate. Check carefully before you assume the job is strictly routine and hold your subs to their target prices. When in doubt, request a bid and have the subcontractor sign an agreement like Figure 19-17.

Controls for Rental Equipment Costs

The objective of controls for rental equipment costs is to keep actual equipment costs in line with planned costs. Controls for equipment costs are similar to other cost controls. The standard is the cost estimate, and measurement may take place either at the point of ordering or of billing. Since total rental cost may not have been predicted accurately, comparing the invoice may be the best method. Comparison of rental time with estimated time can also be made in the office or in the field by the construction superintendent.

Controls for Service and Fees

Services and fees include various direct costs associated with particular jobs, such as permit and inspection fees, taxes, and transfer costs that are direct charges to each job. You have little control over most of these. Control may be put off until after the expense is incurred. The new costs should be used for future estimates.

Controls of Sales Costs

The objective of controlling sales costs is to keep actual costs within planned costs. Some expense items such as sales commissions vary with sales and tend to control themselves. Other expense items are more flexible and may be controlled by various methods.

Relatively stable and predictable items such as sales salaries and sales showroom expense may be controlled through use of a sales budget. At the end of the budgeting period, actual expenses are recorded and compared with budgeted expenses. Large variations should not occur, but an occasional check on sales expenses will help keep them in line. The periodic check is also an appropriate time to consider changes in budgets for future operations.

Because of changed conditions, some expense items considered as reasonably stable and predictable may have increased or decreased during the year. For example, advertising expense may have been stepped up to boost lagging sales. When actual expenses are compared with budgeted expenses at the end of the budget period, significant differences in expenses should be analyzed. You may want to revise plans in the light of changed conditions. This should be done only after careful consideration. Without evaluation, uncontrolled expenses will cut deeply into profits.

Controlling Operating Costs

The objective of controlling operating costs is to keep actual costs within planned costs when operations go according to plan and to make adjustments when actual operations are different from those planned.

As explained earlier, most operating costs may be predicted with a degree of confidence. These costs or expenses are planned in the form of expense budgets. At the end of each budget period, compare the budgeted amount and the actual amount spent for the period. Many items, such as salaries, rent, and utilities, will have been predicted accurately. Such items may be handled by grouping them in broad classifications. If, however, close control is important to you, classifications should be narrow. In most cases, it is enough to compare group totals and investigate individual items only when the group total and budget are obviously different.

Builders Subcontract Agreement

Job Address _____ Date_____, 19____

Owner _____ Address _____

_____   _____
              (Contractor)   (Subcontractor)

Gentlemen:

Subcontractor agrees to furnish all materials and perform all labor necessary to complete the following as per plan and specifications and State and Local Laws and Ordinances:

_____

_____

_____

All of the above materials to be the best of their respective kinds and all work to be completed in a substantial and workmanlike manner.

For the sum of_____ Dollars ($_____)
payable as follows:

_____

_____

_____

Any alteration or deviation from plans whether involving extra cost of materials or labor or not, will be executed only upon written order for same, and will become an extra charge or credit when approved in writing by all parties concerned. It is further agreed that Subcontractor will carry all necessary compensation or liability insurance for the protection of all parties working on the building under the control of or in the employ of the undersigned Subcontractor, also public liability insurance. The above proposal and contract becomes binding when approved and accepted by the Contractor. Subcontractor further agrees to hold the Owner, the Contractor and/or supervisor harmless from any and all Federal and State taxes allocable to the labor or material furnished and/or installed by the undersigned.

Subcontractor agrees to begin work within_____days of written notice of approval and acceptance of this subcontract agreement by the undersigned Contractor, and to continue said work diligently to completion; and if, in the opinion of the Contractor, the job is not proceeding fast enough, the Contractor may give the Subcontractor notice in writing, allowing_____(____) hours in which to supply the necessary labor or material. Should the Subcontractor fail or refuse to comply with the written request, the Contractor may order such labor and material as is necessary to complete the job and the cost of same shall be paid by the Subcontractor, provided, however, the Subcontractor may claim and receive benefit of any excuses for delay to which the Contractor under his original contract with the owner is entitled and to the same extent.

All questions as to the performance of work required under the terms of this contract are subject to arbitration. In case of dispute either party hereto may make a demand for arbitration by filing such demand in writing with the other. One arbitrator may be agreed upon, otherwise there shall be three, one named in writing by each party within five days after demand is given, and a third chosen by the two appointed. Should either party refuse or neglect to appoint said arbitrator or to furnish the arbitrators with any papers or information demanded, he or they are empowered by both parties to proceed exparte. If there be one arbitrator, his decision shall be binding; if there be three, the decision of any two shall be binding. The arbitrators, if they deem that the case demands it, are authorized to award to the party whose contention is upheld such sums as they shall deem proper for the time, expenses and trouble incident to the appeal, and if the appeal was taken without reasonable cause, damages for delay. The arbitrators shall fix their own compensation, unless otherwise agreed upon, and shall assess the cost and charges of arbitration upon either or both parties.

Approved and accepted:

Contractor _____

Subcontractor _____

Date _____

Figure 19-17

Chapter 20
Selling Your Services

To operate a profitable home improvement business you must be able to sell your services effectively. When you are called on to make a proposal for a job, low price alone will decide whether you or your competition gets the contract unless you can sell yourself and your services better than other home improvement contractors. If you have developed the idea of what could be done to improve a home, if you have convinced your prospect that your firm could handle the work efficiently and in a highly professional manner, if you have really created a job where none existed before, you are entitled to your full price, both for the time you spent creating the business and for selling the prospect.

Remodeling is a selling business and anyone who wants to develop a prosperous home improvement firm must know both how to plan for sales and how to sell to potential customers. A new business does not have a reservoir of good will and referral business to call on so it must rely on advertising. This chapter will cover all three parts of selling: the sales plan, advertising methods, and selling the prospect.

The Sales Plan

The object of your plan is to make sure that you have enough sales to reach the profit level you want. By comparing planned sales with your actual sales you can evaluate the effectiveness of your sales efforts and know at an early date when sales are falling seriously behind expectations. When sales lag you may have to take steps promptly to cut overhead or increase the sales budget to keep your company operating profitably. The sooner you know about a problem, the sooner you can do something about it.

The first step in your sales plan is to established a sales goal for the year. Naturally, the goal should be realistic but it should be more than what you feel you could achieve without careful planning and improved sales efforts. You probably will pick a dollar volume figure for the coming year but your goal could be in roofs recovered or

	Sales Per Week				Cumulative Sales				4-Week Running Total 1977	13-Week Running Total 1977
Week	1974	1975	1976	1977 (planned)	1974	1975	1976	1977 (planned)		
1	1	1	1	1	1	1	1	1	4	13
2	1	1	1	1	2	2	2	2	4	13
3	1	2	1	1	3	4	3	3	4	13
4	1	1	1	1	4	5	4	4	4	13
5	1	1	3	1	5	6	7	5	4	14
6	1	1	3	1	6	7	10	6	4	15
7	0	2	2	1	6	9	12	7	4	14
8	1	2	4	1	7	11	16	8	4	13
9	1	2	2	1	8	13	18	9	4	13
10	0	2	2	1	8	15	20	10	4	13
11	0	1	2	2	8	16	22	12	5	14
12	2	1	1	2	10	17	23	14	6	15
13	1	1	0	2	11	18	23	16	7	16
14	1	1	0	2	12	19	23	18	8	17
15	1	1	0	2	13	20	23	20	8	18
.										
.										
.										
47	1	1	2	0	42	47	44	48	2	6
48	1	1	1	1	43	48	46	49	2	7
49	0	1	1	0	43	49	47	49	2	6
50	0	1	1	1	43	50	48	50	2	7
51	0	1	1	1	43	51	49	51	3	7
52	1	1	1	1	44	52	50	52	3	8

Sales chart
Figure 20-1

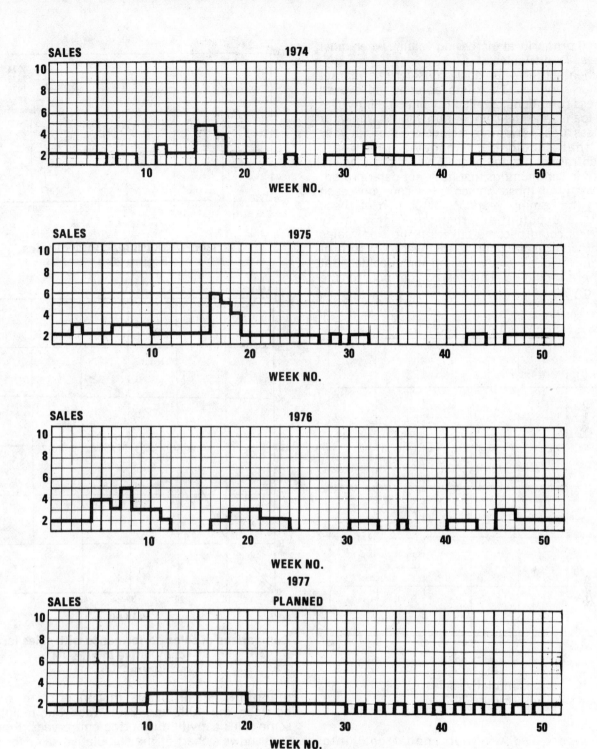

Sales per week
Figure 20-2

bathroom jobs completed. Your goal could be production in a two year period or a six month period, but a year's production is probably about as good a period as any. Next, decide when you will measure each sale: when work is started, when a contract is signed, when money is received, etc. If you have many accounts that pay well after the work is completed, you may want to record the sale when the money becomes due.

Your sales object, naturally, is to increase your volume of profitable business. You may, however, want to increase sales during certain slow months and only maintain your present sales volume during your busy period. This would allow you to keep your best crews busy all year, spread your overhead evenly throughout the year, and stabilize swings in the investment required in your business. On the other hand, you may feel that the market for your services expands greatly during certain seasons and your sales efforts should be

341

directed primarily at increasing volume when most business is available.

When setting your goal, consider the market for your services. Take into account the overall business conditions, availability of home improvement loans, and the competition you have in your business area. How many homes are there in your area that should be remodeled this year? Concentrate on homes over 25 years old that are located in communities that have not deteriorated. How many of these homeowners can you reach during the coming year? What portion of these people are potential prospects? How many prospects could become buyers if the right sales appeal were used? Look at the overall picture of how much work is available and then decide what you share should be. This will be your goal for the coming year.

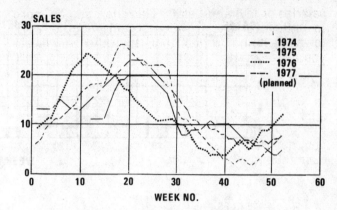

Thirteen-week running total of sales
Figure 20-4

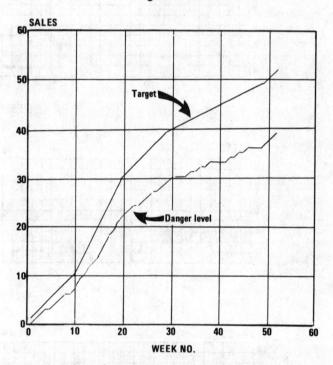

Comparison of sales performance with sales target
[cumulative total]
Figure 20-5

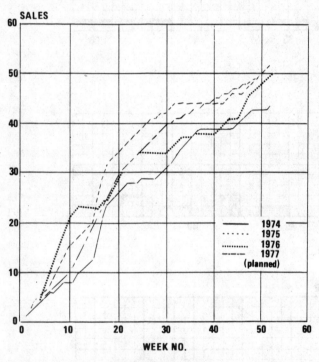

Cumulative sales
Figure 20-3

Now, break down your annual goal into monthly or weekly goals. You should also use a cumulative total of all sales during the period to measure your overall progress with both the annual goal and the history of previous years. Table 20-1 is a sales chart for a kitchen remodeler. The number of jobs sold during each week in the three previous years are laid out against the goal for the current year. Also, cumulative totals and running totals are kept so that overall volume and current rate of sales can be measured. The contractor here wants to spread his sales more evenly over the year but have volume equal to his best year, 1975. Figure 20-2 better illustrates what this contractor is trying to do with his sales

planning. His effort is to concentrate sales between the 10th and 20th week but maintain some sales activity during the entire year. Figure 20-3 shows a chart of the cumulative sales for the year for the same contractor, taken from Figure 20-1. Plotting your sales in this manner will give you an instant picture of where you are and where you should be on your sales plan. Figure 20-4 is also based on Figure 20-1. It gives you a picture of the rate of sales for the year.

You may find it useful to make a chart such as Figure 20-5. As each week passes, add the week's sales to the chart. Having something like Figure 20-5 on the wall above your desk will both help you control sales and motivate everyone in your organization. The danger level is the point at which you, as manager, must take steps to reduce overhead by cutting expenses and reducing salaries or laying off staff. This chart can do more to unite everyone in your organization to work toward a common goal than any other single act you can take.

Advertising Your Services

Setting up a good sales plan is just a good start. It doesn't actually bring in any sales dollars. The next step is to plan your advertising so that you develop prospects that can be turned into customers.

Advertising pays, especially for newer firms that do not have large referral and repeat business to rely on. But how much should you budget for advertising? A figure of 2 percent of gross sales is frequently quoted as a guide. For a home improvement contractor doing $200,000 in business a year, this would provide $4,000 a year for advertising expenses. In practice, though, few contractors grossing $200,000 a year spend that much. There is a strong temptation to scrimp on the advertising and promotion budget particularly for a new company which usually has limited capital. But if yours is a new firm, you will need a well planned campaign to sell your services even more than an established firm does.

There is no sure answer to the question of how much you should spend for advertising. If you are spending less than 2 percent of gross sales you probably could benefit by spending more. The real question is how much business your advertising dollar brings in. If by doubling your advertising you can double your business, you would be foolish not to spend more on advertising. If a budget of four percent of gross does almost no better than a budget of two percent of gross, stick to the smaller budget.

Advertising is simply a means of finding prospects by means of publicity. It makes potential customers familiar with the name and the service featured, associates that service with the name, makes personal sales efforts more productive and creates good will. It proves an economical means of keeping in touch with clients when it would be impossible to do so by personal calls.

Too many of us are inclined to look upon advertising as a means of securing business when we have run short of contracts to be carried through or experience a temporary lull. It is easy to emphasize advertising as one part of selling and overlook its broader advantages — reaching out to thousands of people with the personal message of a salesman and contractor, keeping clients sold, and attracting others. Advertising is often used in a hit-or-miss fashion and the sales results are equally irregular. Regardless of how much or how little advertising you may do as a home improvement contractor, you will find that to be successful, the various forms of advertising must be used with some judgment and must follow a well understood purpose.

Planning Your Advertising

Make a few basic decisions first about what you are offering. What class of work are you anxious to secure? Will you build a reputation for high class work, or high-speed at low cost? Or will you go after any class of work that may come your way? Will you accept repair work, or will you limit yourself to remodeling? In what way does your service differ from that of your competitors?

Considering the class of work you prefer, what are the possibilities in your vicinity? Are your opportunities purely local, or do they extend over a large surrounding area? How will the seasons affect your work locally?

Has the market for home improvement been limited in your territory for any reason in the past? Are there people who can use your services who are not aware of this fact? Are there sections where opportunities can be created for your class of work? What is your competition? How strongly established is it? Does competition offer what you have to offer— the experience, the training, the familiarity with materials and newer construction ideas? How does the quality of their service compare? These and other questions along similar lines will be of material help to you.

Familiarize yourself with any advertising that has been done by other local contractors. Keep a file of their advertising as well as your own. Learn, if possible, what results they had. If unsatisfactory, what was the reason? This alone will open an avenue of inquiry that should prove helpful to you as a guide in your own plans. Advertisements that appear again and again are successful and should tell you something about what works best.

Learn through the business papers, through manufacturers' salesmen, and other channels, what publicity or advertising plans and stunts have been successful for other home improvement companies, no matter where they may be located. Possibly the ideas can be adapted to your own advertising. Consider what the lumber yards in your area are doing. Their problems are similar to your own, and their experience will give you valuable suggestions. The principles behind the advertising of allied lines are closely related.

Forms of Advertising

The focus of your effort should be to find some type of advertising that will consistently pay its own way, even when repeated many times over a long period. Some remodelers have found that a particular advertisement in a local paper will produce inquiries month after month for years. Develop a basic approach that seems to work better than others. Try variations on this theme until you can't seem to improve it. If it is making money for you, it will probably continue to do so for many years, no matter what form of advertising you are using. Experiment with other messages or forms of communication, but keep hitting away at a proven winner. A good ad that makes money year after year will turn any struggling small organization into an established leader in the field, provided of course that the quality of workmanship is high.

Naturally you would hardly expect to use every form of advertising that is available. Select the form of advertising that best suits your community, advertising budget and type of business. Try to visualize who your prospect is and what type of advertising will reach him most economically:

  Newspaper paid ads
  Newspaper press notices and interviews
  Billboards and posters
  Give-away items
  Direct mail
  Telephone
  Yellow pages
  Canvassing
  "Image" advertising

Newspaper Advertising

Newspaper advertising is the mainstay of those whose work is concentrated in a local territory. The value of the newspaper lies in the fact that (a) it carries news and ensures concentrated circulation, (b) it reaches the people often, (c) it reaches into the home. Naturally, its effectiveness depends on what you put into the space you use.

You will find ample material for copy or advertising material in the work you are doing. Be specific in your statements. To announce that your service is "the best that is offered" is not strong advertising. But the fact that you remodeled "five ranch style homes in a particular part of town" would be news of interest to other property owners. What have you been doing for others that would prove interesting? What feature of your service would influence business? What local event or situation presents an opportunity for a tie-in to your services? If you think of nothing else, go back to your analysis of your services.

Writing the copy for your advertising is largely a matter of using your good common sense, a matter of talking to an individual through the printed word instead of by word of mouth. Forget the fine language. Speak the good common English you would use in writing a friendly letter. In fact, advertising experts agree that when writing copy you should try to picture in your mind some one individual, such as a property owner, and write the copy as if you were speaking to that person.

Try to forget that what you are writing is to go into print. Just call to mind what you would like to have those readers know. Just imagine each has a job you would like to handle, and tell them why you should handle it. You'll be surprised at the sincere, straight-from-the-shoulder appeal that results. Brush it up a little later if you wish, but not too much. Remember, it is the friendly, expressive, everyday phrase that gets under the skin and gets action.

Remember that most people you reach don't know anything about your business other than what they see in your ad. Try to make your ad tell your whole story and reflect credit on your organization. Be professional, stress your reliability and emphasize the benefits your customers receive such as prestige, convenience, comfort, security, ego satisfaction, and pride of possession. You can quote prices if you make it clear what is included and what is excluded. Be careful about advertising credit terms. To comply with Regulation Z of the Truth in Lending Act, you can advertise a credit plan or a down payment amount only if you spell out the amount, number and period of payments that are required. If you advertise an interest rate, you must list the annual percentage rate of interest. Take advantage of cooperative advertising plans and ad preparation assistance offered by many building material manufacturers. Keep up to the minute on materials, prices, and labor conditions. Frequently you will be able to feature reductions in costs which will bring many people into the market who have been looking for a timely bargain. This gives you the reputation for being "on the job", well posted, thorough, and a good man to call in when estimates are wanted.

Use enough space in your newspaper ad to tell your story without unnecessary crowding. Use it frequently enough so you will not be forgotten. Otherwise people may get the impression you are out of business. Use copy that appeals to the prospect. Make your advertisements attractive by the use of illustrations, if possible, and by using plenty of white space around the type to set it off. Large space is not necessary. Many a well planned campaign has relied on ads ranging in size from a few lines in the classified section of a single newspaper to ten inches of double column in several papers. Much depends on the way the ad and the copy are laid out.

Figure 20-6 shows several sample newspaper advertisements that might be adapted to your situation. Use all or any portion of these as "copy" when you make up your own advertisement.

Newspaper Publicity

Be friendly with the editor of your local newspaper. Make it a point to give him news items from time to time, but use a little discretion and tact — don't make the reading notice a personal history.

As a home improvement contractor and businessman you are a local institution serving the needs of your community. The public is interested in what goes on there. What are some of the things you are doing that would interest them? A new type of siding material used on some merchant's store, a new roof on a factory building, an old landmark to be restored. If you are active in your local association, let the newspaper know when you attend a convention or meeting. It adds to your reputation — and is news.

THREE
WAYS TO SAVE ON A NEW KITCHEN

1
WE DESIGN IT YOU INSTALL

an original beautifully designed kitchen delivered to your door. You install it yourself.

2
WE DESIGN IT HELP INSTALL

we deliver and arrange the plumbing and electrical work, you do the basic carpentry.

3
WE DESIGN IT AND INSTALL

a beautiful kitchen custom made for your own needs. Completely installed by experienced craftsmen.

(Company Name)
(hours & address)

WE ARE THE BEST REMODELERS IN TOWN!

That's right, we can do highest quality work, at a price any one can afford.

Our staff consists of highly experienced craftsmen, who are trained to serve you in any home remodeling area.

We are the best remodelers in town. Give us a call and let us prove it.

(Company name, phone, address)

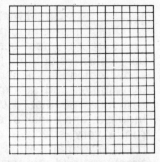

PLAN YOUR OWN ROOM ADDITION!

Then bring or mail it in for a free estimate. It's easy! Just pencil in your floor plan on a gridded piece of paper and let each square represent 1 square foot.

We'll give you a free estimate by phone or mail, to prove ourselves as being able to do the best remodeling work for the lowest price in town.

If you decide that our estimate reaches the price range you had in mind, give us a call and we'll send out an experienced architect who will be happy to sit down and discuss your room plan with you.

(Company Name)
(hours & address)

IS YOUR HOUSE GETTING TOO SMALL?

Maybe it's time for an extra room.

Do you seem to be scraping your elbows on the walls when visitors come over? Are you getting that "cramped in" feeling?

Let us help you solve this space problem by adding a new room or expanding an old room to give you a little more "breathing space."

Adding an addition to your home probably won't cost nearly as much as you had figured on.

Why not give us a call for a free estimate?

We might surprise you.

Company name
Phone
Address

Figure 20-6

Sorry!

Please excuse the noise as we finish remodeling
The _____'s

house at _____

Stop over and see your neighbor's new remodeling job, it's a beauty!

We would be glad to give you a free estimate on any home improvement remodeling that you may have been thinking of. Just fill out the card and mail it in today.

Roof Company
Phone
Address
City, State, Zip

Name _____

Address _____

City _____

Figure 20-7

345

On special occasions there will be an opportunity to give your opinion on matters of a local nature or on general building conditions in relation to the city. An interview has great advertising value, keeping your name prominently before the public. All such advertising is valuable. Some say it's more valuable than paid advertising.

Billboards, Posters and Signs

Moderate sized billboards or signs can be used to good advantage along major highways or elsewhere in districts especially active in building. Before placing such signs along the roads, however, it is best to consult the law of your state and county about restrictions or conditions. In proportion to the number of people who see these signs, this form of advertising is looked upon by many as one of the most economical forms you can use.

Do not overlook the opportunity presented right on the job itself. A sign large enough and attractive enough to be conspicuous should indicate that your company is doing the work. Such signs should be made up by a professional sign painter.

Whenever you do a job, it is a good idea to identify yourself to the neighboring property owners. There is no better endorsement of your service in a neighborhood than the fact that you remodeled the Smith or Jones house. The neighbors of Smith or Jones have houses that are probably in about the same condition as the Smith or Jones house. About the best and easiest way to identify yourself is with a door knob hanger. Place several dozen of these on front door knobs of the adjoining houses. You probably won't get many responses right away. But many of these hangers will be filed away for future reference. Before a year or two have gone by you should get several responses. Figure 20-7 shows a sample layout for a door knob hanger.

Match Books, Pencils, Rulers, Etc.

Despite the ever increasing use of "giveaway" items for advertising purposes, the fact remains that many of these are good advertising. They serve the purpose of keeping your name before homeowner prospects and other users of your services, and permit you to emphasize some feature of your service.

This form of advertising can be distributed through the mail with your regular correspondence or follow-up letters. For example, you could arrange with your bank to supply them with small calendars to enclose with monthly statements or to use on counters. A calendar is effective because it is available for reference when hung up on the wall or placed on a desk. If you use calendars at all, spend enough to get good ones. Remember, a calendar can be practical and attractive at the same time. Calendar manufacturers can furnish remarkably attractive designs at a moderate price.

Direct Mail Advertising

Direct mail selling has real advantages for the home improvement contractor. First, you can be very selective and mail to only homes where you know remodeling is needed. Second, you can mail in volumes and at times which spread out the responses so that you avoid both a deluge of inquiries and a lack of sales leads. Third, you can run a very effective direct mail program at a relatively small cost. Printing 500 envelopes and brochures and mailing them will cost little more than $100. If your message is sound and your names carefully selected, you should develop several good leads from the mailing.

The U. S. Postal Service offers a reduced rate for third class bulk rate mailing. The rate is about one-half of the first class rate and will cut your postage cost considerably. If you are mailing over 500 pieces at a time and prompt delivery is not essential, you should consider using the bulk rate. You have to do some sorting of the mail before mailing but it will be worth the trouble. Apply at the post office for a bulk rate permit. The fee is about $35. Also, get a business reply mail permit so your prospects can send inquiries back to you without affixing postage to the card or envelope you send out with your pitch.

Direct mail advertising can be one of the most productive and economical means of getting leads that you have at your disposal. While there are a great many factors which contribute to the success or failure of direct mail, the single most important factor is the quality of the list. Ideally, the list should be 100 percent homeowners in a strictly-defined area who have a need for remodeling. And it should be a list that's exclusively yours. Few who have gone to the telephone directories, the street address directories, the "Occupant" list, the post office box holder and star route addressing have been successful with direct mail. The only list you can depend on is the one you build yourself. This list is built by writing down the name and pertinent data on every person with whom your firm comes into contact who identifies himself as a prospect. Your advertising, phone solicitors, installers, canvassers, and salesmen all locate prospects. Most don't buy the first time, but you know which remain logical prospects. Look for every conceivable way to identify logical prospects. You'll be surprised how quickly you have 500 names. No matter how large a list you want for your operation, it's only a matter of time, and surprisingly little of that if everyone in your organization adds names to the file.

The best way to keep your prospect list is on 3 x 5 cards. For each prospect fill out three cards with identical information. In reality you are building one list but it will be used three different ways. The first list is kept alphabetically, by the first letter of the prospect's last name. The second list is a duplicate of list number one, except it is separated geographically, according to the territories your several salesmen work or, if you're the only salesman, according to areas you want to work at one time. Within each geographic area the cards are again filed alphabetically, by the first letter of the prospect's last name. List number three is categorized by the assumed interest that prospect has in each of the different product lines you sell.

What information should you record on each of the three cards you make out for each prospect? You'll quickly learn what type of information best suits your needs. But, as a start, list the prospect's name, street, city, state, zip, telephone number, and time of day (and/or day of week) the prospect is at home. Directions to the prospect's home can also be recorded. There should be a space for comments where anything else can be listed that may one day help you make a sale.

In the top left corner of each card should be listed the prospect's geographic location; in the upper right should be listed the home improvement product for which the prospect has the most interest. This will not only facilitate filing each of the three cards in the separate lists, it will also make it easy to locate the cards when you want to make additions or changes. It has proven helpful to use different color cards for each of the three lists. When one (sometimes two) of the cards are out of the file being worked, you'll know by color where to return them.

How To Use The Lists

If you need more business and you don't care what area it comes from or which product line you sell, you mail to your general list (list number one). If one salesman's area is slow and the others are busy, you make a mailing to the names in list number two filed under his area. If your sales of a particular product line are lagging, you give them a push by mailing to the names filed under that product, in list number three.

The format of the three lists suggested here may not exactly fit your particular operation. Don't be hamstrung in your thinking. Understand the idea of the different lists, then modify the system to suit your needs. Just keep in mind, in designing the way you keep your list, the ways in which you intend to use it, and design accordingly.

Your lists, once established, can serve for lead-getting activities besides mailings. There's no reason a canvasser can't take a block of cards from list number two (filed by geographic area) and make calls on these prospects. If you have a phone selling program, your solicitors can periodically go back and rework either geographically (to get work for a particular salesman), or by product (to encourage a better sales mix).

What Should Your Mailing Say

Naturally, the content of your message depends on what you have to sell and what you feel will get the prospect's attention. Some remodelers stress their high degree of reliability, emphasize long lists of satisfied customers and membership in respected trade associations or business groups. Others promote low price specials, unusual designs, the fact that they can handle any type of job, or any other noteworthy feature of their organization. You should, however, remember several points about all advertising and direct mail advertising in particular.

1. Put yourself in the position of the prospect who needs your service. What does he or she really want? Offer something that speaks directly to that need and you will get a response.

2. Make sure what you say is true and that you say it without insulting the intelligence of your prospect.

3. Make it easy to respond. A business reply postcard is about the easiest means available. Your prospect shouldn't have to do any more than check a box, put down their name and address and mail the card.

What you say also depends on the next step of your plan. If you are willing to settle for a large number of prospects that have not yet decided to improve their homes, you can offer a free booklet or an inexpensive but useful gadget that doesn't create any obligation to buy. The line, ''No salesman will call'' will get plenty of responses from people who are interested in home improvement but are not really shopping for a contractor. When you send the free gift, your cover letter offers to set up an appointment, naturally. If you want responses only from people who are looking for a contractor to do the work, offer a premium such as a discount on part of the job. You will get less responses with this method, but you will be able to convert more into buyers. Your salesmen should know what type of prospect they are dealing with. You may find that your salesmen do consistently better on one type of prospect than on the other. Use the method that produces the greatest number of sales at the lowest cost per sales dollar.

Telephone Selling

Many remodeling companies are largely sales companies. Remodeling can be sold over the phone just like Florida land or magazines. The telephone is a relatively inexpensive and very effective means of reaching potential buyers. Generally, remodelers who specialize in phone selling make as many as 100 calls to generate a single lead. Only a small fraction of the leads become buyers. This type of business usually has a fairly large staff of trained and motivated callers and a comparatively small staff of craftsmen. Most work is done through subcontractors.

Most remodelers who don't make thousands of unsolicated calls to find leads do use the phone to stay in touch with old customers (who may offer a good lead) and touch base with prospects who have indicated an interest in improving their home at some time in the future. Every home improvement contractor should use the phone in a very personal way to stay in contact with prospects and old customers. Your direct mail list is also your phone contact list. Use the phone as a supplement to direct mail and in person sales efforts even if you don't use it to find new leads.

Other Advertising Methods

A listing in the yellow-pages of the telephone directory is almost a must for most home improvement contractors. Make sure you are listed under each category that applies to the type of work you want.

Canvassing for leads has been used very successfully by many contractors. It reaches homeowners that obviously can use your services. Concentrate on areas of older homes that are still appreciating in value. Canvassing is especially effective for siding and roofing jobs when you have a job in progress in the immediate vicinity. One job often can serve to sell another if canvassers are used wherever crews are busy.

Don't overlook the advertising value of your trucks. In big, bold letters, identify your company name, address and phone number on both sides of the vehicle and also on the rear if it's a van. Other worthwhile data to include is corporate motto if you have one, types of remodeling you specialize in, corporate identity symbol if you have one, and the seal of the various associations you might belong to. Should local sign painters be too expensive, you can order pressure-sensitive decals and lettering through the mail. One firm which has been doing a good job for remodelers is the VM Corporation, Box 207, 15 Buttonwood Lane, Cinnaminson, NJ 08077. A clean, sharp truck suggests that this is the type of person you are and the type of work you do. Also, an easily identified vehicle helps assure that it is used for business purposes and will discourage thieves. Renew the lettering every two or three years so it stays fresh and clean.

Protecting Your Business Reputation

The people you sell to know that the best form of protection they have as consumers is buying from a reliable business. The truth is that remodeling contractors have less than an untarnished reputation in many parts of this country. You are going to have to overcome consumer reluctance to deal with you as a remodeler before you can even begin to make a sale. The best way to break down these barriers is to be known and active in your community and become a member of your local Better Business Bureau and remodeling or contracting organization. The membership insignia on your truck, letterhead and advertisements gives immediate credibility to your claim to being a reputable firm. Investigate the various membership opportunities available to you and select those which offer genuine benefits to your firm. Better Business Bureau membership is particularly valuable but is offered only to home improvement contractors who conform to Better Business Bureau code of practices. The Better

Business Bureau's pamphlet "Standard of Practice for the Home Improvement Industry" is available from the Council of Better Business Bureaus, Inc., 1150 17th Street, N.W., Washington, D.C. 20036 and your local B.B.B. office. The Better Business Bureau also offers an arbitration program for settling disputes. B.B.B. abritration is much more practical than a court proceeding if the amount in dispute is small. Naturally, association membership is considerably more than a sales tool, but you shouldn't overlook the sales benefits when evaluating which organizations meet your needs. Your prospects want to do business with a company that will be around to stand behind their promises. They want assurance of good professional work without delays and problems. They will pay more for your work if your advertising and sales methods establish your respectability and high professionalism. Membership in organizations such as the National Remodelers Association, 50 East 42nd Street, New York, New York 10017; the National Home Improvement Council, 11 East 44th Street, New York, New York 10017; your local Chamber of Commerce and the Better Business Bureau separate you from the vast majority of home improvement contractors.

Following Up On Leads

You should understand that advertising by itself won't produce sales. It only produces inquires which can be turned into sales. From that point it is up to you or your salesmen to close the sale.

There are several points to remember about handling leads. First, identify the source of the lead. You do this for two reasons; you want to know what advertising is producing results, and you want to collect as much information as you can about your prospect. You know something about your prospect once you know how he happened to contact you. Permanently identify each new client by the source of the lead. On your estimate form, on the file folder, or on some paperwork you generate for that prospect, routinely write in the particular piece of advertising that produced the contact. If it is the result of a direct mail campaign, the lead is usually in the form of a coupon or card. Note the mailing list that card or coupon was used with. Leads from radio or newspaper can be identified by some code number or contact name in your ad. For example, you could insert in your address in a newspaper ad "Department D54" to identify a prospect responding to your April ad in the "Daily Sun." Mail that comes addressed to you at "Dept. D54" is a response to that ad. You could ask prospects to contact "Bob Adams" in your radio spot. All calls for "Bob Adams" (not necessarily the name of anyone in your firm) are the result of that spot. You won't be able to identify prospects produced by referrals, signs, giveaways or the yellow pages so easily. Make a point of asking these prospects how they found out about your company. Most people are not offended by such requests and are willing to reply. If you carefully identify the source of inquiries, you will quickly get a good picture of what advertising works and what doesn't. At the end of the year, review the jobs you have been paid for to find out which advertising dollars produced the most revenue. One additional point. Referrals are your lowest cost advertising. Be sure you send a thank you note to anyone who refers business to you. Referring business to you should be something your past customers get into the habit of doing regularly.

Once the source of the lead is identified, try immediately to qualify the lead. If it comes over the phone, in a very friendly but businesslike manner try to determine if the prospect is serious about home improvement or merely curious about how your business operates. Sending salesmen out on false leads demoralizes your sales staff and delays calls that might be productive.

Follow up every lead within 24 hours of receipt whenever possible. People change their minds, change their circumstances and find your competitors if you don't follow up promptly. One contractor claims that a hot lead is his most valuable asset; a lead over 7 days old is ancient history. Spread out your advertising messages so you avoid a flood of leads one month and an absence of leads the next.

Many building material and appliance manufacturers advertise in national consumer magazines aimed at do-it-yourself remodelers. Leads or product information requests generated from these ads are generally passed to local distributors who may or may not follow up on them. If one of your distributors is receiving and filing these valuable leads, ask him to let you follow up on them. In doing so he would be helping both your business and his. Don't bother with the old leads in his file. Just ask him to let you see the new leads as they come in.

Once you have identified and qualified your lead, assign the salesman who can best handle the job. Salesmen have strengths and weaknesses. Do what you can to match the sales job to the salesman.

Have your salesman call to set up an appointment unless you set up an appointment on the first call. Be as firm but as friendly as possible in insisting that both the husband and wife be present at the appointment if both husband and wife will participate in the decision. It won't do any good to persuade one if both have to agree to the sale. Finally, have your salesman call the prospects 30 minutes before the appointed time. That will save a useless trip on at least 10 percent of your leads.

The Sales Strategy

In the home improvement business, modern sales methods are just as necessary and effective as in any other industry. It is not enough to understand construction and estimating to be successful in the home improvement business. Remodelers must know how to sell. In fact, whether you are in business for yourself or intend to merely sell your services, you should know something of salesmanship.

At the beginning you should know that the old idea that selling is a mysterious gift is not true. While some men appear to be "natural born salesmen," there are certain principles which may be learned and followed by anyone willing to apply himself.

The chief problem in selling is to bring the prospect to feel the way you do about your proposition — to make others think as you do about certain things. The most practical way to do this is to use good common sense, plain honest talk, and give every evidence of a sincere desire to solve the problems and serve the prospect. It is a matter of rendering personal service, and not necessarily a question of a highly persuasive personality.

Selling itself may be divided into routine salesmanship and creative salesmanship. The former is simply a matter of filling an existing want, as when a property owner telephones you and asks the cost of putting on a new 3-tab roof — and your price, without further comment, gets the job.

The creative salesman would drop in on a property owner, show him a modern, attractive colored and textured strip shingle adapted to his property, point out how it would increase the property value and add materially to the appearance. Closing such a sale would be creative salesmanship. In short, it is the difference between handing out what a man asks for, and in educating him to want the thing or service you think best for him.

Thus, the true salesman creates wants — wants that are so insistent and impelling as to overcome any other want at the moment. It is done by talking in terms of the prospect, putting your proposition in his language.

People do not buy things as such, but rather they buy benefits. You don't just buy an automobile. You buy travel, convenience, the open road, ease of operation, low upkeep, and satisfaction of vanity. You do not buy a vacant lot so much as you buy a potential opportunity, enhancing land values, or a prospective homesite. Everyone buys what they expect to derive from the tangible thing secured — the service. And so in selling a remodeling job, the sale is made in the mind of the prospect, not in the mind of the salesman. For the salesman has simply supplied certain ideas which offset previous ideas or objections that were in the mind of the prospect. These newer ideas removed the objections that had been a bar to the action suggested by the salesman.

Before you really can go to a prospect to sell a new roof or to secure a contract for the remodeling of a house or whatever it may be, you should convince yourself that you are really selling him something that you, yourself, would be proud of. Have a high regard for your occupation. Remember that a builder supplies one of the three essentials of life: food, shelter and clothing.

After you have satisfied yourself that you really have something to sell or a service to give that is worth while, you then come to the point of satisfying yourself on the question, "Why can I give this service better than the other man?" or "Is there any way that I can give this service better than my competitor?" If you realize that some other man can build a better room addition and do it cheaper and give better service to the prospect than you can, then it would be up to you to equip yourself so you could give a similar service or get into another line of work where you could give reasonable or better service.

Your Salesmen

Most home improvement contractors have become experienced at selling their own services. As the business grows, salesmen are usually added so the business owner can concentrate on managing and controlling more. When sales reach about $200,000 a year you should have someone begin carrying part of the sales load for you. As sales continue to improve, additional salesmen will be necessary.

Your salesman should be the man who puts the whole job together and handles it to completion. He responds to the inquiry you provide and establishes a relationship with your customer. He listens carefully to your customer's explanation and then makes suggestions about what can and should be done. A background in remodeling work is almost essential here. He must know what your installers can do and what will work well in the prospect's home. He should be able to make a sketch or outline a rough proposal that will attract the prospect's interest. After hearing the client's ideas, the salesman should be able to take down the appropriate dimensions and develop a complete proposal that he can sell to the prospect. Sometimes a draftsman will be needed here, but a salesman with knowledge of plans will be able to make up his own proposal in most cases. Most salesmen do the basic estimating, also. Naturally, you should establish selling prices (within certain limits) and check over the estimate before it is submitted. The salesman should be able to draw up the proposed agreement (on your form), get the prospect's signature on the contract and arrange financing. During construction the salesman is probably your job superintendent, checking on progress, ordering supplies and labor, coordinating the trades and settling disputes, if they come up. He gets the certificate of completion when the work is finished and handles collection problems if there are any. In short, your salesman must be nearly as skilled in the sales and operating ends of your business as you are.

A sales background may be more important than a construction background to your salesman, though most salesmen learn sales after working on home improvement jobs. The sales training and counseling you provide are the keys to developing a good sales staff. You should have a basic training and orientation program for your sales force. Help your salesmen learn your operation and what methods are effective in selling your services. A procedure manual or instruction notebook is an especially good idea here. A new salesman might start by reviewing the office paperwork (estimating, ordering and contract forms) and visiting several jobs in progress (supervision and construction procedures). He should talk with other salesmen about points to emphasize and how to close sales. He should go along on several sales calls before being sent out by himself. Naturally, he should have the energy and drive associated with sales work, but you can go a long way toward motivating him with incentives and your enthusiasm for selling the virtues of your company.

Most sales work is on salary plus commission basis. Generally, a 5 to 10 percent commission is paid to home improvement salesmen upon completion of the work and payment in full. The commission may be more if heavy supervision and paperwork is required or the job is small. Each salesman has a basic quota for each month. His salary is equivalent to the commission due on that quota.

Making The Sale

Most salesmen have or quickly develop good sales techniques that make selling easier. To make the sale, there are five things that must be brought about in the prospect's mind. You must first secure his attention; then you must develop this into interest in your proposition; from here the prospect must be brought to believe in the value of the thing offered; then you must arouse a desire to possess; and finally you must bring this desire to a decision to buy and take the necessary action.

Before this development has taken place, the prospect must be approached and the matter introduced to his attention. With this in mind we can actually divide the complete selling process into the pre-approach, the approach, the interview (or presentation), and the close. Throughout these four steps, various arguments are mingled with proofs, objections, human appeal and countless other points of discussion. Usually at least two meetings are required with the prospect, though several additional meetings may be necessary.

The pre-approach is simply the preparation for the interview. It means getting as much information about the prospect and his side of the interview as possible, in advance of the call. Often a salesman or builder will go to see a prospect "cold turkey" — that is, without any data concerning the prospect. It is best to secure all the information possible in advance. In the case of selling the idea of remodeling a house to a property owner, know something of his circumstances, the type of design he would prefer, whether his wife will be deciding, what is popular in the neighborhood, the names of homeowners in the area that have bought your service, etc. It goes without saying that you will be thoroughly posted on your proposition from your side, ready to modify your arguments, meet objections, and maneuver around a cold turn-down.

Approaching The Prospect

Even your approach is a sale in one sense, for you must "sell" your prospect on the idea of giving you a certain amount of his time. Many people tend to shun the approach of anyone trying to sell them. You will find it almost fatal to say frankly: "My name is Mr. Brown, a builder, and I am here to interest you in remodeling your kitchen." In fact, what answer could a man give such an approach other than "not interested?"

The approach that gets you somewhere in the shortest possible time is one that gives the prospect some idea of what you have to tell him in detail, but in terms of what it will accomplish for him while arousing his desire to know more. A little carefully thinking and planning will help you to do this. A knowledge of the prospect and his business and a knowledge of what your proposition will do for him is needed.

Take the salesman who wants to sell you siding for some jobs you are planning. Does he come at you with, "I'd like to sell you the siding you're going to need on the job"? Not at all. He eases around with the suggestion: "Suppose I could show you a way to save about 25 percent on the cost of that siding job, with a better piece of work and a more lasting job — would you be interested?" Naturally! That is his approach, and he certainly sells you on the idea, for you want to hear more.

To omit this first step — the approach — is fatal to the rest of the interview, for on it is built the first basis of confidence and understanding — the willingness to listen. The approach is based on self-interest. What will this do for me? What will it do for those close to me? So in order to approach your prospect to advantage, ask yourself this question: Why should this man give me the desired few minutes of his time? Your honest answer will give you the key to your opening approach.

Apply this question to the kitchen remodeling project previously referred to. What would be your answer? Wouldn't it be, "Because I know such a kitchen would add to the appearance of this man's property and the type of kitchen I have in mind would increase the resale value of the property far more than the cost of construction." There's your opening approach in your answer. You are going to sell him because what you suggest will benefit him, and not because you are a fellow lodge member, or wear the same color tie. You are going to make his property more useful, attractive and valuable, and you will get him interested by telling him in a way he can believe and that will quickly arouse his interest.

Usually your approach is based on an inquiry received from the prospect. If you have an appointment, your approach is based on that appointment. You then can begin first by listening to your prospect. Listen carefully to his problem and let him explain what he needs. Make suggestions and take notes about what is practical and what is needed. High pressure selling is definitely out. Try out a few ideas if that seems called for. Those ideas that are received well should be incorporated into your proposal. Those that are rejected out of hand should be dropped. Take down the necessary dimensions and make an appointment about a week later to present your proposal. Up to this point you have been receptive and sympathetic as a listener. You have not done any selling. Now, for most home improvement projects, you must form a plan and prepare an estimate. In other words you must come up with a proposal. If you are selling a relatively simple job such as siding or roofing you can eliminate this step and go right on to selling your service as no formal proposal is necessary.

Selling The Proposal

Suppose now that you have completed your plan and estimate and are ready for the presentation. Bear in mind during your presentation that there are two avenues of appeal open to your use. One is an appeal to the prospect's reason and intellect through logic and sound argument; the other is an appeal to the prospect's emotions and imagination by positive suggestions and word pictures.

Few of us decide a matter by logic alone. In fact, there is a mixture of both reasoning and emotion, although it is surprising the number of sales made on the emotional appeal alone. By this we mean the introduction of a suggestion into the prospect's mind around which he constructs a mental picture in which he is the central figure.

Another point to bear in mind is that a human being cannot be influenced except through one of the senses of sight, hearing, taste, smell, and touch. And every idea or feeling that enters the mind creates a reaction. These reactions are expressed in the face, hands, eyes, shoulders, or other parts of the body in some way. These expressions tell the experienced salesman what is going on in the prospect's mind and indicate whether the right line of attack is being followed. So note carefully all reactions to your arguments or suggestions as you go over them with the prospect.

The average man is on strange ground when talking the trade phrases of your business. He picks up your ideas better when the language of seeing and hearing is used — phrases he can picture. People understand pictures more quickly than words or verbal descriptions. They are direct, simple, uninvolved and concrete. It is a word picture that we wish to establish in the prospect's mind. For example, your prospect will get your idea better if you speak of a garage as "just big enough to hold two cars comfortably" instead of "twelve feet three inches deep by twenty-one feet three inches wide". Make use of words and terms familiar to the prospect; talk his language while you are painting him a picture. It may be perfectly clear to you to speak of the "lintel supports", but it would not mean half so much to the prospect as would the phrase "the beam or support across the opening". And while you could instantly picture a "steel casement window" it would be far more clear to your listener were you to say "to insure plenty of light we plan to use three windows set in a large frame."

A plan, sample or sketch is even better than a word picture. It claims your prospect's attention and holds it, allowing his mind to grasp something definite and to picture it. It keeps his mind from wandering — focuses him on one point. When you have laid before your prospect a plan for remodeling his home his mind immediately registers "yes", or "no" to the ideas presented. These ideas immediately become matters of decision, requiring your prospect to make another step toward something definite. Remember, it is for you to do the visualizing for your prospect. Plans are for the workman and the prospect is interested in what his home is going to look like when the work is finished.

Gather your facts and arrange your talking points to lead to a definite end. Each part of your proposal has special features. Point these out. Your prospect may have a fancy for French windows opening out upon a screened-in porch. He may have long dreamed of a room set aside for an honest-to-goodness library — fireplace flanked on either side with built-in bookcases. These are the little touches, the "selling points," that serve to center the attention. Naturally, your proposal has all the standard refinements that one should expect to find. But the other features, the special points give opportunity for special sales appeal.

It is too much to expect that your proposal will be accepted without some modification. The prospect is certain to have some ideas on the matter. It is wise to give due consideration to such suggestions, rather than oppose them. Of course, if the ideas are entirely impractical, they can often be disposed of by a counter suggestion rather than an abrupt criticism. You could say "that would be an idea worth considering, but if we do add it to the plan it may interfere with your ideas on . . . etc." Then when the prospect sees the impracticality of his suggestion he can withdraw gracefully without loss of dignity. Good tact in this respect develops a confidence and respect for you that will serve as a foundation upon which you can eventually lead the prospect to signing the contract.

With this same tact and good judgment you can learn whether the wishes of the man or the woman of the house will be the deciding factor. Salesmen of broad experience will tell you frankly that it is suicidal to ignore the woman of the house, even if the man appears to dominate the interview. It is the woman who will insist on knowing each little detail. She knows by heart every little defect met within her home and has in mind conveniences she will want provided for. And though her decision might not close the deal for you, you can rest assured that if she takes a dislike to your proposal the contract will be vetoed.

Answering Questions and Objections

Throughout the demonstration and selling talk, many questions will arise which must be disposed of to the satisfaction of the questioner. Go over the matter carefully again. Remember, it is not enough that you understand how unimportant the objection is. Your prospect must be made to understand. For it is in his mind that the sale is being made, not in yours. If you do not offset every objection he makes, there is no sale.

These "objections" are often merely honest doubts on points not clear to him. You must supply the facts and truths to offset these ideas in order to bring about the sale. The best way to overcome these objections is to anticipate them whenever possible.

The prospect may object that he cannot finance the type of job you recommend. If you had learned everything possible at the outset regarding your prospect, you would have anticipated this objection by pointing out during the interview that the cash outlay would be approximately so many dollars, while the bank can handle a loan on the remainder.

Have your selling points marshaled and under control; be prepared to answer any questions that arise. In the event of the price question coming up before you are ready for it, try to defer this until later. Some salesmen, answering this point will say: "Why not see first whether this is adapted to your needs, then with the actual requirements before us we can figure more accurately how our costs will run. And I'm sure I can bring it to a basis agreeable to you." Then they will fall back to the point in the demonstration where they left off, and proceed.

Objections offered may embrace price, or service or countless other factors. Frequently, they may not have anything to do with the merit of your service. For example, when a prospect says he was once disappointed in the completion time promised by some builder in the past, this momentarily becomes the leading consideration. Clear this up in its suggested relationship to yourself and then revert back to the main considerations.

Sometime during your presentation the question of your company's reliability may come up. Anyone who has been in the remodeling business for very long knows how valuable testimonial letters can be in making sales. Many remodelers make up a notebook of testimonial letters (including photographs of finished jobs) to use while making sales presentations. The letters should be no more than 6 months to a year old and preferably from customers in the same area as your prospect. Of course, the letters should reflect genuine satisfaction with the work you have done. There is nothing wrong with soliciting testimonial letters from satisfied customers. In fact, as part of your sales pitch you should point out that you hope to receive a testimonial letter on the job you are selling once the work is finished and will work hard to earn it. You should have the permission of past customers before using their testimonial letters. Ask for permission to use the letter when you ask for the letter itself. If you have really done a good job, your customer will usually be glad to recommend your services.

Many prospects ask for references before agreeing to close a deal. This is a legitimate request and should be treated as routine. You should have a bank reference for them, if requested, and a list of satisfied recent customers. The people who wrote your testimonial letters would make good references. Again, you should ask permission before using someone as a reference in your sales work. Get in the habit of asking for a testimonial letter, getting permission to use the customer as a reference and taking a picture every time you finish a job. A stack of testimonial letters, photographs, and a list of satisfied customers costs you little or nothing and is irreplacable when a prospect wants to check your reliability.

Sometime during your sales presentation, if not sooner, your prospect will probably ask if he could do some of the work himself to help cut costs. A flat turn down on this suggestion will be a heavy negative factor in your prospect's decision making process. As a professional in the home improvement business you would rather plan and complete the whole job. You would rather not have an amateur helping on your job. In general, you are going to get more business if your firm is a one-stop supermarket of remodeling services. You should be able to handle planning, code compliance, design, financing, engineering, supervision, carpentry, plumbing, electrical, and all the building specialties yourself or through your subcontractors. But the man who wants only your planning and consultation while he does the work should not be turned down. You can make a good profit charging only for your time in giving advice, if that is all your prospect wants.

Homeowner assistance can cause management and control problems, but many remodeling firms make their willingness to do only what is requested a big selling point. Many homeowners want to do the finishing and demolition themselves but need your help in developing a plan, getting a building permit, and doing the major

part of the construction work. Others may want you to do only the planning and the scheduling of subcontractors. If your prospect wants to do part of the work, don't fight it. Very likely he can't finance the cost of the whole job and there won't be any work unless he handles part of the work himself. There can still be plenty of profit left in the job for you. Concentrate on helping with the parts of the job that are most difficult, planning and coordinating, if that is what your prospect wants. However, if you aren't taking the whole job, build in some controls so you are still in charge of the work you plan to do. Most important, caution the homeowner that the work can be both more difficult and more time consuming than he expects. If the homeowner does decide to do part of the work himself, your contract first should make it clear that you are not responsible for his work or the consequence of his work, or work done under his direction. Second, the contract should make it clear that you can't be responsible for delays caused by the owner or others working under his direction and that if they delay the job unreasonably, you have the right to cancel the contract and recover costs plus your usual profit. Third, agree to do additional work only after the owner has signed a change order and agreed to pay for the additional work at your customary selling price. The contract at the end of this chapter includes these three points to help protect you from an owner who fails to live up to his part of the bargain.

Explain that extensive remodeling is a complex job that requires skill and experience. Point out that you know of homeowners who have spent more money and time trying to do part of the work themselves than if they had hired a professional to do the whole job. Homeowners often pay far too much for materials and waste more materials than necessary. Furthermore, it is fairly easy to pay subcontractors such as plumbers and electricians considerably more than the job is worth. If you handle the whole job you stand behind the whole job. You will be happy to take any part of the job but you can't guarantee what you don't get paid for. That's only fair. When you put it that way most homeowners will see the wisdom of letting you have the whole project. Even if you don't get the whole job, you will probably end up doing more than was originally planned. Your profit on the change orders may be more than the profit you had originally planned into the whole project.

Many salesmen find that they can overcome resistance and objections much easier if the final selling pitch is made in their office or showroom. If you are talking about remodeling bathrooms or kitchens, a bath and kitchen display in your office will be a big selling point. Near the display would be a good place for you to close the sale. When prospects are in their homes there are bound to be occasional interruptions in the flow of your story. Also, for some reason, prospects seem to be more concerned with irrelevant trivia in their homes than in your showroom. Where the dog is going to sleep or where the shower cap can be stored becomes more important than the color of tile or the type of dishwasher. You want to focus attention on the overall beauty and convenience of the new kitchen or bathroom and there is no better way to do this than by having a sparkling new kitchen or bathroom layout right where you are making the sale. If you don't have a showroom, you can use pictures of past jobs, pages torn from magazines or the showroom of a helpful supplier. If your prospect won't leave his home, you have no alternative and must be armed as well as possible with attractive pictures and illustrations when you show up to close the sale. If you have a showroom, spend some time making it perfect down to the last detail. A professional decorator may be well worth his or her fee here.

When the resistance of your prospect begins to ease up, when he has pretty well agreed with the ideas in your proposal, and he starts asking questions of a supplemental nature, it is time to stop your demonstration and get down to closing.

Closing the Sale

It is not the wisest thing to formally ask the prospect to sign a contract at this point. Many secure this same result by taking the willingness for granted and bring it up by test questions. For example: "We are just finishing a job on the west side of town and could get started on this work by the first of the month. Would that be soon enough?"

If he balks, try to find the reason. Then center your attack on this point until it is eliminated. You may wish to show why **now** is the best time for a decision, presenting sound reasons for your advice. Show him why, as a matter of self interest, he cannot afford to delay or arrive at any other decision — that to do so would reflect on his good judgment.

He may want to "think it over." Then point out the result of doing so — often a decision to wait a while results in disappointment later because he failed to decide. Find out if there are any lingering doubts, any information you have failed to supply. If the demonstration has been complete, there is no reason why he should not proceed, if finances permit.

He may want to "talk it over" with someone. If so, you should be present. For how can the man talk it over with anyone in fairness to you if you are not there to supply information? The prospect cannot expect to remember all the facts you have presented — facts you spent hours collecting. Point this out to him and try to arrange to be present when he "talks it over with someone."

Often a man will say "no" to a proposition when he really does not mean it. He is actually waiting for additional facts, further reasons to justify a definite decision. Many salesmen accept "no," only to learn that later the man said "yes" on an identical proposition. Supply facts, test for

the reason behind the refusal. It may simply be a "stall" and can be made a logical starting point for a newer, well-directed sales effort. This is far more satisfactory than being put off, or "stalled."

You may have a prospect who has listened carefully. He may have agreed that what you say and demonstrated are true — but still no sale. "You certainly have an attractive proposition, but I can't go into it right now; drop in Tuesday, will you?" That sounds very nice, but — it doesn't mean a thing more than the fact that the man is putting off a decision. It means you should go over your demonstration and find its weak spot, then find the reason for the delay, and center on that. "Thinking it over", "Talking it over", "Drop in later on"; too many objections at the end of the demonstration and close show a weakness in the presentation — an absence of sufficient facts, forcefullness, earnestness and faith in your proposition. Prevent these "objections" by anticipating them. Eliminate a second or third sales pitch by developing your presentation for a decision, not an opinion.

Suppose your presentation to the prospect proved a decided benefit to him and that he was able to carry it financially. Failure to "close" rests on your own presentation of the proposition. Check it over, picture your interview, call to mind all the objections brought up. Did you satisfy him on all those points? This self-examination is the secret of success in future demonstrations.

Sell yourself absolutely on the merit of your recommendations. It is by faith that you transmit belief to others. Be enthusiastic, optimistic and keep the course of your presentation moving from point to point. It is this enthusiastic action that carries the prospect along, allowing little time to think up reasons for not agreeing with you.

When to Stop Talking

When your prospect agrees to your proposition and gives you your contract, end the discussion. Except for some final decisions relating to your proposition and a few courteous comments, stop talking when you have put the sale across. Don't talk yourself out of it; don't continue to discuss points that might lead to doubts and excuses — and putting off what has been decided. Your sale is concluded — you have finishing touches to put into the plan just discussed — logical reasons to excuse yourself or show your new customers to the door.

These are principles and methods that are being followed by men who stand high in the industry. They apply whether you are developing a sales plan to sell additions or remodel kitchens. The principles only have to be adapted to your own use and requirements.

Plan to Get Paid

No remodeler ever starts a job he doesn't expect to get paid for. Yet, every year thousands of remodelers are stuck with uncollectable accounts or are forced to go to court to collect. The fewer legal problems and controversies you have, the more profitable your organization can become. When a dispute ends up in the offices of two attorneys, there is not going to be any real winner, only a losing side and a side that loses a little less. Any legal dispute over less than one or two thousand dollars is a waste of time. The attorney fees will probably be more than that if one side files suit and will certainly be more than that if the matter goes to trial. Even if you win and your contract provides that you are entitled to collect your attorney fees, you may not be awarded all of your cost, you certainly will not be reimbursed for your time, and you may find that your opponent has no assets or has been able to transfer his assets to someone else so that there is no practical way of collecting without months of delay. It is far better to try to settle smaller disputes by compromise and sue in small claims court if that fails.

You can do a lot to stay out of court when you prepare the contract. Use contract documents that you know and understand and make sure that the people who buy your services understand the contract to mean what you know it actually does mean. Put all of your agreements in writing, even for small jobs if they are to be performed over a period of time longer than a few days or if there are several parties to the agreement. Don't rely on "side agreements" or oral understandings that modify a written agreement. Don't agree to do something you may not be able to do or pay a sum you may not have available.

Don't contract with anyone until you are sure that they have the ability and desire to pay their bills on time. If you have any doubt about prompt payment, put your credit application to use. See Figure 20-8. It is a fact of life that many "sharp" operators and speculators do not pay on time. They realize that even if you have a 100 percent valid claim against a solvent debtor, if you have to use an attorney to collect, you give up at least half the amount owed. The attorney's fee will be between ⅓ and ½ of the debt and you will probably waste between 1 and 3 years getting a judgment against your debtor. Somewhere during that time you will become anxious to settle for something less than the full amount of the debt. The result: you settle for 75 cent on the dollar and your attorney takes about half of that. Your "sharp" operator gets a 25 percent discount and delays payment several months at least. Meanwhile you have to pay your craftsmen and material suppliers.

To avoid disputes you should avoid surprises. No one is surprised if everyone does what is expected. The building contract, whether written or oral, outlines what is expected of the parties contracting. The plans are a part of that contract and impose an obligation on you and the owner and perhaps the subcontractors. These obligations should be as clear and precise as practical and should anticipate as many problems as can reasonably be foreseen. Don't experiment with drafting your own contract and don't sign a contract offered by an owner if you can get the owner to accept your "standard" contract. Some contracts favor contractors and some don't. There are many local and national associations of builders and contractors who offer "standard" contracts that favor and protect the contractor. The contract in Figure 20-9 is weighted in favor of contractors. Have the contract drawn on one of these documents if possible. If this isn't possible, read the contract the owner insists on very carefully. You are held to everything in the contract you sign even if you don't read it. If something seems unfair to you, offer to cross out that portion on all copies of the contract. If the other parties claim that the offensive portion of the contract isn't relevant, you should reply that it is better then to eliminate it by striking it out. If you want something in a contract that isn't there, write it in on all copies and have the parties initial the addition. The addition becomes a perfectly legal part of the agreement! If you don't understand your obligations, you should get the advice of an attorney or pass up the job completely. In any event, don't start the job without a written contract. The Statute of Frauds in most states won't let you collect on some oral contracts. At best you will have to convince a court that there was really a contract; at worst you will only be allowed to collect the reasonable value of your services rather than the full contract price. In either event you may have to go to court to collect what you are owed.

One excellent way to avoid surprises is to make clear what is **not** included. Naturally, you won't be able to list everything that is not included, but a list of items specifically excluded can save a dispute later and may even provide a little extra work for you if the owner wants that item included in your bid. Review Figure 19-2 to find items you may want to exclude specifically.

Your Contract

There are about as many remodeling contracts as there are remodelers. Most of them are good . . . for the remodelers who use them. If you have a good basic contract and it works for you, stick to it. The contract in Figure 20-9 covers some things you may not have thought of or that may help settle disputes in the future. Use it or work part of it into your agreement if you prefer. The important thing is to develop and use a contract that covers as many possible points of conflict as possible. You are way ahead of the game if you can point to specific language in the agreement covering some minor point that your customer says is holding up payment.

Credit Application

Name _____ Date of Birth _____

First _____ Middle _____ Last

First Name of Spouse _____ Social Security Number. ___ — ___ — ___

Home Address _____ City _____ State _____ Zip _____

Home Phone (_____) _____ Years at Present Address _____ Own Home ☐ Rent ☐ ___

Married ☐ Single ☐ Divorced ☐ Widow(er) ☐ Number of Dependents _____

Previous Home Address _____ How Long? _____

Firm Name or Employer's Name _____ Years There _____

Address _____ City _____ State _____ Zip _____

Business Phone _____ Position _____ Nature of Business _____

Previous Employer _____ Years There _____

Address _____ City _____ State _____ Zip _____

College/University (if recent graduate) _____ Year Graduated _____

Your Present Annual Salary _____ List Source & Amount of income other than salary _____

_____

Personal References: Name _____

Address _____ City _____ State _____ Zip _____

Name _____

Address _____ City _____ State _____ Zip _____

Credit References: 1. Name _____ No _____ Open ☐ Closed ☐

Address _____

2. Name _____ No _____ Open ☐ Closed ☐

Address _____

Bank 1. Name _____ Branch _____

_____ Type of Account _____ No _____

Bank 2. Name _____ Branch _____

_____ Type of Account _____ No _____

Finance Company Name _____ Address _____

Street _____ City _____ State _____ Zip _____

Nearest Relative or Friend Not Living with you _____

Address _____ Relationship _____

I hereby certify that the information in this credit application is correct. I hereby authorize you or your agent to investigate the data furnished by me.

X _____

Signature Of Applicant   (Ink)   Date

Figure 20-8

Some home improvement contractors don't use a contract at all. At least they don't use a form they call a "contract." Instead, they draft a letter spelling out in plain language what they plan to do, what they don't plan to do and how much and when they expect to be paid for it. Actually, this letter becomes the contract, whether it is signed and returned by the customer or not, once work begins. The customer agrees to the terms of the contract - letter by letting the work go forward. Your contract could also be verbal, that is, just an oral agreement on what you and your customer are going to do. However, sometimes it is hard to remember exactly what was agreed on and some states won't enforce certain types of oral agreements. Put it in writing unless it is a very small, very simple, very quick job you are doing for a good friend.

Understand Your Contract

You should understand and be able to explain the contracts you offer your client. No matter what contract you use, it will have many of the elements of the sample contract in Figure 20-9. Many of these contract terms are in nearly every building contract you see. Know what they mean, where they apply and you should have no trouble staying out of disputes with your customers.

You should make up at least two copies of all contract documents and materials referred to in the contract documents: one copy for your file, one copy for your customer. A lender should get a third copy if there is a lending institution involved.

The Sample Contract [Figure 20-9]

The front of the contract is the basic contract though the small print on the back is as much a part of the agreement as the terms on the front. Identify your prospect by first and last name (use Mr. and Mrs. if that is appropriate) and date the agreement. Use the blank lines under the first printed line to identify the work you are going to do. Be specific. If you are putting on siding or roofing, installing a few cabinets or doing some minor repairs, identify the maker, style, color, texture, description or design of the material and how many square feet or linear feet or units you are going to apply or install. Avoid language to the effect that you are going to merely reroof, repair or reside "the property." Your customer may think your estimate includes a detached garage or other structure. Instead, spell out how many square feet or linear feet or units you include in your estimate. Include in your description all the necessary associated work such as flashing, molding, trim, etc. Specifically exclude anything that your prospect might think is included but you have not included in your bid. If you need more than five or six lines to describe completely what you are going to do, refer to a sketch, plan, or specifications that show and explain what the job includes. Identify the sketch, plan or specifications clearly in your

reference so there can be no mistake about what the job is.

Next, fill in the job location. A street address is sufficient unless a lender is involved. A bank or saving association will need a legal description.

The next blank space has the payment amount and terms. On most small jobs it is reasonable for you to request payment of one-third as of delivery of the materials and the start of work, one-third at completion and one-third 30 days after completion. Get an agreement to pay something before or at the time work begins. If you are going to have payment problems you should find that out as soon as possible. If the job is over $5,000, you should get 10 percent on signing the contract, 20 percent when the work begins, 20 percent after rough framing, 20 percent when the job is closed in, 20 percent when the carpentry is completed and the last 10 percent upon completion. The idea is to make payments roughly match your expenditures. If a lender is releasing funds to you, they will probably have their own schedule.

Under Regulation Z of the Truth in Lending Act you must notify the buyer of his right to cancel the transaction any time during up to midnight of the third business day following the date the contract was signed. The law requires that a warning appear on the contract near the place for the signature of your customer. You must leave two copies of the separate right to cancel notice and the "Effect of Rescission" paragraph with your customer. You cannot start work until the three days are up (three days excluding Sundays and legal holidays) unless your customer gives up his cancellation right because there is a real emergency that requires repairs necessary to avoid danger to the buyer's family or property. Fill out the separate notice with the date the agreement is signed, the name and address of your company and the date three full days (excluding Sundays and holidays) later when the right to cancel expires.

If there is any finance charge or interest charged on an unpaid balance, you are required to make certain disclosures under Regulation Z of the Truth in Lending Act. If you are planning to charge interest or some finance fee, ask your bank for assistance in complying with Regulation Z. You can get a copy of "What You Ought to Know About Truth in Lending" from most banks and from the Board of Governors, Federal Reserve System, Washington, D.C. 2055l. Any complaint your customer has against you for failing to live up to the contract can be used against whoever holds the right to collect the amount due. An appropriate notice must be included in the loan agreement. Again, your bank or whoever helps you set up credit transactions can provide the necessary forms.

Many states now license construction contractors and require that a notice concerning licensing requirements appear on the contract. Note also that many states require that a notice of the mechanics lien law appear on the contract. Be aware of any such requirements in the states where you do business.

The paragraphs on the reverse of the proposal and contract in Figure 20-9 are numbered. Note especially the following points:

Paragraph 3 sets a standard by which completion can be judged if the owner refuses to agree that you have finished the work. The building inspector's standard will probably be far less exacting than an owner who stubbornly wants you to do some extra work.

Paragraph 4 makes it clear that you can substitute materials when you consider it advisable.

Paragraph 5 requires change orders in writing and specifies that the owner pay your "normal selling price" which is more than the cost of the change. If you have to change the job to comply with building codes, you can collect for the additional work. Additional concrete is a frequent "extra" and your right to collect for concrete not in the bid is guaranteed here.

Paragraph 6 eliminates any responsibility you might have for acts over which you have no control.

Paragraph 7 extends the time you have to complete the contract under a number of conditions and gives you a way out of the contract if the delay goes beyond 30 days.

Paragraph 8 requires that you have workers' compensation and liability insurance and places the burden for other insurance on the owner.

Paragraph 11 means that the American Arbitration Association arbitration system will be used to settle disputes under the contract. The paragraph could have required settlement under the Better Business Bureau's National Program of Consumer Arbitration. Courts usually do not accept cases under contracts requiring arbitration until arbitration has been completed. This means you and your customer will have to arbitrate if you can't agree. Arbitration is much faster and cheaper than a court suit and should solve any problems you have.

Paragraph 12 will help you collect your attorney fees if you have to sue to enforce the contract.

Paragraph 13 should help you collect for most unanticipated expenses that come up after the contract is signed.

Paragraph 15 absolves you of responsibility for a large number of conditions over which you have no control.

Paragraph 22 gives you the important right to stop work if a progress payment isn't made on time.

Paragraph 23 lets you cancel the contract if the credit check you make on your customer reveals a poor payment record or if there was a mistake in your estimate.

Paragraph 24 makes this contract the only agreement you have. In other words, the only promise you make to your customer is what the contract says, not what your salesman may have implied.

Paragraph 25 gives you the right to collect for higher labor and material costs if your labor and material costs go up after the contract is signed.

Paragraph 26 is important where the owner is furnishing materials or working on the job himself.

The contract in Figure 20-9 does not make any warranty or guarantee concerning the work to be done. Naturally, you should pass on to your customer any warranty the manufacturer makes on the products you install. Know how the manufacturer makes good on his warranty and explain that to your client. Most successful home improvement contractors also make good any part of the work they do which fails for the first one or two years after completion. If anything goes wrong with the job in the first year or two, your foreman should fix it, no questions asked. It's good business, costs very little and will pay a big dividend in customer satisfaction. If you feel it is necessary, include $25 or $50 in your bid to cover the cost of service calls.

If you make any warranty of your own you must comply with Federal Trade Commission Rules enacted under the Magnuson — Moss Warranty Act. You must make clear in writing what is covered by your warranty and for how long, what you will do when something fails, and what your customer should do to have the warranty honored. You must also disclose any limitations you place on your warranty and warn that some states do not recognize certain limits on warranties. Finally, you must advise your customer that he has legal rights under the warranty.

Once The Sale Is Made

A day or two after signing the contract your new customers should get a "thank you" letter and the names and phone numbers of key subcontractors and someone in your office who can handle any problem or question they have about their job. Give them an expected beginning date and completion date and reassure them that they will be pleased with the job you are going to do.

PROPOSAL AND CONTRACT

Date _____ 19 _____

To _____

Dear Sir:

We propose to furnish all materials and perform all labor necessary to complete the following:

_____

_____

_____

_____

_____

_____

Job Location: _____

All of the above work to be completed in a substantial and workmanlike manner according to the terms and conditions on the back of this form for the sum of _____

_____ Dollars ($ _____)

Payments to be made as the work progresses as follows: _____

_____

_____

_____

the entire amount of the contract to be paid within _____ days after completion. The price quoted is for immediate acceptance only. Any delay in acceptance will require a verification of prevailing labor and material costs.

By _____

Company Name _____

Address _____

State Licence No. _____

"YOU, THE BUYER, MAY CANCEL THIS TRANSACTION AT ANY TIME PRIOR TO MIDNIGHT OF THE THIRD BUSINESS DAY AFTER THE DATE OF THIS TRANSACTION. SEE THE ATTACHED NOTICE OF CANCELLATION FORM FOR AN EXPLANATION OF THIS RIGHT."

You are hereby authorized to furnish all materials and labor required to complete the work according to the terms and conditions on the back of this proposal, for which we agree to pay the amounts itemized above

Owner _____

Owner _____ Date _____

Figure 20-9

Terms And Conditions

1. The Contractor agrees to commence work hereunder within ten (10) days after the last to occur of the following, (1) receipt of written notice from the Lien Holder, if any, to the effect that all documents required to be recorded prior to the commencement of construction have been properly recorded; (2) the building site has been properly prepared for construction by the Owner, and (3) a building permit has been issued. Contractor agrees to prosecute work thereafter to completion, and to complete the work within a reasonable time, subject to such delays as are permissible under this contract. If no first Lien Holder exists, all references to Lien Holder are to be disregarded.

2. Contractor shall pay all valid bills and charge for material and labor arising out of the construction of the structure and will hold Owner of the property free and harmless against all liens and claims of lien for labor and material filed against the property.

3. No payment under this contract shall be construed as an acceptance of any work done up to the time of such payment, except as to such items as are plainly evident to anyone not experienced in construction work, but the entire work is to be subject to the inspection and approval of the inspector for the Pubic Authority at the time when it shall be claimed by the Contractor that the work has been completed. At the completion of the work, acceptance by the Public Authority shall entitle Contractor to receive all progress payments according to the schedule set forth.

4. The plans and specifications are intended to supplement each other, so that any works exhibited in either and not mentioned in the other are to be executed the same as if they were mentioned and set forth in both. In the event that any conflict exists between any estimate of costs of construction and the terms of this Contract, this Contract shall be controlling. The Contractor may substitute materials that are equal in quality to those specified if the Contractor deems it advisable to do so.

5. Owner agrees to pay Contractor its normal selling price for all additions, alterations or deviations. No additional work shall be done without the prior written authorization of Owner. Any such authorization shall be on a change-order form, approved by both parties, which shall become a part of this Contract. Where such additional work is added to this Contract, it is agreed that all terms and conditions of this Contract shall apply equally to such additional work. Any change in specifications or construction necessary to conform to existing or future building codes, zoning laws, or regulations of inspecting Public Authorities shall be considered additional work to be paid for by Owner as additional work. If the quantity of materials required under this Contract are so altered as to create a hardship on the Contractor, the owner shall be obligated to reimburse Contractor for additional expenses incurred. It is understood and agreed that if Contractor finds that extra concrete is required he is authorized by the Owner to pour the amount of concrete that is required by the building code or site conditions and shall promptly notify Owner of such extra concrete. Owner shall promptly deposit the cost of the required extra concrete with the Contractor. Any changes made under this Contract will not affect the validity of this document.

6. The Contractor shall not be responsible for any damage occasioned by the Owner or Owner's agent, Acts of God, earthquake, or other causes beyond the control of Contractor, unless otherwise herein provided or unless he is obligated by the terms hereof to provide insurance against such hazards. Contractor shall not be liable for damages or defects resulting from work done by subcontractors. In the event Owner authorizes access through adjacent properties for Contractor's use during construction, Owner is required to obtain permission from the owner(s) of the adjacent properties for such. Owner agrees to be responsible and to hold Contractor harmless and accept any risks resulting from access through adjacent properties.

7. The time during which the Contractor is delayed in his work by (a) the acts of Owner or his agents or employees or those claiming under agreement with or grant from Owner, including any notice to the Lien Holder to withhold progress payments, or by (b) any acts or delays occasioned by the Lien Holder, or by (c) the Acts of God which Contractor could not have reasonably forseen and provided against, or by (d) stormy or inclement weather which necessarily delays the work, or by (e) any strikes, boycotts or like obstructive actions by employees or labor organizations and which are beyond the control of Contractor and which he cannot reasonably overcome, or by (f) extra work requested by the Owner, or by (g) failure of Owner to promptly pay for any extra work as authorized, shall be added to the time for completion by a fair and reasonable allowance. Should work be stopped for more than 30 days by any or all of (a) through (g) above, the Contractor may terminate this Contract and collect for all work completed plus a reasonable profit.

8. Contractor shall at his own expense carry all workers' compensation insurance and public liability insurance necessary for the full protection of Contractor and Owner during the progress of the work. Certificates of such insurance shall be filed with Owner and with said Lien Holder if Owner and Lien Holder so require. Owner agrees to procure at his own expense, prior to the commencement of any work, fire insurance with Course of Construction, All Physical Loss and Vandalism and Malicious Mischief clauses attached in a sum equal to the total cost of the improvements. Such insurance shall be written to protect the Owner and Contractor, and Lien Holder, as their interests may appear. Should Owner fail so to do, Contractor may procure such insurance, as agent for Owner, but is not required to do so, and Owner agrees on demand to reimburse Contractor in cash for the cost thereof.

9. Where materials are to be matched, Contractor shall make every reasonable effort to do so using standard materials, but does not guarantee a perfect match.

10. Owner agrees to sign and file for record within five days after the completion and acceptance of work a notice of completion. Contractor agrees upon receipt of final payment to release the property from any and all claims that may have accrued by reason of the construction. If the Contractor faithfully performs the obligations of this part to be performed, he shall have the right to refuse to permit occupancy of the structure by the Owner or anyone claiming through the Owner until Contractor has received the payment due at completion of construction.

11. Any controversy or claim arising out of or relating to this contract, shall be settled by arbitration in accordance with the Rules of the American Arbitration Association, and judgment upon the award rendered by the Arbitrator(s) may be entered in any Court having jurisdiction.

12. Should either party hereto bring suit in court to enforce the terms of this agreement, any judgment awarded shall include court costs and reasonable attorney's fees to the successful party plus interest at the legal rate.

13. Unless otherwise specified, the contract price is based upon Owner's representation that site is level and cleared and is not filled ground or hard rock and that there are no conditions preventing Contractor from proceeding with usual construction procedures and that all existing electrical and plumbing facilities are capable of carrying the extra load caused by the work to be performed by Contractor. Any electrical meter charges required by Public Authorities or utility companies are not included in the price of this Contract, unless included in specifications. If existing conditions are not as represented thereby necessitating additional excavation, blasting, plumbing, electrical, curbing, concrete or other work, the same shall be paid for by Owner as additional work.

14. The Owner is solely responsible for providing Contractor prior to the commencing of construction with such water, electricity and refuse removal service at the job site as may be required by Contractor to effect the construction of the improvement covered by this Contract. Owner shall provide a toilet during the course of construction when required by law.

15. The Contractor shall not be responsible for damage to existing walks, curbs, driveways, cesspools, septic tanks, sewer lines, water or gas lines, arches, shrubs, lawn, trees, clotheslines, telephone and electric lines, etc., by the Contractor, sub-contractor, or supplier incurred in the performance of work or in the delivery of materials for the job. Owner hereby warrants and represents that he shall be solely responsible for the condition of the building site with respect to finish grading, moisture, drainage, alkali content, soil slippage and sinking or any other site condition that may exist over which the Contractor has no control and subsequently results in damage to the building.

16. The Owner is solely responsible for the location of all lot lines and shall identify all corner posts of his lot for the Contractor. If any doubt exists as to the location of such lot lines, the Owner shall at his own cost, order and pay for a survey. If the Owner shall wrongly identify the location of the lot lines of the property, any changes required by the Contractor shall be at Owner's expense. This cost shall be paid by Owner to Contractor in cash prior to continuation of work.

17. Contractor has the right to sub-contract any part, or all, of the work herein agreed to be performed.

18. Owner agrees to install and connect at owner's cost, such utilities and make such improvements in addition to work covered by this contract as may be required by Lien Holder or Public Authority prior to completion of work of Contractor.

19. Contractor shall not be responsible for any damages occasioned by plumbing leaks unless water service is connected to the plumbing facilities prior to the time of rough inspection.

20. The Owner is solely responsible for all charges incurred for grading of lot for level building site, removing all trees, debris, and other obstructions prior to start of construction.

21. Owner hereby grants to Contractor the right to display signs and advertise at the building site.

22. Contractor shall have the right to stop work and keep the job idle if payments are not made to him when due. If any payments are not made to Contractor when due, Owner shall pay to Contractor an additional charge of 10% of the amount of such payment. If the work shall be stopped by the Owner for a period of sixty days, then the Contractor may, at Contractor's option, upon five days written notice, demand and receive payment for all work executed and materials ordered or supplied and any other loss sustained, including a profit of 10% of the contract price. In the event of work stoppage for any reason, Owner shall provide for protection of, and be responsible for any damage, warpage, racking, or loss of material on the premises.

23. Within ten days after execution of this Contract, Contractor shall have the right to cancel this Contract should he determine that there is any uncertainty that all payments due under this Contract will be made when due or that an error has been made in computing the cost of completing the work.

24. This agreement constitutes the entire contract and the parties are not bound by oral expression or representation by any party or agent or either party.

25. The price quoted for completion of the structure is subject to change to the extent of any difference in the cost of labor and materials as of this date and the actual cost to contractor at the time materials are purchased and work is done.

26. The Contractor is not responsible for labor or materials furnished by Owner or anyone working under the direction of the Owner and any loss or additional work that results therefrom shall be the responsibility of the Owner.

27. No action arising from or related to the contract, or the performance thereof, shall be commenced by either party against the other more than two years after the completion or cessation of work under this contract. This limitation applies to all actions of any character, whether at law or in equity, and whether sounding in contract, tort, or otherwise. This limitation shall not be extended by any negligent misrepresentation or unintentional concealment, but shall be extended as provided by law for willful fraud, concealment, or misrepresentation.

28. All taxes and special assessments levied against the property shall be paid by the Owner.

29. Contractor agrees to complete the work in a substantial and workmanlike manner but is not responsible for failures or defects that result from work done by others prior, at the time of or subsequent to work done under this agreement, failure to keep gutters, downspouts and valleys reasonably clear of leaves or obstructions, failure of the Owner to authorize Contractor to undertake needed repairs or replacement of fascia, vents, defective or deteriorated roofing or roofing felt, trim, sheathing, rafters, structural members, siding, masonry, caulking, metal edging, or flashing of any type, or any act of negligence or misuse by the Owner or any other party.

30. Contractor makes no warranty, express or implied (including warranty of fitness for purpose and merchantability). Any warranty or limited warranty shall be as provided by the manufacturer of the products and materials used in construction.

Figure 20-9 [continued]

Notice To Customer Required By Federal Law

You have entered into a transaction on_____which may result in a lien, mortgage, or other security interest on your home. You have a legal right under federal law to cancel this transaction, if you desire to do so, without any penalty or obligation within three business days from the above date or any later date on which all material disclosures required under the Truth in Lending Act have been given to you. If you so cancel the transaction, any lien, mortgage, or other security interest on your home arising from this transaction is automatically void. You are also entitled to receive a refund of any down payment or other consideration if you cancel. If you decide to cancel this transaction, you may do so by notifying

_____
(Name of Creditor)

at_____
(Address of Creditor's Place of Business)

by mail or telegram sent not later than midnight of_____. You may also use any other form of written
                                                     (Date)

notice identifying the transaction if it is delivered to the above address not later than that time.

This notice may be used for the purpose by dating and signing below.

    I hereby cancel this transaction.

_____   _____
         (Date)   (Customer's Signature)

Effect of rescission. When a customer exercises his right to rescind under paragraph (a) of this section, he is not liable for any finance or other charge, and any security interest becomes void upon such a rescission. Within 10 days after receipt of a notice of rescission, the creditor shall return to the customer any money or property given as earnest money, downpayment, or otherwise, and shall take any action necessary or appropriate to reflect the termination of any security interest created under the transaction. If the creditor has delivered any property to the customer, the customer may retain possession of it. Upon the performance of the creditor's obligations under this section, the customer shall tender the property to the creditor, except that if return of the property in kind would be impracticable or inequitable, the customer shall tender its reasonable value. Tender shall be made at the location of the property or at the residence of the customer, at the option of the customer. If the creditor does not take possession of the property within 10 days after tender by the customer, ownership of the property vests in the customer without obligation on his part to pay for it.

Figure 20-9 [continued]

Chapter 21
Financial Controls

Like executives of other manufacturing concerns, home improvement contractors must divide their time among the numerous activities necessary to keep their organizations functioning properly. Unlike managers of many other types of concerns, builders seldom have a staff of highly trained helpers to assist them and are forced to rely to a greater extent on their own judgment and ingenuity. The tasks of overseeing activities in scattered locations, operating in all kinds of weather, devising different operational procedures for different types of jobs, estimating costs, controlling costs, financing operations, determining whether to rent or buy machinery and equipment, and deciding whether to subcontract work or hire personnel to perform certain operations make the job broad and difficult for one person to handle. You should take advantage of all practical aids available to assist you in carrying out these many activities. The decisions you make concerning the problems which you face determines, to a great extent, whether or not operations are profitable. You must have sufficient information available about your operation to make it possible for you to take a businesslike approach to the problems at hand.

Whether you are a former journeyman, trade contractor, or other specialist, accept the fact that you are the director of a complex enterprise and that your time must be divided skillfully among numerous tasks to make the over-all effort profitable. It is entirely possible that a builder who is a former carpenter may continue to do an excellent job as a journeyman and still lose money by neglecting to use good judgment in making financial decisions about his operation.

Of course, it is impossible to set up rules which can be followed to operate every home improvement business successfully. Conditions change so much from day to day that strict rules do not fit the many different situations that arise in the business world. Businessmen, including builders, who can learn to fit practical procedures for operating their organizations into changing surroundings will be the most successful of their group. Whether a person is operating a building concern or any other type of business, the decisions made last week may not apply to this week's problems. In order to grow and operate successfully, you as a builder should recognize that you must become thoroughly familiar with all the parts of your business and learn to make correct decisions when action is required.

A businesslike approach can be only partially the result of the proper use of accounting information. However, to control an operation at all you must have an orderly arrangement of financial data which will provide you with information necessary to make correct decisions. When properly understood, a few records may contain sufficient information to aid in solving many basic problems. Record keeping is your tool to help perform your various tasks efficiently.

The keeping of records is a natural requirement of any operation since the human memory cannot retain all the details relating to business transactions. Records simply represent an orderly arrangement of the financial information that you should naturally feel to be essential in operating an organization. They are just an aid to the memory.

A knowledge of formal bookkeeping methods is not a requirement for the maintenance of adequate records. If you depend on yourself, or your wife or daughter to keep the records, a simplified system using a common-sense interpretation of transactions, such as is discussed in this chapter, is going to be adequate. Where the operation is sufficiently large to warrant employment of a bookkeeper, you may wish to use a more detailed system.

After your books have been set up, you may find that you want a computer service bureau to keep your books for your company. For a cost of about $50 per month a bookkeeping service will process your basic records and make up a profit and loss statement and balance sheet. The service works from the second copies of your checks, records of deposits and your estimate of inventory, receivables and work in progress. Several bookkeeping service companies have programs designed specifically for construction contractors. You still have to pay your bills, make up deposits, and handle your accounts receivable and taxes. Some services and most banks can handle your payroll and payroll tax deposits at a nominal cost.

Why Keep Records

The Internal Revenue Code requires that each taxpayer keep accounting records that will allow him to make a return showing his true taxable income and requires, among other things, that costs be properly classified. Only by keeping proper records will you have the opportunity to take advantage of all possible reductions of taxable income. There will be an additional advantage in that formal records will be available to substantiate any claims for deductions in case they are questioned, for the burden of proof in case of questioning by representatives of the Bureau of Internal Revenue is on the taxpayer. Also, in the case of individuals such as builders who have income from a business, a declaration of estimated tax must be made and the tax must be paid in a lump sum or in installments in advance. Past records will be helpful in estimating the tax. In addition to the preparation of the income tax return, other special reports that the builder must make to governmental units include those for employees' income tax and social security taxes.

Keeping records can help you control costs on your jobs. In considering whether to buy or rent machinery and equipment or whether to subcontract work or hire personnel to do a job, the relative cost in time and money of the alternatives should be considered. If, for example, a builder has been hiring his own carpentry crews he may, by inspection of his carpentry cost records, determine whether or not it would be to his advantage to subcontract carpentry labor. Of course, there are cases where such subcontracting may not be appropriate if the contract carpentry crew would have to be shared with other builders and would not be available when needed. If, as another example, a builder has his own machinery and equipment, he may, by inspection of his cost records, find that he can subcontract excavation work for less than his past cost.

Dependable records not only help you control your specific jobs, but also allow you to plan intelligently the size of your operation in terms of type and number of jobs to take on. That is, you may determine, by examining your records, whether it is more profitable for you to handle a larger number of jobs at a lower margin or a smaller number of jobs at a higher margin. By knowing what types of work make money for you, you know what types of work to sell most aggressively. Most important, as you accumulate financial records, you will see patterns and trends in the figures that help you spot trouble before it becomes critical. A shrinking profit margin or rising expenses could mean a change is due in your way of doing business. There is no better or surer way of checking the health of your company than comparing your financial figures with figures from previous months and with figures from other companies. The whole history of your company can be written in expense statements and balance sheets. The better you understand these documents, the more successful you can become.

Classifying and Recording

Many transactions with which you are concerned consist of the payment for materials, supplies, and labor, and the receipt of cash for draws and settlements. These cash transactions, and a few other occurrences to be discussed later, must be recorded in an orderly arrangement or you will not have the information you need to manage your organization properly.

For the small builder who keeps his own records or has some member of his family do this job, a form such as Form 21-1 will be ideal for recording entries. Most stationery stores have a large number of ruled and three hole punched accounting tablets similar to Form 21-1. If you can't locate a form to suit your needs, remove Form 21-1 and have an "instant printer" produce several hundred copies for you. The examples in this book use a 12 column three page form. The 12 columns are headed as follows (see Form 12-9):
1. Cash on hand
2. Cash in bank
3. Notes, mortgages and other accounts receivable.
4. Materials on hand
5. Cost of jobs under construction
6. Buildings, furniture, equipment, etc.
7. Advances received from lending institutions and purchaser payments
8. Amounts owed suppliers, income taxes withheld from employees, etc.
9. Owner's investment in business
10. Blank
11. Blank
12. Revenue and expense
If you use Figure 21-1 as your Columnar Financial Record, the 12 columns would appear on 12 consecutive pages.

A knowledge of bookkeeping is not required in order to use these records, yet they will provide the same basic information that any more detailed accounting system will provide. The Columnar Financial Record is a framework within which the results of any business transaction may be recorded and from which all information needed for financial statements may be taken. It is only necessary that you have or develop a common sense knowledge of the financial meaning of all transactions to which you are a party to use the Columnar Financial Record efficiently. The information recorded in the Columnar Financial Record, except at the end of an operating period, originates with your checkbook. As the transactions are carried from the checkbook to the Columnar Financial Record, the recording of all transactions involves changes (increases or decreases) in at least two of the classification columns on the forms. This chapter will give the rules for use of the forms and give brief summaries of the essential instructions. Then examples of a number of different transactions covering a period of 2 months will be entered in the record. As you proceed, note that after any transaction has been recorded, the total of the balances of columns 1 through 6 is equal to the total of the balances of columns 7, 8, and 9, and the "net income" sub-column of column 12. This equality provides a means of checking the accuracy of the recorded entries. Keep in mind the

Columnar Financial Record

		Increase		Decrease		Balance	
1							
2							
3							
4							
5							
6							
7							
8							
9							
10							
11							
12							
13							
14							
15							
16							
17							
18							
19							
20							
21							
22							
23							
24							
25							
26							
27							
28							
29							
30							
31							
32							
33							
34							
35							

Form 21-1

No. _121_	$ _105.⁰⁰/₁₀₀_	
DATE _April 5_	19 ___	
To _X Lumber Company_		
FOR _Lumber (Invoice # 1102) Job No. 7_		

	DOLLARS	CENTS
BALANCE BRO'T FOR'D	700.	00
AMOUNT DEPOSITED		
AMOUNT DEPOSITED		
TOTAL	700.	00
AMOUNT THIS CHECK	105.	00
BALANCE CAR'D FOR'D	595.	00

Figure 21-2

fact that this equality represents a natural result of the complete recording of the effects of business transactions rather than the consequence of an attempt to observe formal rules. An accountant would call this double entry bookkeeping but you need not understand it as such. Just note that it helps you reduce errors and actually reflects what is happening to what you have, what you owe and what you are owed.

You are going to have to take certain steps at the outset to make sure your business transactions and only your business transactions are recorded accurately:

1. separate your personal bank account from your business bank account,

2. write an explanation on each check stub indicating the purpose for which each check is drawn

3. number all checks in sequence, and

4. deposit all business receipts in the business bank account.

These procedures will prevent mixing checks for personal expenditures among those for business expenditures and from the mixing of personal and business receipts and deposits. Check stubs or second copies of checks and deposit tickets are used as the basis for entries. Each time a check is written the stub of the check should be filled out completely. The stub should show the number of the check, the date of the check, to whom payment was made, and the purpose for which the check was written, as shown in Figure 21-2. This information must be recorded in the spaces indicated on the stub of the check. This provides the place from which information can be transferred to the Columnar Financial Record.

Deposit tickets should be made out in duplicate so that one copy may be stamped by the cashier and returned to the depositor immediately. The deposit tickets should be numbered. The date of the deposit, the number of the deposit ticket, the individual, company, or lending agency from which the amount was received, and the amount

deposited should be recorded on the deposit ticket as shown in Figure 21-3.

The total of the deposit should be added to the balance shown on the check stub with a note of the deposit number and date as in Figure 21-4.

Before discussing the transfer of information from the check stubs to the forms, the makeup of the Columnar Financial Record should be examined. The forms have 12 columnar sections. These sections are classified, as indicated by the

		No. 1
DEPOSITED WITH THE FIRST NATIONAL BANK		
BY _Frank Pardo_		
April 1		19 ___

PLEASE LIST EACH CHECK SEPARATELY

	DOLLARS	CENTS
CURRENCY		
SILVER		
CHECKS		
Building & Loan, Job #6	375.	00
John Adams, Job #8	125.	00
TOTAL $	500	00

SEE THAT ALL CHECKS AND DRAFTS ARE ENDORSED

Figure 21-3

No. ___	$ ___	
DATE ___	19 ___	
To ___		
FOR ___		

	DOLLARS	CENTS
BALANCE BRO'T FOR'D	595.	00
AMOUNT DEPOSITED No.1	500.	00
AMOUNT DEPOSITED		
TOTAL	1095.	00
AMOUNT THIS CHECK		
BALANCE CAR'D FOR'D		

Figure 21-4

		CLASSIFICATION NUMBER		2			
LINE NO.	DATE	DESCRIPTION OF ENTRY		CASH IN BANK			
				INCREASE	CHECK NO.	DECREASE	BALANCE
1	14 1	Deposit in First National Bank		500.00			500.00
2	2	" " " " "		200.00			700.00
3	5	To pay X Lumber Co. Inv. no. 1102			121	105.00	595.00
4							
5							
6							
7							
8							
9							
10							
11							

Figure 21-5

headings, into columns in which money amounts taken from the check stubs can be written. Under each column heading, except Number 12, subcolumns headed "increase," "decrease," and "balance," are provided, as indicated under classification column 2 as shown in Figure 21-5. Classification column 12 has different subcolumn headings which will be examined later. In all classification columns except 12, when a classification is increased, the amount of the increase is placed in the increase subcolumn and added to the balance subcolumn. When a classification is decreased, the amount of decrease is placed in the decrease subcolumn and subtracted from the balance subcolumn.

The necessity of increasing or decreasing a classification column may be explained by use of an illustration. Assume that $500 has just been deposited in the bank, that the deposit ticket has been filled out completely, and that the amount and date of the deposit have been recorded in the checkbook. Through the act of depositing $500 in the bank account of the business, the bank balance or "cash in bank" has increased by $500. To give effect to this increase, $500 should be written in the increase subcolumn and the balance subcolumn of classification column 2, "cash in bank," increased. This entry has been made on line 1 in Figure 21-5. The date of the deposit should be shown and an explanation of the entry should be written in the column headed, "description of entry." As soon as the $500 amount has been written on the Columnar Financial Record, a check mark should be made after the item on the check stub to show that it has been transferred. A check mark should be made on the check stubs by all amounts which have been transferred so that one can tell by inspection whether or not an amount has been entered on the record.

To carry this example an additional step, suppose another deposit of $200 is made on the following day. After the information on the deposit

ticket has been transferred to the check stub, $200 should be written into the increase subcolumn of classification column 2, "cash in bank," and the balance should be increased by $200. A description of the entry should be written in the column provided for that purpose. This entry has been made on line 2 of Figure 21-5.

When a check is written, the amount should be transferred from the check stub to the Columnar Financial Record and written in the decrease subcolumn of "cash in bank" and the balance subcolumn should be reduced by the amount of the decrease. "Cash in bank" has actually been reduced because money has been spent. Assume that a check for $105 is written on April 5 to pay a supplier. By using the information taken from the check stub, the date, the description of the entry, the number of the check, and the amount of the decrease ($105) should be written on Form 21-1, and the balance of "cash in bank" should be reduced. This entry has been made on line 3 of Figure 21-5.

Another step should now be taken in explaining the use of the Columnar Financial Record. When one classification column is increased or decreased, another classification column must also be increased or decreased. This step can be understood only by recognizing that when money is paid out for any reason, two things actually happen. Assume that a check is written to pay for a typewriter to be used in the office. Money is spent and a typewriter is received. There is (1) a decrease in money in the business and (2) an increase in typewriters. To record this purchase, the date and a description of the action should be written in the explanation column of the Columnar Financial Record; the check number, the amount of the decrease, and the new balance should be written in classification column 2, "cash in bank;" and, since the physical equipment has increased because of the addition of a typewriter, the cost should be entered in the increase subcolumn of the

LINE NO.	DATE	CLASSIFICATION NUMBER DESCRIPTION OF ENTRY	2 CASH IN BANK				6 BUILDING, FURNITURE EQUIPMENT, ETC.		
			INCREASE	CHECK NO.	DECREASE	BALANCE	INCREASE	DECREASE	BALANCE
1	Apr 7	Balance Brought Forward				1,295.00			8,655.40
2	7	Purchased (make) Typewriter Model #		158	145.00	1,150.00	145.00		8,800.40
3									
4									
5									
6									
7									
8									
9									
10									
11									
12									
13									
14									
15									

Figure 21-6

"building, furniture, equipment, etc." classification column, and the balance should be adjusted. This entry is illustrated in Figure 21-6.

When entries are being made in any two of the first six classification columns, the rule applies that if one classification column is increased, the other classification column must be decreased. That is, as long as the first six classification columns are affected, if "cash in bank" is decreased, then either "cash on hand," "materials and lots on hand," "notes, mortgages, and other amounts receivable," "cost of jobs under construction," or "building, furniture, equipment, etc." must be increased. If, for example, a check is written so that one may have cash on hand, then "cash in bank" is decreased and "cash on hand" is increased. There is less money in the bank but more money in the cash box in the office.

If the first six classification columns are considered as one group for purposes of this discussion and classification columns 7, 8, 9, and 12 as another group, then the effect of an increase in any one of the first six classification columns on column 7, 8, 9, or 12 may be determined. For purposes of discussing the relationships, the first six classification columns will be referred to as group A, or assets, and classification columns 7, 8, 9, and 12 will be referred to as group B, or liabilities. Classification columns 10 and 11 are provided for any additional classifications that the builder may wish to use. In those cases where the building concern is owned by partners, these columns may be used for the additional owners. They were not required for use in the individual proprietorship which is illustrated in this chapter. When considering groups A and B, an increase in one, which does not affect another column within the group, calls for an increase in the other. For example, assume a customer makes an initial payment to you before work is started of $500.

Classification column 7, "advances received from lending institutions and purchaser payments," should be increased and "cash in bank" should be increased (assuming the $500 is deposited immediately). Purchaser payments have increased and cash in bank has increased. Or, to take another case, assume the builder takes $1,000 out of his personal bank account and puts it into the business. "Cash in bank" should be increased and "owner's investment in business" should be increased.

Such entries involving columns in group A and group B are different from entries involving columns within either group. When the entry is to be made within the first six columns, it should be observed that if one classification column is increased, the other is decreased. But when an entry involves a column in group A and a column in group B, if the column in group A is increased, the column in group B is increased. Also, when a column in group A is decreased, a column in group B is decreased. Assume, for example, that a builder decreases his owner's investment by withdrawing cash from the bank. Classification column 9, "owner's investment in business" is decreased and classification column 2, "cash in bank," is decreased, since both are reduced.

These steps may be condensed or summarized into the following rules:

1. If an entry in the Columnar Financial Record involves only the classification columns in group A or only the classification columns in group B, when one column is increased the other must be decreased; or, when one column is decreased another column must be increased.

2. If an entry in the Columnar Financial Record involves classification columns in group A and group B, when a column in group A is increased a column in group B must be increased; or

when a column in group A is decreased a column in group B must be decreased.

Entries which do not originate in the checkbook and which should be given consideration are those concerning invoices, depreciation, the transfer of materials and supplies on hand to jobs, and the acceptance of mortgages or notes for work performed.

It is desirable to record purchases made on credit when there is a time lag between receipt of an invoice and payment. In such cases, an appropriate classification column, such as 4, 5, 6, or the "general expense" subcolumn under Number 12, should be increased for the item purchased and classification column 8, "amounts owed suppliers, income taxes withheld from employees, etc." should be increased to indicate the liability. Later, when payment is made, "cash in bank" should be decreased and classification column 8 should be decreased. It is preferable that invoices be entered net of discounts. For example, an invoice for $500 which carries a 2 percent cash discount if payment is made within a limited number of days should be recorded at $490. Since it is to your advantage to take the discount, invoices should be paid within the time limit. If payment is not made within the stipulated number of days, the discount is lost and it will be necessary to increase "general expense" for the lost discount at the time payment of the invoice is made. Where invoices are rendered showing gross prices only, it may prove to be too big a job to convert the price of individual items listed to a net figure, and the invoices may be recorded at gross prices. In such instances, it will be necessary to increase the "purchase discounts, other revenue," subcolumn in classification column 12 when payment of the invoice is made within the discount period and add this increase to the balance of the last subcolumn, "net income."

The entry to record depreciation of buildings, furniture, equipment, etc., is one that should be made periodically. After the amount of depreciation has been determined, the entry should be made by decreasing "building, furniture, equipment, etc." and increasing "general expense."

When materials are used for a job, an entry should be made to decrease classification column 4 and increase classification column 5. Record keeping for materials is discussed in detail later in this chapter.

In those rare cases where a mortgage or delayed payment is accepted as part of the final settlement of a job, classification column 3 should be increased as part of the entry which is made to close a job. In the closing procedure, as illustrated later in this chapter, the amount of the mortgage accepted would be entered in the increase subcolumn of classification column 3 and cash received in the final settlement would be that much smaller.

Space is provided for inclusion of the job number pertaining to various transactions in classification columns 5 and 7. When a transaction involves a specific job, the number of the job should be placed in the space provided in these columns, and the amounts should be transferred to the designated job record (discussed later in this chapter). A check mark may be placed by the job number to indicate that an entry has been transferred to the job record.

Classification column 12 differs from the other columns in arrangement and in the type of information which is provided. Here, the financial results of operating the business are made available. The last column, "net income," reflects the excess of the selling price of jobs sold and other income over costs of jobs completed and general expenses. If costs of jobs and general expenses should exceed revenues from jobs completed and other sources, the amount shown in this last column would be a loss and should be indicated by the use of red ink or parentheses. In the column "selling price of jobs sold" the total of advances received and purchaser payments plus the gross amount of the final settlement for a particular job is entered when the job is completed and payment is due. In those cases where payment is delayed, the "selling price of jobs sold" will consist of advances received and purchaser payments, notes accepted, and the gross amount of the final settlement. The gross amount of the final settlement is made up of the net amount received from the lending institution plus the financing costs. To the column "cost of jobs sold" all direct costs connected with a job are transferred, including the financing costs deducted from the final settlement. The excess of the "selling price of jobs sold" over the "cost of jobs sold" should be added to the balance of the "net income" column. Any expense not classified as a direct cost of the job must be entered in the "general expense" column. When invoices are recorded at gross prices, the discounts taken should be recorded in the "purchase discounts, other revenue" column and added to the balance in the "net income" column. When other revenues, such as interest, forfeited deposits, etc. are realized, these amounts should be entered in the "purchase discounts, other revenue" column and added to the balance in the "net income" column. The explanation column should be used to describe briefly the entries made.

Since classification column 12 differs from the other columns in make-up and since it is included in the group B columns in the statement of rules for making entries, brief comparison of this column with the others should be made. The "net income" column represents the "balance" column for classification column 12. Amounts entered in the "selling price of jobs sold" and "purchase discounts, other revenue" columns increase the "net income" column and amounts entered in the "cost of jobs sold" and "general expense" columns decrease the "net income" column. In comparing classification column 12

		12				
		REVENUE AND EXPENSE				
DATE	EXPLANATION	SELLING PRICE OF JOBS SOLD	PURCHASE DISCOUNTS, OTHER REVENUE	COST OF JOBS SOLD	GENERAL EXPENSE	NET INCOME
19— Mar 14	Balance brought forward					(150.00)
15	Repairs to pickup				200.00	(350.00)
15	Discount Taken		8.00			(342.00)
16	Final Settlement, Job no. 5	12,000.00		11,500.00		158.00
17	Forfeit of Deposit		100.00			258.00

Figure 21-7

with the other classification columns, you may think of the "selling price of jobs sold" and "purchase discounts, other revenue" as being "increase" columns. The columns "cost of jobs sold" and "general expense" could be considered "decrease" columns and the "net income" column could be thought of as a "balance" column. Since the columns in classification column 12 do not have separate "increase" and "decrease" subcolumns, amounts entered in these columns are increases, even though they may either increase or decrease the "net income" column. Thus, if an amount is entered as an increase in the "general expense" column, the "net income" column is decreased; if an amount is entered as an increase in the "purchase discounts, other revenue" column, the "net income" column is increased.

To illustrate the use of classification column 12, assume that a builder has completed job 5, the first job of the year. On March 15, the day before the final settlement, the builder paid general expenses in the amount of $200 (for repairs to pickup) and, in paying an invoice which was recorded at the gross amount, took a discount of $8. Assume further that on March 16 the final settlement on job 5 was completed. Advances received from lending institutions had been $10,000 and the final settlement was $2,000, making the full price $12,000. The direct cost of the job was $11,500. It is assumed also that on March 17 a prospective customer forfeited a $100 deposit which had been made during the latter part of February. On line 1 of Figure 21-7 a balance of $150 appears. Since no jobs have been closed out for the year, this balance of $150 would be a red, or minus, figure. In Figure 21-7 this amount is designated by the use of parentheses. (There has been no revenue recorded to date against which this expense may be applied.) On March 15 the payment of $200 for repairs was recorded by decreasing "cash in bank," entering the $200 in the "general expense" column as an increase, and carrying this amount to the last column, "net income," changing the $150 minus figure to a $350 minus figure. On March 15 the

payment of an invoice originally entered at the gross amount was recorded by decreasing "cash in bank" and "amounts owed suppliers, income taxes withheld from employees, etc.," increasing "purchase discounts, other revenue," and adjusting the "net income" column. On March 16 job 5 was closed. "Cash in bank" was increased by $2,000; "advances received from lending institutions and purchaser payments" was decreased by $10,000; and "selling price of jobs sold," classification column 12, was increased by $12,000. "Cost of jobs sold" (classification column 12) was increased and "cost of jobs under construction" was decreased by $11,500. The difference of $500 was carried to the "net income" column changing the $342 red figure to a $158 black figure. To give effect to the forfeit of the deposit, "advances received from lending institutions and purchaser payments" was decreased by $100, $100 was entered in the "purchase discounts, other revenue" column in classification column 12 as an increase, and the balance in the "net income" column was adjusted to $258.

For purposes of comparing the total of the balances of the first six classification columns with the total of the balances of classification columns 7, 8, 9, and 12, the balance of the last column only of classification column 12, "net income," should be included.

In the examples illustrating the use of the Columnar Financial Record included in this chapter, the balances of classification columns affected by a transaction have been extended after each entry. As a practical matter, these balances, with the exception of that for the "cash in bank" classification column, need not necessarily be extended after each entry. However, you will probably find it to your advantage to extend these balances at least once a week to determine whether the total of the balances of the group A classification columns is equal to the total of the balances of the group B classification columns. It is likely that any remodeler will wish to extend the "cash in bank" classification column balance after each entry in that column in order to know the amount of cash that is available for use.

General Expense Record

Form 21-8

The procedure to be followed in closing out classification column 12 at the end of the operating period or year is discussed under the date of May 31 in the illustrative example which is given later in this chapter.

Form 21-8 is used to classify and group the entries recorded in the "general expense" column of classification column 12, Form 21-1. Note Form 21-12 on page 384.

Two broad classes of expenses are entered in Form 21-8:

1. Indirect building expenses:
   - Tools expense
   - Superintendence
   - Insurance
   - Indirect labor
   - Payroll taxes
   - Depreciation
   - Gasoline and oil
   - Miscellaneous
2. Selling and administrative expenses:
   - Rent
   - Office supplies
   - Telephone and telegraph
   - Advertising
   - Office salaries
   - Depreciation
   - Payroll taxes
   - Miscellaneous

Entries made in the "general expense" column of classification column 12 should be transferred to Form 21-8 since you will want to make an analysis of these expenses. The classifications provided are suitable as a breakdown of expenses for income tax purposes. A check mark should be made in the far left portion of the "general expense" column to indicate that the amount has been transferred to Form 21-8. Transfers to the General Expense Record may be made periodically and need not necessarily be kept up to date at all times.

How The Forms Are Used

Sometimes it is easiest to learn from examples. Forms 21-9, 21-10, 21-11 and 21-12 show how the Columnar Financial Record and the General Expense Record are used. It is assumed that Frank Pardo, a home improvement contractor working out of Ann Arbor, Michigan, adopted the Columnar Financial Record on April 1, 19—. He had one job (No. 43) under construction on which construction costs of $7,136.30 has been incurred. During April he started construction on Jobs 44 and 45, and completed Job 43. During the month of May Job 44 was completed.

April 1.—Frank Pardo set up a formal record of his business on the Columnar Financial Record. He entered balances in classification columns as follows: "cash in bank," $3,000; "notes, mortgages, and other amounts receivable," $25 (amount of loan to employee); "materials and lots on hand," $128; "cost of job under construction," $7,136.30; "building, furniture, equipment, etc.," $10,000; "advances received from lending institu-

tions and purchaser payments," $4,000; "amounts owed suppliers, income taxes withheld from employees, etc.," $302.11; and "owner's investment in business," $15,987.19. The amount entered in classification column 9, "owner's investment in business," was determined by subtracting the total of the balances of classification columns 7 and 8, $4,302.11, from the total of the balances of the first six classification columns, $20,289.30. Notice that the total of the balances of the first six classification columns, $20,289.30, is equal to the total of the balances of column 7, 8, and 9, $20,289.30. This equality of the total of the balances of group A, (asset) columns with the total of the balances of group B (liability and proprietorship) columns will always result after each transaction has been entered if it is handled properly. The Job Record (Form 21-13) for job 43 was opened and the accumulated costs and receipts and draws were recorded. The recording of costs on the Job Record is explained later in this chapter. Check marks were placed in the subcolumns provided on the left hand side of columns 5 and 7 of Form 21-9.

April 2.—A lumber bill was received from Jackson Lumber Co. for $900 (net of cash discount) for lumber delivered to job 43. This invoice was paid. "Cash in bank" was decreased by $900 and the balance was adjusted. "Cost of jobs under construction" was increased by $900 and the balance was adjusted.

Although it has not been specifically mentioned in each transaction which follows, the job number has been entered each time an entry has been made in columns 5 and 7. Check marks have also been entered by the job numbers in these columns indicating that the amounts have been transferred to Job Records. The procedure for transferring such amounts is discussed later in this chapter.

April 5.—The payroll for the week ended April 4 indicated that gross pay was $317.50 and that F.I.C.A. and income taxes withheld totaled $35.76. Check Numbers 104, 105, 106, and 107 was issued in the total amount of $281.74. It was not necessary to indicate the job numbers in column 5 since this information was provided in the payroll from which the labor costs were transferred to the Job Record. For this entry and all succeeding payroll entries, the payroll number has been entered in the job number subcolumn of column 5 for reference purposes only. It has been followed by a check mark to insure that a transfer of this item will not be made from Form 21-9 to the Job Record. Of the gross pay, $40 for clean-up work was charged to general expense. "Cost of jobs under construction" was increased, $277.50; "amounts owed suppliers, and income taxes withheld" was increased, $35.76; $40 was entered in the "general expense" column as an increase and the "net income" column was adjusted; and "cash in bank" was decreased, $281.74.

Form 21-9

Columnar Financial Record

BUILDER _Frank Purdo_   ADDRESS _Ann Arbor Michigan_

PAGE _1_

LINE NO.	DATE	CLASSIFICATION NUMBER / DESCRIPTION OF ENTRY	1. CASH ON HAND INCREASE	DECREASE	BALANCE	2. CASH IN BANK INCREASE	CHECK NO.	DECREASE	BALANCE	3. NOTES, MORTGAGES, AND OTHER AMOUNTS RECEIVABLE INCREASE	DECREASE	BALANCE	4. MATERIALS ON HAND INCREASE	DECREASE	BALANCE
1	Apr 1	To set up record of business							30000 00			25 00			128 00
2	2	To pay Jackson Lumber Co - Inv #534					103	900 00	29100 00						
3	5	Price for work					104-107	251 74	18818 26						
4	5	City of Ann Arbor Building Permit					108	12 50	18805 76						
5	5	Lumber Homes - Plans & Spec					109	15 00	17790 76						
6	5	Apex					110	100 00	17690 76						
7	6	Transfer from personal bank acct				2500 00			19090 76						
8	6	Howard Furnace Co					111	520 00	26900 76						
9	6	Adams & Lumm Accountants					112	100 00	25900 76						
10	9	Draw - Standish Finance Co				3000 00			55900 76						
11	9	Lawndale Hardware Co					113	15 00	55725 76						
12	9	Tappan Oil Company					114	32 65	55743 11						
13	9	Ann Arbor					115	16 32	45431 1				1000 00		1128 00
14	10	Mich Bell Telephone Co					116	20 00	45267 9						
15	10	Ann Arbor Builders Assn					117	150 00	45067 9						
16	10	Office Supply Co - Typewriter Motel					118	13 27	43567 9						
17	11	Maynard Office Supply Co					119	238 15	43435 2						
18	11	Payroll for week					120-122	450 00	41053 7						
19	12	John Luther - Lathing & Plastering					123	250 00	36553 7						
20	14	Howard Furnace Co					124		34053 7						
21	14	Al Eickland					125	112 50	33929 87						
22	14	Ralph Smith					126	480 00	28129 87						
23	15	Eisman Bros					127	85 00	27279 87						
24	15	Purchased new pickup													
25	16	Order Whittemore Standish Ins Co													
26	17	Johnson Supply Co					128	110 00	26129 87						
27	17	Geo Wilson					129	360 00	22359 87						
28	17	Ann Arbor News					130	45 00	23212 87						
29	19	Lazy Ind Company					131	250 00	19629 87						
30	19	Settlement for Lot No. 43							76729 87						
31	19	Payroll for week				5710 00	132-136	337 08	73359 79						
32	20	Collector of Internal Revenue					137	345 11	69969 68						
33	20	To provide cash on hand	20 00		20 00		138	30 00	69709 68						
34	20	Payroll for week					139-143	449 44	65212 4						
35	20	City of Ann Arbor - Bldg Permit					144	14 00	65072 4						

Form 21-9 [continued]
Columnar Financial Record

LINE NO.	5 COST OF JOBS UNDER CONSTRUCTION — JOB NO.	INCREASE	DECREASE	BALANCE	6 BUILDING, FURNITURE, EQUIPMENT, ETC. — INCREASE	DECREASE	BALANCE	7 ADVANCES RECEIVED FROM LENDING INSTITUTIONS AND PURCHASER PAYMENTS — JOB NO.	INCREASE	DECREASE	BALANCE	8 AMOUNTS OWED SUPPLIERS, INCOME TAXES WITHHELD FROM EMPLOYEES, ETC. — INCREASE	DECREASE	BALANCE
1	✓			7136.30			10000.00	✓			4000.00			302.11
2	43✓	900.00		8036.30										
3	PR1✓	277.50		8313.80								35.76		337.87
4	44✓	12.50		8326.30										
5	44✓	15.00		8341.30										
6	44✓	1100.00		9441.30										
7														
8	43✓	500.00		9941.30										
9														
10								43✓	3000.00		7000.00			
11														
12														
13														
14														
15														
16					150.00		10150.00							
17	PR2✓	240.00		10181.30								31.85		369.72
18														
19	43✓	450.00		10631.30										
20	43✓	250.00		10881.30										
21	43✓	112.50		10993.80										
22	43✓	480.00	11516.30	11473.80										
23	44✓	85.00		11558.80	1800.00									
24	43✓	500.00		12058.80		600.00	11350.00	43✓	500.00		7500.00	1200.00		1569.72
25	43✓	500.00		12058.80										
26	43✓	110.00		12168.80										
27	43✓	360.00		12528.80										
28														
29	43✓	250.00		12778.80										
30	43✓	40.00		13025.50						7500.00	0			
31	PR3✓	314.50		11617.00								42.42		1612.14
32													302.11	1310.03
33														
34	PR4✓	503.80		2120.80								64.36		1374.39
35	45✓	14.00		2134.80										

Form 21-9 [continued]
Columnar Financial Record

Line No.	Owner's Investment in Business — Increase	Decrease	Balance	Explanation	Selling Price of Jobs Sold	Cost of Jobs Sold	✓	General Expense	Net Income
1			15987.19						
2									
3									
4				Indirect Labor			✓	40.00	(40.00)
5									
6									
7	2500.00								
8			18487.19						
9				Miscellaneous - S & A			✓	100.00	(140.00)
10									
11				Tools Expense			✓	15.00	(155.00)
12				Gasoline and Oil			✓	32.65	(187.65)
13									
14				Telephone & Telegraph			✓	16.32	(203.97)
15				Miscellaneous - S & A			✓	20.00	(223.97)
16									
17				Office Supplies			✓	13.27	(237.24)
18				Indirect Labor			✓	30.00	(267.24)
19									
20									
21									
22									
23									
24									
25									
26									
27									
28				Advertising			✓	45.00	(312.24)
29									
30				Close Job no 43	13250.00	11516.30			1142.146
31				Indirect Labor			✓	65.00	1356.46
32				Payroll Taxes - I.B.Exp			✓	43.00	1313.46
33									
34				Indirect Labor			✓	10.00	1303.46
35									

A building permit was secured for job 44, $12.50. A Job Record was opened for job 44. Copies of plans and specifications for job 44 were drawn up by Superior Homes, at a cost of $15. A check was written in the amount of $1,100 for the purchase of an easement from the Ajax Company so a driveway could be built across adjoining land owned by Ajax. For each of these transactions "cash in bank" was decreased and "cost of jobs under construction" was increased.

April 6.—Mr. Pardo transferred $2,500 from his personal bank account into the business bank account. "Cash in bank" was increased and "owner's investment in business" was increased. Where a deposit includes amount received from a single source, as in this case, the entry may be made directly from the check stub. Where a deposit includes amounts received from several different sources, it may be necessary to make the entry on the Columnar Financial Record from the duplicate deposit ticket as space would not be available on the check stub to analyze the deposit.

Mr. Pardo paid Howard Furnace Co. $500 on the heating contract for job 43. "Cash in bank" was decreased and "cost of jobs under construction" was increased.

A check for $100 was written to Adams & Turner, accountants, for audit performed in March. "Cash in bank" was decreased and $100 was entered in the "general expense" column as an increase and the "net income" column was adjusted.

April 9.—Mr. Pardo received a check as a draw (second inspection) on job 43 from Standish Finance Co., $3,000. This check was deposited. "Cash in bank" was increased and "advances received from lending institutions and purchaser payments" was increased.

Checks were issued to the Lawndale Hardware Co. to cover the purchase of a lantern and other small tools, $15, and to Tappan Oil Co. for gas and oil, $32.65. For each of these transactions "cash in bank" was decreased and the amounts were entered in the "general expense" column as increases and the "net income" column was adjusted.

April 10.—A range-oven was purchased and paid for in the amount of $1,000. "Cash in bank" was decreased and "materials on hand" was increased.

Mr. Pardo paid the Michigan Bell Telephone Co. $16.32. A check was sent to the Ann Arbor Builders' Association for Mr. Pardo's annual dues, $20. For each of these transactions "cash in bank" was decreased and the amounts were entered in the "general expense" column as increases and the "net income" column was adjusted.

April 11.—A check was written to pay for the purchase of a typewriter, $150. "Building, furniture, equipment, etc.," was increased and "cash in bank" was decreased. You may find it helpful to use a Fixed Asset Record to list the cost and description of business equipment.

A check was sent to the Maynard Office Supply Co., $13.27. "Cash in bank" was decreased and $13.27 was entered in the "general expense" column as an increase and the "net income" column was adjusted.

April 12.—The payroll for the week ended April 11 indicated that gross pay was $270 and that F.I.C.A. and income taxes withheld totaled $31.85. Check Numbers 120, 121, and 122 were issued in the total amount of $238.15. Of the gross pay, $30 for work in the storeroom was charged to general expense. "Cost of jobs under construction" and "amounts owed suppliers, income taxes withheld from employees, etc." were increased; $30 was entered in the "general expense" column as an increase and the "net income" column was adjusted; and "cash in bank" was decreased.

April 14.—A check was written for work completed by Loren Little for lathing and plastering job 43, $450. Howard Furnace Co. was paid in full for heating contract on job 43, $250. Payment in full was made to Al Ecklund for electrical work on job 43, $112.50, and to Ralph Smith for plumbing contract on job 43, $480. For each of these transactions "cash in bank" was decreased and "cost of jobs under construction" was increased.

April 15.—Elkins Bros. were paid in full for excavation work completed for job 44, $85. "Cash in bank" was decreased and "cost of jobs under construction" was increased.

Mr. Pardo purchased a used pickup truck for $1,800. He traded in an old truck with a depreciated cost of $600 and received a $600 trade-in allowance. To take the cost of the old truck out of the records, classification column 6, "building, furniture, equipment, etc.," was decreased by $600. This same column was increased by $1,800, the cost of the purchased pickup, and "amounts owed suppliers, income taxes withheld from employees, etc." was increased by $1,200, the amount which remained to be paid.

If Mr. Pardo had traded in an old truck with a depreciated cost of $800 and received a trade-in allowance of $600, the amount to be added to the "building, furniture, equipment, etc.," column would then be $2,000. The entry would be made as follows: Classification column 6, "building, furniture, equipment, etc.," would be decreased by $800, the depreciated cost of the old truck, and increased by $2,000, the book value of the purchased pickup truck, and "amounts owed suppliers, income taxes withheld from employees, etc." would be increased by $1,200.

If Mr. Pardo had traded in an old truck with a depreciated cost of $400, the amount to be added to the "building, furniture, equipment, etc." column would then be $1,600. The entry would be made as follows:

Classification column 6, "building, furniture,

equipment, etc.'' would be decreased by $400, the depreciated cost of the old truck, and increased by $1,600, the book value of the purchased pickup, and ''amounts owed suppliers, income taxes withheld from employees, etc.'' would be increased by $1,200.

Although the procedure suggested above does not conform with good accounting theory, it is suggested in order to simplify the preparation of tax returns from the records. The Internal Revenue Service will not permit a gain or loss to be recognized for tax purposes on the exchange of a fixed asset which is used in the business.

April 16.—An order was written on Standish Finance Co. to pay Art Pilsner for completion of painting contract on job 43, $500. ''Cost of jobs under construction'' was increased and ''advances received from lending institutions and purchaser payments'' was increased.

Received materials for stock. An invoice was not received and an entry was not made.

April 17.—Johnson Supply Co. was paid in full for floor covering, job 43, $110. A check for $360 was mailed to Jim Wilson for tile work on job 43. For each of these transactions, ''cash in bank'' was decreased and ''cost of jobs under construction'' was increased.

Ann Arbor News was paid $45 for advertising. ''Cash in bank'' was decreased and $45 was entered in the ''general expense'' column as an increase and the ''net income'' column was adjusted.

April 18.—Mr. Pardo paid Lotz Seed Co. in full for landscaping work on job 43, $250. ''Cash in bank'' was decreased and ''cost of jobs under construction'' was increased.

April 19.—Mr. Pardo received final settlement for job 43 from the Standish Finance Co. The net amount received was $5,710. Forty dollars had been deducted by Standish to cover tax and interest, making the gross settlement $5,750. ''Cash in bank'' was increased by $5,710, ''cost of jobs under construction'' was increased by $40 (covering tax and interest), ''advances received from lending institutions and purchaser payments'' was decreased by $7,500, and this latter amount, plus the gross final settlement, $5,750, or $13,250 was entered under ''selling price of jobs sold'' in classification column 12. ''Cost of jobs under construction'' was decreased by $11,516.30, the total cost indicated on the Job Record, and this cost was entered under ''cost of jobs sold'' in classification column 12. The difference of $1,733.70 was added to the balance in the column headed ''net income'' in classification column 12.

The total of the balances of the first six columns was found to agree with the total of the balances of columns 7, 8, and 9, and the ''net income'' subcolumn of classification column 12, as follows:

1.	Cash on hand	0
2.	Cash in bank	$7,672.87
3.	Notes, mortgages, and other amounts receivable	25.00
4.	Materials on hand	1,128.00
5.	Cost of jobs under construction	1,302.50
6.	Building, furniture, equipment, etc	11,350.00
7.	Advances received from lending institutions and purchaser payments	0
8.	Amounts owed suppliers, income taxes withheld from employees, etc	$1,569.72
9.	Owner's investment in business	18,487.19
12.	Net income	1,421.46
		$21,478.37 $21,478.37

The payroll for the week ended April 18 indicated that gross pay was $379.50 and that F.I.C.A. and income taxes withheld totaled $42.42. Checks Numbers 132, 133, 134, 135 and 136 were issued in the total amount of $337.08. Of the gross pay, $65 for inventory work was charged to general expense. ''Cost of jobs under construction'' and ''amounts owed suppliers, income taxes withheld from employees, etc.'' were increased; $65 was entered in the ''general expense'' column as an increase and the ''net income'' column was adjusted; and ''cash in bank'' was decreased.

April 20.—A check in the amount of $345.11 was written to pay F.I.C.A. and income taxes; $302.11 of this amount had been withheld from the employee's payroll and $43 represented Mr. Pardo's share of the F.I.C.A. taxes. These amounts cover withholding from employees for F.I.C.A. and income tax and the employer's share of F.I.C.A. taxes for the first quarter of the year. ''Cash in bank'' was decreased by $345.11; ''amounts owed suppliers, income taxes withheld from employees, etc.,'' was decreased by $302.11; and $43 was entered in the ''general expense'' column as an increase and the ''net income'' column was adjusted.

A check was written in the amount of $20 to provide Mr. Pardo with cash for miscellaneous business expenses. ''Cash in bank'' was decreased and ''cash on hand'' was increased.

April 26.—The payroll for the week ended April 25 indicated that gross pay was $513.80 and that F.I.C.A. and income taxes withheld totaled $64.36. Checks Numbers 139, 140, 141, 142, and 143 were issued in the total amount of $449.44. Of the gross pay, $10 for yard work was charged to general expense. ''Cost of jobs under construction'' and ''amounts owed suppliers, income taxes withheld from employees, etc.,'' were increased; $10 was entered in the ''general expense'' column as an increase and the ''net income'' column was adjusted; and ''cash in bank'' was decreased.

April 29.—A building permit was obtained for job 45, $14. Clyde Collins was paid one-half of the contract for concrete work on job 44, $482.50. For each of these transactions, ''cash in bank'' was decreased and ''cost of jobs under construction'' was increased.

Form 21-10
Columnar Financial Record

PAGE 2
BUILDER: Frank Pardo
ADDRESS: Ann Arbor, Michigan

LINE	DATE	DESCRIPTION OF ENTRY	CASH ON HAND INCREASE	DECREASE	BALANCE	CASH IN BANK INCREASE	CHECK NO.	DECREASE	BALANCE	NOTES, MORTGAGES, ETC. INCREASE	DECREASE	BALANCE	MATERIALS ON HAND INCREASE	DECREASE	BALANCE
1	Apr 28	Brought forward from previous page			20 00				6507 24			125 00			1128 00
2	29	Clyde Collins					145	482 50	6024 74						
3	30	Johnson Supply Co					146	58 50	5966 24						
4	30	Frank Pardo Drawing					147	500 00	5466 24						
5	30	To record Cash on hand Agent for gravel oil		8 00	12 00										
6	30	To record depreciation													
7	May 1	Transfer cost of lot to job no. 45												1000 00	128 00
8	3	Payroll for week					148-156	592 36	4873 88						
9	5	Mobil Plant					157	20 00	4853 88						
10	6	Bank Service Charge						1 05	4852 83						
11	6	Drew - Standard Finance Co				3000 00			7852 83						
12	7	Jackson Lumber Co					158	1345 72	6507 11						
13	7	Ann Arbor Building Matls Co					159	498 62	6008 49						
14	7	Purchased w/ small tools from Cash on Hand		10 00	2 00										
15	8	Eklund Bros					160	90 00	5918 49						
16	8	John Percy					161	645 00	5273 49						
17	8	Al Eklund					162	105 00	5168 49						
18	10	Payroll for week					163-169	615 49	4553 00						
19	10	Detroit Furnace Co					170	453 33	4099 67						
20	10	Midwest Finance Co					171	80 00	4019 67						
21	12	Clyde Collins					172	490 00	3529 67						
22	12	Mich. Bell Telephone Co					173	14 95	3514 72						
23	12	Tappan Oil Company					174	40 79	3473 93						
24	14	Jackson Lumber Co - 2nd Apr 30					175	600 00	2873 93				588 00		716 00
25	14	Ann Arbor Bank - Deposit for Co. of I.R.					176	196 60	2677 33						
26	15	Boland Glass Co					177	37 90	2639 43						
27	16	Steven Little					178	400 00	2239 43						
28	16	Johnson Supply Co					179	60 00	2179 43						
29	16	General Insurance Agency					180	200 00	1979 43						
30	16	Maynard Office Supply Co					181	15 02	1964 41						
31	17	Payroll for week					182-189	720 33	1244 08						
32	18	Johnson - Standard Finance Co				3000 00			4244 08						
33	20	John Percy					190	430 00	3814 08						
34	20	Transfer Matls to job no. 45												159 26	556 74
35	21	Al Eklund					191	105 00	3709 08						

Column headings:

- **5 — COST OF JOBS UNDER CONSTRUCTION** (Job No. | Increase | Decrease | Balance)
- **6 — BUILDING, FURNITURE, EQUIPMENT, ETC.** (Increase | Decrease | Balance)
- **7 — ADVANCES RECEIVED FROM LENDING INSTITUTIONS AND PURCHASER PAYMENTS** (Job No. | Increase | Decrease | Balance)
- **8 — AMOUNTS OWED SUPPLIERS, INCOME TAXES WITHHELD FROM EMPLOYEES, ETC.** (Job No. | Decrease | Balance)

Line	5 Job No.	5 Increase	5 Decrease	5 Balance	6 Increase	6 Decrease	6 Balance	7 Job No.	7 Increase	7 Decrease	7 Balance	8 Job No.	8 Decrease	8 Balance
1				2134 80			11350 00				0			1374 39
2	44✓	4829 50		2617 30										
3	44✓	58 50		2675 80										
4														
5														
6	45✓	1000 00		3675 80		66 25	11283 75							
7	PR5✓	615 50		4291 30										
8	45✓	20 00		4311 30								73 14		1447 53
9														
10														
11	44✓	1345 72		5657 02				44✓	3000 00		3000 00			
12	44✓	498 62		6155 64										
13														
14	45✓	90 00		6245 64										
15	44✓	645 00		6890 64										
16	44✓	105 00		6995 64										
17	PR6✓	650 00		7645 64								74 51		1522 04
18	44✓	453 33		8098 97										
19	45✓	490 00		8588 97										
20													80 00	1442 04
21														
22														
23														
24														
25	44✓	37 90		8626 87									174 39	1267 65
26	44✓	400 00		9026 87										
27	45✓	60 00		9086 87										
28														
29														
30				9900 87										
31	PR7✓	814 00										103 67		1371 52
32								44✓	3000 00		6000 00			
33	44✓	430 00		10330 87										
34	45✓	159 26		10490 13										
35	44✓	105 00		10595 13										

Form 21-10 [continued]
Columnar Financial Record

Line No.	Owner's Investment in Business — Increase	Decrease	Balance	Explanation	General Expense (✓)	Net Income	Line No.
1			18487.19		✓	1303.46	1
2							2
3							3
4		500.00	17987.19				4
5				Gasoline and Oil 33.75 IB / Depreciation 32.50 S&A	✓ 8.00	1295.46	5
6					✓ 66.25	1229.21	6
7				Indirect Labor	50.00		7
8				Miscellaneous – S&A	✓	1179.21	8
9					1.05		9
10				Tools Expense	✓	1178.16	10
11							11
12							12
13				Indirect Labor	10.00		13
14					✓	1168.16	14
15							15
16							16
17				Indirect Labor	40.00	1128.16	17
18					✓		18
19							19
20							20
21				Telephone and Telegraph	✓ 14.95	1113.21	21
22				Gasoline and Oil	✓ 40.79	1072.42	22
23				Miscellaneous – S&A	✓ 12.00		23
24				Payroll Taxes – I.B. Exp	✓ 22.21	1060.42	24
25						1038.21	25
26							26
27							27
28				Miscellaneous – S&A	✓ 200.00		28
29				Office Supplies	✓ 15.02	838.21	29
30				Indirect Labor	✓ 10.00	823.19	30
31						813.19	31
32							32
33							33
34							34
35							35

Column sections: 9 — Owner's Investment in Business; 10 — Increase / Decrease / Balance; 11 — Increase / Decrease / Balance; 12 — Revenue and Expense (Selling Price of Jobs Sold, Purchase Discounts. Other Revenue, Cost of Jobs Sold, General Expense, Net Income)

April 30.—Johnson Supply Co. was paid for an I-beam purchased for job 44, $58,50. "Cash in bank" was decreased and "cost of jobs under construction" was increased.

Mr. Pardo withdrew $500 for personal use. "Cash in bank" was decreased and "owner's investment in business" was decreased.

Eight dollars of the cash on hand had been spent for gasoline and oil. "Cash on hand" was decreased and $8 was entered in the "general expense" column as an increase and the "net income" column was adjusted.

Mr. Pardo records depreciation monthly. Depreciation of $66.25 was recorded by decreasing "building, furniture, equipment, etc.," and entering the $66.25 in the "general expense" column as an increase and adjusting the "net income" column. Depreciation for one-half month on the old pickup traded in and for one-half month on the new pickup amounted to a total of $38.75. If Form 21-12 is used, depreciation on the pickup should be classified as an "indirect building expense" and the remainder, $27.50, covering depreciation on the building, office furniture, and typewriter, should be classified as a "selling and administrative expense."

The column balances as of April 30 were:

1.	Cash on hand	$12.00
2.	Cash in bank	5,466.24
3.	Notes, mortgages, and other amounts receivable	25.00
4.	Materials on hand	1,128.00
5.	Cost of jobs under construction	2,675.80
6.	Building, furniture, equipment, etc	11,283.75
7.	Advances received from lending institutions and purchaser payments	0
8.	Amounts owed suppliers, income taxes withheld from employees, etc	$1,374.39
9.	Owner's investment in business	17,987.19
12.	Net income	1,229.21
		$20,590.79 $20,590.79

May 1.— Mr. Pardo transferred the cost of the range purchased on April 10 to job 45, $1,000. "Materials on hand" was decreased and "cost of jobs under construction" was increased.

May 3.—The payroll for the week ended May 2 indicated that gross pay was $665.50 and that F.I.C.A. and income taxes withheld totaled $73.14. Check Numbers 148 through 156 were issued in the total amount of $592.36. Of the gross $50 for loading and unloading materials was charged to general expense. "Cost of jobs under construction" and "amounts owed suppliers, income taxes withheld from employees, etc." were increased; $50 was entered in the "general expense" column as an increase and the "net income" column was adjusted; and "cash in bank" was decreased.

May 5.—A check in the amount of $20 was paid to Model Plans for plans for job 45. "Cash in bank" was decreased and "cost of jobs under construction" was increased.

May 6.—The bank statement for the month ended April 30 indicated service charges of $1.05. "Cash in bank" was decreased and $1.05 was entered in the "general expense" column as an increase and the "net income" column was adjusted.

Mr. Pardo received $3,000 as a draw from Standish Finance Co., job 44. "Cash in bank" was increased and "advances received from lending institutions and purchaser payments" was increased.

May 7.—Mr. Pardo issued a check in the amount of $1,345.72 to Jackson Lumber Co. in payment of an April 30 invoice covering job 44. He also paid Wilson Building Materials Co. for masonry materials, $498.62 (April 30 bill), for job 44. For each of these transactions "cash in bank" was decreased and "cost of jobs under construction" was increased.

Mr. Pardo purchased a putty knife, screwdriver, and other small tools, for a total cost of $10. He used cash on hand. "Cash on hand" was decreased and $10 was entered in the "general expense" column as an increase and the "net income" column was adjusted.

May 8.—Mr. Pardo paid Elkins Bros. in full for excavation work completed on job 45, $90. He also paid John Percy 60 percent of the plumbing contract on job 44, $645, and Al Ecklund 50 percent of the electrical contract on job 44, $105. For each of these transactions "cash in bank" was decreased and "cost of jobs under construction" was increased.

May 10.—The payroll for the week ended May 9 indicated gross pay of $690 and that F.I.C.A. and income taxes withheld totaled $74.51. Check Numbers 163 through 169 were issued in the total amount of $615.49. Of the gross pay, $40 for work in storeroom was charged to general expense. "Cost of jobs under construction" and "amounts owed suppliers, income taxes withheld from employees, etc.," were increased; $40 was entered in the "general expense" column as an increase, and the "net income" column was adjusted; and "cash in bank" was decreased.

Mr. Pardo paid Howard Furnace Co. $453.33 on heating contract for job 44. "Cash in bank" was decreased and "cost of jobs under construction" was increased.

Mid-West Finance Co. was paid $80 on amount owed for pickup truck purchased on April 15. "Cash in bank" was decreased, and "amounts owed suppliers, income taxes withheld from employees, etc.," was decreased.

May 12.—Mr. Pardo paid Clyde Collins $490 on the contract for concrete work for job 45. "Cash in bank" was decreased and "cost of jobs under construction" was increased.

PAGE 3

Form 21-11
Columnar Financial Record

BUILDER Frank Pardo

ADDRESS Ann Arbor, Michigan

LINE NO.	DATE	DESCRIPTION OF ENTRY	CASH ON HAND			CASH IN BANK				NOTES, MORTGAGES, AND OTHER AMOUNTS RECEIVABLE			MATERIALS ON HAND		
			INCREASE	DECREASE	BALANCE	INCREASE	CHECK NO.	DECREASE	BALANCE	INCREASE	DECREASE	BALANCE	INCREASE	DECREASE	BALANCE
1	May 31	Brought forward from previous page			2.00				3709.08			25.00			556.74
2	31	Carl Pelani					192	430.00	3279.08						
3	31	John Piroy					193	800.00	2479.08						
4	31	Howard Furnace Co					194	500.00	1979.08						
5	31	Ann Arbor News					195	30.00	1949.08						
6	31	Clyde Collins					196	510.00	1439.08						
7	31	Howard Furnace Co					197	226.67	1212.41						
8	31	Draw Standard Finance Co				2000.00			3212.41						
9	31	Payroll for week					198-206	764.39	2448.02						
10	31	Flint Floor Covering Co					207	99.25	2348.77						
11	31	Jackson Lumber Co					208	564.00	1784.77						
12	31	City Steel Company					209	200.00	1584.77						
13	31	Settlement for job no 44				5030.00			6614.77						
14	31	Payroll for week					210-217	637.58	5977.19						
15	31	Depreciation for month													
16	31	To close net income to owner's invest													
17															
18															
19															
20															
21															
22															
23															
24															
25															
26															
27															
28															
29															
30															
31															
32															
33															
34															
35															

Form 21-11 [continued]
Columnar Financial Record

LINE NO.	(5) COST OF JOBS UNDER CONSTRUCTION — JOB NO.	INCREASE	DECREASE	BALANCE	(6) BUILDING, FURNITURE, EQUIPMENT, ETC. — INCREASE	DECREASE	BALANCE	(7) ADVANCES RECEIVED FROM LENDING INSTITUTIONS AND PURCHASER PAYMENTS — JOB NO.	INCREASE	DECREASE	BALANCE	(8) AMOUNTS OWED SUPPLIERS, INCOME TAXES WITHHELD FROM EMPLOYEES, ETC. — INCREASE	DECREASE	BALANCE
1				10595 13			11293 75				6000 00			1371 52
2	44	430 00		11025 13										
3	45	800 00		11825 13										
4	45	500 00		12325 13										
5														
6	44	510 00		12835 13										
7	44	226 67		13061 80				45	2000 00					
8	PR8	843 00		13904 80		65 00	11218 75				8000 00	113 61		1484 93
9	44	99 25		14004 05										
10	44	564 00		14568 05										
11	44	200 00		14768 05										
12	44	450 00		14908 36				44		9000.				
13	PR9	670 00	9903 69	5579 36							3000 00	72 42		1557 35
14														
15														
16														
17														
18														
19														
20														
21														
22														
23														
24														
25														
26														
27														
28														
29														
30														
31														
32														
33														
34														
35														

Form 21-11 [continued]
Columnar Financial Record

LINE NO.	OWNER'S INVESTMENT IN BUSINESS (9)			(10)			(11)			REVENUE AND EXPENSE (12)					
	INCREASE	DECREASE	BALANCE	INCREASE	DECREASE	BALANCE	INCREASE	DECREASE	BALANCE	EXPLANATION	SELLING PRICE OF JOBS SOLD	PURCHASE DISCOUNTS, OTHER REVENUE	COST OF JOBS SOLD	GENERAL EXPENSE	NET INCOME
1			17987 19												8131 19
2															
3															
4															
5										Advertising				30 00	7983 19
6															
7															
8															
9										Indirect Labor				35 00	7948 19
10			19801 69												
11															
12															
13										Close Job no.44	11075 00		9903 69		1979 00
14										Indirect Labor 37.50 + 8				40 00	1979 50
15										Depreciation 27.50 518				65 00	1814 50
16	18 14 50									Transfer to owners invest.					0
17															
18															
19															
20															
21															
22															
23															
24															
25															
26															
27															
28															
29															
30															
31															
32															
33															
34															
35															

Form 21-12
General Expense Record

Line	Date	Fol	Line No.	Explanation	Tools Expense	Superintendence	Insurance	Indirect Labor	Payroll Taxes	Depreciation	Gasoline and Oil	Misc.	Rent	Office Supplies	Telephone and Telegraph	Advertising	Office Salaries	Depreciation	Payroll Taxes	Misc.	Total
1	Apr 5	1	1	Labor - Cleanup				40.00													40.00
2	6	1	2	Audit Fee																100.00	140.00
3	7	1	3	Small Tools	15.00																155.00
4	9	1	4	Gas and Oil							32.65										187.65
5	10	1	5	Telephone + Telegraph											16.32						203.97
6	10	1	6	Dues to Builders Assn																20.00	223.97
7	11	1	7	Office Supplies										13.27							237.24
8	12	1	8	Labor - Storeroom				30.00													267.24
9	17	1	9	Advertising												45.00					312.24
10	19	1	10	Labor Taking Inventory				65.00													377.24
11	20	1	11	Employer's Share					43.00												420.24
12	26	1	12	Labor - Yard Work				10.00													430.24
13	30	2	13	Gas and Oil							8.00										438.24
14	30	2	14	Depreciation						38.75								27.50			504.49
15	May 3	2	15	Labor - Loading Materials				50.00													554.49
16	6	2	16	Cash Service Charge																1.05	555.54
17	7	2	17	Small Tools	10.00																565.54
18	10	2	18	Labor - Storeroom				40.00													605.54
19	12	2	19	Telephone + Telegraph											14.95						620.49
20	12	2	20	Gas and Oil							40.79										661.28
21	14	2	21	Purchase Discount Lost																12.00	673.28
22	14	2	22	Employer's Share					22.21												695.49
23	16	2	23	Insurance																200.00	895.49
24	16	2	24	Office Supplies										15.02							910.51
25	17	2	25	Labor - Materials Yard				10.00													920.51
26	21	3	26	Advertising												30.00					950.51
27	24	3	27	Labor - Cleanup				35.00													985.51
28	31	3	28	Labor - Materials Yard				40.00													1025.51
29	31	3	29	Depreciation						37.50								27.50			1090.51
30			30																		
31			31																		
32			32																		
33			33																		
34			34																		
35			35																		

Mr. Pardo wrote checks to Michigan Bell Telephone Co., $14.95, and to Tappan Oil Co., $40.79. For each of these transactions "cash in bank" was decreased and the amounts were entered in the "general expense" column as increases and the "net income" column was adjusted.

May 14.—It was discovered that an invoice which had been received on May 2 from Jackson Lumber Co. covering materials purchased for stock on April 16 had not been paid. The net amount of this invoice was $588. Since it was not paid by May 12 the purchase discount of $12 was lost. "Cash in bank" was decreased by $600, "materials on hand" was increased by $588, and $12 was entered in the "general expense" column as an increase, and the "net income" column was adjusted.

Since F.I.C.A. and income taxes withheld during April exceeded $100 (income taxes withheld plus employee's share of F.I.C.A. totaled $174.39, and the employer's share of the F.I.C.A. amounted to $22.21), a check for $196.60 was sent to the Federal Reserve Bank in Detroit and a receipt was obtained to attach to the quarterly report to the Internal Revenue Service. "Cash in bank" was decreased by $196.60, "amounts owed suppliers, income taxes withheld from employees, etc.," was reduced by $174.39, and $22.21 was entered in the "general expense" column as an increase and the "net income" column was adjusted.

May 15.—Mr. Pardo paid Bogland Glass Co. $37.90 for glass work on job 44. "Cash in bank" was decreased and "cost of jobs under construction" was increased.

The column balances as of May 15 were:

1.	Cash on hand	$2.00	
2.	Cash in bank	2,639.43	
3.	Notes, mortgages, and other amounts receivable	25.00	
4.	Materials on hand	716.00	
5.	Cost of jobs under construction	8,626.87	
6.	Building, furniture, equipment, etc	11,283.75	
7.	Advances received from lending institutions and purchaser payments		$3,000.00
8.	Amounts owed suppliers, income taxes withheld from employees, etc		1,267.65
9.	Owner's investment in business		17,987.19
12.	Net income		1,038.21
		$23,293.05	$23,293.05

May 16.—Loren Little was paid in full for the lathing and plastering contract on job 44, $400. Mr. Pardo sent a check to Johnson Supply Co. to pay an invoice covering an I-beam purchased for job 45, $60. For each of these transactions, "cash in bank" was decreased and "cost of jobs under construction" was increased.

The General Insurance Agency was paid $200 and Maynard Office Supply Co. was sent a check for $15.02. For each of these transactions "cash in bank" was decreased and the amounts were entered in the "general expense" column as an increase and the "net income" column was adjusted.

May 17.—The payroll for the week ended May 16 indicated gross pay of $824 and that taxes withheld totaled $103.67. Check Numbers 182 through 189 were issued in the total amount of $720.33. Of the gross pay, $10 for work in the materials yard was charged to general expense. "Cost of jobs under construction" and "amounts owed suppliers, income taxes withheld from employees, etc.," were increased; $10 was entered in the general expense" column as an increase and the "net income" column was adjusted; and "cash in bank" was decreased.

May 18.—Mr. Pardo received $3,000 as a draw from Standish Finance Co., job 44. "Cash in bank" was increased and "advances received from lending institutions and purchaser payments" was increased.

May 20.—Mr. Pardo paid John Percy in full for plumbing work on job 44, $430. "Cash in bank" was decreased and "cost of jobs under construction" was increased.

Materials were transferred from stock to job 45, $159.26. "Cost of jobs under construction" was increased and "materials on hand" was decreased.

May 21.—Al Ecklund was paid the remaining amount due on the electrical contract on job 44, $105. Art Pilsner was paid the remaining amount due on the painting contract on job 44, $430. John Percy, the plumbing contractor, was paid $800, and Howard Furnace Co., the heating contractor, was paid $500 for work on job 45. "Cash in bank" was decreased and "cost of jobs under construction" was increased for each of these transactions.

Ann Arbor News was paid $30 for advertising. "Cash in bank" was decreased and $30 was entered in the "general expense" column as an increase and the "net income" column was adjusted.

May 23.—Mr. Pardo paid Clyde Collins in full for the concrete contract on job 44, $510. Howard Furnace Co. was paid in full for heating contract on job 44, $226.67. "Cash in bank" was decreased and "cost of jobs under construction" was increased for each of these transactions.

May 24.—Mr. Pardo received a draw from Standish Finance Co. for job 45, $2,000. "Cash in bank" was increased and "advances received from lending institutions and purchaser payments" was increased.

The payroll for the week ended May 23 indicated that gross pay was $878 and that F.I.C.A. and income taxes withheld totaled $113.61. Check Numbers 198 through 206 were issued in the total amount of $764.39. Of the gross pay, $35 for clean-up work was charged to general expense. ''Cost of jobs under construction'' and ''amounts owed suppliers, income taxes withheld from employees, etc.'' were increased; $35 was entered in the ''general expense'' column as an increase and the ''net income'' column was adjusted; and ''cash in bank'' was decreased.

May 25.—Filbert Floor Covering Co. was paid $99.25 for the contract on job 44. ''Cash in bank'' was decreased and ''cost of jobs under construction'' was increased.

May 28.—Mr. Pardo paid Jackson Lumber Co. $564 for materials on job 44, and Lotz Seed Co. $200 for landscaping contract on job 44. ''Cash in bank'' was decreased and ''cost of jobs under construction'' was increased for each of these transactions.

May 29.—Final settlement was received from the Standish Finance Co. for job 44. The net amount received was $5,030. Forty-five dollars had been withheld by Standish to cover tax and interest, making the gross settlement $5,075. ''Cash in bank'' was increased by $5,030, ''cost of jobs under construction'' was increased by $45 (covering tax and interest), ''advances received from lending institutions and purchaser payments'' was decreased by $6,000, and this latter amount plus the gross settlement of $5,075, or $11,075, was entered under ''selling price of jobs sold'' in classification column 12. ''Costs of jobs under construction'' was decreased by $9,903.69, the total cost indicated on the Job Record, and this cost was entered under ''cost of jobs sold'' in classification column 12. The difference of $1,171.31 was added to $748.19, the amount shown as net income in the ''net income'' subcolumn of classification column 12. The sum of these two amounts, $1,919.50, was then entered in the ''net income'' subcolumn of classification column 12.

May 31.—The payroll for the week ended May 30 indicated that gross pay was $710 and that F.I.C.A. and income taxes withheld totaled $72.42. Check Numbers 210 through 217 were issued in the total amount of $637.58. Of the gross pay $40 for work in the materials yard was charged to general expense. ''Cost of jobs under construction'' and ''amounts owed suppliers, F.I.C.A., income taxes withheld from employees, etc.'' were increased; $40 was entered in the ''general expense'' column as an increase and the ''net income'' column was adjusted; and ''cash in bank'' was decreased.

Depreciation for the month of May was $65. ''Building, furniture, equipment, etc.'' was decreased and ''general expense'' was increased and the ''net income'' column was adjusted. Mr. Pardo entered $37.50 of the depreciation as ''indirect building expense'' and $27.50 as ''selling and administrative expense'' on Form 21-12.

For illustrative purposes, the net income for the months of April and May was transferred from the ''net income'' column in classification column 12 to ''owner's investment in business'' (classification column 9). This transfer (1) left a zero balance in the column ''net income'' and (2) increased the balance showing the ''owner's investment in business'' to $19,801.69. (A net loss would decrease the owner's investment.) This transfer may be accomplished monthly or at the end of each year. If the ''net income'' balance is transferred to the ''owner's investment in business'' column monthly, or at other intervals during the year, the total of such balances transferred will represent net income for the year. In comparing the balances of the columns immediately after transferring the net income figure, the first nine columns only are used, as follows:

1.	Cash on hand		$2.00
2.	Cash in bank		5,977.19
3.	Notes, mortgages, and other amounts receivable		25.00
4.	Materials on hand		556.74
5.	Cost of jobs under construction		5,579.36
6.	Building, furniture, equipment, etc		11,218.75
7.	Advances received from lending institutions and purchaser payments	$2,000.00	
8.	Amounts owed suppliers, income taxes withheld from employees etc		1,557.35
9.	Owner's investment in business		19,801.69
		$23,359.04	$23,359.04

The Job Record

The recording of costs of jobs under construction is one of the most important parts of any accounting system for a home improvement contractor. You must be able to observe the progress of your jobs in relation to the costs accumulated and determine whether costs are ''out of line'' or not. The Job Records, used on the Pardo jobs (Forms 21-13 and 21-14) should be used to provide the builder with job cost information.

Below each column heading on the Job Record is a space provided for the entry of estimated costs for each classification. These amounts should be taken from the Estimate Summary form and entered when the Job Record is set up before any costs are incurred. As material and labor costs are incurred and entered in the ''increase'' subcolumn of column 5, ''costs of jobs under construction'' in the Columnar Financial Record, they should be copied in the correct classification column of the Job Record for each job. By such a procedure the actual costs of individual jobs are accumulated in the Job Records below the estimate of the total costs for each of the classifications. Pencil totals may be inserted at frequent intervals in order to make the comparison with estimated costs in each classification. Column 28 is used to keep a running total of the costs incurred on each job. When a job is completed comparisons of actual and estimated costs may be made on the Job Record below the last line used and the completed form becomes a permanent record of the job.

It is natural for you to think of your total operation in terms of individual projects. This involves both the cost accumulations and the financing arrangements. Columns 29 and 30 of the Job Record should be used to show the advances received from lending institutions, down payments from purchasers, and final settlements upon completion of the job. "Receipts and draws" should be entered first in the Columnar Financial Record as an increase in the "cash in bank" column and an increase in column 7 headed "advances received from lending institutions and purchaser payments," with the job number shown at the left in the space provided. The amount should then be copied in column 29 of the Job Record and column 30 should be increased by the amount received. This procedure makes it possible to have a record in column 31 headed "balance" which represents for each job the excess of costs incurred over receipts and draws. If the costs recorded have all been paid for in cash (if there are no amounts owed to suppliers on a particular job), this balance shows the amount of the builder's own funds tied up in each job. In some cases, particularly during the early stages of the job, the advances received may exceed the costs incurred. The amount in column 31 should then be shown in red or in parentheses. After a job has been closed and after all receipts and draws and the final settlement have been entered, the amount in column 31 represents the margin on that job.

During April Mr. Pardo completed job 43 and started jobs 44 and 45. In May he completed all of job 44. His financial transactions for these 2 months are entered in the Columnar Financial Record shown in Forms 21-9, 21-10 and 21-11. All entries in the sections for "cost of jobs under construction" and "advances received from lending institutions and purchaser payments" are transferred to the Job Records indicated at the left of the amounts. These transfers should be made at frequent intervals (daily if possible) so that the Job Records will be up to date at all times.

When the Columnar Financial Record was put into use on April 1, job 43 was in progress and costs incurred to that date were analyzed into the various cost classifications and entered on the first line of the form directly under the estimated figures. Adding across, the "total cost to date" was found to be $7,136.30. This figure agrees with the beginning amount in the "balance" sub-column of the section for "cost of jobs under construction" on the Columnar Financial Record since job 43 was the only job in progress. If more than one job had been under construction, the total of all the "total cost to date" figures on all the jobs would be equal to the amount shown in the Columnar Financial Record. Prior to April 1, Mr. Pardo had received $4,000 on job 43 from his customer and the lending institution. This amount

agreed with the "balance" subcolumn of the section for "advances received from lending institutions and purchaser payments" on the Columnar Financial Record since job 43 was the only job in progress on April 1. The "balance," column 31, on the Job Record for job 43 showed $3,136.30 as the excess of costs incurred over the amounts drawn.

Final settlement on job 43 was made on April 19. Mr. Pardo received $5,710 in cash after Standish Finance Co. deducted $40 for financing charges. The $40 was copied in the "miscellaneous" column of the Job Record from column 5 of the Columnar Financial Record. This made the Job Record complete as far as the direct costs were concerned. The $5,710 plus $40 or $5,750 was the gross amount of the settlement which, when added to the $7,500 already drawn, agreed with the contract price of $13,250. The $5,750 was, therefore, entered in column 29 of the Job Record and added to the total in column 30. The "balance" column was changed to reflect the difference in columns 28 and 30 or by an amount of $1,733.70. This was the margin on job 43.

The Job Record for job 44 is shown in Form 21-14. The illustrative entries are similar to those of job 43 with the final margin on the job being $1,171.31. After the completion of each job it is helpful to show after the total of each classification column a comparison of actual cost with estimated cost. This has been done for jobs 43 and 44 with the excessive actual costs shown in parentheses. These amounts point out possible areas for question or investigation. On job 43 the actual direct cost exceeded the estimate by $191.30. This is shown at the bottom of column 28. Several of the classifications contributed to this excess but the largest contributor was lumber, millwork, and other carpentry materials. This excess was only $60, however, or less than 3 percent of the estimate and may not be large enough to cause much concern. On the other hand, on job 44 the carpentry labor in classification 6 exceeded the estimate by $180, or 16 percent of the estimate. This should cause the builder to make a study of the cause for such a high cost in this category and either to take the action necessary to prevent its recurrence or revise his estimates on future jobs. The masonry cost on job 44 was $126.02, or more than 9 percent less than the estimate. This fact should give the builder some satisfaction. It should cause him to determine the reason for the difference and to examine the estimate on job 44 and masonry estimates on houses to be remodeled in the future.

Job Progress Reports

The accounting records compiled by Mr. Pardo will be exceptionally useful in examining his whole business operation as well as examining each individual job. The Job Progress Report, Form 21-15 is a useful way of summarizing the status of all jobs under construction.

Form 21-13
Job Record

JOB NO. ___43___

DESIGNATION _____

NAME ___B. L. Dunn___

LOCATION ___11 Lookout Heights___

LINE NO.	DATE	EXPLANATION	FOL.	1 PERMITS, PLANS, SPECIFICATIONS, ETC.	2 EXCAVATION, BACKFILL, & FINISH GRADING	3 MASONRY	4 CONCRETE AND CEMENT FINISHING	5 STRUCTURAL STEEL	6 CARPENTRY LABOR	7 LUMBER, MILL WORK, AND OTHER CARP. MAT'LS	8 PLUMBING	9 ELECTRICAL	10 HEATING & AIR CONDITIONING	11 SHEET METAL	12 ROOFING	13 INSULATING
		ESTIMATED COST (FROM SUMMARY COST ESTIMATE)		50.00	120.00	1,500.00	1,100.00	75.00	1,275.00	2,100.00	1,200.00	225.00	750.00	—	—	—
1	Apr 1	Totals to date	√	47.50	120.00	1,520.50	1,146.70	72.25	898.75	1,260.00	720.00	112.50				
2	2	Jackson Lumber Co	—							900.00						
3	5	Payroll	PR1						277.50							
4	6	Howard Furnace Co (⅔)	—										500.00			
5	9	Dean-Standard Finance Co	—													
6	12	Payroll	PR2						150.00							
7	#	John Little (In full)	—													
8	14	Howard Furnace Co (In full)	—										250.00			
9	14	Al Eklund (In full)	—									112.50				
10	14	Ralph Smith (In full)	—								480.00					
11	16	Order on Standard Fin. & Pay Loan	—													
12	17	Johnson Supply Co (In full)	—													
13	17	Fred Wilson	—													
14	18	City Lbr Company (In full)	—													
15	19	Final Settlement	—													
16																
17		Underline Own Estimate	√	47.50	120.00	1,520.50	1,146.50	72.25	1,906.25	2,160.00	1,200.00	725.00	750.00			
18				2.50	—	(20.50)	(46.20)	2.75	(31.25)	(60.00)	—	—	—			
19																
20–33																

Form 21-13 [continued]
Job Record

JOB NO. __43__

DESIGNATION

NAME _B. L. Brown_
LOCATION _11 Lookout Heights_

LINE NO.	14 LATHING AND PLASTERING	15 PAINTING AND PAPER-HANGING	16 FLOOR COVERING	17 FLOOR FINISHING	18 GLASS AND GLAZING	19 TILE AND MOSAIC	20 BLINDS, APPLIANCES, AND FURNITURE	21 SCREENS, STORM DOORS AND STORM WINDOWS	22 LAND-SCAPING	23 EXTRAS	24 MISCEL-LANEOUS	25	26	27 LAND	28 TOTAL COST TO DATE / EST. TOTAL DIRECT JOB COSTS	29 RECEIPTS AND DRAWS	30 TOTAL RECEIPTS AND DRAWS TO DATE / PROPOSED SELLING PRICE	31 BALANCE
(est.)	450.00	500.00	120.00	—	60.00	350.00	—	—	250.00					1,200.00	11,325.00		13,250.00	
1					58.50									1,200.00	7,136.30	4,000.00	4,000.00	3,136.30
2															8,036.30		4,000.00	4,036.30
3															8,313.80		4,000.00	4,313.80
4															8,813.80		4,000.00	4,813.80
5															8,813.80	3,000.00	7,000.00	1,813.80
6	450.00														8,963.80		7,000.00	1,963.80
7															9,413.80		7,000.00	2,413.80
8															9,663.80		7,000.00	2,663.80
9															9,776.30		7,000.00	2,776.30
10			110.00												10,236.30		7,000.00	3,256.30
11		500.00													10,756.30	500.00	7,500.00	3,256.30
12															10,866.30		7,500.00	3,366.30
13						360.00									11,226.30		7,500.00	3,726.30
14									250.00		40.00				11,476.30		7,500.00	3,976.32
15															11,516.30	5,750.00	13,250.00	(1,733.70)
16																		
17	450.00	500.00	110.00		58.50	360.00			250.00		40.00			1,200.00	11,516.30			
18	—	—	10.00		1.50	(10.00)			—		(40.00)				(119.30)			
19																		
20																		
21																		
22																		
23																		
24																		
25																		
26																		
27																		
28																		
29																		
30																		
31																		
32																		
33																		

Form 21-14
Job Record

JOB NO. ___44___

DESIGNATION _____

NAME _O. L. Underwood_

LOCATION _273 Logan Heights_

LINE NO.	DATE	EXPLANATION	FOL.	1 PERMITS, PLANS, SPECIFICATIONS, ETC.	2 EXCAVATION, BACKFILL, & FINISH GRADING	3 MASONRY	4 CONCRETE AND CEMENT FINISHING	5 STRUCTURAL STEEL	6 CARPENTRY LABOR	7 LUMBER, MILLWORK, AND OTHER CARP. MAT'LS	8 PLUMBING	9 ELECTRICAL	10 HEATING & AIR CONDITIONING	11 SHEET METAL	12 ROOFING	13 INSULATING
		ESTIMATED COST (FROM SUMMARY COST ESTIMATE)		50.00	90.00	1,374.84	965.00	60.00	1,125.00	1,867.36	1,075.00	210.00	680.00	—	—	—
1	Apr 5	Building Permit	1	12.50												
2	5	Copies of Plans & Spec	1	15.00												
3	5	Lot 273 Logan Heights														
4	12	Virgil	PR2		85.00											
5	15	Akins Bros (in full)	1						90.00							
6	19	Payroll	PR3			102.00			212.50							
7	26	Payroll	PR4			193.80			31.00							
8	29	Clyde Collins (½ of contract)	2				482.50									
9	30	Johnson Supply (1 team)	2			275.80	482.50	58.50	61.50							
10	May 3	Payroll	PR5	27.50	95.00	207.00			287.50							
11	6	Wilson from Standish Iron Co	2													
12	7	Jackson Lumber Co (Apr 30 bill)	2							1,345.72						
13	7	Wilson Building Mat'ls (Apr 30 bill)	2			498.62										
14	8	John Percy (60%)	2								645.00					
15	9	Gil Eckland (50%)	2									105.00				
16	9	Payroll	PR6			249.90			195.00							
17	10	Holland Furnace Co (⅓)	2										453.33			
18	15	Logan Glass Co	2													
19	16	Wren Little (in full)	2													
20	17	Payroll	PR7						190.00							
21	18	Wilson from Standish Iron Co	2													
22	20	John Percy (in full)	2								430.00					
23	21	Gil Eckland (in full)	2									105.00				
24	21	Art Planer (in full)	3													
25	23	Clyde Collins (in full)	3				51.00									
26	23	Holland Furnace Co (in full)	3										226.67			
27	24	Payroll	PR8						20.00							
28	25	Gilbert Floor Covering	3													
29	28	Jackson Lumber Co (May 23 bill)	3							564.00						
30	28	Stopfort Company (in full)	3													
31	29	Final Settlement	3	17.50	85.00	1,248.32	492.50	58.50	1,305.00	1,909.22	1,075.00	210.00	680.00			
32		Under (over) Estimate	✓	22.50	5.00	126.02	(27.50)	1.50	(180.00)	(142.36)			—			
33																

Form 21-14 [continued]
Job Record

JOB NO. __44__

DESIGNATION _____

LOCATION _____

NAME: O. L. Underwood
273 Logan Heights

LINE NO.	14 LATHING AND PLASTERING	15 PAINTING AND PAPER-HANGING	16 FLOOR COVERING	17 FLOOR FINISHING	18 GLASS AND GLAZING	19 TILE AND MOSAIC	20 BLINDS, APPLIANCES, AND FURNITURE	21 SCREENS, STORM DOORS, AND STORM WINDOWS	22 LAND-SCAPING	23 EXTRAS	24 MISCEL-LANEOUS	25	26	27 LAND	28 TOTAL COST TO DATE	29 RECEIPTS AND DRAWS	30 TOTAL RECEIPTS AND DRAWS TO DATE	31 BALANCE
(est./proposed)	400.00	430.00	85.00	—	35.00	—	—	—	200.00	—	—			1,100.00	EST. TOTAL DIRECT JOB COSTS 9,746.70		PROPOSED SELLING PRICE 11,075.00	
1															12 50			12 50
2															27 50			27 50
3														1100 00	1127 50			1127 50
4															1217 50			1217 50
5															1302 50			1302 50
6															1617 00			1617 00
7															2120 80			2120 80
8															2603 30			2603 30
9	400 00														2661 80			2661 80
10															3153 30			3153 30
11															3153 30	3000 00	3000 00	153 30
12															4499 02		3000 00	1499 02
13															4997 64		3000 00	1997 64
14															5642 64		3000 00	2674 64
15															5747 64		3000 00	2747 64
16															6192 54		3000 00	3192 54
17															6645 87		3000 00	3645 87
18					37 90										6683 77		3000 00	3683 77
19															7083 77		3000 00	3683 77
20															7273 77		3000 00	4083 77
21															7273 77	3000 00	6000 00	4273 77
22															7703 77		6000 00	1273 77
23															7808 77		6000 00	1703 77
24		430 00													8238 77		6000 00	1808 77
25															8748 77		6000 00	2238 77
26									200 00						8975 44		6000 00	2748 77
27															8995 44		6000 00	2975 44
28			99 25												9094 69		6000 00	2995 44
29											45 00				9658 69		6000 00	3094 69
30															9858 69		6000 00	3658 69
31	400 00													1100 00	9903 69	5075 00	11075 00	3858 69
32	—	430 00	99 25		37 90				200 00		45 00				9903 69			(1713)
33		—	(14 25)		(2 90)				—		(45 00)				(156 99)			

Job Progress Report

Form 21-15

JOB PROGRESS REPORT

BUILDER *Frank Pardo*

MONTH OF *April* 19___

1 JOBS	2 COSTS ACCUMULATED AT BEGINNING OF MONTH	3 COSTS ADDED DURING MONTH	4 TOTAL COSTS TO DATE	5 TOTAL RECEIPTS AND DRAWS TO DATE	6 EXCESS OF COSTS OVER RECEIPTS AND DRAWS	7 ESTIMATED TOTAL COST	8 COST TO COMPLETE - ORIGINAL ESTIMATE	9 PER CENT COMPLETE	10 COST TO COMPLETE - ADJUSTED ESTIMATE
No. 43	$7136.30	$4380.00	$11516.30	Job Closed	Job Closed	Job Closed	Job Closed	100%	
No. 44	—	2661.80	2661.80	None	2661.80	9746.00	7084.20	26%	7580.00
No. 45	—	14.00	14.00	None	14.00	12000.00	11986.00	— —	—
A	5500.00	4000.00	9500.00	10000.00	(500.00)	12000.00	2500.00	70%	4000.00
B	1200.00	5000.00	6200.00	5000.00	1200.00	10000.00	3800.00	65%	3300.00
Totals	13836.30	16055.80	29892.10						
Less Cost of Jobs Completed			11516.30						
Cost of Jobs Under Construction			18375.80						

Form 21-16
Job Progress Report

The Job Progress Report shown in Form 21-16 includes data covering Mr. Pardo's operations for April and, in order to make the illustration more complete, two other jobs not mentioned in the chapter are added.

Data for the Job Progress Report should be compiled as follows:

1. Amounts shown in column 2, "Costs accumulated at the beginning of the month," were taken from the individual Job Records and represent the accumulated costs in the "total cost to date" column for individual jobs as of the beginning of April. The total of these accumulated costs is equal to the amount in the "balance" subcolumn of classification column 5, "cost of jobs under construction" on the Columnar Financial Record as of the beginning of April.

2. Amounts entered in column 3 of the Job Progress Reports, "costs added during month," were taken from the individual Job Records and represent the difference between the "total costs to date" as of the end of April and the beginning of April. The total of "costs added during the month," column 3 of the Job Progress Report, is equal to the total of amounts entered in the "increase" subcolumn of classification column 5, "cost of jobs under construction," on the Columnar Financial Record.

3. Amounts entered in column 4 of the Job Progress Report, "total costs to date," represent the totals of amounts entered in columns 2 and 3 of the Job Progress Report. The total of column 4 less the cost of jobs completed during the month is equal to the "balance" subcolumn of "cost of jobs under construction," classification column 5 on the Columnar Financial Record.

4. Amounts entered in column 5, "total receipts and draws to date," were taken from the individual Job Records. By excluding from column 5 the receipts and draws for jobs closed during the period, the total of this column will be equal to the "balance" subcolumn of classification column 7 on the Columnar Financial Record, "advances received from lending institutions and purchaser payments" at the end of April.

5. Amounts recorded in column 7, "estimated total cost," of the Job Progress Report represent total direct costs as estimated on the Summary Cost Estimate.

6. Amounts entered in column 8, "cost to

complete—original estimate," were determined by subtracting amounts entered in column 4 from the amounts entered in column 7 of the Job Progress Report.

7. In column 9, "percent complete," the percent of completion was recorded. This percentage is only a rough estimate of the physical activity which has taken place on the individual jobs.

8. Amounts entered in column 10, "cost to complete—adjusted estimate," were determined by subtracting from adjusted estimates of total costs for jobs the actual costs which have been incurred.

The purpose of compiling the Job Progress Report is, as has been mentioned, to give you an over-all view of your operation. Although interpretation of some of the columns is obvious, certain portions of the data may require explanation.

Amounts entered in column 6, "excess of cost over receipts and draws," which represent the difference between the amounts entered in columns 4 and 5, should be carefully interpreted. Except for Job A on the Job Progress Report in Form 21-16, this difference represents costs to the builder which have not been reimbursed by lending institutions or covered by purchaser payments. This amount is significant in that it reflects the net investment of the builder in jobs under construction (or will reflect the net investment when amounts owed suppliers are paid). If you enter a great number of invoices on the Columnar Financial Record and if you want to know the net amount of your own cash invested in jobs, deduct the amounts owed suppliers for materials, etc., included in job costs from the balances shown in column 6.

In those instances where the receipts from purchaser payments and from lending institutions exceed costs incurred to date, as is the case for job A in the Job Progress Report, you should not view this excess as profit. Only in those cases where all costs have been incurred on a job will this excess represent gross margin. Far from being a profit if costs are yet to be incurred, this excess may represent in fact your customer's funds which have been used to pay for labor or buy materials that have been applied to other jobs. Where prepayments have been received, you have a liability to your customer until the final settlement has been made.

Although the percent of physical completion listed in column 9 of the Job Progress Report is an estimate, it may be useful as a means of helping you to control your operation. It is indicated that job A on the Job Progress Report is 70 percent complete. Without taking such factors as changes in wage rates, material prices, or the effect of bad weather on job progress into consideration at this point, it appears that the cost estimate needs to be revised from $12,000 to approximately $13,500 ($9,500 ÷ 0.70). If this were the case, then the "estimated cost to complete" would not be $2,500

but approximately $4,000. On the other hand, indications are, at least at the end of April, that job B will not cost $10,000. Such interpretations based on estimates are, of course, only approximations. However, this means of investigating job progress will call to attention cost differences which may otherwise be overlooked.

Financial Statements

An Income Statement (profit and loss statement) and a Statement of Financial Condition (balance sheet) can be compiled easily from information provided by the Columnar Financial Record. Lending institutions sometimes require both of these statements before approving loans. An Income Statement should be included on the personal income tax return of an individual proprietor and both an income statement and a statement of financial condition should be included on the partnership income tax information return. From a managerial point of view, you should always have a current copy of these statements available.

The income statement shows an orderly arrangement of revenue from jobs and other sources, direct costs, expenses and losses, and net income or net loss.

Income statement, (profit and loss) for the months of April and May 19___

FRANK PARDO

Revenues			
Revenues from building activities			$24,325.00
Other revenues			00
Total revenues			$24,325.00
Less cost of jobs sold and expenses:			
Cost of jobs sold			$21,419.99
Indirect building expenses:			
Total expense		$25.00	
Indirect labor		320.00	
Payroll taxes		65.21	
Depreciation		76.25	
Gasoline and oil		81.44	
Total indirect building expenses			567.90
Selling and administrative expenses:			
Office supplies		$28.29	
Telephone and telegraph		31.27	
Advertising		75.00	
Depreciation		55.00	
Miscellaneous:			
Auditing	$100.00		
Dues to builders' association	20.00		
Bank service charges	1.05		
Purchase discounts lost	12.00		
Insurance	200.00		
		333.05	
Total selling and administrative expenses			522.61
Total of cost of jobs sold and expenses			$22,510.50
Net income			$1,814.50

Figure 21-17

Data for preparation of the income statement can be taken from classification column 12 of the Columnar Financial Record. Data for the income statement shown in Figure 21-17, was taken from the Columnar Financial Records used by Mr. Frank Pardo during the months of April and May.

The amount shown as "revenues from building

activities'' under the heading of ''revenues'' in Figure 21-17 represents the sum of amounts entered in the ''selling price of jobs sold'' subcolumn of classification column 12 of the Columnar Financial Records for April and May. If any of Mr. Pardo's transactions had required entry in the ''purchase discounts, other revenue'' subcolumn of classification column 12, the sum of such amounts entered would have been shown as ''other revenues.''

The amount shown as ''cost of jobs sold'' under the heading of ''less cost of jobs sold and expenses'' in Figure 21-17 represents the sum of amounts entered in the ''cost of jobs sold'' subcolumn of classification column 12 for the months of April and May. The ''indirect building expenses'' and ''selling and administrative expenses'' represent a detailed breakdown of amounts entered in the ''general expense'' subcolumn of classification column 12 for the months of April and May. Since Mr. Pardo used Form 21-12, the General Expense Record, the expenses were classified there and the expense column totals were used in compiling the income statement.

The amount of ''net income'' shown in Figure 21-17 is the same as the last amount shown in the ''net income'' subcolumn of classification column 12 at the end of May.

The statement of financial condition shows in an orderly arrangement the items owned, the amounts owed, and the investment of the owner in the business.

The data necessary for the preparation of this statement can be taken from the Columnar Financial Record. The items owned are represented by the ''balance'' subcolumns of the first six classification columns, the amounts owed by the ''balance'' subcolumns of classification columns 7 and 8, and the investment of the owner in the business by the ''balance'' subcolumn of classification 9. Since classification column 12 is not used in the preparation of the statement of financial condition, the amount shown in the ''net income'' subcolumn of classification column 12 should be transferred to the ''balance'' subcolumn of classification column 9. A net income will increase and a net loss will decrease the ''balance'' subcolumn of classification column 9.

Mr. Pardo prepared the statement of financial condition, shown in Figure 21-18, using the ''balances'' in his Columnar Financial Record as of the end of May.

Using Financial Statements

Many builders have not schooled themselves to the significance of financial statements in their businesses. As an operating builder, you very likely tend to concentrate on your income account in seeking ways to increase profits or reduce losses. Nevertheless, a careful review of your balance sheet is a worthwhile related procedure. A knowledge of the distribution of your assets and

Statement of financial condition (balance sheet) May 31, 19___

FRANK PARDO

Items owned (assets)

Cash on hand	$2.00
Cash in bank	5,977.19
Notes, mortgages, and other amounts receivable	25.00
Materials and lots on hand	556.74
Cost of jobs under construction	5,579.36
Building, furniture, equipment, etc	11,218.75
Total of items owned	$23,359.04

Amounts owed (liabilities)

Advances received from lending institutions and purchaser payments	$2,000.00	
Amounts owed suppliers, income taxes withheld from employees, etc	1,557.35	
Total of amounts owed		$3,557.35
Owner's investment in business		$19,801.69

Figure 21-18

liabilities and an appreciation of the typical ratios for the more successful concerns in your business can be of great value. It is particularly so in judging whether the financial structure of your company should be altered or redesigned so as to improve operating performance.

If you don't have an operating statement to guide you, it would be easy to rationalize a bad situation which could be corrected by saying, ''We don't have enough capital,'' or even ''let the creditors carry us for awhile; look at the business we give them.''

All too often, slow payments are an end result of unhealthy underlying conditions that may ultimately endanger your business. You might find, for example, excessive collection periods, poorly thought out expansion programs, or too large investments in fixed assets. In the course of the existence of most construction companies there are peaks in business activity, followed by valleys. In the ensuing fluctuation in prices and sales, some unbalanced concerns are unable to make the adjustment. There is always a chance of trouble developing as the result of some unforeseen event, such as the introduction of new techniques or a shift in style. If liabilities are heavy, real difficulties will certainly be faced.

Brief descriptions of key balance sheet ratios appear below. Average ratios for building contractors and subcontractors are also listed. The examples are taken from the Balance Sheet of ABC Remodelers, Inc., Figure 21-19, and the Income and Expense Statement, Figure 21-20.

1. **Current assets to current liabilities**. Widely known as the ''current ratio,'' this is one test of solvency, measuring the liquid assets available to meet all debts falling due within a year's time. For ABC Remodelers:

$$\frac{\text{Current assets}}{\text{current liabilities}} = \frac{288,122}{163,008} = 1.76 \text{ times}$$

Current assets are those normally expected to flow into cash in the course of one year. Ordinarily these include cash, notes and accounts receivable,

BALANCE SHEET

ABC REMODELERS, INC.

December 31, 19 __

ASSETS

Current assets:
Cash in bank		$15,000.41
Petty cash		50.94
Accounts receivable	$170,000.43	
Less: Reserve for bad debts	9,000.39	161,000.04
Retainage due		18,870.70
Inventory		93,200.72
Total current assets		**$288,122.81**

Fixed assets:
Autos and trucks	$68,875.03	
Buildings and improvements	4,259.67	
Furniture and fixtures	15,740.34	
Shop equipment	9,006.89	
Communications equipment	2,554.46	
Total fixed assets	$100,436.39	
Less: Reserve for depreciation	62,514.08	
Book value of fixed assets		**$37,922.31**

Other assets:
Cash surrender value of life insurance	$8,280.16	
Deferred expenses	3,440.59	
Unexpired insurance	3,800.85	
Investments	24,200.00	
Deposits	530.00	
Total other assets		**$40,251.60**

TOTAL ASSETS: $366,296.72

LIABILITIES AND STOCKHOLDERS' EQUITY

Current liabilities:
Accounts payable	$122,592.45
Payroll taxes	4,914.76
Accrued expenses	10,085.63
*Notes payable to the bank (due this year)	10,495.36
Federal income taxes	14,920.52
Total current liabilities	**$163,008.72**

Long term liabilities:
None

Stockholders' equity:
Common stock	$20,000.00
Retained earnings as of 12/31	150,000.00
Net profit for the 12 months	33,288.00
Total stockholders' equity	**$203,288.00**

Total liabilities and stockholders equity $366,296.72

*NOTES:

The balance of the company's bank loan is scheduled to be paid off in full this year.

Figure 21-19

ABC REMODELERS, INC.

STATEMENT OF INCOME AND EXPENSES AS OF DECEMBER 31, 19__

TOTAL SALES	$1,000,000	100.0

Sales by product & service group
Kitchens	105,000	10.5
Baths	170,000	17.0
Room additions, other	380,000	38.0
Recreation rooms	100,000	10.0
Siding	100,000	10.0
Miscellaneous	145,000	14.5
Total Sales	$1,000,000	100.0

Cost of sales:
Material	354,000	35.4
Labor	321,000	32.1
Subcontract	35,000	3.5
Other direct costs	14,000	1.4
Total Direct Job Costs	$724,000	72.4
Gross Margin	$276,000	27.6

Overhead expenses:
Sales:
Salaries, salesmen	$11,450	1.1
Commission, salesmen	3,650	0.3
Promotion	4,230	0.4
Travel and entertainment	1,237	0.1
Auto expense	3,740	0.3
Sub Total	$24,307	2.4

Truck & Shop:
Truck	$7,240	0.7
Maintenance	1,412	0.1
Tools and supplies	1,935	0.1
Miscellaneous	1,004	0.1
Sub Total	$111,591	1.1

Administrative:
Salaries, managers	$ 49,715	4.9
Salaries, clerks	71,000	7.1
Rent	14,500	1.4
Telephone	3,640	0.3
Insurance	8,653	0.8
Depreciation	3,247	0.3
Payroll taxes	42,059	4.2
Miscellaneous	9,000	0.9
Sub Total	$201,814	20.1
Total Overhead	$237,712	23.7
Net Profit From Operations	$ 38,288	3.8
Provision For Income Tax	$ (5,000)	(0.5)
Net Profit After Taxes	$ 33,288	3.3

Figure 21-20

current retainage, inventory, and the value of work in progress. While some concerns may consider items such as cash-surrender value of life insurance as current, the tendency is to post the latter as noncurrent.

Current liabilities are short term obligations for the payment of cash due on demand or within a year. Such liabilities ordinarily include notes and accounts payable for merchandise, open loans payable, short term bank loans, taxes, and contract advances received but not earned. Other sundry short term obligations, such as maturing equipment obligations and the like, also fall within the category of current liabilities.

Most small building contractors have a "current ratio" of at least 1.5 (current assets 1½ times current liabilities). More solvent firms have about twice as much in current assets as current

debts (a ratio of 2 times). Current assets should be more than 60 percent of total assets and cash should make up about 6 percent of assets. Subcontractors should have a current ratio of 1.5 to 1 and 2.5 to 1 would not be excessive. Current assets should be 60 to 80 percent of total assets and cash should be about 10 percent of total assets for subcontractors. ABC Remodelers, Inc. has a good current ratio of 1.76 times. They have maintained this ratio by keeping accounts payable and inventory relatively low.

2. Current liabilities to tangible net worth. Like the "current ratio," this is another means of evaluating financial condition by comparing what is owned to what is owed. For ABC Remodelers:

$$\frac{\text{Current liabilities}}{\text{tangible net worth}} = \frac{163,008}{203,288} = 80\%$$

Tangible net worth is the total asset value of a business, minus any intangible items in the assets such as goodwill, trademarks, patents, copyrights, leaseholds, treasury stock, organization expenses, or underwriting discounts and expenses. In a corporation, the tangible net worth would consist of the sum of all outstanding capital stock — preferred and common — and surplus, minus intangibles. In a partnership or proprietorship, it could comprise the capital account, or accounts, less the intangibles.

A word about "intangibles". In a going business, these items frequently have a great but undeterminable realizable value. Until these intangibles are actually liquidated by sale, it is difficult to evaluate what they might bring. In some cases, they have no commercial value except to those who hold them — for instance, the item of goodwill. To a profitable business up for sale, the goodwill conceivably could represent the potential earning power over a period of years, and actually bring more than the assets themselves. On the other hand, another business might find itself unable to realize anything at all on goodwill.

Many building contractors have current liabilities which are 140 percent of tangible net worth. A more reasonable figure would be about 80 percent. Contractors who subcontract less of their work and most subcontractors usually have current liabilities at about 70 percent of net worth. ABC Remodelers are well within reasonable limits in the ratio of their current liabilities to tangible net worth.

3. Turnover of tangible net worth. Sometimes called "net revenues to tangible net worth," this ratio shows how actively invested capital is being put to work by indicating its turnover during a period. It helps measure the profitability of the investment. Both overwork and underwork of tangible net worth are considered unhealthy. For ABC Remodelers:

$$\frac{\text{Net sales (year)}}{\text{tangible net worth}} = \frac{1,000,000}{203,288} = 4.9 \text{ times}$$

Turnover of tangible net worth is determined by dividing the average tangible net worth into net revenues for the same periods. The ratio is expressed as the number of times the turnover is obtained within the given period. Builders should "turn" their net worth about 5 to 5½ times a year and subcontractors should average 6 to 6½ times. ABC Remodelers is a little low in turnover of its tangible net worth. Some remodelers turn their net worth 6 or 7 times per year, especially remodelers who maintain only minimum inventory, rent as much of the equipment they need as possible and keep fixed assets as low as possible.

4. Turnover of working capital. Known also as the ratio of "net sales to net working capital," this ratio also measures how actively the working cash in a business is being put to work in terms of sales.

Working capital or cash are assets that can readily be converted into operating funds within a year. It does not include invested capital. A low ratio shows unprofitable use of working capital; a high one, vulnerability to creditors. For ABC Remodelers:

$$\frac{\text{Net sales (year)}}{\text{working capital}} = \frac{\text{net sales (year)}}{\text{current assets-current liabilities}} = \frac{1,000,000}{125,114} = 8.0 \text{ times}$$

Deduct the sum of the current liabilities from the total current assets to get working capital, the business assets which can readily be converted into operating funds. A builder with $100,000 in cash, receivables, and work in progress and no unpaid obligations would have $100,000 in working capital. A business with $200,000 in current assets and $100,000 in current liabilities also would have $100,000 working capital. Obviously, however, items like receivables and retainage cannot usually be liquidated overnight. Hence, most businesses require a margin of current assets over and above current liabilities to provide for stock and work-in-progress, and also to carry ensuing receivables after the work is finished until the receivables are collected. Home improvement contractors should "turn" their working capital about 10 times a year while subcontractors average 6 to 8 times a year. ABC Remodelers shows a fairly low sales volume for the working capital in the company. The company should be able to handle some additional sales without straining the working capital available.

5. Net profits to tangible net worth. As the measure of return on investment, this is increasingly considered one of the best criteria of profitability, often the key measure of management efficiency. Profits after taxes are widely looked upon as the final source of growth. If this return on capital is too low, the capital invested could be better used elsewhere. For ABC Remodelers:

$$\frac{\text{Net profits (after taxes)}}{\text{tangible net worth}} = \frac{33,288}{203,288} = 16.4\%$$

This ratio relates profits actually earned in a given length of time to the average net worth during that time. Profit here means the revenue left over from sales income and allowing for payment of all costs. These include costs of goods sold, writedowns and chargeoffs, Federal and other taxes accruing over the period covered, and whatever miscellaneous adjustments may be necessary to reduce assets to current, going values. The ratio is determined by dividing tangible net worth at a given period into net profits for a given period. The ratio is expressed as a percentage.

Home improvement contractors should show a profit after taxes of at least 12 percent of net worth. A 20 percent profit is a reasonable expectation and the top 10 percent of the industry shows about a 40 percent profit on net worth. Subcontractors should show a profit of at least 15 percent of net worth. However, many types of subcontracting operations seem to fall close to or below 15 percent while others reach 35 percent or more. Small concrete and electrical contractors have averaged above 20 percent but contractors in the mechanical trades have averaged less than a 15 percent profit on net worth in many years. ABC Remodelers had a good year and made back 16.4 percent of their net worth. A good goal for management during the following year would be to increase that 16 percent closer to 20 percent.

6. Fixed assets to tangible net worth. This ratio, which shows the relationship between investment in land, office and equipment and the owner's capital, indicates how liquid is net worth. The higher this ratio, the less the owner's capital is available for use as working capital or to meet debts. For ABC Remodelers:

$$\frac{\text{Fixed assets}}{\text{tangible net worth}} = \frac{37,922}{203,288} = 18.7\%$$

Fixed assets means the sum of assets such as land, buildings, leasehold improvements, fixtures, furniture, machinery, tools, and equipment, less depreciation. The ratio is obtained by dividing the depreciated fixed assets by the tangible net worth. Builders should not have more than 25 percent of their tangible net worth invested in land, office, equipment, tools, and other fixed assets. Heavy construction contractors that need heavy equipment on a daily basis may often have 50 percent of their net worth invested in fixed assets. Subcontractors who do not need expensive equipment should have no more than 25 percent of their tangible net worth invested in fixed assets. ABC Remodelers has been successful in keeping fixed assets a small part of tangible net worth. You can see the advantage of having only $4,259.67 less depreciation invested in "Buildings and Improvements." If ABC owned its own building or had made extensive improvements in its leased building, fixed assets would be way above the 25 percent figure. Most remodelers agree that their resources should be held in working capital rather than tied up in land and buildings.

7. Total debt to tangible net worth. This ratio also measures "what is owed to what is owned". As this figure approaches 100 percent, the creditors' interest in the business assets approaches the owner's. For ABC Remodelers:

$$\frac{\text{Total debt}}{\text{tangible net worth}} = \frac{\text{current debt + fixed debt}}{\text{tangible net worth}} = \frac{163,008}{203,288} = 80\%$$

Total debt is the sum of all obligations owed by the company such as accounts and notes payable and mortages payable. The ratio is obtained by dividing the total of these debts by tangible net worth. Speculative builders often have a debt to worth ratio of 300 to 400 percent. Most general contractors have a debt to worth ratio of about 150 percent. Subcontractors should average 100 percent to 140 percent for most types of work. ABC Remodelers has no long term or fixed debt so the 80 percent figure is the same as the "current ratio" under paragraph 2.

8. Net profit on net sales. This ratio measures the rate of return on net sales. The resultant percentage indicates the number of cents of each sales dollar remaining after considering all income statement items and excluding income taxes. A slight variation of the above occurs when net operating profit is divided by net sales. This ratio reveals the profitableness of sales—i.e., the profitableness of the regular operations of a business.

Many builders think a high rate of return on net sales is necessary for successful operation. This view is not always sound. To evaluate properly the significance of the ratio, consideration should be given to such factors as the value of sales and the total capital employed. A low rate of return compared with rapid turnover and large sales volume, for example, may result in satisfactory earnings. For ABC Remodelers:

$$\frac{\text{Net profits}}{\text{net sales}} = \frac{33,288}{1,000,000} = 3.3\%$$

Most contractors and subcontractors show a net profit on sales of between 1 percent and 2 percent. An exceptionally profitable operation would show about a 4 percent profit. These figures may be misleading because "profit" in a small company often disappears into the owner's pocket before the final figures are prepared for each year. ABC Remodelers had a good profit per sales dollar.

You should have an operating statement (income and expenses) and a balance sheet for your company and make an analysis of the key elements on those documents every month. The history of your company is being written each month in these figures. To understand the strengths and weaknesses of your company, you have to be able to read and understand these figures. Compare each month with the previous months, the same month in previous years, and the standards in this chapter. Try to improve your performance in key areas where you do not measure up to standard. Set goals for each year and handle your sales, estimating, production and investment so that you meet these goals. Your analysis will reveal areas where specific actions will help overall company performance. Take the action needed and watch your profits and financial soundness grow. If you watch your financial ratios carefully you probably won't spend as much on fancy, expensive (and unnecessary) automobiles, office furnishings, and copy machines as some other builders. Your overhead will probably be lower and you won't own a big, expensive office and showroom. In fact, less informed competition may consider your operation "small time" or marginal. But you may rest assured that your profit and the salary you pay yourself will be consistently higher than most of the "fast and loose" operators. More important, you will have the controls and the resources to operate profitably long after a business slowdown has driven others to insolvency.

Capital Requirements

It has been pointed out that working capital (the excess of current assets over liabilities) is the lifeblood of your company. Your own experience will probably confirm that there are three types of home improvement companies: the marginal, the limited, and the growing company. Quite often, the amount of capital a company has determines which of these three classifications it falls into.

Lack of working capital has caused many firms to fail. In some cases, insufficient capital has kept small building businesses from growing. But even so, many men have started their own home improvement businesses with a minimum of capital, and have succeeded.

You may find that certain equipment can help to cut your labor costs. However, such equipment represents working capital. If you buy too much of it too soon you may be tying up dollars you will need later — perhaps to tide your company over a period of slow sales. Sometimes it may be better to rent equipment. Or it may be cheaper to subcontract work that requires expensive equipment. In other instances, you may be able to lease equipment with an option to buy after you have used it a certain length of time. Such an arrangement allows you to spread equipment costs out without tying up a large amount of capital.

Home improvement contracting requires the smallest amount of working capital per dollar of sales of any construction business. This is because contract jobs are generally small, and the customer pays shortly after the work is finished or while the work is being done.

If you are operating a modernization company with **minimum capital**, you might try to finish jobs more quickly to increase the turnover of your working capital. In time, though, this could lead to trouble. Rushing through a job so you can get paid for it could result in sloppy workmanship. Then too some customers might resent your demands for immediate payment. In trying to get new work, you might use high pressure salesmanship with more emphasis on price than on performance. Here you will be damaging your company's reputation. In fact, with such practices your chances of going broke would be good.

However, if you have some margin over the minimum amount of capital, you can run a better home improvement company. You will be able to supervise the work so as to help cut costs while giving your customers a quality job. You will buy more advertising as a supplement to the word-of-mouth advertising you get from satisfied customers.

Adequate capital will allow you to
(1) Emphasize quality work and build a reputation for reliability
(2) Carry on more extensive advertising and merchandising
(3) Prepare detailed estimates which should help control costs
(4) Possibly offer financing to customers (You will probably sell their notes to financial institutions).

Growing With Working Capital

Notice that the rate at which you accumulate working capital probably determines your rate of growth. Very often, the money you make as a result of operations (the profit left in the company after taxes are paid) will not be enough to finance the growth you want. Refer to Figure 21-20. ABC Remodelers made $33,288 on sales of $1 million. Assume management projected sales of $1.3 million for the new year, a reasonable goal for the company. How much working capital is needed for additional material, labor and other costs to carry the addition $300 thousand in sales? Check the percentage of direct costs, 72.4 percent on the income statement. Assuming costs for the additional $300,000 in net sales would remain the same, ABC will have to find $217,210 to finance the new business (72.4 percent x $300,000). For every additional dollar of sales, ABC needs 72.4 cents for added direct costs. Of course, this money is part of sales dollars. The problem is, how long does it take to collect these dollars? If it takes 60 days, they are short $49,314.60 in additional working capital. $300,000 divided by 365 days = $821.91 per day. $821.91 x 60 days = $49,314.60.

Faced with this problem, most home improvement contractors turn to suppliers for additional or extended terms. Accounts payable start to rise and payment dates must be stretched out. The result: an otherwise healthy company may begin trying to operate on inadequate capital, and everyone they do business with probably senses this. It is far better to grow within the limits of your working capital and let your analysis of the balance sheet tell you how much growth is reasonable.

Reports to Governmental Units

Builders are required to make periodic reports and payments to Federal, State, and in some cases, city governing units for unemployment compensation, Federal Insurance Contributions Act (F.I.C.A.), Medicare, income taxes, and worker's compensation insurance. You must be familiar with the exact requirements of Federal, State, and city regulations in order to set up procedures to facilitate the preparation of necessary reports and the making of proper periodic payments.

Most builders are subject to three payroll taxes, namely:

1. Approximately twelve percent of about the first $16,000 of taxable wages paid each employee during each calendar year for F.I.C.A. and Medicare. One-half of this amount of the gross wage is paid by the employer and the other half is paid by deduction from the employee's wages.

2. A small percentage of the first several thousand dollars of taxable wages paid to each employee during the calendar year for Federal unemployment tax (F.U.T.A.). This tax is paid by the employer alone; however, a credit is allowed for contributions to State unemployment funds (including credits under merit rating plans). Most builders will pay a net Federal unemployment tax of only six tenths of 1 percent.

3. Federal income tax withholding which is deducted from the employee's pay. The amount depends on the number of exemptions claimed by the employee and the amount of wages earned.

A summary of the laws and regulations governing the withholding, deposit, payment, and reporting of these taxes is found in the Employer's Tax Guide which is issued by the Internal Revenue Service and is revised periodically to provide the employer with up-to-date information. Every builder should have a copy of this booklet. It includes precise instructions and statements of employers' responsibilities and a complete set of income tax withholding tables.

On page 2 of the Employer's Tax Guide is a calendar of employer's duties. The following paragraphs in this section include partial excerpts from the "guide" with some additional comments on the procedures involved.

Upon hiring new employees you should require that each employee submit a completed Withholding Exemption Certificate, Form W-4. It is necessary that you keep in your files a copy of the W-4 form for each employee. Each form should be signed by the employee and should show the name and address of the employee, his social security number, and the number of tax exemptions claimed by the employee. These forms provide data that you must use to determine the amount of income tax withholding from each employee's wages and that you will need to prepare the quarterly reports of F.I.C.A. (Federal Insurance Contributions Act) taxes, Medicare taxes, and F.U.T.A. (Federal Unemployment Tax Act) taxes.

From each payment of wages to an employee you must withhold an amount for income taxes in accordance with the Employee's Withholding Exemption Certificate and the appropriate withholding rate as shown in the tables included in the Employer's Tax Guide (Circular E).

By the 15th day of each month other than the first month of a quarter, you must deposit income tax withheld and the employees' and employer's F.I.C.A. taxes for the previous month in a Federal Reserve or other authorized bank. Taxes for the third month of any quarter may be deposited by the 15th of the succeeding month or paid with the quarterly return, Form 941, which is mentioned below.

On or before each April 30, July 31, October 31, and January 31 you must file a report on the Employer's Quarterly Federal Tax Return, Form 941, with the Internal Revenue Service and pay the total amount of taxes due for the previous quarter for the total of the income taxes withheld from wages and employee's and employer's F.I.C.A. taxes. If you have made monthly deposits, you should deduct the amounts deposited from the total tax to determine the amount to be paid with the quarterly return.

When an employee's withholding exemption changes, he should be requested to file a new certificate, Form W-4, so that future withholdings may be in accordance with his changed status.

On or before each January 31 and at the termination of employment, each employee should be given a Withholding Statement, Form W-2, in triplicate showing (1) the total wages subject to income tax withholding and the total amount of income tax withheld and (2) the amount of wages subject to F.I.C.A. taxes and the amount of such taxes withheld from the employees. You should retain the last copy, Form W-2A, which must be submitted with the annual report, Form W-3.

On or before February 28 of each year you should file a Transmittal of Income and Tax Statements, Form W-3, together with the collector's copies (Form W-2A, mentioned above) of all withholding statements furnished employees for the preceding calendar year. You must also file by January 31 Form 940, showing the total wages subject to F.U.T.A. tax, the credits for contributions paid into State funds and for merit rating credits, and the net amount owed. The full amount owed must be paid on or before January 31. If you deposited the full amount of tax when due, you are permitted an additional 10 days to file the return.

The Internal Revenue Service will automatically send to you Federal Tax Deposit Forms 501 (for income and social security taxes) and 508 (unemployment taxes) and blank Forms 940 and 941 once you apply for an employer identification number. Your bank will probably handle all these Federal tax deposits and returns for no additional charge if you use their payroll service (usually at a cost of about $15 per month plus a small fee per employee).

Requirements of State Governing Units

All States have enacted legislation providing for unemployment compensation funds to which employers must contribute in accordance with merit rating plans of assessment. A few States require employers to withhold from the employees' pay and make payments to State unemployment insurance funds. For the most part, these regulations require payments of varying percentages of the first several thousand dollars earned by each employee during any calendar year. Special forms for the filing of such information to accompany payments of these amounts may be obtained from the State commissions whose offices are usually found in major cities and the State capitals. Some States have enacted withholding legislation to accompany the State income tax regulations and in such cases builders are required to submit periodic reports. Some States have also enacted worker's compensation insurance funds to which employers may be required to make periodic reports and payments.

Requirements of City Governing Units

Certain cities have enacted legislation providing for the withholding from employees' wages for city income taxes. Builders should investigate their particular situation and see if any such procedures and reports are required.

Income Taxes

Since a proprietorship's net income is in the form of business profits rather than payments from which your income tax may be withheld, a self-employed builder must file a Declaration of Estimated Tax, Form 1040-ES, on or before March 15 if the calendar year is used or on or before the 15th of the third month of the builder's fiscal year if other than the calendar year. This report provides a basis for paying currently (pay as you go) any tax liability being incurred. You must pay with your declaration at least one-fourth of the total estimated tax on the date of filing and equal payments on or before June 15, September 15, and January 15, or corresponding months if a fiscal year is being used. If, after filing a declaration, you find that your estimated tax is in error by a substantial amount, you should file a new declaration, marking it "Amended." This should be done on or before one of the payment dates mentioned above which follows the recognition of a need for a change in the estimated tax. Subsequent payments of the tax should be amended accordingly with the total remaining tax divided equally among the remaining payment dates.

You must file a Form 1040, Individual Income Tax Return, by the 15th of the fourth month following the close of the fiscal or calendar year and make payment of the excess of the tax, if any, over the payments made on the estimated tax. If you are operating as a partnership, your share of the earnings of the partnership should be entered in Schedule E on Form 1040, and no detailed analysis of the amount is required since the partnership information return will provide the necessary detail. If you are the owner of a corporation engaged in building, your personal income from the activity will be the amount of salary and dividends received during the taxable year and the detail of the results of the building operation will be reported only on a corporation tax return. If you operate as an individual enterprise, the detailed results of operations should be shown in Schedule C on Form 1040 or on a separate sheet attached to the form. Figure 21-17 shows how the income and expense items could be listed on Schedule C.

If your operation is being carried on as a partnership, a Partnership Tax Return, Form 1065, must be filed by the 15th day of the third month following the close of the taxable year. This return is designed for information purposes only. Space is provided in the return for the detailed results of operations. It may, however, be preferable to attach a statement similar to Figure 21-17. Such an arrangement and computation with schedules for any items shown on Form 1065 is satisfactory. On Schedule H the partnership should show balance sheets (statements of financial condition) as of the beginning and end of the taxable year being reported. These may be filled in using the "balance" columns shown on the Columnar Financial Record, or statements similar to the one shown in Figures 21-17 and 21-18 may be attached to the return. No payment is made with the partnership return, but in Schedule I of the return a division of the partnership net income must be indicated and these amounts must agree with those shown in the "income from partnership" schedule of Form 1040 for each partner. The partners must pay income taxes on their respective shares of the net income without regard to the amounts they have withdrawn. Salaries to partners must not be considered as an expense in determining net income. The year in which each partner must report his share of the partnership net income is the year in which the partnership year ended.

If your building operation is being carried on by a corporation, a Corporation Income Tax Return, Form 1120, must be filed on or about the 15th of the third month following the close of the taxable year. The corporation tax must be paid at the date of filing the return or in two equal installments by the 15th of the third month and the sixth month following the close of the taxable year. Instructions for the preparation of Form 1120 are published separately by the Government Printing Office and may be obtained from any office of the Internal Revenue Service. Every corporation also must estimate the taxes due for each year and make deposits (Form 503) toward the estimated tax by the 15th of the fourth, sixth, ninth and twelfth month of that year.

Chapter 22

Production Control

Most home improvement contractors have worked on home improvement jobs enough to know the importance of production control. The construction process should flow smoothly without delay, each trade or part of the job being completed just before the next stage is scheduled to begin. Unfortunately, it doesn't always work this way. Delays, avoidable or unavoidable, may drag the job out while your client suffers the inconvenience of an uninhabitable home and your labor costs go well beyond estimates. Fortunately, though remodeling a home can be a fairly complex process, a little planning and supervision can avoid most major production problems. Your aim should be to balance the advantages of prompt completion against the need to equalize work flow, to avoid wasted labor and materials and to ensure sound construction.

Scheduling

Scheduling production is the central step in coordinating all the varied elements of your enterprise. The process of coordination extends back to getting enough contracts so that all your resources will be occupied all the time, but avoiding over-commitment that would stain your staff, crews and financial resources. It includes seeing that you have the necessary labor, crews, materials, and subcontractors lined up and that they are on the right site at the right time. It extends forward to completing each contract on time and having new ones to take its place.

Coordination and scheduling will be among your most important overall responsibilities as a manager. As your work increases, you will not be able to keep all the work in progress in your head and you will begin using a calendar or checklist to remind you of critical dates. As your work becomes more complex and timing becomes important, you should begin laying out work schedules and charting progress.

Principles of Scheduling

The development of a schedule is governed by four principles which remain inflexible even though the type of schedule is changed.

Scheduled operations cannot exceed the capability to accomplish the work. For example, if only two carpenters are available, the maximum possible number which can be scheduled at any given time cannot be greater than two.

Scheduled operations must follow the sequence of work required for the particular job. For instance, a roof cannot be put on a room addition before the walls are up.

These limitations may appear so obvious as to be not worth mentioning, yet the most common errors in scheduling involve violations of these two principles. If a schedule is made which does not violate either of these principles it will be a workable schedule, but it will not necessarily be a good schedule. Two other principles must be observed to develop a good schedule.

Critical items must be scheduled as soon as possible. Maximum speed of completion of a project is important when home interiors are exposed to the weather or a kitchen is unusable. This is usually achieved by scheduling those items which take longest, or upon which many other operations are based, to begin as early as possible.

Scheduling must ensure continuity of work effort. Maximum efficiency in accomplishing a particular work item is best achieved if the time for accomplishing that item is as continuous as possible. For example, a carpenter framing an addition should not be required to install cabinets or siding or work on some other job until the framing is completed. To do so would reduce the output, and require more supervision. These last two principles are often in opposition to one another and a balance must be made between them to obtain the best schedule.

Work schedules can be prepared for your company as a whole and for each project you have. Manpower and equipment schedules are normally prepared at the same time because the information they contain is required for preparation of the work schedules.

In scheduling a job, the first procedure is to list the work elements. Next, determine the construction sequences; obviously, excavating must come before foundation placement, wall construction before the installation of finish door and window frames, and so on. Prepare a list of the work elements in the proper order for the work you do most frequently. This list is your work schedule. The starting date for each job is, of course, the starting date for the work element which is first in construction sequence. A work schedule for a large remodeling and room addition job might look like Figure 22-1.

The time required for each work element is determined by dividing the estimated man-days required by the number of men expected to be assigned to constructing that element. Each work element is scheduled in its proper construction

Address _____

Promised Completion Date _____

	Type of Work	Working Day	
		Start	Finish
1.	Layout and leveling	1	1
2.	Foundation forms	1	2
3.	Foundation pour	3	3
4.	Rough grading	4	4
5.	Exterior walls	5	5
6.	Decking, paper	6	6
7.	Roofing and concrete	6	10
8.	Plumbing, rough	7	8
9.	Heating, rough	8	10
10.	Wiring, rough	8	10
11.	Exterior paint prime	10	10
12.	Slab pour	11	11
13.	Brickwork	11	14
14.	Grade garage floor and porch	12	12
15.	Interior walls	13	13
16.	Garage floor and porch	13	13
17.	Heating, rough	14	15
18.	Plumbing, rough	14	15
19.	Electrical, rough	14	15
20.	Electrical inspection	15	16
21.	Garage door	15	15
22.	Wall insulation	17	17
23.	Septic tank and laterals	17	20
24.	Guttering	18	18
25.	Drywall board	18	18
26.	Concrete flatwork	18	21
27.	Tile rough-in	19	19
28.	Drywall tape	19	19
29.	Drywall 2	20	20
30.	Drywall 3	21	21
31.	Ceiling insulation	21	21
32.	Drywall 4	22	22
33.	Clean brick and stucco	22	22
34.	Furnace	23	23
35.	Exterior paint 2	23	23
36.	Light fixtures, finish wiring	23	25
37.	Tile finish	23	25
38.	Carpentry trim	23	26
39.	Finish grading, landscaping	23	27
40.	Drywall sand	26	26
41.	Plumbing finish	27	28
42.	Painting	27	29
43.	Exterior paint finish	28	28
44.	Paperhanging	29	30
45.	Hardwood floors	31	32
46.	Light fixture glass	32	32
47.	Floors resilient	32	32
48.	Shower enclosure, tile cleaning	32	32
49.	Cleaning	32	33
50.	Hardware and shoe	33	33

Work schedule
Figure 22-1

sequence, showing starting and completion dates. Often, of course, it is not wise or even possible to wait until one element is finished before starting another. For example, roughing in electrical and plumbing fixtures may occur on the same day, and perhaps at the same time.

Work finished should be recorded daily. In some types of work, it is more convenient to report work quantities as portions are completed, rather than to attempt to report partial completion of portions. Major problems affecting progress should be described, and any unusual construction met-

hods should be reported in detail, with sketches included if necessary. If progress is behind schedule, the report should describe what measures are being taken to bring it back on schedule, or explain why the promised completion date cannot be met and what additional time is needed for completion. Your work schedule for each job should be used when scheduling and estimating similar jobs in the future. File the completed work schedule with your estimate for the job.

The Critical Path Method

In recent years, a network analysis system of project planning, scheduling, and control, called the Critical Path Method (CPM), has come into existence and widespead use in the construction industry. The object of CPM is to combine all the information relevant to the planning and scheduling of project functions into a single master plan—a plan that coordinates all of the many different efforts required to accomplish a single objective, that shows the interrelationships of all of these efforts, that shows which efforts are critical to timely completion, and that promotes the most efficient use of equipment and manpower.

There is a lot to know about CPM and you may not find all the information you need on the subject in this handbook. Sufficient information is provided, however, to assist you in preparing arrow diagrams, interpreting critical path schedules drawn up for jobs under your control, and in developing critical path schedules for future construction projects.

Arrow Diagrams

An arrow diagram must be drawn to identify, in a graphic form, the individual items of work, services, or tasks—referred to in CPM as activities—that are involved in the project. Also, of equal importance, the arrow diagram must show how each activity depends upon others during the sequence of construction. The arrow diagram graphically describes the sequence of activities as well as the interrelationship of activities within the project.

In an arrow diagram, both arrows and circles are used to describe the sequence of work. An arrow represents an activity, and a circle represents an event. An event is the starting point of an activity and occurs only when all the activities preceding it (which means all the arrows leading to the circle) have been completed.

In Figure 22-2, the starting point for the arrow marked activity is the occurrence of event number (2). Event number (4) does not occur until the work represented by the arrow (2) to (4) and the work represented by the arrow from (3) to (4) has been completed—and this means entirely completed. This means, then, that the work represented by the activity from (4) to (6) cannot start until (2) - (4) and (3) - (4) have been finished. If this does not

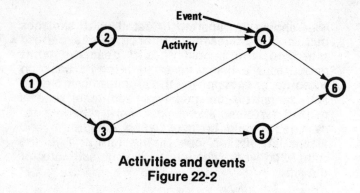

Activities and events
Figure 22-2

accurately describe the situation, the arrow diagram must be redrawn.

Because everything that happens in CPM is based on dependency situations (i.e., one activity dependent upon others), the arrow diagram must be a meaningful description of the project. If it is not, less than satisfactory results will be obtained from CPM. In almost every case of difficulties or dissatisfaction with CPM, the cause can be traced to a faulty or unrealistic arrow diagram. Everything in an arrow diagram has significant meaning, and for this reason the basic principles must be understood and applied completely.

Principle Number 1

The first principle of arrow diagram development that must be understood is that everything in the diagram has meaning. Within this principle, the following rules must be learned and applied.

1. Every arrow represents an item of work and is referred to as an activity.

2. An event is the starting point of an activity, shown as a circle.

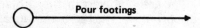

3. An activity depends upon and cannot begin until the completion of all preceding activities.

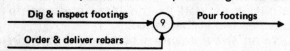

As illustrated above, **pour footings** depends upon the completion of **dig & inspect footings** and **order & deliver rebars**.

4. All activities that start with the same event depend upon and cannot begin until the completion of all activities that enter that event.

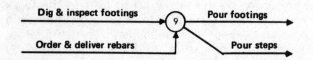

In the case illustrated here, **pour footings** and **pour steps** depend upon the completion of the two activities that enter their common starting event. In other words, it is impossible to **pour footings** or to **pour steps** until **dig & inspect footings**, and **order & deliver rebars** have both been completed. The diagram indicates that all the footings (not just some) must be dug and inspected, and all necessary rebars must be on hand, before either of the two activities starting with event (9) can begin.

Principle Number 2

A second principle is that an activity has a single definite starting point and a single definite ending point. Placing an arrow in a diagram must satisfy two basic questions:

1. "What activities must be completed before this one can start?" This indicates the event from which to start the activity.

2. "What activities cannot be started if this one is not completed?" This indicates into which event the activity should enter.

Suppose, for example, Figure 22-3 had been drawn. As shown, excavating for the footings and pads for the air conditioner pad is the first activity, followed by placing the concrete footings and pads. At event (3) and as a result of the completion of the previous activities, several independent work items then can commence.

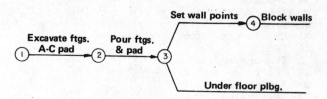

Diagram development
Figure 22-3

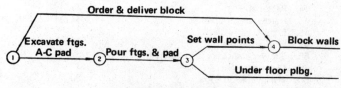

Adding an activity
Figure 22-4

Suppose, however, that it is desired to add an activity to indicate delivery of concrete block for a sight screen enclosure wall around the pad. The first question asked about this new activity should be, "What must be finished before the block can be ordered and delivered?" Actually, there is nothing in the diagram that, if not accomplished, would hold up the ordering and delivery of block. The starting point for this activity would then be event (1).

The second question to be asked about the new activity is, "What cannot proceed until this activity is completed?'' The answer is, of course, block walls. The termination point for this new activity then is event (4) and the results of the analysis described above would appear as illustrated in Figure 22-4.

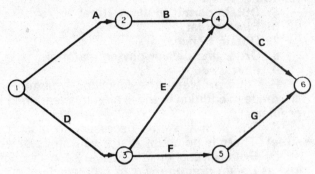

Dependency relationships
Figure 22-5

Principle Number 3

A third principle is that the arrow diagram does not describe time relationships, but rather dependency relationships. Generally, the arrow diagram is not drawn on a time scale. That is, the length and direction of an activity arrow has no relationship to the amount of time required to accomplish the work represented by it. Likewise, two activities starting with the same event do not necessarily occur at the same time. In Figure 22-5, the only thing that is actually known about activity A & D is that they are independent. They may or may not go on at the same time. The time that an activity takes place is decided in the scheduling phase, not by the arrow diagram. The arrow diagram merely defines the dependency situations that exist. In the illustration involving the concrete block for the air conditioner pad, for example, the activity order & delivery block starts with event (1), as does excavate footings & pad, etc. This does not mean that both activities must be conducted at the same time. They might, but probably won't. The only thing indicated is that these two activities are independent.

Principle Number 4

Because the schedule produced from the arrow diagram must be followed by several people, anyone who has an important role in the job should be consulted when creating the arrow diagram. Crew leaders should be asked to review the arrow diagram carefully to make certain that the activities relating to their work are accurately described.

Other Rules and Conventions

Notice in the illustrations thus far that the events are all numbered. This makes it possible to identify an activity and its position in the diagram. An activity is identified by using the event number at its tail and the event number at its head. In the

air conditioner pad illustration, pour footings & pad could be referred to as activity (2) - (3) and identified in that fashion. Each activity should have its own arrow. To observe this rule, it is sometimes necessary to use a "connector" type of activity that doesn't really represent work, but merely helps to observe the rule. This type of activity is usually referred to as a "dummy". This special activity is drawn as a dotted line and indicates that no work is involved in that activity.

Figure 22-6 illustrates this rule and the use of the dotted-line (dummy) activity. The dotted-line activity represents the dummy and involves no duration and no cost. It serves only as a dependency connector or sequence indicator.

Dummies have purposes in addition to the one mentioned above, and will be discussed in more detail later.

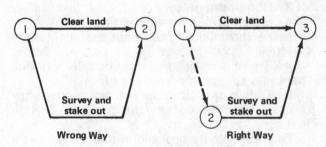

Activity identification
Figure 22-6

When you produce your schedule, the numbering of the events must be watched closely. Extreme care must be taken to make sure of two things:

1. Every number must be used, and
2. The number at the tail of any arrow must be less than the number at the head of that arrow.

It is always wise to delay numbering the diagram until it has been completed.

Another rule governing the creation of the arrow diagram is that each job has only one starting event and only one ending event. When nothing must be done prior to the start of an activity, the arrow representing that activity starts with the project's starting event. When nothing depends upon the accomplishment of an activity, its arrow ends with the project's ending event. You always know, because of this rule, where an activity belongs in the network.

It must be kept firmly in mind that restricting the arrow diagram to one starting event and one ending event, does not at all limit the number of starting or ending activities.

For a fairly complex job, your arrow diagram is the single most important planning document. It will help you spot potential bottlenecks before they develop. To make a good arrow diagram you must have a good understanding of the work you do and what your crews can accomplish.

Sample Problems

You will find that, with a little practice, you can make simple arrow diagrams in a few seconds to solve most scheduling problems. The problems that follow will give you practice working with C.P.M. scheduling.

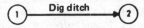

Problem 1

To illustrate the arrow diagram with an actual problem, suppose you want to install a drainage ditch in the backyard of a house. Assume for this problem that only one man was going to work on this project. Draw a diagram for this one-man project including only the following activities:

1. **Dig ditch**
2. **Place gravel in bottom of ditch**
3. **Obtain draintile**
4. **Install draintile and backfill**

Make the following assumptions about these activities:

A. There is no need to order or deliver gravel. There is a supply in the driveway.

B. **Obtain draintile** does not require the effort of the one man doing this project. It is a "delivery" type activity.

The first activity possible in this project is **dig ditch**-so an activity indicating this would be drawn.

After the ditch had been dug, the placing of the gravel would be next and the diagram would describe this as follows:

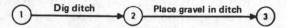

After the gravel is in, the draintile could be installed. However, in order for this to happen, the draintile must have been obtained. This means that the activity **obtain draintile** must end at the beginning of **install draintile and backfill**. This will be at event (3) and the diagram would look like this:

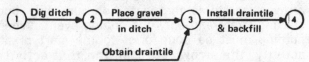

The question remaining is where does **obtain draintile** start? One must further ask what activities must be accomplished before this one can start. It turns out that nothing need to be done in order to start this activity and it thus starts with event (1). The completed solution is below.

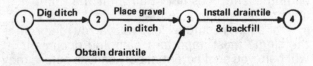

Problem 2

To create a slightly more complicated problem,

consider the situation that would exist if the work described in problem 1 had been performed in the street in front of the house rather than in the backyard.

If this were the case, it would be necessary to add to the list of activities given for 1, the following:

5. **Obtain permit to block street**
6. **Block street**
7. **Obtain trencher**
8. **Order and deliver paving material**
9. **Pave street**

For this problem, the following assumptions are made in addition to those already mentioned in problem 1:

C. The permit is only necessary to block the street. There is no doubt that it will be obtained.

D. **Obtain trencher** is a delivery type activity just as **obtain draintile** was in problem 1.

The solution to this problem would start with the first activity that can take place. In this problem, four independent activities can be shown as starting the project. (See view A of Figure 22-7.) Which one goes where, is not at all important. In the beginning, it may appear that all four activities are expected to start or take place at the same time. This is definitely not the case. The person drawing the diagram must realize and remember at all times that the arrow diagram does not describe time relationships but rather only dependencies. This diagram, so far, states only that **obtain trencher, obtain permit, obtain draintile,** and **order and deliver paving material** are independent, that is, can start at any time whether other activities have started or not. "When" these activities take place will be determined in the scheduling phase.

The next item that can be placed into the diagram is **block street**. The only thing that must be done before this activity can start is **obtain permit**. This new activity then starts with event (2) as shown in view B of Figure 22-7.

The assumption that obtaining the permit was only a formality makes it possible to show **obtain trencher, obtain draintile** and the **paving material** activity as starting with event (1). If this assumption had not been made, a prudent decision would have been to start these activities at event (2) so as to avoid spending money without being sure of the results.

After the street had been blocked, **dig ditch** could start. However, because the ditch is to be dug through the pavement, the trencher would have to be on hand. The arrow diagram indicating this would appear as shown in view C of Figure 22-7.

From this point until the final activity, the diagram would be constructed as it was for problem 1, as shown in view D of Figure 22-7.

The final activity would be repair of the damage caused by tearing up the street. In order to pave, however, the paving materials would have to be on hand. This situation is described by

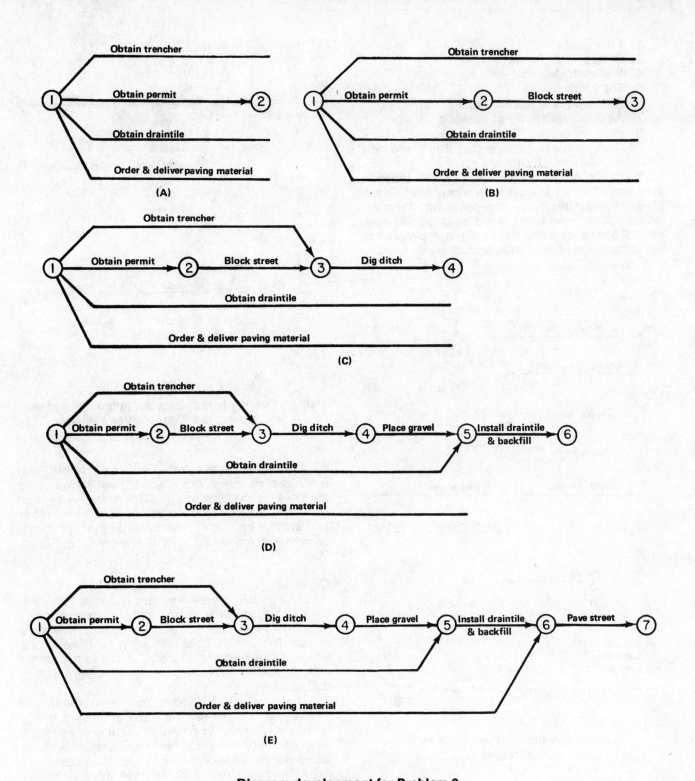

Diagram development for Problem 2
Figure 22-7

making the event that starts **pave street** (6), and making the same event the end point for **order and deliver paving material**. The final solution to this problem is illustrated in view E of Figure 22-7.

An important point to remember about arrow diagrams is that the assumptions made about the activities are as important as the activities themselves. They should always be written down, so as to be remembered.

Problem 3

Create an arrow diagram for a reinforced concrete foundation to be built partially below ground level. Assume that all necessary tools, equipment and materials (including concrete) are on the job site, and that there is no limit to the number of workers. A rented backhoe is used to excavate. Use only the following activities:

1. Lay out and excavate

2. **Fine grade**
3. **Prefabricate forms**
4. **Prefabricate rebar**
5. **Set forms**
6. **Set rebar and anchor bolts**
7. **Advise availability of backhoe**
8. **Pour concrete**

For this problem, assume that **fine grade** must be done before the forms are set.

Start the arrow diagram with three independent activities as indicated in view A of Figure 22-8. Remember that the diagram does not indicate that these three items go on at the same time. It only shows that they are not dependent on the completion of any other activities.

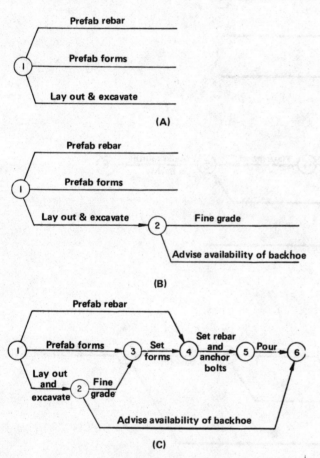

(A)

(B)

(C)

Diagram development for Problem 3
Figure 22-8

After the layout and excavation had been completed, two activities could be started. **Fine grade** would depend upon the completion of this activity, and, of course, the backhoe could not be returned until the completion of this phase of work. At this point the diagram would appear as indicated in view B of Figure 22-8.

Again, the diagram is not stating that **fine grade** and that **advise availability of backhoe** occur simultaneously, but rather that they both depend upon—cannot start until—the completion of **lay out and excavate.**

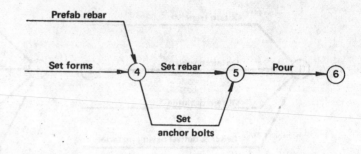

(A)

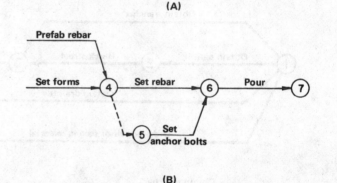

(B)

Using a dummy to maintain the uniqueness of the header and tail identification system
Figure 22-9

After **fine grade** had been completed, and when **prefab forms** was finished, **set forms** could start. At its conclusion, and after **prefab rebar** had been completed, it would be possible to **set rebar** and **anchor bolts** and then finally **pour.** The final arrow diagram would look like view C of Figure 22-8.

Dummies

Occasionally, it is necessary to use a "connector" type activity to indicate a dependency relationship without causing confusion. This type of activity, which does not represent work and which has a duration of zero, is called a **dummy** activity and is shown on the arrow diagram as a dotted line.

There are two reasons for using dummies. These two reasons can be illustrated by altering problem 3 above. Suppose that instead of having a single activity called **set rebar and anchor bolts**, it was desired to make two activities: one called **set rebar** and the other called **set anchor bolts.** Suppose that both depended upon **prefab rebar**, and **set forms**, and that both had to be completed before the **pour.** The affected part of the diagram would appear as illustrated in view A of Figure 22-9.

From a dependency point of view, there is nothing wrong with this kind of description. However, confusion results from having more than one activity with the same I - J numbers. It is not clear which is activity (4) - (5). Earlier in this chapter a rule was established to solve this problem. It was stated that not more than one activity may have the same arrow head and the same arrow tail. So as not to break this rule, one of the two "non-unique" activities must be changed into two, one of which is a dummy. The affected part of the diagram would now appear as illustrated in view B of Figure 22-9. If view B of Figure 22-9 is examined closely, the following points become clear:

1. Event (6) occurs when **set rebar** and **set anchor bolts** have been completed.

2. Since the **dummy** from (4) to (5) activity has zero duration, it is completed at the same point in time that event (4) occurs.

3. For this reason, event (5) occurs when **set forms** is finished and **prefab rebars** is finished, and **set anchor bolts** can start at this time.

4. This is exactly the same dependency statement that was made in the incorrect solution illustrated in view A of Figure 22-9 in which two activities from (4) - (5) appeared.

The first reason for using a dummy then is to maintain the uniqueness of the arrow head and tail identification system.

The second reason for using a dummy is somewhat more complex. A connector type activity is sometimes needed to describe dependencies in such a way that non-dependent activities are not shown as dependent. This can be illustrated also with the diagram from problem 3.

If the problem were changed by eliminating the assumption that **fine grade** had to precede the setting of the forms, one would be tempted to produce an arrow diagram that looked like view A of Figure 22-10.

This approach, however, is incorrect because it is not possible to **set forms** unless at least the excavating has been completed. Another solution—which is also incorrect—might be to combine event number (2) and event (3). Event (3) would then appear as in view B of Figure 22-10.

The error in this arrow diagram exists because **fine grade** and **advise availability of backhoe** do not depend upon the completion of **prefab forms** as indicated, but rather only on **lay out and excavate**.

The correct solution is one in which **set forms** is indicated as depending upon both **prefab forms** as well as on **lay out and excavate,** and in which **fine grade** and **advise availability of backhoe** depend only on **lay out and excavate**. The proper way to show this situation is by using the first solution with a **dummy** from (2) to (3), as illustrated in view C of Figure 22-10.

Because the **dummy** from (2) to (3) has a zero duration, it is finished when event (2) occurs. It merely transfers, then, the dependency relationship—the sequence—desired to event (3).

The second reason for using a **dummy** is to establish a dependency or sequence without confusing non-dependent activities.

Dummy Problem

Go back to the very first problem in this section. There it was assumed that only one man

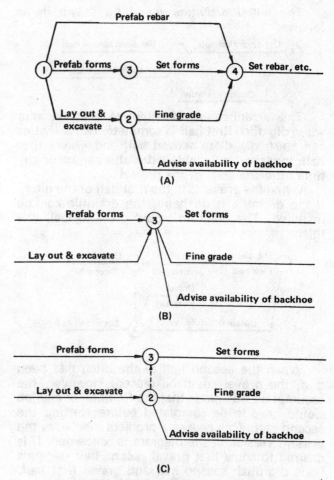

Using a dummy to establish a dependency or sequence without confusing non-dependent activites
Figure 22-10

would work on the job. If the job is broken into segments to show how several crews would get the job done, the activity list could be almost doubled. Assume for this problem that each of those original activities will now be done each in two phases. That is, approximately half the ditch will be dug before the "gravel crew" will start placing gravel, and the gravel will be in half the ditch before the crew installing the draintile can start work. The activity list would then consist of the following items:

1. **Obtain draintile**
2. **Dig ditch first half**
3. **Dig ditch second half**
4. **Place gravel first half**
5. **Place gravel second half**
6. **Install draintile first half**
7. **Install draintile second half**

It must be assumed for this problem that the first half of each job will be finished before the second half starts. Assumed also is that **obtain draintile** is a delivery type of activity, and that the ditch will be backfilled as part of the **install draintile** operation.

The initial activities would be drawn in as follows:

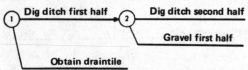

This arrangement of activities indicates that when **dig ditch first half** is complete, two activities can start: **dig ditch second half** and **gravel first half**. Independently, **obtain draintile** can occur any time after the start of the project.

When the gravel is in the first half of the ditch, if the draintile is on hand, the draintile can be installed. The diagram would grow to look like this:

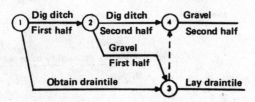

When the second half of the ditch had been dug, the gravel could be placed. However, the assumption was made that the first half of any job would have to be completed before starting the second half. This causes a problem insofar as the present version of the diagram is concerned. It is desired to show that **gravel second half** depends upon **dig ditch second half** and **gravel first half**. Tying this new activity to the ditch digging operation is no problem but difficulties arise when **gravel first half** is considered. The problem is caused by the fact that **gravel first half** has been tied into event (3). If the end of **gravel first half** is connected to the beginning of **gravel second half**, and a dummy is about the only way this could be done, a serious sequence error is committed.

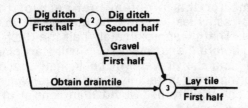

The diagram, as it appears above, indicates that **gravel second half** cannot start until **obtain draintile** has been completed. That is, the diagram states that it is impossible to place gravel unless draintile is on hand. No such assumption was made, so this is incorrect.

The problem is solved by rearranging the diagram as originally created and splitting **gravel first half** into two paths. Dummy activities are, of course, required to do this. The final solution to this expanded drainage ditch problem appears at the top of the next column.

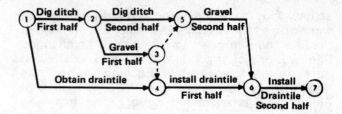

Duration Estimates

The next stage of input development involves the element of time. This requires the planner to estimate the amount of time that will be necessary to complete each activity, or the duration time of each activity. This is necessary in order to produce the CPM schedule for the project. The arrow diagram describes how the activities fit together; the duration estimates indicate how long each will take.

Duration estimates can be made in a variety of ways. The simplest way is to determine the "normal" amount of time needed to finish the activity with a normal sized crew or with the normal amount of equipment. In addition to this simple approach, the CPM user has the ability to examine a range of durations for the activities so as to produce more than one schedule for the project, starting with the longest duration for each activity and then "speeding up" those that will shorten the project.

It must be emphasized that when duration estimates are made, they are made on an individual activity basis. That is, no consideration of other activities can be made. Remember that the planning phase—in which the duration estimates are made—is entirely separate from the scheduling phase.

The duration estimate is used to calculate the schedule for a project. It is also used to find those activities that are controlling the amount of time needed to get the project done. These are the critical activities and collectively they make up "The Critical Path".

The critical path is the longest path—in time—through the network. Since these critical activities are added to determine the total duration of the project, any delay of one of these activities will proportionately delay the job—and conversely, any speed-up will decrease the total duration time.

Look at the example illustrated in Figure 22-11. In this arrow diagram, durations have been assigned each of the activities. The diagram indicates that activity A-a twelve-day item—must be finished before the four-day activity B[(2) - (4)] can begin and that B must be finished before activity C[(4) - (7)] can start, etc.

This diagram contains three paths. A first path consists of activities A, B, and C. The amount of time required to complete this set of activities is the total of activities A, B, and C, or 24 days.

A second path through this network is made up

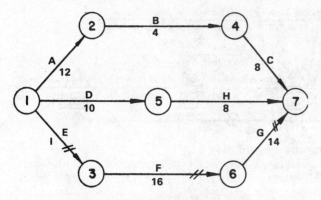

Determining the critical path
Figure 22-11

of activities E, F, and G. This path requires a total of 31 days to finish, (1 + 16 + 14).

A third path in the network is also possible. It is made up of activities D and H. The duration of this path is 10 + 8, or 18 days.

The longest path in time through the network illustrated is thus E, F and G. This then is the critical path and is indicated by small double slants on the arrows. If it is desired or required that the amount of time needed to complete the project be shortened, these are the activities upon which to concentrate. Non-critical activities are strictly dependent upon the completion of the critical items, so speeding up non-critical activities is of no value at all, in terms of project duration.

It turns out generally that a very small number of activities make up the critical path—usually less than thirty per cent of the total activities. This means that a large percentage of the activities in a project have extra time available—since they are, in a sense, waiting for the critical items to be completed. A project manager can adjust his non-critical activities to take best advantage of weather conditions, manpower and equipment availabilities, and other items, without delaying the project.

An important point to keep in mind when using CPM is that there is not necessarily only one critical path. There may be several. Also, remember that if a critical path is shortened, a

new "parallel" path will most likely become critical. In the simplified example illustrated in Figure 22-12, the critical path goes through the top activities. If these are shortened by only one day, however, the bottom path becomes a parallel critical path. Any further reducing of the amount of time required to get the project finished would have to be done on both critical paths.

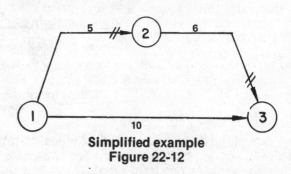

Simplified example
Figure 22-12

CPM Scheduling

When a project has been planned on an arrow diagram, the next step is to schedule it—that is, to place it on a working timetable. When this has been done, it will be possible to determine when each of the various activities must be performed, when deliveries must take place, how much (if any) spare time there is for each activity, and when completion of the whole operation may be expected. It will also be possible to determine which activities are critical, and to what extent a delay in one activity will affect succeeding activities.

The earliest time at which an event can occur is the sum of the durations of the activities on the longest path leading up to the event. This time is entered in a box next to the event on the arrow diagram, as shown in Figure 22-13.

The times shown are, of course, project times—that is, successive working days, not successive calendar days, reckoned from 0 at the tail of the first arrow. The duration of the first activity in Figure 22-13 is 2 days; therefore, event (2) occurs at project time 2. The time for event (3)

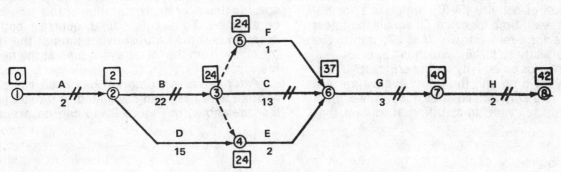

Earliest event times on arrow diagram
Figure 22-13

411

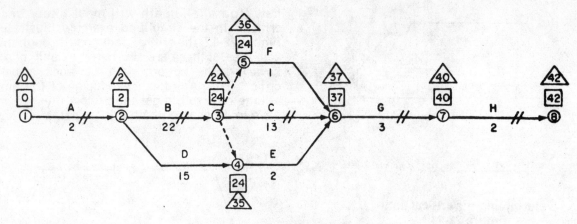

Latest event times on arrow diagram
Figure 22-14

is the sum of the duration times of activities (1)-(2) and (2)-(3), or 24. However, there are two paths leading to event (4): one from event (1) through (2) for a total of 17, the other from event (1) through (2) and (3) for a total of 24. Following the rule of selecting the longest path, the earliest event time for event (4) is 24. Similarly, three paths lead to event (6), and the longest from event (1) through (2) and (3) is selected, giving an earliest event time for event (6) of 37.

You also need to know the latest time at which an event can occur. To determine this, you begin at the end of the project and work backward. To calculate the latest time at which an event can occur, subtract the duration of the immediately following job from the immediately following latest event time. The latest event time is entered in a small triangle adjacent to the box containing the earliest event time, as shown in Figure 22-14.

In that figure, the latest event times for events (6), (7), and (8) are the same as the earliest event times. This follows from application of the rule given. The latest event time for event (7), for example, equals the latest event time for event (8), which is 42, minus the duration of activity (7)-(8), which is 2. The remainder is 40.

However, the latest event time for event (4) equals the latest time for event (6), which is 37, minus the duration of activity (4)-(6), which is 2, or 35. The latest event time for event (5) equals the latest event time for event (6), which is 37, minus the duration of activity (5)-(6), which is 1, or 36. The latest event time for event (3) equals the latest event time for event (6), which is 37, minus the duration of activity (3)-(6), which is 13, or 24.

Note that for an activity on the critical path the earliest event time and the latest event time are the same; it is only for activities not on the critical path that these event times differ. It follows that

identical earliest and latest event times are another means of identifying activities on the critical path.

The spare time available to perform a task is called **float**. Properly controlled, the manipulation

Activity	Project Days		Calendar Days		Float
	Start	Duration	Start	Finish	
(1, 2)	1	2	March 1, Thurs.	March 2, Fri.	0
(2, 3)	3	3	March 5, Mon.	March 7, Wed.	0
(2, 4)	3	3	March 5, Mon.	March 7, Wed.	16
(3, 5)	6	4	March 8, Thurs.	March 11, Sun.	0
(4, 6)	6	3	March 8, Thurs.	March 12, Mon.	16
(5, 7)	10	6	March 12, Mon.	March 19, Mon.	0
(6, 8)	9	5	March 13, Tues.	March 17, Sat.	16
(7, 9)	16	6	March 20, Tues.	March 25, Sun.	0
(8, 11)	14	1	March 19, Mon.	March 19, Mon.	16
(9, 10)	22	3	March 26, Mon.	March 28, Wed.	0
(10, 12)	.	0			0
(10, 13)	25	3	March 29, Thurs.	April 2, Mon.	0
(10, 17)	25	1	March 29, Thurs.	March 29, Thurs.	11
(11, 15)	.	0	--	--	0
(11, 18)	15	2	March 20, Tues.	March 21, Wed.	20
(12, 17)	25	2	March 29, Thurs.	March 30, Fri.	10
(13, 14)	28	5	April 3, Tues.	April 7, Sat.	0
(14, 15)	33	2	April 9, Mon.	April 10, Tues.	0
(15, 16)	35	3	April 11, Wed.	April 13, Fri.	0
(16, 17)	.	0			0
(16, 18)	38	1	April 16, Mon.	April 16, Mon.	2
(17, 18)	38	3	April 16, Mon.	April 18, Wed.	0
(18, 19)	41	2	April 19, Thurs.	April 20, Fri.	0

Project timetable
Figure 22-15

of float is valuable in determining the most efficient use of manpower, equipment, and materials. The existence of float allows latitude in the timing of the jobs with which it is associated. On the other hand, a job having no float is inflexible; it must start and end precisely at specific times, or the completion of the project will be affected. To calculate float, subtract both the duration and the earliest event time at the tail of the arrow from the latest event time at the head of the arrow.

After the arrow diagram has been completed and the float has been calculated, a timetable like the one shown in Figure 22-15 can be prepared.

Index

Practical References For Builders

From Craftsman Book Company, 542 Stevens Avenue, Solana Beach, California 92075

Roofers Handbook

The journeyman roofer's complete guide to wood and asphalt shingle application on both new construction and reroofing jobs: When and how to use shakes, shingles, and T-locks to full advantage. How professional roofers make smooth tie-ins on any job. The right way to cover valleys and ridges. An excellent chapter on handling and preventing leaks is by itself worth the price of the book. Chapters 12 and 13 show you how to prepare an estimate (including man hour requirements), how to set up and run your own roofing business and how to sell your services as a professional roofer. A sample roofing contract is included. It's all here with easy-to-follow explanations, over 250 illustrations and hundreds of inside-trade tips available nowhere else. If you install shingle roofing this will be your most valuable working reference.

| 192 pages | 8½ x 11 | $7.25 |

Practical Rafter Calculator

Cut every rafter right the first time and know it is perfect. This book gives you rapid, accurate, 100% error-free answers . . . the exact, actual lengths for common, hip, valley and jack rafters for every span up to 50 feet and for every rise from 1/2 in 12 to 30 in 12. Everything is worked out to give you the correct rafter length at a glance—to the nearest 1/16 inch! Angle, plumb and level cuts are included so that you have all the information you need to do the job right the first time—everytime. If you know the pitch and you know the roof span, this practical reference will give you everything else. Every framing professional needs this time-saver.

| 124 pages | 3 x 7 | $3.00 |

National Construction Estimator

Material prices for every commonly used building material, the proper labor cost for installing the material, hundreds of time saving rules of thumb, square foot costs and estimating tables. If you plan to estimate building costs, you should have the current edition at your fingertips. The National Construction Estimator covers the construction estimating field to give you figures for common as well as hard to estimate items. Over 15,000 material and labor costs are included . . . covering every phrase of residential, industrial and commercial construction. Just about every construction cost you will need is here — detailed, accurate, clear cost data designed to give you the information you need when you need it.

Every page is loaded with data that would take you hours or days to develop. All costs are organized and presented to meet your working needs and to solve tough cost problems. You work up your bid or figure the cost of changes in minutes. This popular reference will save you hours in gathering material and labor costs and will help you avoid costly errors. If you only have one book of construction costs, this is the book to have — an accurate, easy to use construction cost guide.

Order your copy of the National Construction Estimator now. Use it for 10 days. You will be pleased or you may return the book for a refund.

| 304 pages | 8½ x 11 | $7.50 |

Finish Carpentry

Money-making know-how for the carpentry "pro". Sure answers to practical carpentry problems. This new handbook has the time-saving methods, inside trade information and proven shortcuts you need to do first class finish carpentry work on any job: cornices and rakes, gutters and downspouts, wood shingle roofing, asphalt, asbestos and built-up roofing, prefabricated windows, door bucks and frames, door trim, siding, wallboard, lath and plaster, stairs and railings, cabinets, joinery, and wood flooring. Here you have all the information you need to figure the materials and labor required, lay out the work and cut, fit and install the items required. Over 350 man-hour tables, charts and clear illustrations make this the practical, step by step handbook every carpentry "pro" needs.

| 192 pages | 8½ x 11 | $5.25 |

Wood Frame House Construction

From the layout of the outer walls, excavation and formwork, to finish carpentry, sheet metal and painting, every step of construction is covered in detail with clear illustrations and explanations. Here the builder will find everything he needs to know about framing, roofing, siding, plumbing, heating, insulation and vapor barrier, interior finishing, floor coverings, millwork and cabinets, stairs, chimneys, driveways, walks . . . complete step by step "how to" information on everything that goes into building a frame house. Many valuable tips are included: What you should know about building codes; ways to reduce costs without cutting quality; when concrete block should be used; the advantages of slabs; how to use reinforcing properly. This new book is carefully written, well illustrated and worth several times the price.

| 240 pages | 8 x 10 | $3.25 |

The Successful Construction Contractor

Can you succeed at construction contracting? Thousands do every year and these two volumes are filled with the working knowledge contractors need and use to succeed.

Volume I covers plan reading, working with specifications and practical construction. It is written to solve problems and answer practical questions: How to understand working drawings and specifications (actual plans and complete specifications are included for residential, commercial and industrial structures). How to work with concrete and steel. Selecting the right lumber grades and framing system for your job. How to handle masonry, drywall, lath and plaster to best advantage. Over 100 practical pages on carpentry alone will help you ensure that carpentry work on your jobs is professional and efficient. With over 450 pages and more than 600 illustrations, tables and charts, this volume puts at your fingertips the time and money saving construction knowledge you need on your next job.

Volume II has the essential estimating, selling and mangement information contractors need: A complete estimating system for excavation, concrete, reinforcing steel, carpentry, masonry, lath and plaster and much more. How to take off materials and compile an estimate including profit, contingency and overhead. How to be sure your estimate is correct and control costs once work is started. Hundreds of actual man-hour estimates are included (important reference material you will use again and again). One chapter examines actual case histories of contractors who made it and contractors who didn't and isolates 8 key elements that you can use to build your own operation. Another chapter shows you how to get started—licensing requirements and permits. How to incorporate or form a partnership (complete with forms) and meet state and Federal tax requirements. How you can make sure you get the loan you request (including 26 Federal loan sources for builders). How to plan and schedule your job (using modern CPM techniques). The chapter on selling has a complete sales plan for remodelers and custom builders, from initial advertising to closing the sale. A very important section shows you how to check the financial health of your company, whether large or small, and protect it so that it can grow and prosper. A final chapter covers construction law and builders insurance and includes important information on how to get the bonds you need to move into more profitable types of work. If you want to develop a strong, money-making construction business, you should have these practical manuals.

| Volume I | 8½ x 11 | 452 pages | $8.75 |
| Volume II | 8½ x 11 | 496 pages | $9.50 |

Rough Carpentry

Modern construction methods, labor and material saving tips, the facts you need to select the right lumber grade and dimension for every framing problem. All rough carpentry is covered in detail: sills, girders, columns, joists, sheathing, ceiling, roof and wall framing (over 60 pages on roofs alone), roof trusses, dormers, bay windows, furring and grounds, stairs (42 illustrations of stair work) and insulation (including how much is needed for your area). Many of the 24 chapters explain practical code approved methods for saving lumber and time without sacrificing quality . . . important information every cost conscious builder should have. Chapters on columns, headers, rafters, joists and girders show you how to use simple engineering principles to select the right lumber dimension for whatever species and grade you are using. This new handbook will open your eyes to faster, money-saving, more professional ways to handle framing, sheathing, and insulating and will be the most useful carpentry reference in your library.

| 288 pages | 8½ x 11 | $6.75 |

Home Builder's Guide

The "how to" of custom home contracting explained by a successful professional builder. Clearly explains what you should be doing, avoiding and watching out for during each phase of the job. How to anticipate problems, eliminate bottlenecks, keep the work going smoothly, and end up with a finished home that puts profits in your pocket and pleases your customers. Includes what you need to know about working with subcontractors, lenders, architects, municipal authorities, building inspectors, tradesmen and suppliers. Here you will find simple but effective ways of avoiding design problems, getting the right kind of financing, making sure your building permit is issued promptly, avoiding excavation mistakes, preventing delays if a subcontractor's work doesn't pass inspection, coordinating framing with the other trades, developing a flexible but effective construction schedule, and getting the work done without the hundreds of problems that often delay even highly experienced builders.

| 359 pages | 8½ x 5½ | $7.00 |

Stair Builders Handbook

Modern methods, proven techniques and precise tables that guarantee professional results on every stairway you build. If you know the floor to floor rise, this handbook will give you everything else: the number and dimension of treads and risers, the total run, the correct well hole opening, the angle of incline, the quantity of materials and settings for your framing square. Accurate tables give you over 3,500 code approved rise and run combinations—several for each 1/8 inch interval from a 3 foot to a 12 foot floor to floor rise. Simple step by step instruction with big, clear illustrations help you build the right stairway for your job. Anyone who designs, lays out or builds stairways needs this time-saving, money-making handbook. There is nothing else like it available. It's your key to perfect stairs—the first time, every time—from now on.

413 pages   8½ x 5½ $5.95

Construction Industry Production Manual

Reliable man hour tables developed by professional estimators from hundreds of jobs and all types of construction. Detailed, accurate answers to the question "How much labor is required?" Thousands of carefully researched figures, accurate charts and precise tables to give the estimator the information he needs to compile an accurate estimate: How much soil will your D8 dozer move in the terrain and soil type you are figuring? How many man hours should you allow for forming up the columns and beams on your next job? For framing the roof? For hanging the lighting fixtures? If you have only one book of labor tables, this is the book to have — a one volume library with about all the labor figures you will ever need.

176 pages   5½ x 8½ $6.00

Work Items For Construction Estimating

This book has the computer generated and verified estimating and scheduling data you can use on every commercial, industrial and heavy construction project. Man-hour requirements for all frequently used construction work items, crew composition data, work scheduling information and equipment requirements. Over 10,000 individual data records catalogued into the C.S.I. system of account codes. This is the only reference of its kind with both practical data and effective estimating techniques.

208 pages   11 x 8½ $20.00

Carpentry by H. H. Siegele

This book illustrates all the essentials of residential work: layout, form building, simplified timber engineering, corners, joists and flooring, rough framing, sheathing, cornices, columns, lattice, building paper, siding, doors and windows, roofing, joints and more. One chapter demonstrates how the steel square is used in modern carpentry. A whole generation of journeymen and apprentices have learned carpentry from H. H. Siegele. This reference has the essential knowledge you need to become a skilled professional carpenter.

219 pages   8½ x 11 $6.95

National Repair And Remodeling Estimator

The complete pricing guide for dwelling reconstruction costs. Reliable, specific data you can apply on every remodeling job. Up-to-date material costs and labor figures based on thousands of repair and remodeling jobs across the country. Professional estimating techniques to help determine the material needed, the quantity to order, the labor required, the correct crew size and the actual labor cost for your area. This new volume is a complete file of repair and remodeling costs and will become your working file for all pricing. Revised annually.

128 pages   11 x 8½ $6.50

Concrete And Formwork

All the information you need to select and pour the right mix for the job, lay out the structure, select the right form materials, design and build the forms and finish and cure the concrete. This is the handbook for the man on the job who needs sure answers to practical problems: What type of mix is best? What admixtures are needed? How deep should the footing be? What is the best way to lay out and design the forms? How much concrete and form material are needed? Nearly 100 pages of step-by-step instructions cover the actual construction and erecting of all forms in common use. The most useful single volume for anyone working with site fabricated wood concrete forms.

176 pages   8½ x 11 $3.75

Mechanical Estimators Handbook

The indispensable guide to estimating pipe and fittings for residential and commercial construction, power plants, refineries, water works, booster stations, sewer plants and storage facilities. Side by side on two pages you find the labor required to handle, cut, fit and install the pipe you are figuring. The following page has the weight, diameter, wall thickness and test pressures you need. Full information on planning and layout, figuring overhead and profit and professional cost estimating techniques. If you estimate pipe and fittings, this book will be your working companion for many jobs to come.

288 pages   5½ x 8½ $8.95

Process Plant Estimating, Evaluation And Control

The most comprehensive assembly of cost data yet offered to the process plant industries. The carefully selected, well organized estimating data cost engineers, chemical process designers and project analysts need to evaluate the cost of the next generation of refineries and chemical process plants. Based on cost data developed from the actual field construction records of 50 major capital projects. Includes labor, material and equipment costs for every type of refinery and chemical process plant, indirect costs, pipe estimating data, project control techniques, computer cost systems and much more. Over 1,000 charts, diagrams and displays to help you go from proposal to definitive cost— with assurance that you have the most timely, most complete cost information available anywhere.

608 pages   8½ x 11 $25.00

Craftsman Book Company
542 Stevens Avenue
Solana Beach, California 92075

Name _____

Company _____

Address _____

City _____ State _____ Zip _____

10 DAY FULL MONEY BACK GUARANTEE

☐ Roofers Handbook $7.25
☐ Practical Rafter Calculator 3.00
☐ National Construction Estimator 7.50
☐ Finish Carpentry 5.25
☐ Wood-Frame House Construction 3.25
☐ The Successful Construction Contractor Vol. 1 ...... 8.75
☐ The Successful Construction Contractor Vol. 2 ...... 9.50
☐ Rough Carpentry 6.75
☐ Home Builder's Guide 7.00
☐ Stair Builders Handbook 5.95
☐ Construction Industry Production Manual 6.00
☐ Work Items For Construction Estimating 20.00
☐ Carpentry 6.95
☐ National Repair and Remodeling Estimator 6.50
☐ Concrete and Formwork 3.75
☐ Mechanical Estimators Handbook 8.95
☐ Process Plant Estimating, Evaluation and Control ... 25.00
☐ Remodelers Handbook 12.00

_____ **Total Enclosed**

_____ **In California add 6% tax**